T0336201

Hillslope Hydrology and Stability

Landslides occur when hillslopes become mechanically unstable, because of meteorological and geologic processes, and pose a serious threat to human environments in their proximity. The mechanical balance within hillslopes is governed by two coupled physical processes: hydrologic or subsurface flow and stress. The stabilizing strength of hillslope materials depends on effective stress, which is diminished by rainfall, increasing the risk of gravity destabilizing the balance and causing a landslide.

This book presents a cutting-edge quantitative approach to understanding hydro-mechanical processes in hillslopes, and to the study and prediction of rainfall-induced landslides. Combining geomorphology, hydrology, and geomechanics, it provides an interdisciplinary analysis that integrates the mechanical and hydrologic processes governing landslide occurrences, across variably saturated hillslope environments. Topics covered include a historic synthesis of hillslope geomorphology and hydrology, total and effective stress distributions, critical reviews of shear strength of hillslope materials, and different bases for stability analysis. Exercises and homework problems are provided for students to engage with the theory in practice.

This is an invaluable resource for graduate students and researchers in hydrology, geomorphology, engineering geology, geotechnical engineering, and geomechanics, and also for professionals in the fields of civil and environmental engineering, and natural hazard analysis.

Ning Lu, F. GSA, F. ASCE, is Professor of Civil and Environmental Engineering at Colorado School of Mines, and his primary research in the past decade has concentrated on hillslope hydrology and slope stability. He is the senior author of *Unsaturated Soil Mechanics* (John Wiley & Sons, 2004), and has also published extensively in peer-reviewed journals on unifying effective stress in variably saturated porous media. Professor Lu is a recipient of the Norman Medal and the Croes Medal from the American Society of Civil Engineers for his seminal work on defining suction stress in variably saturated soils.

Jonathan W. Godt is a Research Physical Scientist with the United States Geological Survey and has worked on landslide hazard problems for more than 15 years, both in the United States and around the world. His research focuses on monitoring and understanding landslide processes to improve tools for landslide hazard assessment and forecasting. He has published numerous reports, maps, and journal articles on the subject of landslide hazards.

Hillslope Hydrology and Stability

NING LU
Colorado School of Mines

JONATHAN W. GODT
United States Geological Survey

CAMBRIDGE
UNIVERSITY PRESS

CAMBRIDGE
UNIVERSITY PRESS

University Printing House, Cambridge CB2 8BS, United Kingdom

One Liberty Plaza, 20th Floor, New York, NY 10006, USA

477 Williamstown Road, Port Melbourne, VIC 3207, Australia

4843/24, 2nd Floor, Ansari Road, Daryaganj, Delhi - 110002, India

79 Anson Road, #06-04/06, Singapore 079906

Cambridge University Press is part of the University of Cambridge.

It furthers the University's mission by disseminating knowledge in the pursuit of education, learning and research at the highest international levels of excellence.

www.cambridge.org
Information on this title: www.cambridge.org/9781107021068

First published 2013
Reprinted 2015

A catalogue record for this publication is available from the British Library

Library of Congress Cataloging in Publication data
Lu, Ning, 1960–
Hillslope hydrology and stability / Ning Lu, Jonathan W. Godt.
p. cm.
ISBN 978-1-107-02106-8 (hardback)
1. Mountain hydrology. 2. Slopes (Physical geography) 3. Soil erosion. 4. Soil mechanics. 5. Landslides. 6. Groundwater flow. I. Godt, Jonathan W. II. Title.
GB843.5.L8 2012
551.43′6 – dc23 2012024783

ISBN 978-1-107-02106-8 Hardback

To
Connie, Vivian, and Shemin
and
Neva and Laura

Contents

The color plates can be found between pages 216 and 217

Foreword

Even a cursory inspection of *Hillslope Hydrology and Stability* by Lu and Godt will impress most professionals interested in processes at the interface between geotechnical engineering and hydrology. This unique textbook represents an attempt to systematically unify concepts from vadose zone hydrology and geotechnical engineering into a new hydro-geo-mechanical approach with special emphasis on quantifying natural mechanisms for the onset of hydrologically induced landslides. Professionals will particularly appreciate the comprehensive coverage of concepts ranging from fundamentals of geomechanics and soil properties to the state-of-the-art concepts of hillslope hydrology, with explicit treatment of soil heterogeneity, layering, and vegetation mechanical and hydrologic functions. The authors have been able to weave a coherent picture based on the cutting-edge state of knowledge regarding landslides as natural geomorphological processes and as ubiquitous natural hazards in mountainous regions.

Students will appreciate the lucid coverage of topics offering a systematic introduction to key ingredients essential for understanding the occurrence of landslides in their broader natural context (often missing in technical textbooks). Students are guided through aspects of precipitation with its instantaneous to inter-annual patterns, as well as aspects of soil types and the geomorphological context of landslides. This provides a solid foundation for introduction of more specific technical aspects of infiltration, hillslope hydrology, and hydro-mechanical properties, and assembles the roles of these factors on a hillslope mechanical state. Students will find clear explanations of fundamental concepts inspired by numerical examples to help them develop appreciation for the orders of magnitude for the quantities involved. Numerous motivating homework problems further promote self-study.

Hillslope Hydrology and Stability helps chart the boundaries of the emerging interdisciplinary field of soil hydromechanics. The authors offer a rigorous link between hydrology and soil mechanics by providing a unified treatment of effective stress (suction stress) under variably saturated conditions (Chapter 6). The authors also provide a fresh look at well-established concepts found in textbooks from hydrology and geotechnical engineering fused together using new crucial aspects typically glossed over in standard texts, thereby providing a unique new perspective. For example, the interplay between hillslope subsurface flows and soil layering (forming hydrologic barriers), a critical mechanism for abrupt landslide triggering, has rarely been previously discussed in a quantitative hillslope hydro-mechanical context as done in Chapter 3. The quantitative treatment of root reinforcement and the role of plants in the mechanical picture of natural hillslopes (Chapter 7) is another example of the conceptual integration in the basis of the book. The wealth of information on numerical values of key parameters and the instructive use of case studies described in

Chapters 9 and 10 make *Hillslope Hydrology and Stability* an outstanding resource for students, researchers, and practitioners alike. No doubt the test of time would add refinement to this labor of love that contains numerous new concepts – I hope students and researchers would be challenged and inspired by the breadth and depth offered in this unique treatise on hydro-mechanical hillslope processes.

Professor Dani Or
ETH Zurich

Preface

We strive to provide a thorough description on the cutting edge of the spatial and temporal occurrence of rainfall-induced landslides by quantifying the hydro-mechanical processes in hillslopes. Landslides are a pervasive natural phenomenon that constantly shapes the morphology of the earth's surface. Over geologic time, landslides are the result of two episodic, and broadly occurring geologic processes; tectonics and erosion. At human scale, the former operates at a uniform rate barely sensed by humans except during earthquakes. However, the latter is entirely sensible and is driven largely by rainfall. The results of these dynamic geologic processes are the infinite variety of landforms that vary remarkably in geometry; from flat plains to rolling hills, to vertical or even overhanging cliffs, and to shapes that test human's imagination.

Understanding of how landslides occur is vital to the well being of human society and our environment and has been a research focus for many disciplines such as geomorphology, hydrology, geography, meteorology, soil science, and civil and environmental engineering. While each of these disciplines tackles landslide problems from quite different perspectives, a common thread is the mechanics of landsliding. From the vantage of mechanics, no matter how complicated the morphology of the land surface, it is the mechanical balance within hillslopes that determines if they are stable or not. Two coupled physical processes govern the mechanical balance; hydrological or subsurface flow process and stress equilibrium process.

Understanding and quantifying the hydro-mechanical processes provide the key link to the knowledge gained from different disciplines and pathways for predicting the spatial and temporal occurrence of landslides. In each hillslope, driving and resisting forces dictate the state of stability. The driving or destabilizing forces are mainly provided by gravity and the resisting or stabilizing forces are mainly provided by the strength of hillslope materials. This mechanical balance is mediated by the presence of water, which varies dramatically over climatic, seasonal, and shorter time scales and has both a stabilizing and destabilizing effect. The effect of water on the stability of hillslopes is quantified using the concept of effective stress, which provides a connection between subsurface hydrologic and mechanical processes under variably saturated conditions.

In this volume, we present quantitative treatments of rainfall infiltration, effective stress, their coupling, and roles in hillslope stability. An overall introduction to landslide phenomena, their classification, and socio-economic impacts is provided in Chapter 1. The settings where landslides occur are described in Chapter 2: slope geomorphology. Subsurface hydrologic process under variably saturated conditions is systematically described in the forms of steady infiltration (Chapter 3) and transient infiltration (Chapter 4). The background stress or total stress fields driven by gravity in hillslopes are quantified under the theory of linear elastostatics in Chapter 5. A unified effective stress framework linking soil suction to effective stress is provided in Chapter 6. The pertinent material properties, both the strength of soil and vegetation roots, and hydrologic constitutive laws, are provided in Chapters 7 and 8. Integration of slope geomorphology, hydrology, and soil mechanics leads to a rigorous treatment of slope stability analysis that is described in Chapters 9 and 10. Chapter 9 provides an in-depth introduction to the classical or conventional slope stability methodologies as well as expansions to include environments under variably saturated conditions by the unified effective stress principle. Chapter 10 presents a framework departing from the conventional slope stability paradigm by employing scalar fields of suction stress and factor of safety, which has potential to reveal spatial and temporal occurrence of rainfall-induced landslides in variably saturated hillslopes. The effectiveness of the proposed hydro-mechanical framework is examined through two case studies in these chapters. The first case study is an analysis of a shallow landslide induced by rainfall and is based on a multi-year field-monitoring program where the reduction of a few kPa of suction stress eventually led to slope failure. The second case study applies the hydro-mechanical framework to analyze a deep-seated landslide that moves each year in response to melting snow.

The book is truly the journal of our joint endeavor to advance the understanding of occurrence of landslides. The materials covered here have been grown out of a course, *Hillslope Hydrology and Stability*, taught at Colorado School of Mines, USA, EPFL-Lausanne, Switzerland, and University of Perugia, Italy over the past 6 years. From teaching, we gained much from our interactions with students and professionals. The major part of NL's contribution to the book was written while he was on sabbatical as the Shimizu Visiting Professor at Stanford University and a visiting scientist at the U.S. Geological Survey campus in Menlo Park, California office in 2010–2011. His hosts, Ronaldo Borja at Stanford and Brian Collins at the USGS provided an intellectually stimulating and productive environment. The authors benefitted greatly from contributions from the following colleagues who provide insightful, critical, and thorough reviews of parts of the manuscript: Rex Baum, Brian Collins, Richard Healy, Richard Iverson, and Mark Reid of the U.S. Geological Survey, Dalia Kirschbaum of NASA Goddard Space Flight Center, Giovanni Crosta of the University of Milano-Bicocca, William Likos of the University of Wisconsin-Madison, John McCartney of the University of Colorado-Boulder, Dani Or of ETH Zurich, Ricardo Rigon of the University of Trento, Diana Salciarini of the University of Perugia, Alexandra Wayllace of the Colorado School of Mines, and Raymond Torres of the University of South Carolina-Columbia. We extend special thanks to Rex Baum for looking at the entire proof of the book. Nonetheless, all errors and bias remain ours. Başak Şener-Kaya prepared the figures and tables for the total stress distributions in hillslope in Chapter 5. Finally, the authors would like to express our gratitude to Peter Birkeland who acts as Pe(te)casso for illustrating the essentials of our thoughts in art form at the beginning of each part.

Symbols

Symbol	Description	Units
A	Skempton pore pressure parameter for isotropic loading	–
A_L	landslide area	m^2
A	area; cross sectional area	m^2
a_1	root tensile strength parameter	MPa m^{-a2}
a_2	root tensile strength parameter	–
B	Skempton pore pressure parameter for deviator loading	–
$\mathbf{b}$	body force vector	N/m^3
$\boldsymbol{b}$	parameter for inter-grain friction angle	–
b_1	root shear strength parameter	MPa
b_2	root shear strength parameter	MPa m^3/kg
b_n	width of the *n*th slice in a method of slices	m
b_o	parameter for cumulative rate of root mass with depth	–
b_i	body force components	N/m^3
$C(\psi)$	specific moisture capacity as function of suction	1/kPa
$C(h)$	specific moisture capacity as function of head	1/m
c	cohesion	kPa
c	solute concentration	mol m^3
c_c	cohesion mobilized by cementation bonds	kPa
c_d	mobilized or developed cohesion along failure surface	kPa
c_o	cohesion due to grain inter-locking	kPa
c_s	cohesion mobilized by suction stress	kPa
c_u	undrained shear strength	kPa
c'	cohesion in terms of effective stress	kPa
D	diffusivity	m^2/s
D_o	free vapor diffusivity in air	m^2/s
D_v	free vapor diffusivity in porous media	m^2/s
D_r	maximum depth of landslide body	m
D_r	relative density	–
D_v	diffusion coefficient for water vapor	m^2/s
D_{10}	10% finer particle diameter	m
D_{50}	50% finer particle diameter	m
d	diameter of capillary tube	m
d	root diameter	mm

Symbol	Description	Units
d	shear strength parameter defined by cohesion and friction angle	kPa
d_1, d_2, d_3	root shear strength growth parameters	kPa
d_4	root shear strength growth parameter	y^{-1}
d_5	root shear strength decay parameter	y^{-d_5}
d_6	root shear strength decay parameter	y^{-d_6}
E	Young's modulus	kPa
E	inter-slice normal forces in method of slices	kN
e	void ratio	–
e_{max}	void ratio in loosest state	–
e_{min}	void ratio in densest state	–
e_s	saturation vapor pressure	hPa
FS	factor of safety for a hillslope	–
FS_s	shear strength based factor of safety	–
f	infiltration capacity	cm/hr
$f(u_a - u_w), f(S)$	suction stress characteristic function	kPa
f_c	minimum steady constant infiltration capacity	cm/hr
f_0	initial infiltration capacity	cm/hr
F_{ij}	force components	N
G	elasticity modulus	kPa
G_s	specific gravity of soil solids	–
g	acceleration due to gravity	m/s^2
$\mathbf{g}$	acceleration vector due to gravity	m/s^2
H	Kirchhoff integral transformation	m^2/s
H_{max}	maximum slope height of a finite slope	m
H_{ss}	depth of sliding surface from ground surface	m
H_{wt}	depth of water table from ground surface	m
h	height of capillary rise; head	m
h_a	air-entry head	m
h_c	maximum height of capillary rise	m
h_d	applied increment in matric suction head	m
h_g	total gravitational head	m
h_i	initial suction head in a soil column	m
h_m	matric suction head	m
h_n	height of the water table from the failure surface for slice n	m
h_o	suction head at wetting front	m
h_o	osmotic suction head	m
h_t	total head	m
h_{vap}	potential head of water vapor	m
h_v	kinetic or velocity head	m

Symbol	Description	Units
h_w	applied decrement in matric suction head	m
$I_{1\sigma}$	first stress invariant	kPa
i	hydraulic gradient	–
i	initial root orientation with respect to failure plane	deg
i, j, m, s	series indices	–
K	bulk elastic modulus	kPa
K	hydraulic conductivity	m/s
$\mathbf{K}$	hydraulic conductivity tensor	m/s
K^*	dimensionless hydraulic conductivity in Laplace space	–
K_f	permeability-dependent constant for infiltration capacity	hr^{-1}
K_o	hydraulic conductivity at wetting front	m/s
K_o	horizontal to vertical stress ratio under no horizontal displacement condition	–
K_{eq}	equivalent hydraulic conductivity of soil-HAE ceramic stone system	m/s
K_s	saturated hydraulic conductivity	m/s
K_{sat}	saturated hydraulic conductivity	m/s
K_s^d	saturated hydraulic conductivity for drying state	m/s
K_s^w	saturated hydraulic conductivity for wetting state	m/s
K_s^c	saturated hydraulic conductivity of HAE ceramic stone	m/s
K_x, K_y, K_z	hydraulic conductivity in the x, y, and z directions	m/s
L	diversion width for capillary barrier	m
L	soil layer thickness	m
L	length of soil body in infinite-slope model	m
L	depth of the water table from ground surface	m
L_r	length of the surface of rupture of a landslide body	m
l	sample height plus thickness of HAE ceramic stone	m
l_1, l_s	sample height	m
l_2, l_c	thickness of HAE ceramic stone	m
l_n	length of the base of slice n	m
M	shear strength parameter defined by internal friction angle	–
M_r	cumulative mass fraction in depth z	–
m	total number of slices in a method of slices	–
m	slope stability number for assessing stability of finite slope	–
m_r	root mass per unit volume of the reinforced soil	kg/m^3
m_s	mass of solid	kg
N	index variable	–
N	normal force	N
N_n	normal reacting force	N
n	Corey's 1954 hydraulic conductivity model parameter	–
n	porosity	–

Symbol	Description	Units
n	SWCC modeling constant	–
n^d	SWCC modeling constant for drying state	–
n^w	SWCC modeling constant for wetting state	–
n	series index	–
$\mathbf{n}$	unit directional vector on boundary	–
n_x, n_y, n_z	components of unit directional vector on boundary	–
n_a	air-filled porosity	%
n_p	porosity	–
P	annual precipitation	mm
PET	annual potential evaporation	mm
p	landslide probability density	m^{-2}
Q	dimensionless flow variable	–
Q	diversion capacity for capillary barrier	m^2/s
Q	total cumulative infiltration	m
q	fluid flow velocity	m/s
$\hat{q}_d(l,t)$	simulated outflow rate during drying	m/s
$\hat{q}_d^{\exp}(l,t)$	experimental outflow rate during drying	m/s
$\hat{q}_w(l,t)$	simulated inflow rate during wetting	m/s
$\hat{q}_w^{\exp}(l,t)$	experimental inflow rate during wetting	m/s
q_{in}	total inflow rate of water into a unit cell	kg/s
q_{out}	total outflow rate of water out of a unit cell	kg/s
q_v	vapor flow velocity	m/s
$\mathbf{q}$	fluid velocity vector	m/s
R	universal gas constant	J/mol K
R	radius of Mohr circle	kPa
R	resultant force	N
RDD	relative dry density	–
R_{max}	maximum resultant force	N
R_r	root shear strength conversion factor	–
REV	representative elementary volume	m^3
r	radius of circular failure surface	m
r	equivalent or mean pore radius	μm
r_u	pore-water pressure parameter in infinite-slope model	–
S	degree of saturation	%
S	cross section	m^2
S_{xy}	cross section perpendicular to z axis	m^2
S_{xz}	cross section perpendicular to y axis	m^2
S_{yz}	cross section perpendicular to x axis	m^2
S	shear force	N
S_{max}	maximum shear force	N
S_e	effective degree of saturation	%
S_n	mobilized shear resistance along the base of the nth slice	N

Symbol	Description	Units
S_r	residual degree of saturation	%
S_s	specific storage	1/m
s	sorptivity	$m/s^{1/2}$
T	absolute temperature	K
T	dimensionless time	–
T_s	surface tension	N/m
t	time	s
t_x	traction or stress component in the x direction at boundary	Pa
t_y	traction or stress component in the y direction at boundary	Pa
t_z	traction or stress component in the z direction at boundary	Pa
u	pore-water pressure	kPa
u_x, u_y, u_z	displacement components	m
$\bar{u}_x$, $\bar{u}_y$, $\bar{u}_z$	displacement components at boundary	m
u_a	pore air pressure; air pressure	kPa
u_b	air-entry (bubbling) pressure	kPa
u_c	pore pressure due to isotropic stress loading	kPa
u_d	pore pressure due to deviatoric stress loading	kPa
u_{ij}	displacement components	m
u_{sat}	saturated vapor pressure	kPa
u_{v0}	saturated vapor pressure	kPa
u_w	pore-water pressure; water pressure	kPa
$(u_a - u_w)$	matric suction	kPa
V	volume of landslide body	m^3
V_t	volume of soil specimen	m^3
v	discharge velocity	m/s
v_v	volume of void in REV	m^3
v_s	volume of solid in REV	m^3
v_w	volume of water in REV	m^3
v_w	molar volume of water	m^3/kmol
W	virtual work due to effective stress	J
W	weight of soil body	N
W_n	weight of slice n	N
W_σ^s	virtual work due to suction stress stress	J
W_v	weight of soil column per unit cross section area	N/m^2
X	maximum displacement of failure zone along failure surface	m
X	inter-slice shear forces in method of slices	kN
x, y, z	Cartesian coordinate directions	m
x_*, z_*	Cartesian coordinate aligned with sloping direction	m
Z	dimensionless distance	–

Symbol	Description	Units
Z	wetting from position	m
Z	thickness of failure zone	m
z_w	depth of loose or weathering zone	m
α	rotational angle on Mohr circle	deg
α	local topographic slope	–
α	pore size distribution index; SWCC modeling constant	1/kPa
α^d	pore size distribution index for drying state	1/kPa
α^w	pore size distribution index for wetting state	1/kPa
α_n	angle of slice n	N
α_s	bulk compressibility of soil	m^2/N
β	rotational angle on Mohr circle	deg
β	angle of failure surface with respect to horizontal direction	deg
β	pore size distribution index; SWCC modeling constant	1/m
β_w	compressibility of water	m^2/N
χ	coefficient of matric suction	–
$\chi(u_a - u_w)$	suction stress (capillary stress)	kPa
ε	strain	%
$\varepsilon_x, \varepsilon_{xy}$	strain components	%
ϕ	angle of internal friction	deg
ϕ	angle of dip for capillary barrier	deg
ϕ_c	inter-grain friction angle	deg
ϕ_d	angle of internal friction at dry state	deg
ϕ_0, ϕ_{100}	angle of internal friction at 0 and 100% relative dry density	deg
ϕ'	effective angle of internal friction	deg
ϕ'_d	developed or mobilized effective friction angle	deg
ϕ_{CU}	friction angle under consolidation undrained condition	deg
ϕ'_{NC}	effective friction angle under normal consolidation condition	deg
ϕ'_{OC}	effective friction angle under overly consolidation condition	deg
γ	bulk (total) unit weight	kN/m^3
$\gamma_{dmax}, \gamma_{dmin}$	maximum or minimum dry unit weight	kN/m^3
γ	slope angle	deg
γ_{xy}	strain components for angle of distortion	radian
γ_w	unit weight of water	kN/m^3
Λ_n	nth positive root of pseudoperiodic characteristic equation for K^*	–
λ	latent heat of vaporization	J/kg
λ	slope stability number for assessing soil unit weight	–

Symbol	Description	Units
λ	Boltzmann transformation variable	–
λ	Lamé elastic constant	kPa
δ_{ij}	identity tensor	–
ν	Poisson's ratio	–
μ_t	total chemical potential	J/kg
μ_0	chemical potential of reference state	J/kg
μ_v	chemical potential of water vapor	J/kg
π	osmotic pressure	kPa
Θ	effective water content (effective degree of saturation)	%
θ	volumetric water content	%
θ	mobilized friction angle	deg
θ	angle of shear distortion with respect to initial root orientation	deg
θ	maximum mobilized friction angle	deg
θ	angle of potential failure surface with respect to horizontal direction	deg
θ_1	angle of distortion for element Δx	radian
θ_2	angle of distortion for element Δz	radian
θ_{cr}	critical angle of potential failure surface	deg
θ_r	residual volumetric water content	%
θ_r^d	residual volumetric water content for drying state	%
θ_r^w	residual volumetric water content for wetting state	%
θ_s	saturated volumetric water content	%
θ_s^d	saturated volumetric water content for drying state	%
θ_s^w	saturated volumetric water content for wetting state	%
θ_i	initial volumetric water content	%
θ_o	volumetric water content at wetting front	%
ρ_v	density of water vapor (absolute humidity)	kg/m^3
ρ_w	density of water	kg/m^3
σ	total normal stress	kPa
σ_o	normal stress	kPa
σ_{cap}	suction stress component due to capillarity	kPa
σ_{pc}	suction stress component due to physico-chemical forces	kPa
σ_C	counterbalance stress to suction stress due to Born repulsion	kPa
σ_c	cementation bonding stress	kPa
σ'	effective stress	kPa
σ_r	root tensile strength	kPa
σ_{ri}	root *i*'s tensile strength	kPa
σ^s	suction stress (capillary cohesion)	kPa
σ_{xi}	stress components	kPa
σ_1	major principal stress	kPa

Symbol	Description	Units
σ_2	intermediate principal stress	kPa
σ_3	minor principal stress	kPa
σ_n	total normal stress	kPa
σ'_n	effective normal stress	kPa
σ_{tia}	isotropic tensile strength	kPa
σ_{tua}	uniaxial tensile strength	kPa
$(\sigma - u_a)$	net normal stress	kPa
$(\sigma_f - u_a)_f$	net normal stress on failure plane at failure	kPa
τ	shear stress	kPa
$\tau_{xy}, \tau_{xz}, \tau_{zy}$	shear stress components	kPa
τ_{max}	maximum shear stress at a point	kPa
τ_d	mobilized or developed shear stress along failure surface	kPa
τ_f	shear stress at failure	kPa
τ_{rs}	root shear strength mobilized by root tensile strength	kPa
ω_w	molecular mass of water	kg/mol
ω_v	molecular mass of water vapor	kg/mol
ω	capillary barrier efficiency	–
ω	parameter for defining van Genuchten's SWRC model	–
ψ	composite contact or dilation angle	deg
ψ	suction	kPa
ψ	rupture root orientation with respect to failure plane	deg
ψ_m	matric suction	kPa
ψ_o	osmotic suction	kPa
ψ_o	matric suction beyond wetting front	kPa
ψ_t	total suction	kPa
RH	relative humidity	%
HCF	hydraulic conductivity function	
SSCC	suction stress characteristic curve	
SWCC	soil water characteristic curve	
SWRC	soil water retention curve (also called SWCC)	

PART I

INTRODUCTION AND STATE OF THE ART

1 Introduction

1.1 Landslide overview

Landslides are one of the most widespread and effective agents in sculpting the earth's surface (Eckel, 1958, p.1). They are ubiquitous in mountainous and hilly environments in all parts of the world and are an important mechanism for moving earth materials from uplands to river systems. The general term "landslide" is used to describe a wide range of gravity-driven mass movements both on the land surface and beneath bodies of water. Landslides include diverse slope movements such as rock fall and debris flows, which are described in more detail in Section 1.2.

Landslides are the failure of sloping earth materials. A hillslope fails when forces or stresses acting upon it overcome the strength of the earth materials. Some of the forces acting on a hillslope include gravity, pore-water pressure, tectonic uplift, and earthquake shaking. These forces act over time scales ranging from geologic to essentially instantaneous and over spatial scales that range from continental to the soil grain. The strength of hillslope materials is a function of geologic composition and stress state, and is modified by past movement, weathering, vegetation, and hydrologic processes. These concepts are discussed in detail in Chapters 3, 4, 5, 6, 7, and 8.

Processes leading to landslide occurrence are separated into "causes" and "triggers." Landslides can be caused by morphologic, geologic, and other factors that set the stage for a landslide to occur. Landslide triggers are the events that initiate landslide motion. The difference between a cause and a trigger is the time scale over which the processes take place. This range of time scales is obviously a continuum and often it is impossible to determine the precise trigger for a given slide. In other cases the trigger is easily identified, such as heavy rainfall, earthquake shaking, or volcanic eruption.

Landslides are also among the most costly natural hazards in terms of human life and economic loss. Because landslides occur over much of the land surface, and are generated by a range of processes, they frequently intersect human activities and the built environment, often with disastrous consequences. Statistics on losses associated with landslides are difficult to compile, in part because landslides often result from large earthquakes or coincide with large-scale flooding or tropical cyclones, which tend to capture the attention of the media and official inquiries. However, as human activities continue to expand into landslide-prone environments, the recognition of the scope and the magnitude of the hazard has increased.

To understand how landslides occur, it is useful to examine the geometry of a "typical" landslide body. Figure 1.1 is an idealized sketch showing the various parts of a rotational

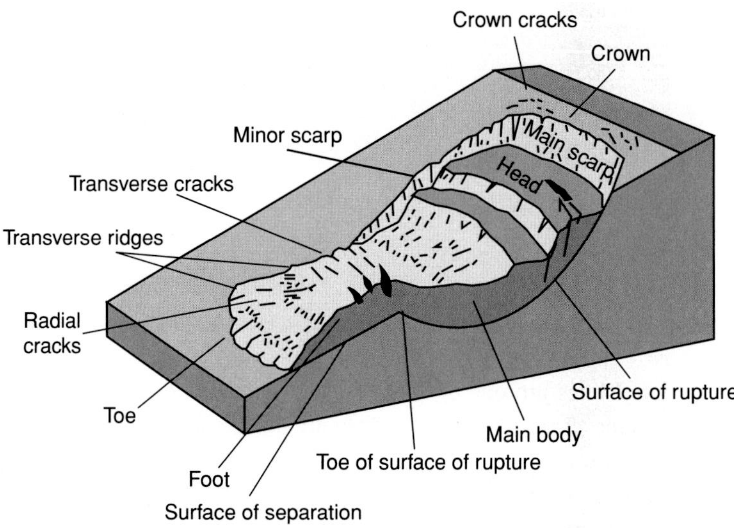

Figure 1.1 Diagram showing the location of the various parts of a landslide (after Varnes, 1978).

landslide, which are described in Table 1.1 (Varnes, 1978). A characteristic that distinguishes landslides from other mass movement processes such as saltation or grain-by-grain transport is the presence of a rupture surface. The rupture surface is the boundary between the relative motions of the landslide body and undisturbed ground around and beneath the slide. At the scale of the overall landslide, the shape of the rupture surface may range from planar to roughly circular. The upslope extent of the landslide is bounded by the main scarp, which is often a vertical or sub-vertical exposure of the hillside materials. The downslope extent of the landslide is referred to as the toe, and the lateral margins of the landslide are called flanks. Morphologic features, such as internal scarps, cracks, and ridges, are an expression of deformation in the landslide body and underlying topography (e.g., Baum *et al.*, 1998; Coe *et al.*, 2009).

While in general the geometry of a landslide body is complicated, it can be approximated as part of an ellipsoid such that the volume of a landslide shown in Figure 1.1 can be estimated by the following equation:

$$V = \frac{\pi}{6}\,(\text{length})\,(\text{width})\,(\text{depth}) \tag{1.1}$$

1.2 Landslide classification

The two most prominent English-language landslide classifications are by Hutchinson (Hutchinson, 1968, 1988; Skempton and Hutchinson, 1969; Hungr *et al.*, 2001) and Varnes (Varnes, 1958, 1978; Cruden and Varnes, 1996). The two systems are generally similar but treat flows of earth materials somewhat differently. Hutchinson's classification emphasizes the results of movement whereas Varnes' tends to emphasize the conditions of slope

Table 1.1 Description of landslide parts (after Varnes, 1978)

Main scarp	A steep surface on the undisturbed ground around the periphery of the slide, caused by the movement of slide material away from undisturbed ground
Minor scarp	A steep surface on the displaced material produced by differential movements within the displaced mass
Head	The upper parts of the slide material along the contact between the displaced material and the main scarp
Toe	The margin of displaced material most distant from the main scarp
Main body	Part of the displaced material of the landslide that overlies the surface of rupture between the main scarp and toe of surface of rupture
Original ground surface	The surface of the hillslope that existed before the landslide occurred
Surface of rupture	Surface that forms the lower boundary of displaced material
Toe of surface of rupture	Intersection (usually buried) between the lower part of surface of rupture and original ground surface
Foot	The part of the displaced material that lies upslope from the toe of the surface of rupture
Crown	Practically undisplaced material adjacent to highest parts of main scarp
Flank	Undisplaced material adjacent to sides of surface of rupture, left and right refer to flanks as viewed from crown

Table 1.2 Abbreviated classification of slope movements (after Varnes, 1978)

			Type of material		
				Engineering soils	
Type of movement			Bedrock	Mostly coarse grained	Mostly fine grained
Falls			Rock fall	Debris fall	Earth fall
Topples			Rock topple	Debris topple	Earth slump
Slides	Rotational	Few units	Rock slump	Debris slump	Earth slump
	Translational	Many units	Rock slide	Debris slide	Earth slide
Lateral spreads			Rock spread	Debris spread	Earth spread
Flows			Rock flow	Debris flow	Earth flow
Complex combination of two or more types of movement					

failure (Crozier 1986; Hungr *et al.*, 2001). The purpose of this book is the analysis of the mechanics and hydrology of slope failure, so the Varnes system is used. The Varnes (1958, 1978) scheme classifies landslides based on the type of movement and the material involved (Table 1.2).

The type of movement is separated into falls, topples, slides (rotational and translational), spreads, and flows. Another category of "complex" movements is used to describe a

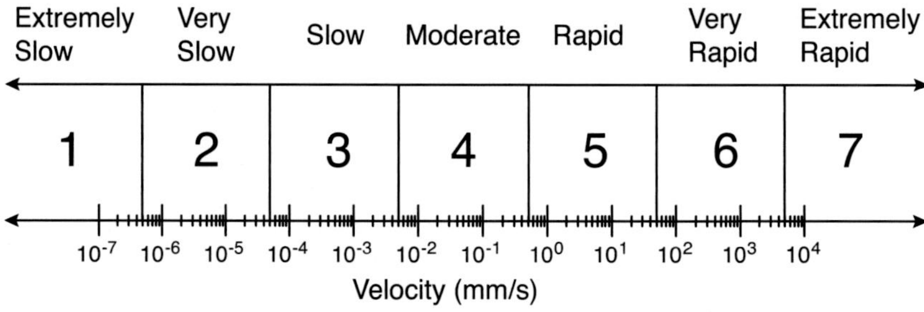

Figure 1.2 Landslide velocity scale (after Cruden and Varnes, 1996).

combination of any of these types of movements. Materials are separated into two classes, soil (in the engineering usage) and rock. Soil is differentiated from rock in that it is unconsolidated earth materials typically at the ground surface. Soil is subdivided into debris, which is predominantly coarse-grained in composition, and earth, which is predominantly fine-grained. The demarcation between debris and earth is arbitrary; debris is material in which 20–80% of the particles are greater than 2 mm in size, earth is material in which 80% of the particles are less than 2 mm (Shroder, 1971).

1.2.1 Landslide velocity

Landslides move at speeds ranging from a few millimeters per year to tens of kilometers per hour. The movement of the slowest slides is imperceptible to visual observation while the fastest may travel down mountain slopes in minutes or seconds. The state of landslide activity is classified as active, inactive, or fossil (Varnes, 1978). Active landslides are those that are currently moving or movement has been suspended, but they have moved within the last cycle of seasons. Deformation features such as scarps are typically distinct and easily identified. Inactive landslides are those for which no evidence of movement is detectable within the last cycle of seasons. Deformation features may be subdued by weathering and vegetation growth. Dormant slides are those in which movement has ceased, but changes in conditions may lead to renewed deformation. Reactivation of movement of dormant landslides is generally possible within the current climate. In contrast, fossil or relict landslides are slides in which movement has ceased and reactivation is not generally possible under the current precipitation climate unless human activities, such as reservoir construction, alter topographic or hydrologic conditions.

The destructiveness or hazard associated with landslides is proportional to their velocity. Cruden and Varnes (1996) provide a velocity ranking of landslides inspired by the Mercalli scale for earthquake damage (Figure 1.2). Extremely slow landslides that move less than a few tens of millimeters per year are likely to be undetected without instrumental observations. Properly engineered structures can be built on many such slides if the movement is recognized. Very slow landslides may move a few meters per year, and slow landslides move as much as 13 meters in a month (5×10^{-3} mm/s). Temporary and remedial structures can be maintained and built on such slides if episodic movements are limited. Landslides with moderate velocity, typically a few meters in an hour, are generally not a threat to human safety, but damage to buildings and other structures is common. Rapid landslides may move a few meters in minutes and potentially threaten human safety. Escape is generally possible; however, movement at this rate typically destroys buildings and property. Very rapid landslides move a few meters in a second. Lives are often lost if people are in the path of these landslides. Extremely rapid landslides move more than 5 meters per second. Escape from such landslides is unlikely and structures and buildings in their path are typically destroyed.

Figure 1.3 shows idealized sketches of various landslide types. The names describe either or both the style of movement and the materials involved. Styles of movement include *falls*, *topples*, *slides*, *spread*, and *flows* (Cruden and Varnes, 1996). *Falls* are a mass of rock or soil that is detached from a steep slope or cliff with little or no shear displacement along the failure surface. Under gravity, the mass then descends very rapidly to extremely rapidly by falling, bouncing, or rolling. *Topples* are the extremely slow to extremely rapid rotation of a mass of soil or rock out of the slope face. *Slides* are the downslope movement of soil or rock, in which the extremely slow to extremely rapid motion occurs primarily along a discrete surface of rupture. Sliding can occur in a variety of modes defined primarily by the shape of the rupture surface and the relative motion between the landslide body and the surrounding ground. Two distinct modes are translational and rotational slides. *Spread* is defined as the extension of a soil or rock mass combined with general subsidence of underlying material. The surface of rupture is not a zone of shear in such style of movement and movement rates are typically extremely slow to moderate. *Flows* are defined as spatially continuous movement in which shear surfaces are transient and short-lived and the moving mass takes the appearance of a viscous liquid. Flows can be extremely slow to extremely rapid and are often a secondary style of movement of a mass of soil or rock that initially fails as a fall or slide.

Because many flows move at great speed, they are potentially very destructive. This destructive potential has spurred the study of mass flows and driven the development of methods for hazard assessment and mitigation, as well as a rich vocabulary. Hungr *et al.* (2001) updated Hutchinson's (1968, 1988) classification of "debris movements of flow-like form" to further categorize the broad range of flows and conserve long-used terms. Post-failure movement is emphasized and classification of "landslides of the flow type" is based on the origin, character, and moisture condition of materials.

A widely used term from this classification is "flow slide." Flow slides are very rapid to extremely rapid flows of loose sorted or unsorted granular material involving excess pore pressures or liquefaction of material originating from the landslide source (Hutchinson, 1988; Hungr *et al.*, 2001). The term "flow slide" was introduced by Casagrande (1936) and redefined by Hutchinson (1988). A flow slide is characterized by the collapse of the internal soil structure during sliding or as a result of earthquake shaking that reduces pore space and elevates pore pressures in moist materials. These landslides can be particularly hazardous in that they tend to travel at high speeds over long distances. This process is effective at generating debris flows (Iverson *et al.*, 1997).

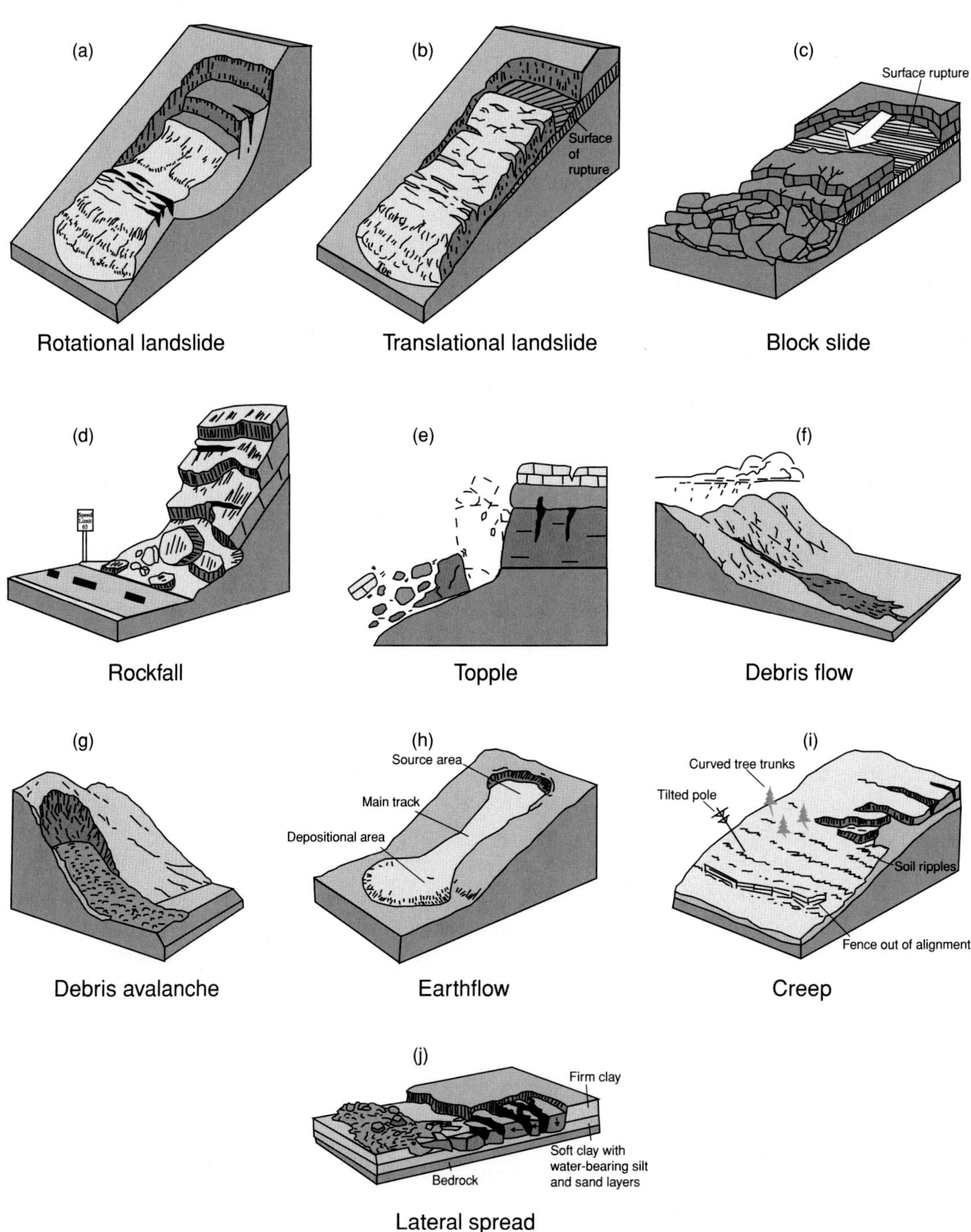

Figure 1.3 Idealized diagrams showing styles of landslide movement (after Varnes, 1978).

1.2.2 Illustration of landslide classification

The classification based on the type of movement and landslide materials is further illustrated below.

Rotational landslides (Figure 1.3a) move along an upwardly concave curved surface of rupture. In soil, the ratio of the depth (D_r) to length (L_r) of the surface of rupture D_r/L_r is typically between 0.15 and 0.33 (Skempton and Hutchinson, 1969).

Translational landslides (Figure 1.3b) move out and down a planar or undulating surface of rupture, often with a channel-shaped cross section that is parallel to the undisturbed ground. Rupture surfaces often coincide with discontinuities in earth materials such as bedding surfaces or the contact between rock and overlying soil. Translational slides are generally shallower than rotational landslides with D_r/L_r ratios of less than 0.1 (Skempton and Hutchinson, 1969). Translational landslides often mobilize as debris flows if slide velocity and pore-water pressures are sufficient (Iverson *et al.*, 1997).

Block slides (Figure 1.3c) or planar slides (Hoek and Bray, 1981) are translational landslides that move on a single discontinuity in rock masses. With sufficient displacement, block slides may break up into debris or transform into rock avalanches (Hutchinson, 1988).

Rockfall (Figure 1.3d) is the detachment of particles from a rock mass typically along a steep or vertical surface. The particles then descend by falling, bounding, or rolling.

Topple (Figure 1.3e) is the forward rotation of a rock or soil mass out of the slopes that may result in falls or slides.

Debris flows (Figure 1.3f) are poorly sorted slurries of rock, soil, and mud that are saturated with water. Debris flows are distinguished from other slope movement processes such as rock avalanches in that both fluid forces and particle interactions affect motion (Iverson, 1997).

Debris avalanches (Figure 1.3g) or rock avalanches are distinguished from debris flows in that the debris or rock is not saturated with water.

Earthflows (Figure 1.3h) are slope movements in which the moving mass resembles a viscous liquid. The mass may be bounded by discrete shear surfaces (Hutchinson and Bhandari, 1971; Keefer and Johnson, 1983). These landslides are also known as mudslides (e.g., Hutchinson, 1988).

Creep (Figure 1.3i) is any extremely slow movement that tends to be diffuse rather than movement along a distinct surface of rupture.

Lateral spreads (Figure 1.3j) are the extension of cohesive soil or rock mass combined with subsidence of the moving mass into underlying material. Spreads may result from liquefaction of underlying material.

1.3 Landslide occurrence

1.3.1 Landslide triggering mechanisms

A wide range of geologic and meteorologic processes may trigger landslides including volcanic eruptions, earthquakes, and heavy precipitation (Wieczorek, 1996). Human activities,

Figure 1.4 Debris flow on the flank of Mt. St. Helens triggered by an eruption in March 1982 (photo by Tom Casadevall, USGS). See also color plate section.

such as slope excavation, reservoir operation, and irrigation, may also initiate landslides. In this section some of the more common natural landslide triggers are described.

Landslides initiated by volcanic eruptions are among the largest and most destructive. The 18 May 1980 eruption of Mt. St. Helens in Washington State was coincident with the occurrence of a 2.8 km^3 rock slide–debris avalanche from the north flank of the edifice that travelled 22 km down the North Fork of the Toutle River (Voight *et al.*, 1983). This landslide and other debris flows initiated by the eruption destroyed homes and transportation infrastructure in the area (Schuster, 1981). Debris flows from volcanoes are often described using the Indonesian term "lahar." Lahars can be initiated by volcanic eruption or other mechanisms. Figure 1.4 shows a lahar that swept down the flank of Mt. St. Helens in March of 1982 following one of the 17 eruptive episodes in the 6-year period following the 1980 event.

Strong ground shaking from earthquakes has triggered all types of landslides in a variety of physiographic settings. Large earthquakes in mountainous areas can generate landslides over very large regions and are often responsible for a significant part of the societal consequences (Keefer, 1984). Examples from Kashmir in 2005 and Sichuan, China, in 2008 highlight their effects in terms of human loss and damage to the built environment. Figure 1.5 shows rockslides triggered by the M7.9 Wenchuan, China, earthquake and the resulting damage to the town of Beichuan. Perhaps as many as 20% of the nearly 100,000 lives lost in the earthquake were the result of landslides (Yin *et al.*, 2009).

Heavy precipitation, either rainfall or melting snow, is the most common landslide trigger. Landslides triggered by heavy precipitation occur in all parts of the world in a wide variety of climatic, geologic, and topographic settings. Large storm systems or sequences of storms

Figure 1.5 Rockslides triggered by the 2008 M7.9 Wenchuan, China, earthquake in Beichuan town (photo by Yueping Yin, China Geological Survey). See also color plate section.

can generate landslides over broad geographic areas. Some have caused tremendous loss of life and property, and damage to transportation infrastructure often hampers relief efforts. For example, Figure 1.6 shows landslides that were triggered by heavy rainfall in January of 2011 in the Novo Friburgo region of southeastern Brazil. More than 400 fatalities were reported from this disaster.

1.3.2 Frequency and magnitude of landslide events

Landslides that occur in response to individual heavy rainstorms or earthquakes may be distributed over regions that extend from a few to tens of thousands of square kilometers (e.g., Coe and Godt, 2001; Dai *et al.*, 2011). The largest of these landslide "events" may comprise many thousands of landslides (e.g., Bucknam *et al.*, 2001) and can dominate sediment production from hillslopes (Hovius *et al.*, 1997; Parker *et al.*, 2011). Such events are also often responsible for great loss of life and destruction of property (e.g., Guzzetti, 2000). Maps of the location and boundaries of landslides interpreted from aerial photography or other remotely sensed imagery are referred to as landslide inventories and provide critical empirical data for landslide hazard assessment (e.g., Harp *et al.*, 2010). Examination of landslide frequency and magnitude using inventories reveals a consistent relation between landslide area, which is assumed to be a proxy for volume or magnitude, and their relative frequency. Figure 1.7 shows the relation between the normalized probability density and landslide area for three landslide inventories (Malamud *et al.*, 2004). The probability density or relative frequency of landslides of a given size is similar despite variation in the type of triggering event, or geologic or climatic setting. The relative frequency of landslide size

Figure 1.6 Landslides triggered by heavy rainfall in Brazil in January 2011 (AP photo). See also color plate section.

increases with decreasing landslide area in an inverse power-law relation up to a limit of a few hundred square meters, below which the relative frequency tends to decrease again. Thus, very large and very small landslides tend to be relatively rare compared to landslides of more modest size. Similar relations appear to hold for submarine landslides as well (ten Brink *et al.*, 2006).

1.4 Socio-economic impacts of landslides

The social and economic impact of landslides worldwide is significant. Landslides associated with extreme events (e.g., tropical cyclones and earthquakes) have killed tens of thousands and their destructiveness can lead to loss of life that ranks with any natural disaster. For example, the 1786 Sichuan Province earthquake in China killed perhaps as many as a 100,000 people when a dam created by an earthquake-induced landslide subsequently failed. The 1920 earthquake in the Gansu Province in China triggered thousands of landslides in loess hillslopes, again killing perhaps 100,000 people.

Figure 1.8 shows the distribution and numbers of deaths from landslides in 2007 (Petley, 2008). In this year fatalities were concentrated in South Asia, Latin America, and the Indonesian archipelago. This reflects the physical conditions of steep topography and heavy rainfall in these regions as well as the human conditions of dense and expanding population and associated land-use activities such as deforestation.

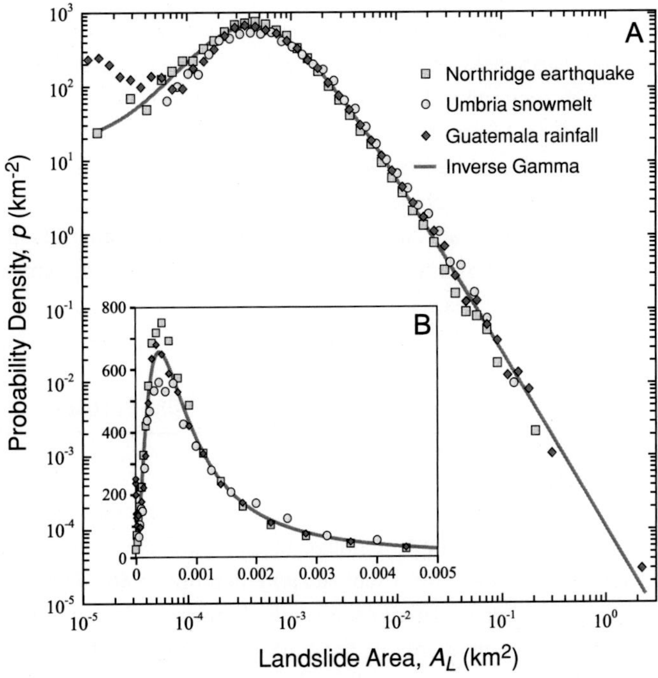

Figure 1.7 Relation between landslide area and probability density in log-log (A) and linear (B) space for three landslide inventories (from Malamud *et al.*, 2004). Landslide inventories were mapped after the 17 January 1994 Northridge earthquake in California (Harp and Jibson, 1995), a snowmelt-triggered event in Umbria in January 1997 (Cardinali *et al.*, 2000), and after Hurricane Mitch in Guatemala in late October and early November of 1998 (Bucknam *et al.*, 2001). The best-fit line is a three-parameter inverse-gamma function.

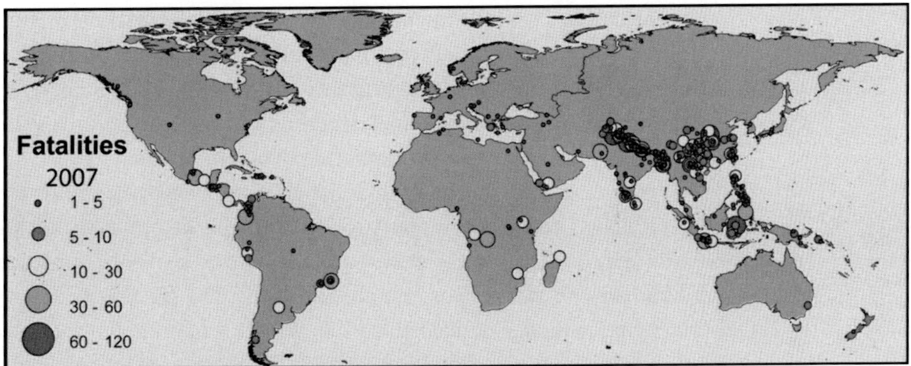

Figure 1.8 Distribution of landslide fatalities in 2007 (from Petley, 2008). See also color plate section.

In the United States, landslides are estimated to be responsible for between 25 and 50 deaths and more than US$2–3 billion in damage per year (Schuster and Highland, 2001; National Research Council, 2004). Despite the large economic impact of landslides, no systematic catalog of their occurrence or impact is maintained in the United States, and no insurance scheme exists to spread losses associated with landslides. Landslide damage is poorly documented because it is often considered part of the triggering event (e.g., earthquake, hurricane, or volcanic eruption), and while they may occur over broad regions their individual impact is often isolated. However, the damage from landslides often exceeds that from the other effects. For example, most of the damage in Anchorage, Alaska, associated with the 1964 M9.2 Good Friday Earthquake was the result of landslides (Hansen, 1965). In the developed world, the number of lives lost due to landslides is decreasing (Sidle and Ochiai, 2006); however, this is not generally the case in the less-developed world (Global South) due to population expansion into areas susceptible to landslides and land-use activities.

1.4.1 Types of costs

Economic losses resulting from landslides can be categorized as either direct or indirect, public or private. Direct economic loss is defined as the costs of "replacement, repair, or maintenance due to damage to installations or property within the boundaries of the responsible landslide" (Schuster and Fleming, 1986). These costs include material and labor to rebuild, repair, or replace roads, homes, buildings, sewers, water supplies, pipelines, and other components of the built environment. Maintenance costs to remove landslide debris from roadways, walkways, parks, and other areas are also included. All other costs associated with landslides can be classified as indirect. Such costs can be substantial and are difficult to quantify, and include the interruption of utility and transportation services. Costs of increased travel time, lost wages and tax revenue, decreases in property values, litigation expenses and settlements, and mitigation efforts aimed at preventing damage from future landslides are also classified as indirect costs.

Public costs are defined as those borne by local, municipal, county, state and federal government entities, and ultimately by taxpayers. Public costs are largely related to the maintenance and repair of roadways damaged by landslides (Fleming and Taylor, 1980). Other public costs include those associated with the response to emergency and relief needs by fire, police, and medical personnel. Building inspections, hazard evaluations, warnings, and evacuations are also typically supported by public funds.

Private costs are those borne by individuals and businesses and result from damage to homes, buildings, and other property. Condemnation of property and the requirement of demolition may compound the costs of damage to a residence. In some cases the value of the land may be reduced or completely lost. Costs for construction, demolition, and real estate can generally be determined; however, litigation and fear of condemnation of property may make this information difficult to obtain in practice. Other private costs are much harder to define, but should not be discounted. The loss of a home or loved one may cause social and psychological damage that far outweighs the financial cost.

Although landslide losses are difficult to obtain and estimates are subject to error, Table 1.3 provides a comparison of the annual losses from landslides for three countries in the developed Global North: Italy, Japan, and the United States. The annual landslide costs for both Japan and Italy are about US$4 billion, about 1.6 times that for the United States. However, the population of both countries is considerably smaller than that of the

Table 1.3 Average annual costs for three nations in the developed Global North (from Sidle and Ochiai, 2006)

Country	Italy	Japan	United States
Annual cost (BUS$)	4.00	4.00	2.50
Population (M)	58.15	127.40	301.00
GDP (BUS$)	1,760.00	4,218.00	13,130.00
Per capita (US$/person)	68.79	31.40	8.31
%GDP	0.23	0.09	0.02

United States. Thus, landslide costs constitute about a quarter of one percent of Italy's gross domestic product (GDP) whereas in the United States those costs are a few hundredths of one percent.

1.4.2 Historical examples of widespread landslide events in North America

Table 1.4 describes several rainfall-induced widespread landslide events in North America and includes estimates of loss of life and economic costs. The most destructive event in this list by far was the category 5 Hurricane Mitch, in October of 1998. The stalled remnants of Hurricane Mitch impacted much of Central America and resulted in the deadliest hurricane to hit the region in more than 100 years. Many areas received more than 1.5 m of rainfall over a several day period, which caused widespread flooding and hundreds of thousands of landslides. For example, in Honduras, more than half a million landslides occurred (Harp *et al.*, 2002), more than 7,000 people were killed, and about one-half of the population of the country was impacted financially. One in three farms lost crops or livestock, and one in ten businesses were impacted. More than 200 bridges were damaged or destroyed and more than 70% of the road network was rendered impassable (Morris *et al.*, 2002).

1.4.3 Direct economic loss in the San Francisco Bay region in 1997–8

Godt and Savage (1999) contributed to a catalog of the direct economic loss from landslides in the San Francisco Bay region in California following the unusually wet winter season of 1997 and 1998. Landslides during this period were related to the warm phase of the El Niño Southern Oscillation (ENSO). The region experienced near-record rainfall (600–1,000 mm) with some areas receiving more than twice the normal average during the period. Particularly intense rainfall struck the region during the first week of February and triggered thousands of shallow landslides and debris flows (Coe and Godt, 2001; Coe *et al.*, 2004).

Landslides considered in the damage catalog included any failure of hillside materials, both natural and engineered, that impacted the built environment. More than 300 damaging landslides were documented (Figure 1.9); the slides ranged in size from a 25 m^3 failure of engineered material to the reactivation of the massive (13 million m^3) Mission Peak earthflow complex (Rogers, 1998).

The direct economic impact of landslide damage was assessed throughout the San Francisco Bay region through interviews with emergency managers and public works employees,

Table 1.4 Examples of landslide disasters in North America

Location	Date	Description	Loss
Greater Los Angeles, California, USA	January and February, 1969	Widespread landslides and debris flows	18 deaths; 175 homes destroyed; losses > $40M
Central Virginia, USA	August 1969	Widespread landslides and debris flows triggered by Hurricane Camille	~150 deaths thought to mainly result from debris flows
San Francisco Bay region, California, USA	January 1982	>18,000 landslides and debris flows	14 deaths; >7,000 homes and businesses damaged or destroyed; direct and indirect losses >$4,000M
Wasatch Front, Utah, USA	Spring 1983	Widespread landslides and debris flows	Destruction of the town of Thistle; >1,300 jobs lost, losses > $600M
Honduras, Guatemala, Nicaragua, El Salvador	October 1998	Widespread (hundreds of thousands) of landslides of all types; destructive debris flows from Hurricane Mitch	Perhaps as many as 30,000 deaths; development set back by decades
Caraballeda, Venezuela	December 1999	Intense localized landslides and debris flows	Perhaps as many as 30,000 deaths
Western North Carolina, USA	September 2004	Widespread landslides and debris flows triggered by Hurricanes Ivan and Frances	5 deaths; many homes damaged or destroyed

private engineering consultants, and the general public who owned or managed property impacted by landslides in the area. More than US$158 million in direct losses to both public and private property were tabulated (Table 1.5). The distribution of public and private costs varied significantly by county (Figure 1.9). Where landslide damage was widespread (e.g., San Mateo, Alameda, and Santa Cruz counties) losses were about evenly distributed between public and private entities. However, intense localized landslide activity tended to bias the loss estimates. For example, a large portion of the damage in Sonoma County resulted from the localized landslide disaster in the community of Rio Nido. Three homes were completely destroyed and 32 other properties condemned. Thus it is difficult to infer private direct losses from more easily obtained public data.

1.5 Rainfall-induced landslides

Shallow landslides are typically translational slope failures a few meters thick of unlithified soil mantle or regolith (Cruden and Varnes, 1996; Sidle and Ochiai, 2006). They may occur wholly or partly in the unsaturated zone, and may dominate mass-movement processes

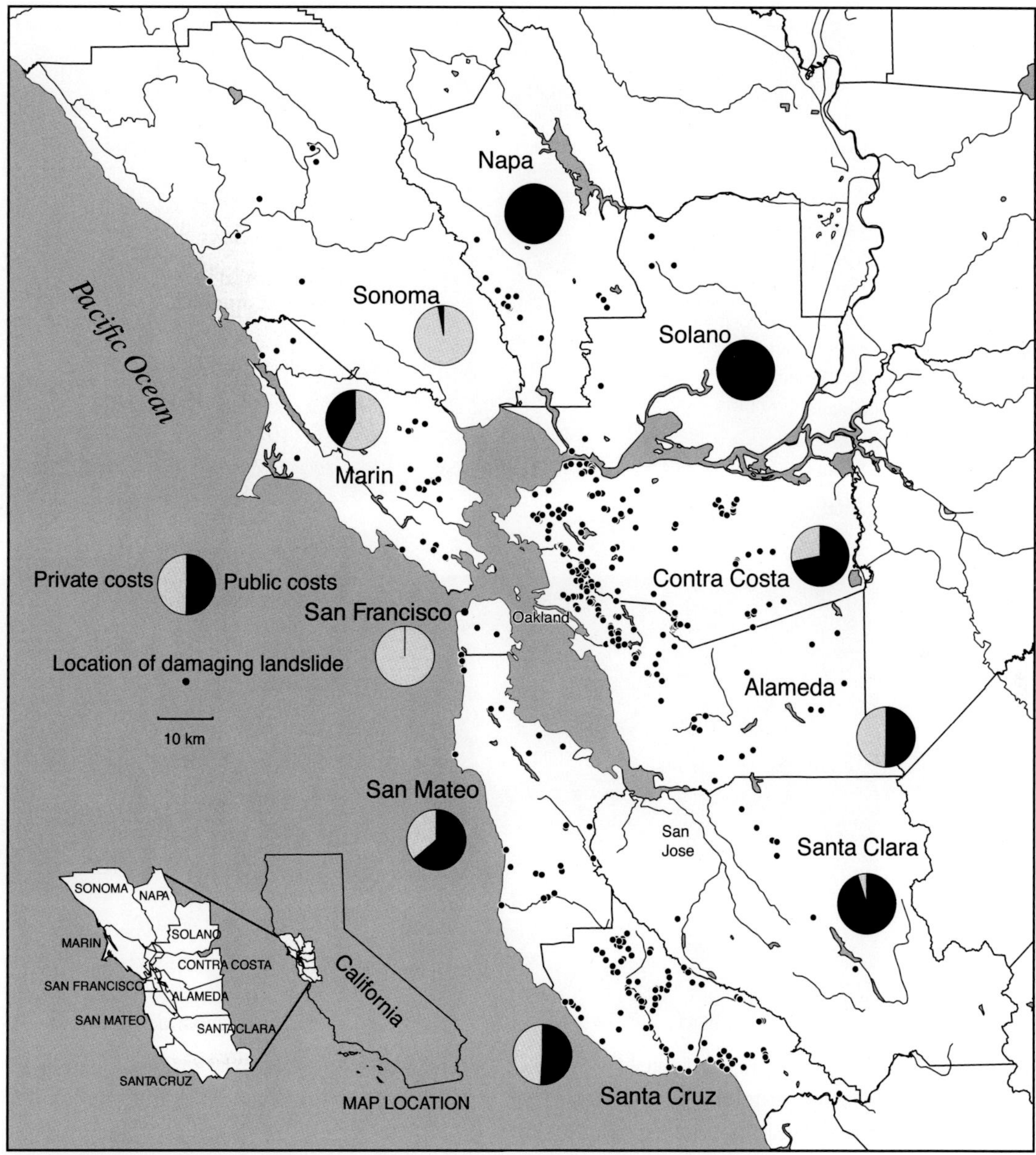

Figure 1.9 Map showing the location of damaging landslides in the San Francisco Bay region during the winter season of 1997–8. Pie charts show the distribution of public (black) vs. private (grey) costs (from Godt and Savage, 1999). See also color plate section.

in hillslope environments (e.g., Trustrum *et al.*, 1999). They are particularly destructive when they initiate or coalesce to form debris flows (Iverson *et al.*, 1997). According to a recent survey, about half of the 40 most destructive landslide disasters worldwide in the past century resulted from prolonged or intense rainfall (Sidle and Ochiai, 2006). Although

Table 1.5 Direct costs of landslides in the San Francisco Bay region resulting from the El Niño rainstorms in the winter season of 1997–8. Per-capita losses are estimated based on population statistics from the California Department of Finance, City and County Populations Estimates, May 1998 (from Godt and Savage, 1999).

County	Landslide losses (millions US$)	Per-capita loss (US$)
Alameda	20.02	14.22
Contra Costa	27.00	29.98
Marin	2.54	10.33
Napa	1.12	9.08
San Mateo	55.00	76.88
Santa Clara	7.60	4.50
San Francisco	4.10	5.19
Solano	5.00	13.03
Sonoma	21.00	48.04
Santa Cruz	14.68	58.67
Total	158.06	22.76

Figure 1.10 Photograph showing abundant shallow landslides and debris flows near Valencia, Los Angeles County, California. The light, un-vegetated scars are shallow failures of hillslope material caused by heavy rainfall during the winter of 2005 (photo by the authors). See also color plate section.

the volume of individual shallow landslides is often small, typically less than 1000 m^3, extensive areas are often affected. Figure 1.10 shows an example of widespread shallow landslides that resulted from heavy rainfall during the winter of 2005 in Los Angeles County, California.

Recent advances in understanding of the initiation of precipitation-induced shallow landslides can be broadly divided using two conceptual models on the occurrence of slope failure above the water table. The first is within the context of classical soil mechanics and it states that the failure surface is saturated and has compressive pore-water pressures acting on it. Field and experimental results have identified vertical infiltration and shallow water table accretion above a permeability contrast (Sidle and Swanston, 1982; Reid *et al.*, 1988; Johnson and Sitar, 1990), transient development of positive pore pressures (Reid *et al.*, 1997), and exfiltration of groundwater from bedrock (Montgomery *et al.*, 1997) as some of the hydrologic processes that can generate shallow landslides in this manner. Few of the field studies provide conclusive support for any single hydrologic process, but rather emphasize the spatial and temporal variability of the near-surface groundwater response to precipitation and the implied consequences for slope stability (e.g., Fannin and Jaakkola, 1999). In contrast, experimental study in controlled settings has shown that a variety of hydrologic processes can generate conditions sufficient to cause shallow landslides (Reid *et al.*, 1997).

The second conceptual model asserts that the state of stress of the soil or regolith is modified by infiltration and changes in soil matric suction and that these changes can lead to slope failure without complete saturation and compressive pore-water pressures along the failure surface. However, to date, understanding shallow landslide generation mechanisms rests primarily with observations and analytical and numerical results from geologic and climatic settings where precipitation-induced shallow landslides occur in soils or regolith that overlie permeable substrates (e.g., Lumb, 1975; Brand, 1981; Day and Axten, 1989; Wolle and Hachich, 1989; Fourie *et al.*, 1999; Cho and Lee, 2001; Springman *et al.*, 2003; Collins and Znidarcic, 2004; Francois *et al.*, 2007). Further evidence for this mechanism can be found in the theoretical understanding of the contribution of the reduction of soil suction to the stability of slopes (e.g., Morgenstern and de Matos, 1975; Rahardjo *et al.*, 2007) and the effective stress reduction in partially saturated soil masses (e.g., Khalili *et al.*, 2004; Lu and Likos, 2004; Lu and Likos, 2006; Godt *et al.*, 2009). However, the mechanics of slope failure and behavior under unsaturated conditions have principally been investigated using numerical sensitivity studies (e.g., Ng and Shi, 1998; Cho and Lee, 2001; Rahardjo *et al.*, 2007), and quantitative physical confirmation from field or laboratory experiments on pore-water conditions of the failure surface remains elusive.

Recent work in the Serra do Mar, Brazil, provides a vivid example of landslides above the water table, and demonstrates that the loss of soil suction during rainstorms can trigger shallow failures on steep slopes without development of positive pore pressures (Abramento and Carvalho, 1989; Wolle and Hachich, 1989). In the Serra do Mar, shallow landslides are typically about a meter thick and fail in colluvial soils underlain by more permeable fractured granitic and gneissic rocks and saprolite. Field observations over several years indicated that groundwater levels are consistently 20–30 m below the ground surface and that the permeable bedrock drains the overlying saprolite and soil layers. Thus, it was unlikely that perched water bodies above the water table contributed significantly to the slope failures.

1.5.1 Evidence of shallow landslide occurrence in the unsaturated zone

Monitoring efforts by the USGS and co-workers in the Puget Sound region in Washington captured shallow landslide occurrence above the water table under nearly saturated conditions for the first time (Godt *et al.*, 2009; Figure 1.11). Shallow landslides are a common

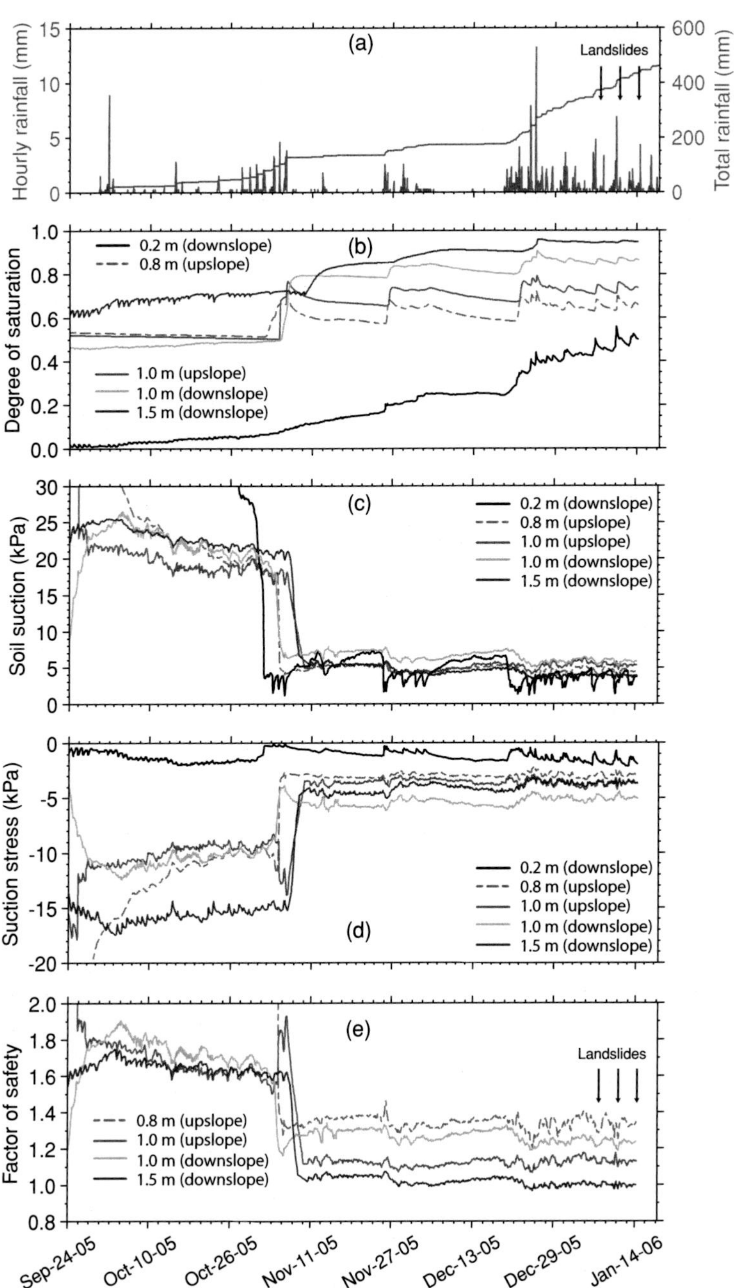

Figure 1.11 Shallow landslides under partially saturated conditions at the Edmonds, Washington, site. (a) Hourly and cumulative rainfall, (b) hourly soil saturation, (c) hourly soil suction, (d) suction stress, and (e) factor of safety for the period 24 September 2005 to 16 January 2006 at various depths from an upslope and downslope instrument array. Arrays are separated by about 3 m (from Godt *et al.*, 2009). See also color plate section.

and often hazardous occurrence on the steep coastal bluffs in the Puget Sound Lowland during the wet winter season (Miller, 1991) and typically occur during extended wet periods lasting several days (Godt *et al.*, 2006). The steep (>30°) 50 to 100 m high coastal bluffs in the Seattle, Washington, area are the result of Pleistocene age glaciation, wave attack at the shoreline, and mass movement processes (Shipman, 2004). In such a geologic, hydrologic, and climatic setting, shallow landslides are generally less than a few meters thick and typically occur in the loose, sandy, colluvial deposits derived from the glacial and non-glacial sediments that underlie the bluffs (Galster and Laprade, 1991). The instrumented hillslope is a steep coastal bluff with a thin (<2.0 m) colluvial cover near Edmonds, Washington. Measured profiles of soil saturation and suction are shown in Figure 1.11b and c (Baum *et al.*, 2005). The bluffs along this section of coastline are subject to a variety of landslide processes that periodically disrupt rail services between Seattle and Everett (45 km north of Seattle).

Changes in effective stress can be captured by suction stress (Lu and Likos, 2006; also Chapter 6). Suction stress is calculated as the product of soil saturation (Figure 1.11b) times soil suction (Figure 1.11c). Factors of safety were calculated using the classical infinite-slope model (e.g., Duncan and Wright, 2005); however, pore pressure was replaced by suction stress (Lu and Godt, 2008) for measured values of slope angle of 45°, internal friction angle of 36.1°, true soil cohesion of 0.93 kPa, a saturated moisture content of 40%, and a residual moisture content of 5%. Periods when the factor of safety was less than or equal to 1.0, indicating unstable conditions: (1) 29 Nov. 2005 to 7 Dec. 2005: rainfall 25.6 mm, no landslides reported; (2) 23 Dec. 2005 to 28 Dec. 2005: rainfall 84.3 mm, no landslides reported along the Seattle–Everett corridor. However, landslides were reported in Seattle and south in Tacoma on 25 Dec. 2005; (3) 30 Dec. 2005 to 4 Jan. 2006: rainfall 42.5 mm, no landslides reported; (4) 5 Jan. 2006 to 8 Jan. 2006, rainfall 31.2 mm, several landslides reported on 6 Jan. 2006, rail corridor closed between Edmonds and Everett; (5) 9 Jan. 2006 to 14 Jan. 2006: rainfall 58.6 mm, several landslides reported on 10 Jan. 2006 between Seattle and Edmonds, instruments damaged and buried under unsaturated landslide debris on 14 Jan. 2006. When landslides occurred during the middle of January 2006 (Figure 1.11e where factor of safety remains less than 1.0), all monitoring stations show that soil was unsaturated (Figure 1.11b) and negative pore pressure (soil suction) existed (Figure 1.11c).

Experimental results from controlled, large-scale landslide experiments also indicate that shallow landslide failures may occur in the absence of widespread compressive pore pressures (Iverson *et al.*, 1997; Reid *et al.*, 1997). In Experiment II (Iverson *et al.*, 1997), a slope failure was initiated by artificial rainfall, which initially occurred in a mostly unsaturated part of the experimental hillslope. Traditional limit-equilibrium analysis of this slope failure assuming saturated conditions produced unsatisfactory results yielding a factor of safety much greater than unity. Failure was triggered in Experiment III by high-intensity rainfall resulting in near zero pore pressures throughout the soil mass without the formation of a water table (Reid *et al.*, 1997). This style of failure was also difficult to explain using conventional slope stability analyses. Reid *et al.* (1997) point to pressure perturbations induced by the high-intensity rainfall as a possible triggering mechanism

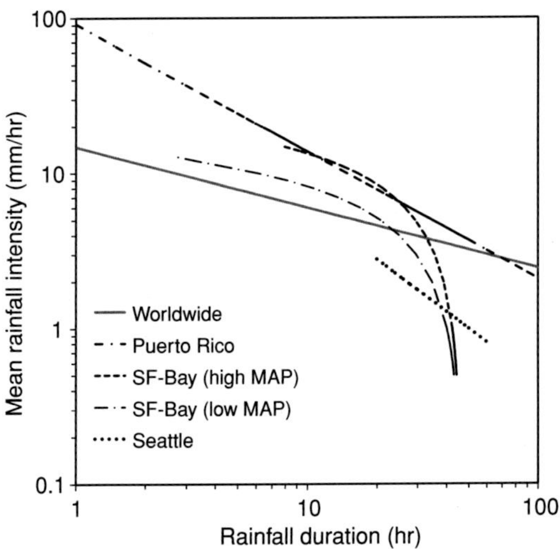

Figure 1.12 Rainfall intensity–duration thresholds for Seattle (Godt *et al.*, 2006), Puerto Rico (Larsen and Simon, 1993), San Francisco Bay region for areas of high and low mean annual precipitation (MAP) (Cannon and Ellen, 1988), and worldwide data sets (Caine, 1980).

for failure, similar to the abrupt water-table rise caused by rainfall in tension-saturated materials (Gillham, 1984; Torres and Alexander, 2002).

1.5.2 Role of precipitation characteristics in triggering shallow landslides

The forecasting of precipitation-induced shallow landslides has traditionally relied on empirical correlations between landslide occurrence and antecedent and storm rainfall to define a threshold condition above which shallow landsliding can be expected (Caine, 1980; Keefer *et al.*, 1987; Hong *et al.*, 2006). This methodology of forecasting landslides is illustrated in Figure 1.12. The application of rainfall thresholds for forecasts or warnings rests on the assumption that past rainfall conditions associated with shallow slope failure are likely to trigger landslides in the future. In practice, the use of rainfall thresholds requires real-time precipitation monitoring and a decision protocol for issuing warnings. Several approaches have been used to account for the local climatic, topographic, and geologic effects on the intensity and duration of threshold rainfall (Cannon and Ellen, 1988; Wilson, 1997). Typically, these approaches use a regional climatic variable such as mean annual precipitation to extend the application of thresholds developed for a specific locale (often chosen based on data availability) to a broader region.

An intensity–duration threshold is an attempt to identify the minimum rainfall conditions that yield destabilizing pore-water conditions at depth in a hillslope. The pore-water response at depth to infiltrating rainfall that leads to hillslope destabilization is a transient process, which for a given depth in the hillslope is controlled by the hydraulic properties of the hillslope materials and its initial moisture content as well as the intensity and duration of rainfall (Reid, 1994; Iverson, 2000). This implies that as the moisture content of the colluvium on steep hillslopes increases, rainstorms of lower intensity or duration may cause shallow failures, or in probabilistic terms, the likelihood that a given storm will cause shallow failure increases with antecedent moisture (Crozier, 1986, 1999).

Both the approach to determine the antecedent or initial rainfall conditions and the approach to define the triggering rainfall intensity and duration rely on empirical correlations between rainfall characteristics and shallow landslide occurrence. The quality of forecasts or predictions of landslide occurrence using either empirical method are highly dependent on the length and quality of the historical rainfall and landslide records. These records are unavailable for most areas. Therefore, these methods are generally empirical qualitative or semi-quantitative analysis and are difficult to generalize for prediction of precipitation-induced shallow landslides over broad regions.

1.5.3 Role of infiltration and unsaturated flow within hillslopes

The role of infiltration of water into shallow soils and the subsequent pore-pressure response at depth is critical to the understanding of transient conditions that lead to shallow slope failure. There is a strong correlation between the timing of intense precipitation and dynamic variation in soil pore-water pressure.

Quantification of rainfall infiltration into hillslopes has been an active research area for the previous three decades. Infiltration of water into soils is a transient process that is highly dependent on the hydraulic properties of the material and the initial moisture conditions. Numerical solutions to the governing equation for partially saturated flow have been widely applied to agricultural, water supply, and contaminant transport problems, beginning with the work of Rubin and Steinhardt (1963). More recently these models have been applied to evaluate the hydrologic response of steep hillslopes to natural and applied rainfall (e.g., Ebel *et al.*, 2007) and conditions that cause landsliding in a wide variety of geologic and climatologic settings. For example, Reid *et al.* (1988) applied a two-dimensional variably saturated flow model to the fine-grained soils of a shallow landslide location in the Santa Cruz Mountains of California and showed that a downslope variation in hydraulic conductivity apparently led to mounding of a perched groundwater table and resultant failure.

Sustainable land use is one of the driving forces for research and prevention of landslides around the world. For example, in Hong Kong, landsliding presents a significant hazard to densely populated areas in steep terrain, and numerical modeling of unsaturated flow has been applied to a variety of slope stability problems. For example, Anderson and Howes (1985) developed a one-dimensional, coupled, unsaturated flow-slope stability model (CHASM) to evaluate the role of soil matric suction on the strength of the deep colluvial materials on steep hillslopes in the Mid-Levels of Hong Kong. CHASM is an integrated slope hydrology/slope stability model that solves the governing flow equations using a finite-difference numerical approach. This model has been extended to two dimensions to incorporate non-planar failure surfaces (Anderson *et al.*, 1988), and has been applied to evaluate the effect of climate change and pedogenesis on shallow slope failures in Scotland over the Holocene (Brooks and Richards, 1994; Brooks *et al.*, 1995). It has also been used to examine the effectiveness of vegetation as a mitigation strategy for shallow failures in the humid tropics (Collison *et al.*, 1995). Recently, the CHASM model has been used to investigate the changes in hillslope susceptibility to shallow landsliding following forest removal and depletion of the soil and regolith by erosion processes in New Zealand (Brooks *et al.*, 2002).

1.6 Scope and organization of the book

1.6.1 Why does rainfall cause landslides?

Here is a simple story. Rainfall leads to transient infiltration and redistribution of soil moisture in hillslopes. Depending on how the process is conceptualized, soil moisture change or infiltration induces changes in either soil strength or effective stress. When the state of stress in a hillslope reaches its limit of strength, failure of the hillslope material occurs, leading to the possibility of landslide movement. Thus, it is important to understand how water infiltrates into the soil and how the water is redistributed in hillslope environments (Part II: Chapters 3–4). It is equally important to understand how slope geometry, dimensions, infiltration (Part III: Chapters 5–6), and material properties (Part IV: Chapters 7–8) affect the state of stress and the stability of hillslopes in response to hydrological and mechanical changes (Part V: Chapters 9–10).

1.6.2 Organization of the book

The authors intend to systematically present the current knowledge on geomorphology, hydrology, and geomechanics pertinent to landslide problems. The book is written in five parts in ten sequential chapters; Part I: 1 Introduction and 2 Slope geomorphology, Part II: 3 Steady infiltration and 4 Transient infiltration, Part III: 5 Total stresses in hillslopes and 6 Effective stress in variably saturated media, Part IV: 7 Strength of hillslope materials and 8 Hydro-mechanical properties, and Part V: 9 Failure surface based stability analysis and 10 Stress field based stability analysis. The first two chapters introduce the problem by describing landslides, their impact on human activities, and the role of landslides in sculpting the earth's surface. Chapters 3 and 4 are a quantitative description of water movement through the near-surface environment. Chapters 5 and 6 provide systematic descriptions of stress fields including total stress and effective stress in shallow hillslope environments. Chapters 7 and 8 describe mechanical properties or strength and hydro-mechanical properties of hillside materials. These chapters build a foundation for the final two chapters (Chapters 9 and 10). In these chapters the authors introduce failure surface based or limit-equilibrium slope stability and stress field based slope stability analyses, and rainfall-induced landslides are examined.

1.7 Problems

1. What are the two major geologic processes that determine the morphology of the earth's surface?
2. What is the main reason for erosion?
3. Name several disciplines that are highly related to landslides.
4. What kind of physical processes govern the stability of slopes?
5. What is the role of water in the stability of hillslopes?
6. In a few sentences, describe what you expect to learn from this book after reading the preface.
7. Name at least three countries that are severely affected by landslides. What is the typical annual cost of landslides in those countries?
8. Why are the impacts of landslides becoming more pronounced worldwide?
9. Name the areas (regions) in the United States that are most affected by landslides.
10. What are the two most important criteria used in landslide classification? And how do we describe those criteria?
11. What are the major causes of landslides? From a mechanical perspective, what is the common trigger of landslides?
12. A landslide is measured with the following dimensions: the width of rupture surface is 50 m, the depth of rupture is 5 m, and the length of rupture is 100 m. Estimate the volume of the landslide.
13. Name the four factors that contribute to landslide occurrence in mountainous regions.
14. What geologic units are particularly susceptible to landsliding?
15. Why are areas along plate margins particularly susceptible to landslides?
16. List three ways industrial forest practices may affect landslide frequency.
17. Describe and give examples of the types of costs associated with landslides.
18. Name three major factors that have increased the loss due to landslides.
19. In today's environment, why is vulnerability to landslide disasters increasing?
20. List three of the most common physical reasons for landslides.
21. What is the main difference between landslide "causes" and "triggers?"
22. Would you expect to have more frequent landslides in the future, and why?
23. What is the main failure mode in the crest region? What is the main failure mode along the rupture surface? And what is the main failure mode near the toe region?
24. Briefly explain why there are different failure modes in different locations of a landslide?
25. For landslides with different velocities, which one has the highest energy? And where does the energy come from?
26. What are the major differences between topple and rockfall?
27. What are the likely geologic and hydrologic conditions for translational landslides?
28. What are the major triggering mechanisms for landslides?
29. Is there an insurance scheme for landslide losses in the United States, and why?
30. Which region had the greatest number of landslide fatalities in 2007?

31 In the United States, which region(s) tends to have large landslide damages?
32 What makes shallow landslides particularly destructive?
33 What triggered the most destructive landslide disasters over the past century?
34 What are the two possible conditions under which landslides may occur?
35 What two quantities, measured in the Edmonds, Washington, case study, were used to calculate suction stress?
36 Describe the traditional, empirical method used to forecast precipitation-induced landslides.

2 Hillslope geomorphology

2.1 Hillslope hydrologic cycle

2.1.1 Global patterns of precipitation and evaporation

Hydrologic processes describe the form of water movement. Water movement in hillslopes, either in liquid or vapor form, is dynamic. It provides the most common physical mechanism that drives the spatial and temporal variation in stress in hillslopes and thus landslide occurrence. As shown in Figure 2.1, the movement of water in and over a hillslope typically involves several physical processes, namely precipitation, evaporation, transpiration, runoff, infiltration, and saturated groundwater flow. Furthermore, these physical processes are often inter-related and are greatly controlled by a hillslope's physical location, morphology, climatic setting, and the geologic materials of the hillslope.

At a particular location, the rate, amount, and seasonal distribution of precipitation and evaporation are greatly influenced by the global climate. The amount of annual precipitation is largely a function of latitude, the proximity to the moisture source of the ocean or sea, the global atmospheric circulation, and local wind patterns, which are greatly influenced by topography. The dependence of precipitation over land on latitude and proximity to the ocean is illustrated in Figure 2.2. The lightest-shaded land areas indicate the driest regions, which receive an average annual precipitation of less than 25 cm. This zone roughly lies between the latitudes of 17° and 30° N and 17° and 30° S as well as the zone at latitudes higher than 70° in both the southern and northern hemispheres. Examples of locations that lie in these zones of low precipitation are much of northern and southern Africa, southern and western Australia, and much of central Asia. The darker-shaded land areas indicate the wettest regions, which receive more than 200 cm of precipitation annually. These areas of high precipitation are located in the equatorial latitudes roughly between 10° N and 10° S, and in middle latitudes where global and smaller-scale circulation patterns interact with topography to produce abundant precipitation. Examples of such regions are located in the northeast part of the South Asian subcontinent at about 30° N latitude where monsoonal flow, described in more detail in Section 2.1.4, interacts with the Himalaya upland, and the Pacific coasts of North and South America at latitudes between 40° and 60° where westerly flow interacts with coastal mountains.

The locations of precipitation maxima and minima are a consequence of the latitudinal variation in the movement of air over the globe. At the global scale, the circulation pattern is divided into "cells" that span broad latitudinal bands. The primary cell, known as the Hadley cell describes the movement of warm moist air that rises from the equatorial

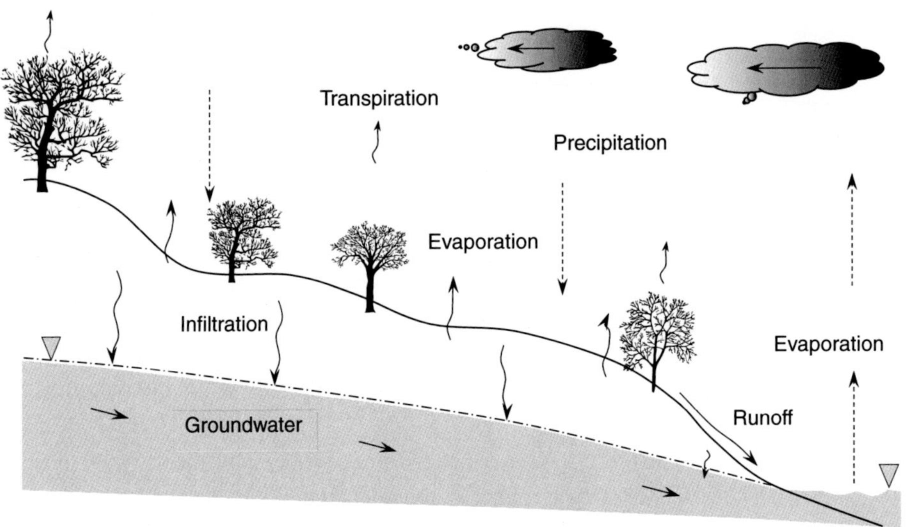

Figure 2.1 The hydrologic cycle in and around a hillslope (from Lu and Likos, 2004a).

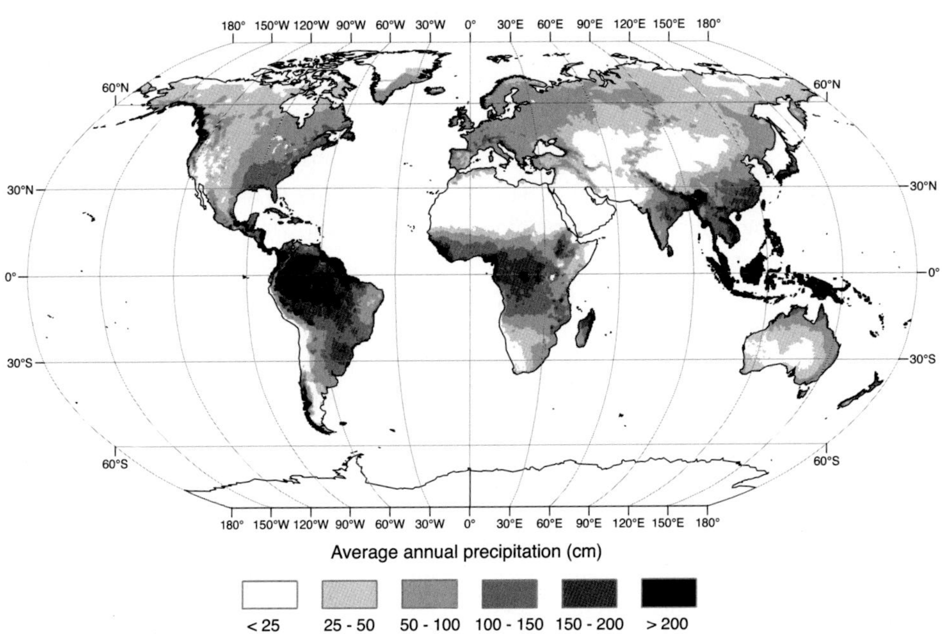

Figure 2.2 Global average annual precipitation (data from the Global Precipitation Climatology Centre, Meyer-Christoffer *et al.*, 2011).

ocean, which then moves poleward both north and south, descends at about 30° N and S latitudes, and finally moves back towards the equator. Along this pathway, the air cools and precipitation results, but the amount that falls diminishes with distance from the moisture source of the equatorial ocean. This results in dry zones between about 17° and 30° latitude

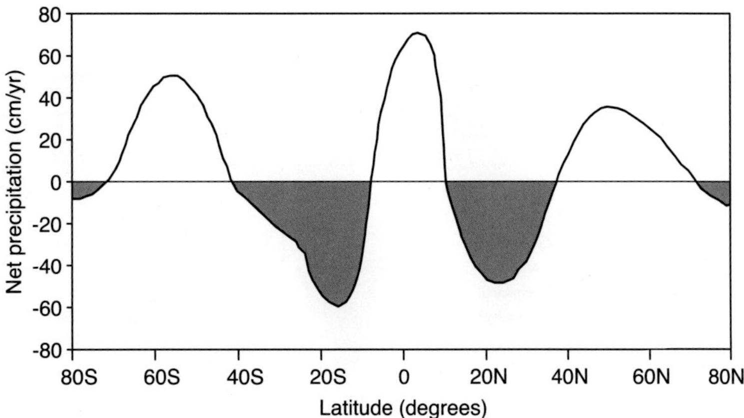

Figure 2.3 Net precipitation as a function of latitude (data from Peixoto and Kettani, 1973).

in both hemispheres. Another circulation pattern (mid-latitude cell) occurs between 30° and roughly 50° latitude, which gives rise to the precipitation patterns in those regions: characteristically moderate precipitation between 30° and 60° latitudes, but extremely dry zones in regions poleward of 60° in both hemispheres. The general circulation cells interact with the rotation of the earth, the ocean circulation, and the land mass to produce the complex distribution of precipitation over the earth's land area.

The evaporation process is controlled by the thermodynamic variables of pressure, relative humidity, and temperature of the atmosphere immediately adjacent the hillslope ground surface and the thermodynamic variables of the hillslope materials and vegetation. In general, gradients in these variables between the atmosphere and soils and vegetation provide the driving forces for vapor exchange; the higher the gradient, the higher the evaporation rate will be.

The global latitudinal pattern of net precipitation (annual precipitation corrected with evaporation) is shown in Figure 2.3. The pattern is characterized by net evaporation in zones between 10° and 35° N latitude, between 10° and 40° S latitude, and in polar regions. Precipitation exceeds evaporation in equatorial regions at latitudes lower than 10° and in the mid-latitude zones between about 36° and 72° N latitude and between 41° and 75° S latitude.

Climate has a great influence on the occurrence of landslides as it influences the degree of weathering, the availability of moisture, and the type and structure of vegetation. The ratio of precipitation to evaporation has been used to classify climate conditions. The global humidity index (UNESCO, 1984) is a widely used classification system that is based on the ratio of average annual precipitation and potential evaporation (P/PET). If an area has $P/PET < 0.05$, it is called a hyper-arid zone. If an area has $0.05 < P/PET < 0.2$, it is called an arid zone. If an area has $0.2 < P/PET < 0.5$, it is called a semi-arid zone. If an area has $0.5 < P/PET < 0.65$, it is called a dry–semihumid zone. If an area has $P/PET > 0.65$, it is called a humid zone. Based on these criteria, a global humidity index map can be created as shown in Figure 2.4.

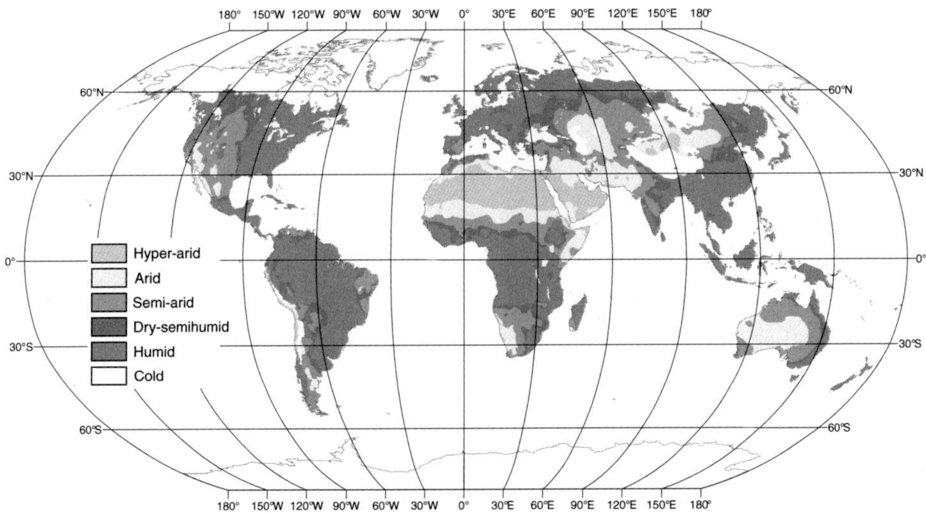

Figure 2.4 Global humidity index map (adapted from GRID/UNEP, Office of Arid Land Studies, University of Arizona). See also color plate section.

By using the same criteria as those defined by UNESCO's global humidity index, a humidity index map of North America is shown in Figure 2.5.

Because the humidity index, together with hillslope geometry and materials, and vegetation, has a direct impact on the distribution of soil moisture and soil suction, it describes the general state of soil stress (suction stress) in the unsaturated zone of hillslopes. While the correlation between humidity index and soil moisture and suction distribution is straightforward (i.e., a high humidity index generally promotes wet soil and low suction conditions), the correlation between humidity index and soil stress is quite complicated and material dependent. The exact nature of the interplay among precipitation, hillslope materials, and soil suction stress is a focus of this book and will be addressed quantitatively throughout the subsequent chapters.

A qualitative illustration of the interplay between precipitation and landsliding can be drawn using three landslide events that occurred in January 2006 close to a USGS monitoring site near Seattle, Washington, USA.

The location of Seattle is shown in Figure 2.5. The hillslopes where the landslides occurred are steep ($>35°$) west-facing coastal bluffs, generally less than 100 m high, that form the coastline of the Puget Sound. The hillslopes are composed of a glacial sequence of outwash and lacustrine deposits and are mantled with a one to several meters thick sandy colluvium. Figure 2.6 shows the hourly and cumulative rainfall measured at the site. In the winter wet season of 2005 and 2006 a series of storms brought nearly 400 mm of rainfall to the location between about 1 October 2005 and 1 January 2006 when the first landslide was reported. During the two-week period leading up to the time of the occurrence of the slides (shown as arrows) rainfall was recorded nearly continuously. Finally, in the hours prior to the occurrence of the slides, moderately intense rainfall was recorded. Over the

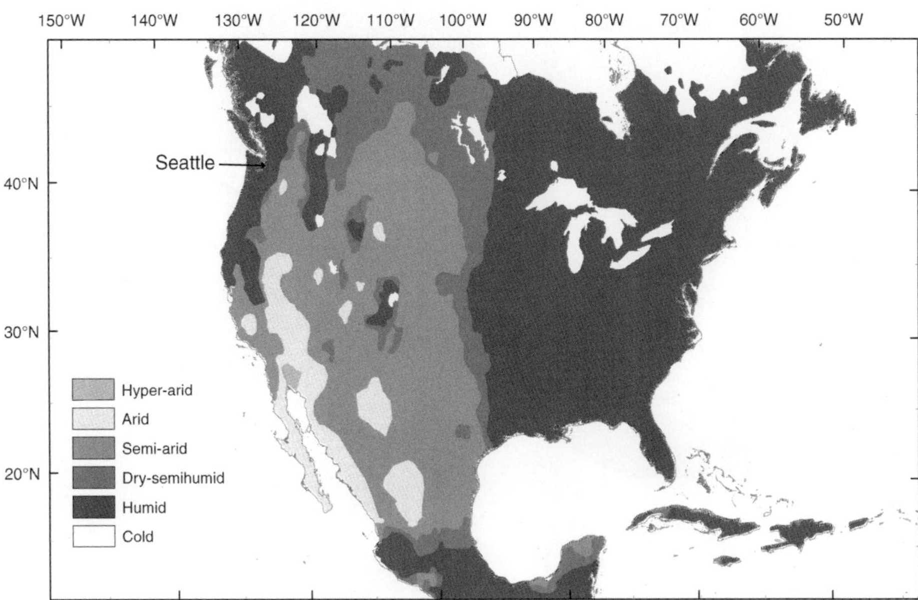

Figure 2.5 Humidity index map of North America (adapted from GRID/UNEP, Office of Arid Land Studies, University of Arizona). See also color plate section.

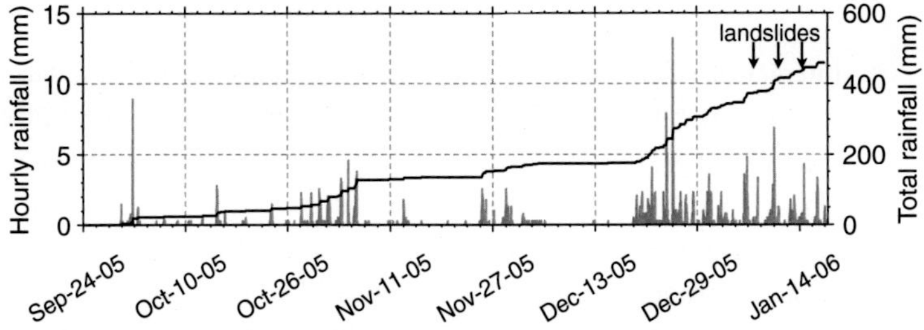

Figure 2.6 Rainfall during the winter wet season and landslide occurrence in January 2006 (arrows) close to the USGS monitoring site near Seattle, Washington. The time series indicates a correlation between four precipitation characteristics (total rainfall, short-term intensity, antecedent precipitation, and storm duration) and landslides occurrence (from Godt *et al.*, 2009).

past several decades, empirical relations have been developed to correlate precipitation characteristics with landslide initiation, and quantitative physical frameworks have been developed to predict shallow landsliding. However, a general quantitative framework that explicitly incorporates the effects of the unsaturated zone is still lacking. Any reliable predictive framework should consider the physical processes that occur between precipitation and landslide occurrence. Specifically, these physical processes include runoff, infiltration, the consequent changes in soil moisture, suction, and stress, as well as the geometry and

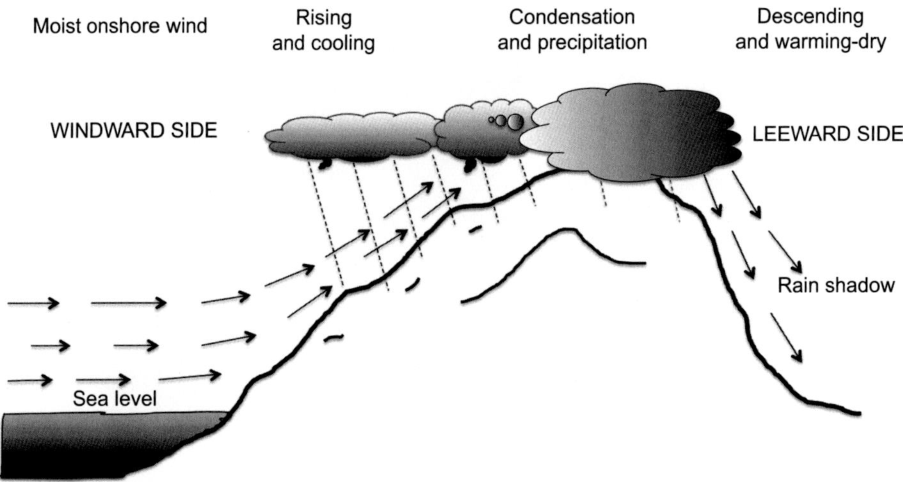

Figure 2.7 Orographic enhancement of precipitation.

stratigraphy of geologic materials and their hydrologic and mechanical properties. In the subsequent sections, these processes and factors will be described.

2.1.2 Orographic precipitation enhancement

Interaction of atmospheric flow with orographic features such as mountains or hills creates spatial variability in precipitation amount and intensity (e.g., Bergeron 1960). Figure 2.7 shows a simple schematic diagram that illustrates the basic processes that can enhance precipitation in topographically complex environments. Moist airflow rises when it encounters topography leading to cooling and the condensation of water into clouds. In a thermally stable environment these clouds typically produce little precipitation on their own. However, rain droplets or ice crystals that fall out of clouds higher in the atmosphere can drive a positive feedback creating precipitation (the so-called "seeder-feeder" mechanism) that falls on the windward side of the topographic rise. A wide variety of additional physical mechanisms influence precipitation rates and amounts in complex topography (e.g., Smith, 2006).

Over geologic time scales, the interactions of atmospheric circulation and topography control erosion rates and landscape form (e.g., Douglas, 1976; Roe, 2005). Over human time scales, orographic enhancement of precipitation can lead to flooding and landslide initiation (e.g., Neiman *et al.*, 2008b).

2.1.3 Atmospheric rivers

Atmospheric rivers are narrow plumes of moisture that typically extend over a few thousand kilometers and are responsible for the great majority of the poleward transport of water

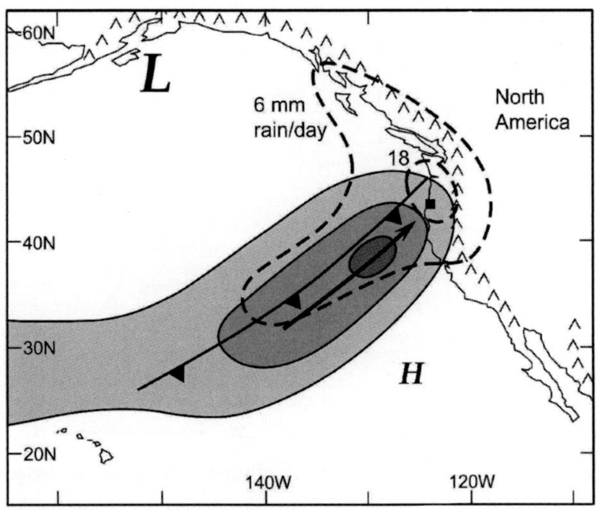

Figure 2.8 Conceptual sketch showing the general setting of atmospheric rivers during the winter season in the North Pacific impacting North America (from Neiman *et al.*, 2008a). Shaded contours indicate relative availability of moisture in the atmosphere. Dashed contours show typical rainfall rates.

vapor from the tropics into the mid-latitudes (Zhu and Newell, 1998; Neiman *et al.*, 2008a). When this moisture flow interacts with topography, heavy precipitation falls that often leads to flooding and landslides (e.g., Ralph *et al.*, 2006; Neiman *et al.*, 2008b). The impact of atmospheric rivers on water supply and flooding on the west coast of North America is well established (e.g., Dettinger *et al.*, 2011) and recent work points to their influence on flooding and landslide occurrence in Norway as well (Stohl *et al.*, 2008). Figure 2.8 shows a conceptual map of atmospheric rivers that impact western North America in the winter season. The circulation pattern consists of a strong low-pressure system centered at about 55° N latitude in the Gulf of Alaska and a high-pressure system off shore of Baja California. Low-level flow of moisture extends from near the Hawaiian Islands to the west coast of North America along a frontal boundary. When this flow interacts with topography, precipitation rates of 10 mm/day or more are possible. The so-called "Pineapple Express" storms are a particularly strong set of atmospheric rivers that impact western North America, where the plume of moisture extends to the Hawaiian Islands. On average, the west coast of North America receives precipitation from atmospheric rivers about 16 times a year, of which about six can be classified as Pineapple Express storms (Dettinger *et al.*, 2011).

2.1.4 Monsoons

Monsoons are seasonal shifts in airflow direction accompanied by rainfall that occur in Asia, West Africa, and the Americas. Where the flow interacts with topography, heavy rainfall, flooding, and landslides are common. Figure 2.9 is a conceptual diagram showing the general pattern of circulation that drives the North American or Mexican monsoon. Water vapor is transported from both the Gulf of California and the Gulf of Mexico (Adams and Comrie, 1997). The moisture flow interacts with topography driving convection. This

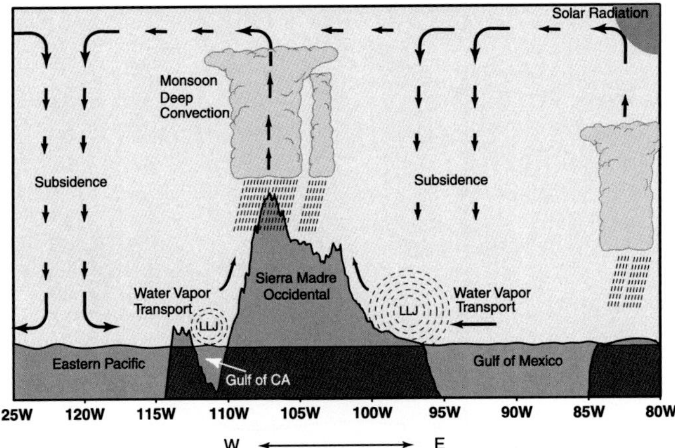

Figure 2.9 Conceptual sketch showing the transport of water vapor by the low-level jet (LLJ) and circulation that drive convection and precipitation of the North American or Mexican monsoon (from the World Climate Research Programme).

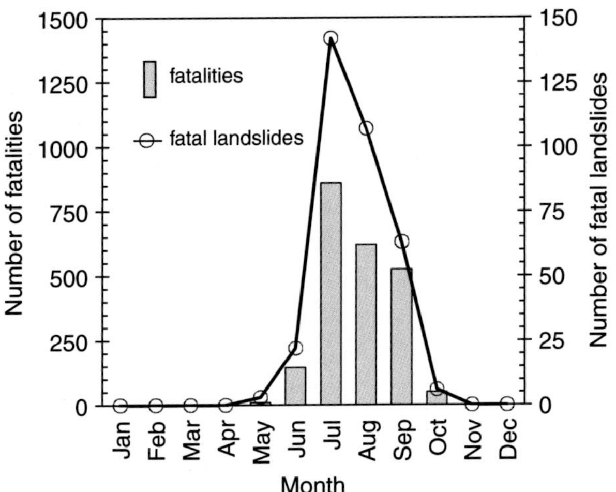

Figure 2.10 Graph showing the number of deaths resulting from landslides (bars, left hand scale) and the number of landslides causing those deaths (lines, right hand scale) by month for the period of 1978–2005 in Nepal (from Petley *et al.*, 2007).

convection provides a large part of the summer season precipitation over the southwestern United States and Mexico. Strong storms related to the monsoon are capable of producing flooding and landslides (e.g., Godt and Coe, 2007; Griffiths *et al.*, 2009).

The monsoon climate of the Middle Himalaya of Nepal leads to a marked seasonality in landslide occurrence and related fatalities. Figure 2.10 shows the numbers of deaths and associated landslides for Nepal during the period from 1978 to 2005. The peak occurs in July related to the peak of rainfall associated with the monsoon. Interestingly, years with strong atmospheric monsoonal conditions (i.e., the shifts in wind direction and speed are pronounced) tend to be those with lower rainfall intensities and durations and resulting decreases in landslide activity (Shrestha *et al.*, 2000; Petley *et al.*, 2007).

2.1.5 Tropical cyclones

Tropical cyclones, hurricanes, and typhoons are capable of delivering copious rainfall amounts and when they impact mountainous regions they often generate a large number of landslides. For example, in 1996, Typhoon Herb hit the northern part of the island of Taiwan and dropped nearly 2 m of rain over a three-day period. Thousands of landslides and debris flows occurred resulting in the loss of 73 lives and more than US$1 billion in damage (Lin and Jeng, 2000). In the United States, tropical systems that impact the Appalachian Mountains often generate copious rainfall, flooding, and occasionally widespread landslides. For example, Hurricanes Frances and Ivan impacted western North Carolina bringing daily rainfall in excess of 200 mm for some locations in September 2004. This sequence of storms triggered widespread landslides and flooding and associated loss of life and damage to property in the region (Wooten *et al.*, 2008).

2.1.6 El Niño and La Niña

El Niño is the periodic warming of the tropical Pacific Ocean that persists for 12 or more months and occurs every three to seven years (McPhaden, 2002). The opposite pattern, a cooling of sea-surface temperatures in the Pacific is referred to as La Niña. El Niño and La Niña conditions are dynamically linked to the differences in surface atmospheric pressure over Australia and the eastern tropical Pacific. This pressure oscillation is called ENSO (El Niño Southern Oscillation). Both El Niño and La Niña conditions are associated with changes in weather patterns and impacts such as drought, flooding, and landslides. Figure 2.11 shows a conceptual drawing of the normal and El Niño conditions in the Pacific. Under normal (ENSO neutral) conditions the warm pool of water resides in the western Pacific giving rise to deep convection and rainfall. The trade winds in both the northern and southern hemispheres are easterly. Under El Niño conditions, the trade winds weaken and the pool of warm water expands and extends eastward, shifting the deep convection and rainfall along with it. The impacts of the change in the location of deep convection and trade wind shift are most consistent in the tropical regions that border the Pacific and generally diminish in both intensity and consistency at higher latitudes, as shown in Figure 2.12. During El Niño, drought conditions often develop in Southeast Asia and Australia, whereas the west coasts of both South and North America often experience heavy rainfall, flooding, and landslides. The spatial distribution of the impacts of La Niña is generally the same as that for El Niño, but of opposite sign.

The El Niño of 1997–8 was the largest and strongest El Niño of the twentieth century and brought heavy precipitation to large parts of the southern and western United States (Chagnon, 2000). Record rainfall in southern and central California triggered widespread shallow landslides and debris flows over many parts of the state (Godt and Savage, 1999; Gabet and Dunne, 2002). El Niño conditions have also been linked with severe flooding

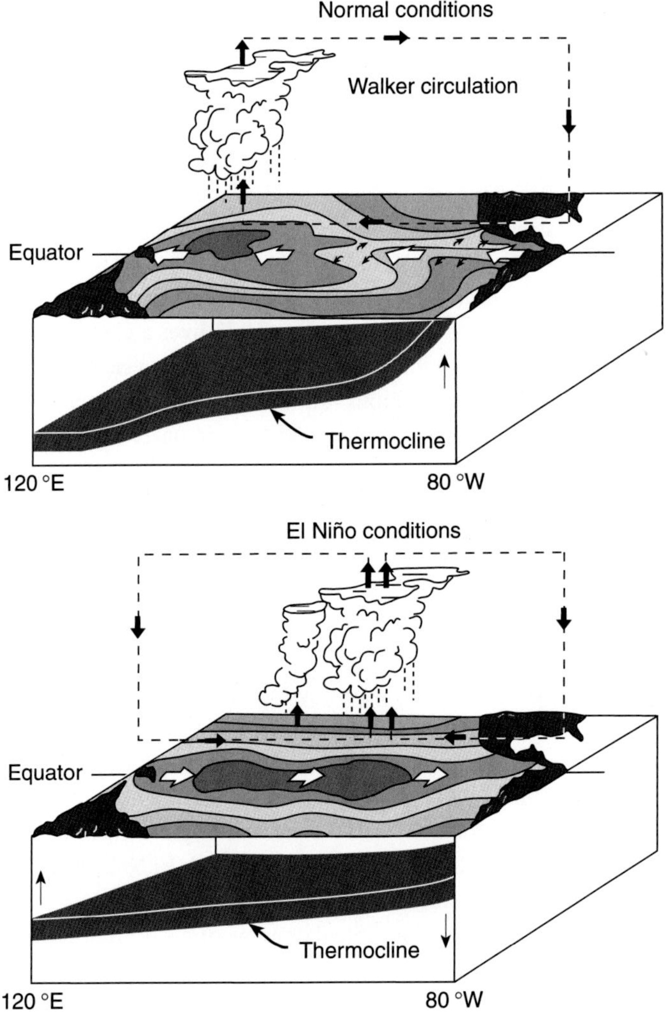

Figure 2.11 Schematic drawing of normal and El Niño conditions in the equatorial Pacific. Dark grey colors indicate relatively warm sea-surface temperatures (from McPhaden, 2002).

and debris flow events in other parts of the Pacific Americas, such as Peru (Keefer *et al.*, 2003) and Argentina (Moreiras, 2005), and in Kenya where landslides associated with the 1997 and 1998 El Niño caused US$1 billion in damage (Ngecu and Mathu, 1998).

2.1.7 Trends in extreme precipitation

Extreme precipitation episodes or heavy downpours have become more frequent and more intense in recent decades than at any other time in the historical record. These extreme events also account for an increasingly larger percentage of the total annual precipitation. These changes have been significant over much of North America.

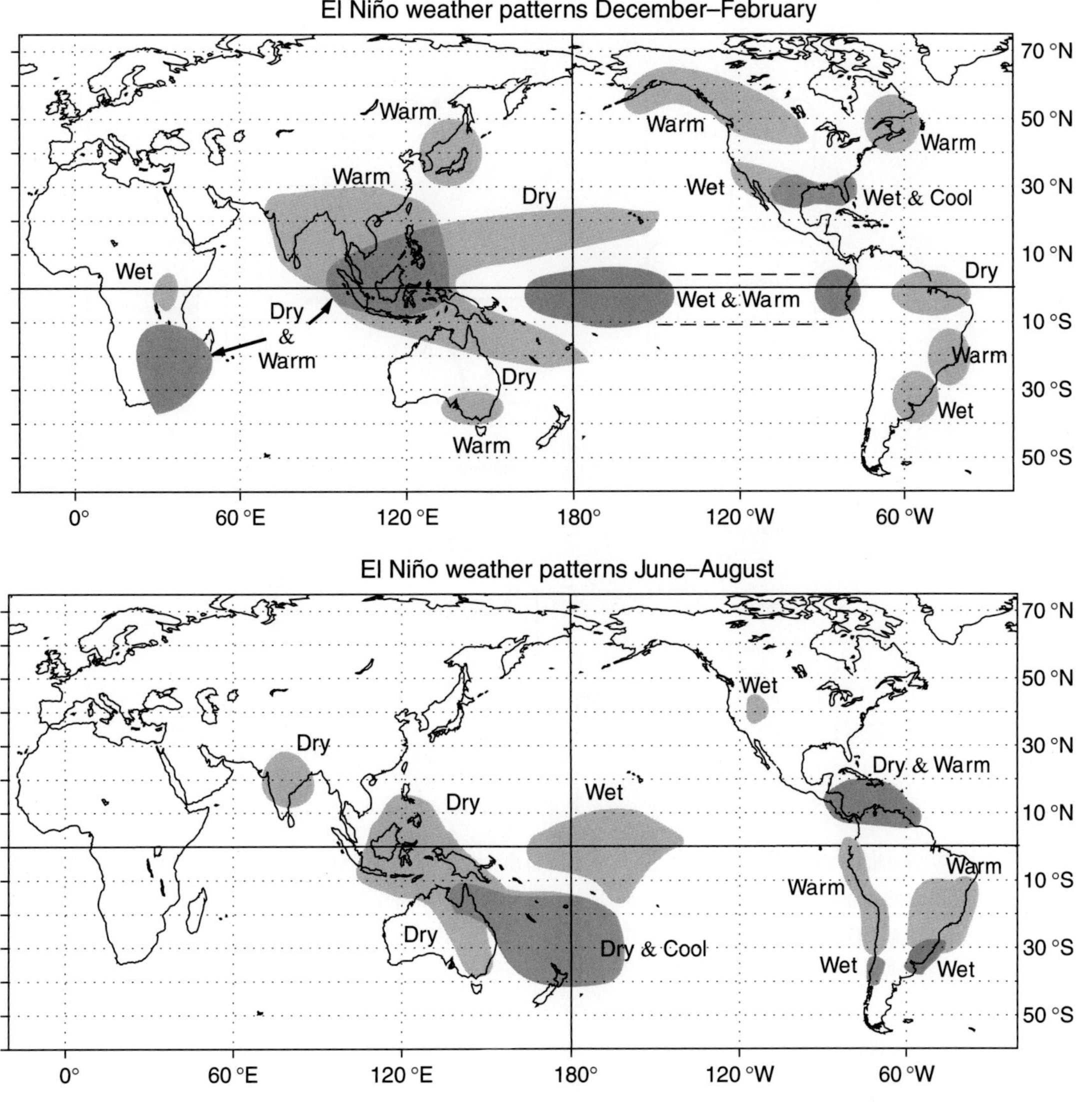

Figure 2.12 Maps showing temperature and precipitation anomalies associated with El Niño (from McPhaden, 2002).

Trenberth (1999) has proposed a conceptual model based on the assumption that an increase in trace gas concentrations in the atmosphere increases the radiative forcing, resulting in an enhancement of the hydrologic cycle (Figure 2.13). Since a large part of surface heating is used to evaporate surface moisture, an increase in the global mean temperature will increase the water-holding capacity of the atmosphere and the available moisture in the atmosphere. While the model simplifies the process and does not take into account several feedbacks, it provides a schematic way to illustrate the mechanism by which global precipitation may increase in a warmer climate.

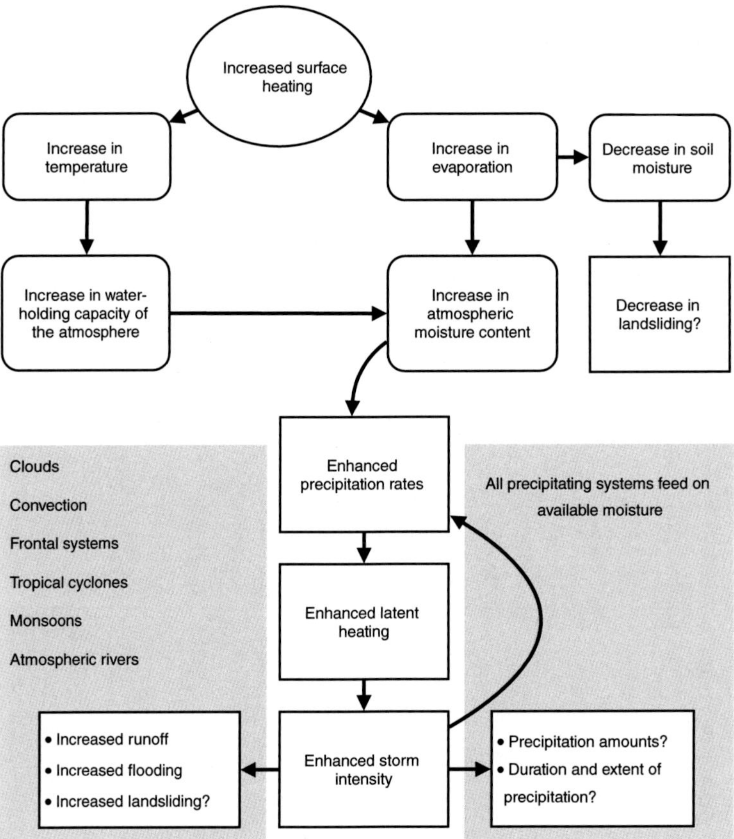

Figure 2.13 Outline of the sequence and processes involved in climate change and how they influence the moisture content of the atmosphere, evaporation, and precipitation rates. All precipitating systems feed on available atmospheric moisture leading to enhanced precipitation rates and potential changes in hydrologic extremes, such as flooding and landsliding (modified from Trenberth, 1999).

The Clausius–Clapeyron relation defines the phase-change boundary between liquid and gas phases of matter and can be used to describe the water-holding capacity of the atmosphere as a function of temperature and pressure (also see Section 3.1.5). A useful approximation for describing the effects on rainfall is known as the August–Roche–Magnus approximation:

$$e_s(T) = 6.1049\exp\left(\frac{17.625T}{T + 243.04}\right) \tag{2.1}$$

where $e_s(T)$ is the saturation vapor pressure in hPa, and T is temperature in degrees Celsius and provides a means to quantitatively estimate the saturation vapor pressure of the atmosphere and thus the amount of water available for precipitation as the atmosphere heats or rises over topography (e.g., Roe, 2005). For every one degree Celsius of temperature increase the atmosphere can hold an additional ~7% of water by volume.

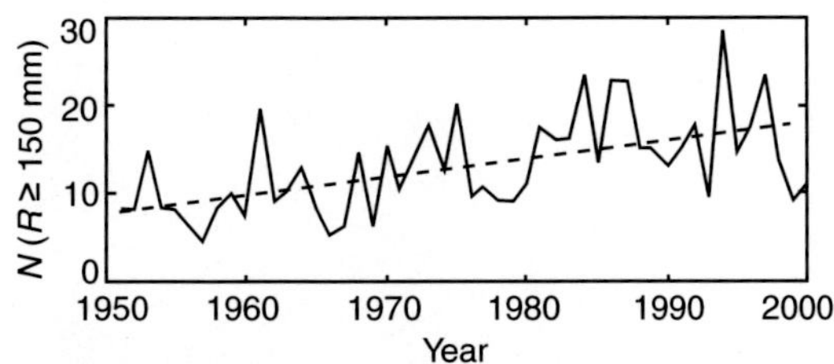

Figure 2.14 Trend in the number of extreme (> 150 mm) daily rainfall events during the monsoon season (1 June–30 September) in central India for the period 1951 to 2000 (from Goswami *et al.*, 2006).

The Clausius–Clapeyron relation generally holds for climate model predictions of future changes in extreme daily precipitation (e.g., Allen and Ingram, 2002). However, rainfall intensities over shorter periods (hourly) may increase by more than the 7% per degree described by the relation (Lenderink and Meijgaard, 2008).

Observational evidence indicates that short-term (daily) precipitation extremes have increased over much of North America, with the most significant increases in southern Alaska, western British Columbia, Arctic and southeastern Canada, and the central United States (Groisman *et al.*, 2005). The changes in heavy precipitation are greater than any change in precipitation totals. Increases in heavy precipitation occur even where annual totals are decreasing. Time series analysis shows an increase in the frequency of extended wet periods (90 days or longer) in the United States (Kunkel *et al.*, 2003).

Goswami *et al.* (2006) examined daily rainfall anomalies during the monsoon season (1 June to 30 September) in Central India and found a significant increasing trend in the variance during the period 1951 to 2000. Rainfall was gridded on a 1 degree by 1 degree box and heavy rainfall events were defined as more than 100 mm/day and very heavy rainfall as more than 150 mm/day. The trend in a larger number of extreme events came at the expense of light to moderate rainfall, yielding no overall trend in annual rainfall. The heaviest four rainfall events in each monsoon season have increased about 10% per decade over the 50-year record (Figure 2.14).

Documentation of tropical cyclone intensity is patchy. The availability of tropical cyclone data is uneven and data are not collected in a standardized manner (Kunkel *et al.*, 2003). Global counts of tropical storms can only be considered accurate for the era of earth-orbiting satellites, which begins in the middle 1970s. Estimates of storm intensity are much less reliable, even today. Holland and Webster (2007) found no trend in the mean intensity of Atlantic basin storms. No consistent trend in the number of tropical storms in the North Atlantic basin can be separated from natural fluctuations in number and intensity of tropical storms (Webster *et al.*, 2005). However, Emmanuel (2005) and Webster *et al.* (2005) show an increase in the frequency of category 4 and 5 cyclones in the North Pacific basin with potential impact on Taiwan. Furthermore, even in areas where the frequency of tropical storms is predicted to decrease, rainfall rates within 100 km of the tropical cyclone center are expected to increase (Knutson *et al.*, 2010).

This section discussed several factors that control the spatial distribution of precipitation at global, continental, and regional scales. Emphasis was placed on both the effects of

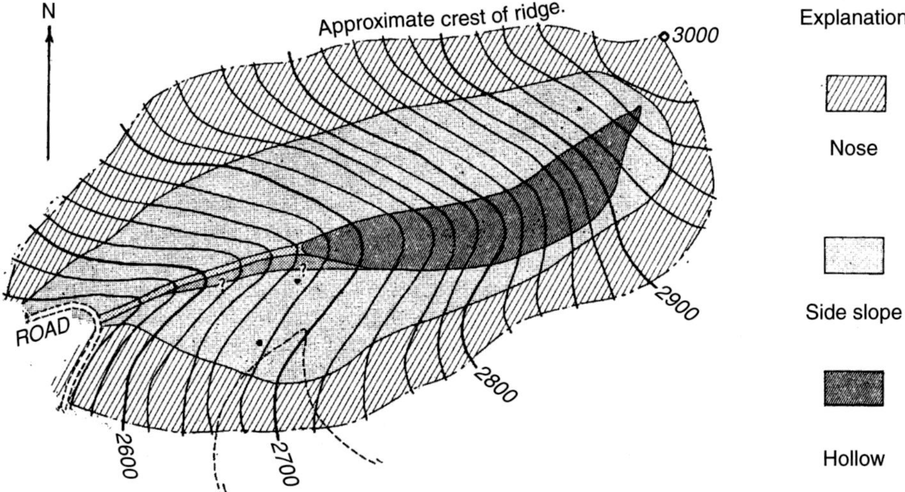

Figure 2.15 Map showing the parts of a hillslope (from Hack and Goodlett, 1960).

topography on precipitation and the circulation patterns such as El Niño that influence the occurrence of heavy precipitation. The next section explores the role of topographic form in the distribution of water at the hillslope scale.

2.2 Topography

2.2.1 General topographic features

Much of the current work in identifying the spatial location and extent of areas susceptible to landslides has focused upon the topographic characteristics of areas where slides have been observed (Borga *et al.*, 1998; Montgomery *et al.*, 1998; Wu and Sidle, 1995). Obvious geomorphic requirements are sufficiently steep slopes overlain with material available for transport. Many workers have identified finer scale topographic characteristics that favor debris flow initiation (e.g., Hack and Goodlett, 1960; Reneau and Dietrich, 1987).

Hack and Goodlett (1960), in their study of hillslope morphology in the Appalachians, divided the hillslope into three zones based upon topographic characteristics and the resultant behavior of surface water flow (Figure 2.15). Side slopes consist of planar topography with little or no curvature. Noses are defined as areas of divergent topography or ridge that divide small catchments. Conversely, hollows are areas of convergence. Often subtle, these depressions at the heads of stream channels were identified as the primary source areas for many of the shallow landslides investigated.

In soil-mantled landscapes, mass-wasting processes from side slopes contribute to the colluvial deposition in topographic hollows. This colluvium is the source material for

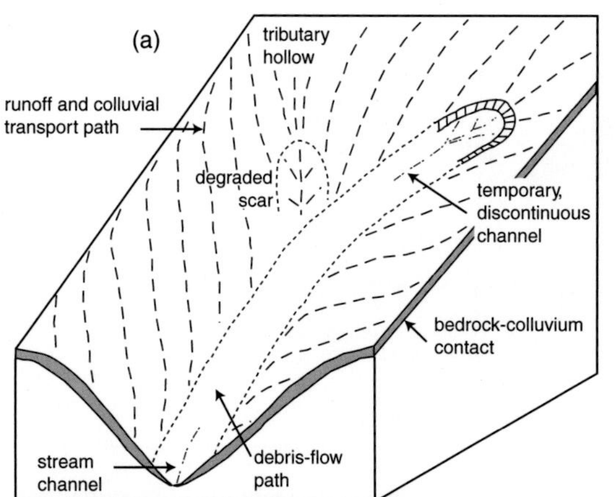

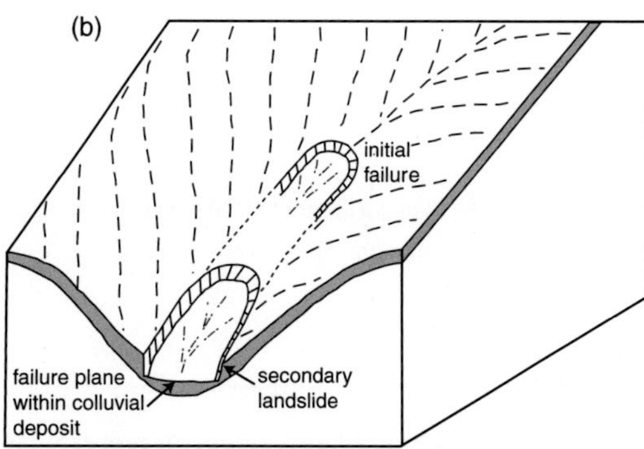

Figure 2.16 Conceptual sketches of typical landslide locations in soil-mantled landscapes. (a) Debris flow mobilized from a landslide in a topographic hollow. (b) Debris flow originating in a hollow that triggered another landslide downslope (from Reneau *et al.*, 1990)

shallow landslides and debris flows (e.g., Reneau and Dietrich, 1987; Reneau *et al.*, 1990) (Figure 2.16).

In addition to collecting colluvium, topographic hollows are also areas of concentrated flow through the soil (Dietrich and Sitar, 1997). Water near the surface is routed by terrain features and concentrated in areas of topographic convergence, which increases saturation and the likelihood of positive pore-water pressures during a storm event (Montgomery and Dietrich, 1994b). Side slopes and noses, where subsurface flow does not converge, are less susceptible to erosion (Horton, 1945; Montgomery and Dietrich, 1994a). Spatial variability of subsurface properties may perturb the flow path and residence times, but convergence of flow in hollows dominates (Ijjasz-Vasquez and Bras, 1995). By providing a concentration point in space for both subsurface flow and material that may be mobilized, various empirical relationships between topographic hollows and debris flows have been observed.

Reneau and Dietrich (1987), in a study of damaging shallow landslides in Marin County, have proposed an inverse relationship between area occupied by side slopes in a valley and the importance of hollows as a debris flow source area. Following a widespread debris flow event in the San Francisco Bay region in January 1982, Ellen *et al.* (1988), also in Marin County, identified about two-thirds of mapped debris flow scars as originating in hollows.

There are other studies that indicate a less concrete relationship between the location of debris flow source areas and hollows. In an intensive field study following the disastrous rainstorm of 5 June 1995, in Madison County Virginia, Wieczorek *et al.* (1997) found that planar slopes were more likely to be areas of initiation. They concluded that the thin soils overlying the planar slopes were more susceptible to failure. Conversely, Montgomery and others (1998) reported poor performance of their susceptibility model that relies on the identification of hollows in areas overlain with thick glacial deposits in the Pacific Northwest of the United States.

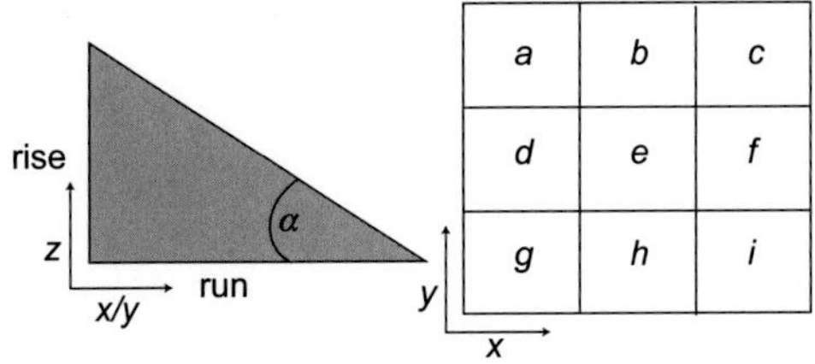

Figure 2.17 Diagram illustrating the calculation of topographic slope from a 3 × 3 evenly spaced grid with elevation values (*a* – *i*).

2.2.2 Digital landscapes

Digital landscapes are representations of the earth's surface, collected using a wide variety of photogrammetric and remote sensing methods. These data can be manipulated in a geographic information system (GIS) to address a range of hydrologic and geomorphic problems (e.g., Moore *et al.*, 1991; Tarboton, 1997). Digital topographic data are typically evenly spaced grids of surface elevation values, although other data structures are used (Maune, 2007). Remotely sensed data using lidar, or light detection and ranging technology, are particularly useful for landslide studies in that they potentially provide very high spatial resolution (<1 m) and information on the surface elevation below vegetation (e.g., McKean and Roering, 2004).

Local topographic slope is commonly derived from grid-based digital elevation models (DEMs) and can be computed by fitting an inclined plane to the elevation values of a 3 × 3 evenly spaced array. Local slope s as a percentage is (Figure 2.17)

$$\alpha = \frac{\text{rise}}{\text{run}}100 \tag{2.2}$$

and in degrees is

$$\alpha = \arctan\left(\frac{\text{rise}}{\text{run}}\right) \tag{2.3}$$

Given a 3 × 3 grid with cells of equal size and elevation values, $a - i$, the local slope at the center cell is (Figure 2.17)

$$\frac{\text{rise}}{\text{run}} = \sqrt{\left(\frac{dz}{dx}\right)^2 + \left(\frac{dz}{dy}\right)^2} \tag{2.4}$$

where

$$\frac{dz}{dx} = \frac{(a + 2d + g) - (c + 2f + i)}{8\,(\text{cellsize})} \tag{2.5}$$

and

$$\frac{dz}{dy} = \frac{(a + 2b + c) - (g + 2h + i)}{8\,(\text{cellsize})} \tag{2.6}$$

2.2.3 DEM methods for landslide analysis

Models used to identify landslide source areas are a subset of physically based, distributed hydrologic models. These models attempt to integrate physical interactions known to occur at a point in space over a defined spatial extent. To identify potential landslide locations, hydrologic models are often coupled with a limit-equilibrium slope stability model and a DEM (Borga *et al.*, 1998; Montgomery and Dietrich, 1994b). The DEM is used to disaggregate the drainage basin of interest into like topographic elements. Each element is assigned topographic parameters defined by local slope, contour length, and upslope contributing area (Montgomery and Dietrich, 1994b). A simple form of the one-dimensional limit-equilibrium equation assumes an infinite slope and relates the frictional strength, pore-water pressure, and gravity for a cohesionless soil of constant thickness (Montgomery and Dietrich, 1994b). The hydrologic models are essentially routing models that utilize the DEM to identify flow paths and compute the depth of subsurface flow for each cell (Montgomery and Dietrich, 1994b; Wu and Sidle, 1995). Calculations are performed for each cell in the drainage basin and a relative measure of slope stability for a given hydrologic input is assigned (Borga *et al.*, 1998; Montgomery *et al.*, 1998b).

Several coarse assumptions must be made to calibrate these models for application to actual hillslopes. Many of the subsurface characteristics are assumed to be spatially and vertically uniform and are assigned values based upon a representative sample (Wu and Sidle, 1995; Borga *et al.*, 1998; Montgomery *et al.*, 1998). Hydrologic inputs are assumed to be spatially and temporally constant (Montgomery and Dietrich, 1994b). Assumptions of homogeneity are justified on practical and theoretical grounds. Practically, subsurface characteristics are, by their hidden nature, essentially unknowable (Beven, 1996). Theoretically, the reliance on topography as a controlling mechanism provides a convenient alternative. Subsurface characteristics can be ignored because of their limited influence on the spatial location of initiation events (Montgomery *et al.*, 1998). However, these assumptions may not be valid because geologic and soil formation processes create heterogeneity in soil mechanical and hydrologic properties. Some of these processes are examined in the following sections.

2.3 Soil classification

2.3.1 Soil stratigraphy

Over time, infiltration plays a key role in soil formation. Consequently, soil hydrologic and mechanical properties have great influence on soil stability or landslides. Figure 2.18 illustrates typical residual soil horizons and their hydrologic properties.

Near the ground surface the *O horizon* is characterized by accumulation of organic material with varying degrees of decomposition. Saturated hydraulic conductivity is typically very high, up to 10^{-3} m/s. Cracks commonly exist in this horizon, providing pathways for

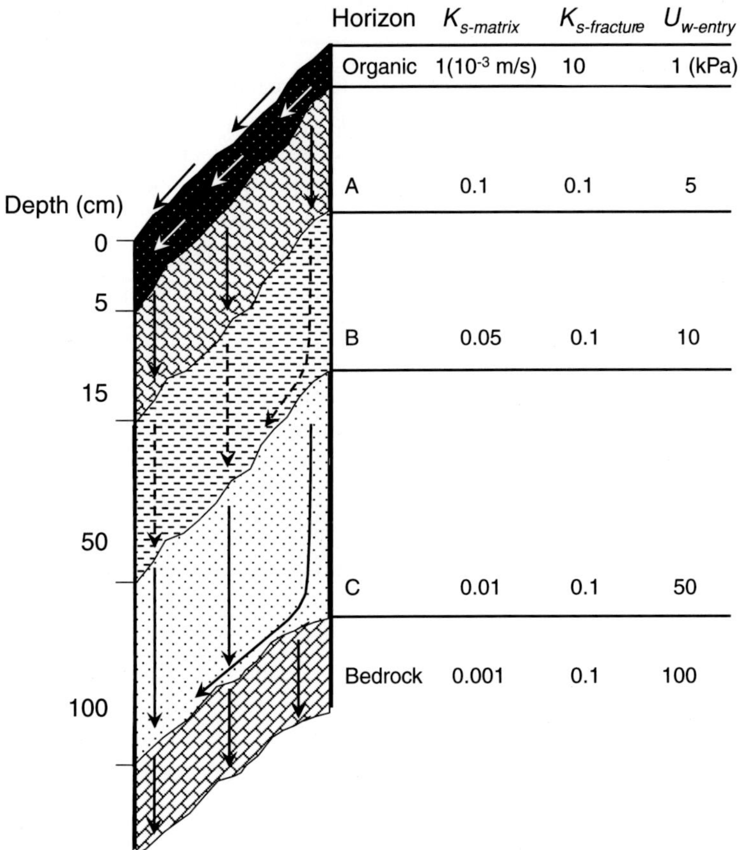

Figure 2.18 Illustration of hypothetical residual hillslope soil horizons and their hydrologic properties.

water to flow into this horizon. The hydrostatic pressure at which water can flow into soil is called water-entry pressure. This hydrologic parameter is indicative of the initial infiltration of water into soil. Water at the surface under pressure below this value will flow over the ground surface rather than infiltrate. Water entry pressure is an intrinsic soil property that mainly depends on soil pore size distribution, and soil wetness (water–solid contact angle and prevailing soil water content). Although this parameter has been correlated to air-entry pressure, accurate mathematical description has yet to be established. The O horizon typically has the lowest water-entry pressure, on the order of 1 kPa. The thickness of this horizon varies greatly, mainly depending on vegetation and climatic conditions, and slope gradient.

Below the O horizon, the *A horizon* is characterized by accumulation of humidified organic materials mixed with a dominantly mineral fraction. Due to mixing, the hydrologic properties of this zone are often homogeneous. The saturated hydraulic conductivity is typically smaller than that in the O horizon, and the water-entry pressure is typically several times greater than that of the O horizon. The thickness of this horizon is typically several times that of the O horizon.

Below the A horizon is typically a *B horizon* with no or little evidence of original sediment or rock structure, but with various materials illuviated into it and residual dissolvable concentration of materials. For example, alkaline earth carbonates, mainly calcium carbonate, can be found in this horizon. They are typically moved to this horizon and precipitated there by either downward infiltration or upward capillary rise flow. Due to the weathering and mineral precipitation, saturated hydraulic conductivity is much smaller that the overlying A horizon. Fractures and holes often exist in this horizon, causing sharp contrast in hydrologic properties between fractures and soil matrix. The thickness of this horizon varies from a few centimeters to several meters.

Soil below the B horizon represents weathered materials that lack the properties of A and B horizons. This is the *C horizon*. Depending on the degree of weathering, the characteristics of parent rocks (typically consolidated bedrock underlying the C horizon) can be identified in this horizon.

2.3.2 Commonly used classification systems

Hillslope materials are mostly weathered rocks called soils. Depending on the geographic location, climate conditions, and the origin and history of soils, chemical compositions, and mechanical and hydrologic properties vary from one place to another. Very rare is the case that soils from different locations will have identical chemical compositions, and mechanical and hydrologic properties. And often soils a few meters apart have different compositions and very different hydro-mechanical properties. Thus it is important to classify them such that their hydro-mechanical properties could be similar within the same classification.

A number of soil classification systems are available, depending on the disciplines and their general engineering or scientific objectives. Some of the most widely used classification systems in the United States are: U.S. Department of Agriculture (USDA), International, Unified Soil Classification System (USCS) by the American Society for Testing Materials (ASTM), and American Association for State Highway and Transportation Organization (AASHTO). The commonality among these systems is the use of soil particle sizes in dividing soils into different categories of clay, silt, sand, gravel, and larger sized materials, although the boundaries for each of these systems are somewhat different, as illustrated in Figure 2.19. For clay, both the USDA and International use the same particle size of 0.002 mm to distinguish it from silt, AASHTO uses 0.006 mm, while USCS does not explicitly distinguish particles sizes between clay and silt. The boundaries between silt and sand among these four systems are somewhat similar but different: USCS and AASHTO use 0.075 mm, USDA uses 0.05 mm, and International uses 0.02 mm. The boundary between sand and gravel is 2 mm in all four systems, except 4.75 mm is used in the USCS. Subdivisions within sand are employed in all four systems, but the boundaries are different between those subdivisions (Figure 2.19). The boundary between gravel and cobbles/stones/boulders is the same for USDA, USCS, and AASHO: 76 mm; 20 mm is used in the International system.

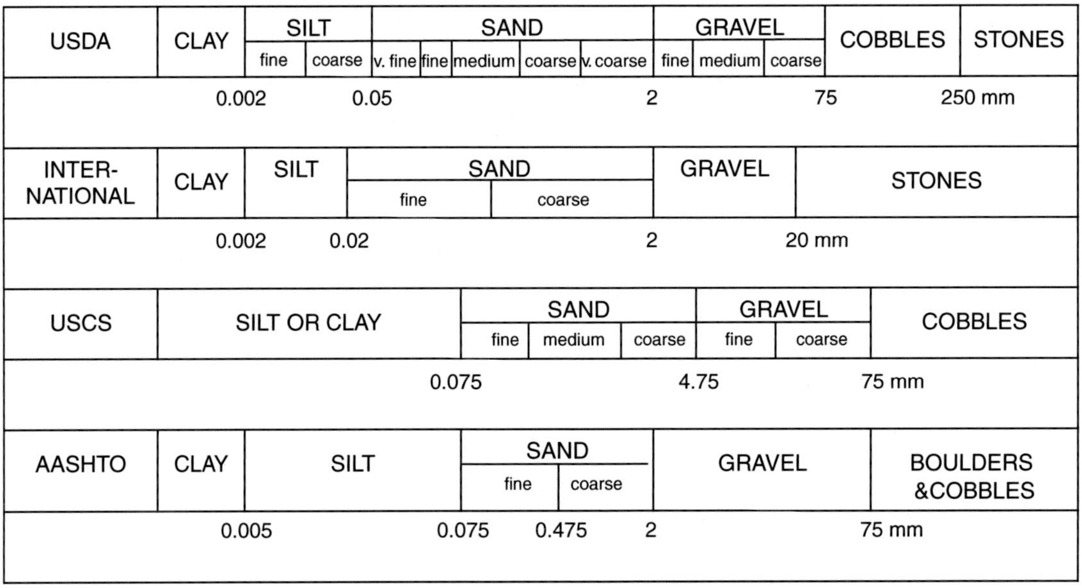

USDA	CLAY	SILT		SAND					GRAVEL			COBBLES	STONES
		fine	coarse	v. fine	fine	medium	coarse	v. coarse	fine	medium	coarse		
	0.002		0.05					2			75		250 mm

INTER-NATIONAL	CLAY	SILT	SAND		GRAVEL	STONES
			fine	coarse		
	0.002	0.02		2	20 mm	

USCS	SILT OR CLAY	SAND			GRAVEL		COBBLES
		fine	medium	coarse	fine	coarse	
	0.075			4.75		75 mm	

AASHTO	CLAY	SILT	SAND		GRAVEL	BOULDERS &COBBLES
			fine	coarse		
	0.005	0.075	0.475	2	75 mm	

Figure 2.19 Chart comparing the particle-size categories of the four widely used soil classification schemes (from Soil Survey Division Staff, 1993).

It is important to note that none of the above soil classifications are based on soil chemical compositions, i.e., mineral structures or material types. Thus, a soil classified as clay does not mean it is clay mineral and could be silicon or other materials.

The USDA system assigns soil names based on a soil's percentages of sand, silt, and clay, as illustrated in ternary diagram in Figure 2.20. The presence of organic matter, humus, is explicit in the term "loam" and used in the soil names. Soils whose names contain the word "loam" are generally good for growing plants. Loamy soils generally are mixtures of sand, silt, and clay. The details of each of these systems can be found in the literature. For an example of the USDA system see Eswaran *et al.* (2002). For the AASHTO system see Hogentogler and Terzaghi (1929). For the International system see Buol *et al.* (2003), and for the USCS see ASTM (1985).

2.4 Hillslope hydrology and stream flow generation

2.4.1 Runoff and infiltration

The movement of water from precipitation to discharge to streams or rivers is a dynamic process. It involves many physical processes such as overland flow, infiltration into the unsaturated zone, evapo-transpiration, and saturated groundwater flow. A simplified schematic illustration of the route water may take from the atmosphere to the stream channel is illustrated in Figure 2.21.

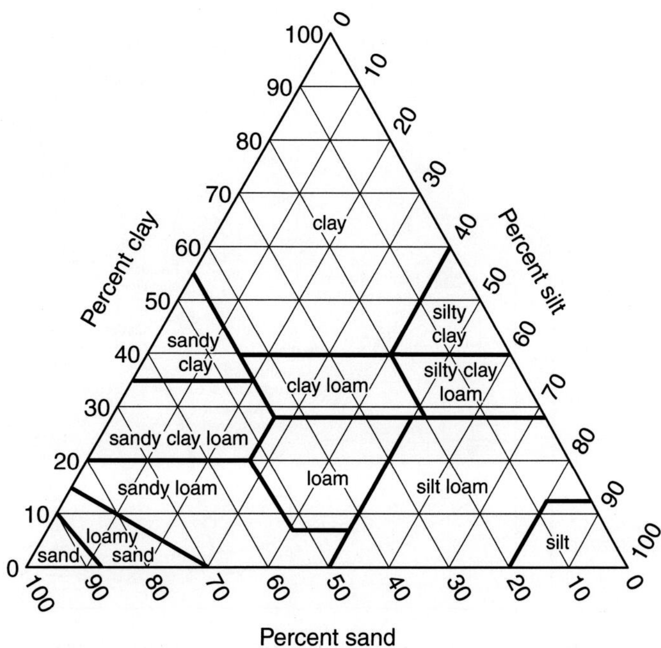

Figure 2.20 Ternary diagram showing the soil classification as relative percentages of sand, silt, and clay in the basic textural classes.

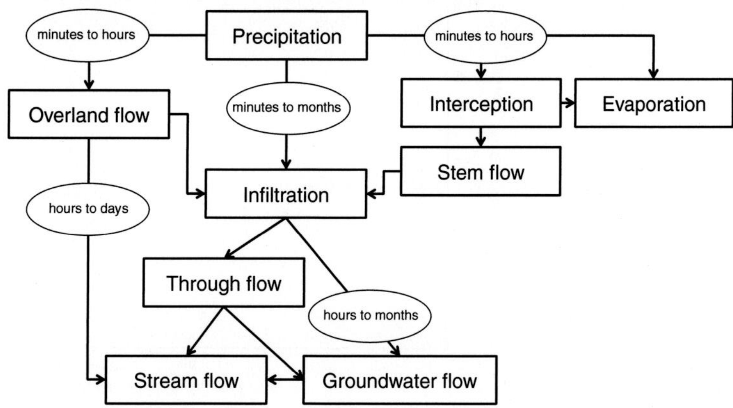

Figure 2.21 Relations among the major hydrologic processes in hillslopes.

Direct runoff is the water that is discharged to streams shortly following rainfall or snowmelt. The source of runoff may be precipitation that moves over the ground surface (overland flow) or through surficial materials, or water stored in the vadose or saturated zones. The complexity of infiltration and runoff processes is illustrated in Figure 2.22. Infiltration is the portion of precipitation falling on the ground surface that moves into surficial materials. Infiltration may find its way to the stream (through flow) or may join

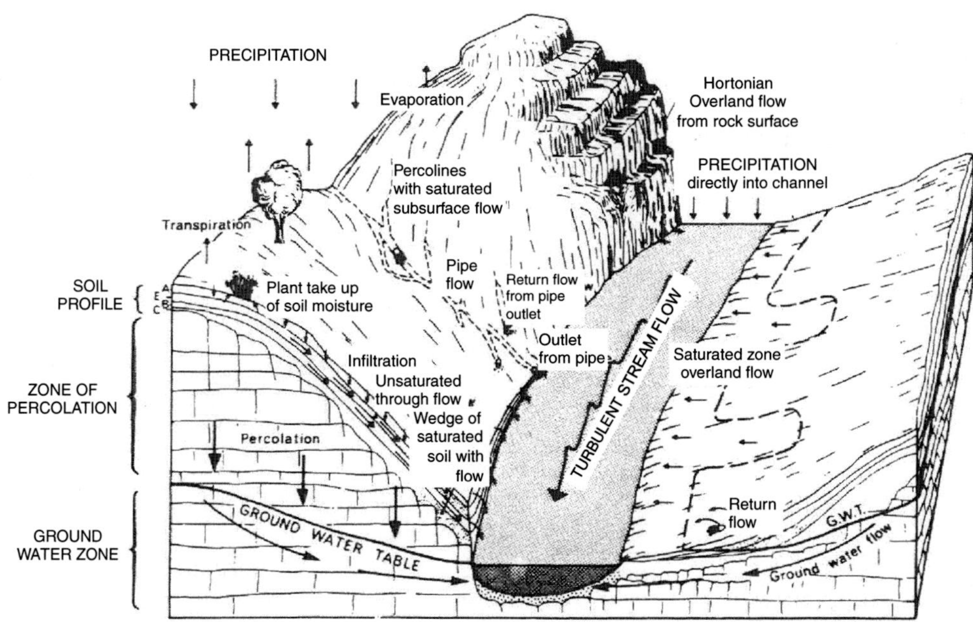

Figure 2.22 Illustration of the mechanisms that generate stream flow (after Selby, 1993).

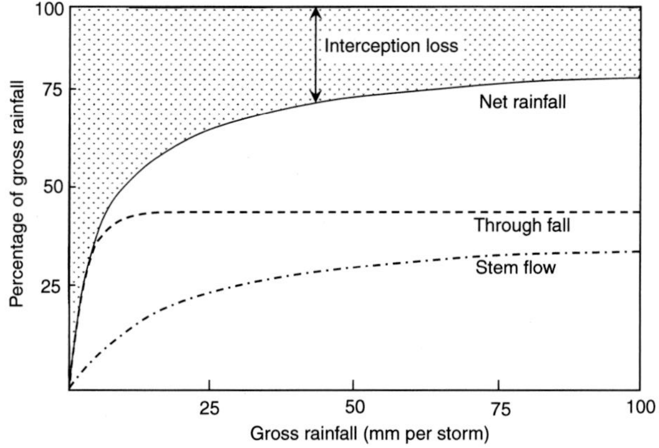

Figure 2.23 Illustration of the effect of vegetation on rainfall (from Aldridge and Jackson, 1968).

the groundwater flow. Some part of the precipitation that falls on vegetated landscapes is intercepted or retained for some period and may evaporate back to the atmosphere, whereas the rest (through flow) moves into or over the ground surface (Figure 2.23). The part of runoff that moves downhill is commonly called overland flow. Water intercepted by plants may either evaporate back to the atmosphere or move downward along the stem of the plant (stem flow) and infiltrate into the soils or become part of overland flow. All water that moves into soil is called infiltration, which typically passes through the unsaturated zone via small pores in the soil matrix or through fissures or other preferential pathways before it recharges the groundwater at the water table.

Modern surface water hydrology in general, and hillslope hydrology in particular, originated in part from the practical needs of reservoir design and flood control (e.g., Kirby, 1978). Of great interest were methods that could provide reliable predictions of stream flow. Hillslope hydrology was greatly influenced and arguably founded by Robert E. Horton. Horton, in a series of papers, beginning in 1933, described the role of soil and its ability to absorb water on the generation of stream flow (Chorley, 1978; Beven, 2004). In Horton's stream flow generation model, the soil mantle acts as a sieve or diversion that partitions rainfall into overland flow, which eventually makes its way to stream channels, and infiltration, which becomes groundwater. A key concept was that the soil mantle has an "infiltration capacity." The infiltration capacity is the volume of water that will flow into surficial materials per unit time, and it varies over seasonal and shorter time scales dependent on initial soil moisture conditions and pore structure. Horton's (1939) equation for the infiltration capacity f in centimeters per hour at time t, is

$$f = f_c + (f_0 - f_c)\exp^{-K_f t} \tag{2.7}$$

where f_c is the minimum steady constant infiltration capacity, f_0 is the infiltration capacity at time $t = 0$, K_f is a constant related to the permeability of the soil. For the limiting case of initially dry, well-drained materials, the infiltration capacity is relatively large at the onset of rain of a constant rate and decreases exponentially towards a constant with time. On a sloping soil surface, if the infiltration capacity is exceeded by the rainfall rate, the excess water will travel downhill on the ground surface. This water that flows over the surface was termed "rainfall excess" overland flow (Horton, 1933). Overland flow of water will continue to move downhill until it either enters a stream channel, infiltrates at some other location where the infiltration capacity has not been exceeded, or is stored in depressions where it either evaporates or infiltrates some time later. The water that infiltrates into the soil moves to the stream channel at a much lower rate and is effectively isolated from overland flow. When overland flow is produced it occurs over extensive hillslope areas. Thus the source of water that controls the flood peak is overland flow. This mechanism for generating stream flow is frequently referred to as "Hortonian overland flow" (Kirkby and Chorley, 1967).

Horton (1945) later extended the model of surface runoff to include surface erosion as well as overland flow generation. He proposed that downslope of a threshold distance from the drainage divide, the depth of overland flow becomes sufficient to generate shear stress adequate to entrain soil particles (Figure 2.24). Early field observations supporting the Hortonian view of stream flow generation and erosion came primarily from poorly vegetated environments with thin soil cover and low infiltration capacities (e.g., Schumm, 1956; Chorley, 1978).

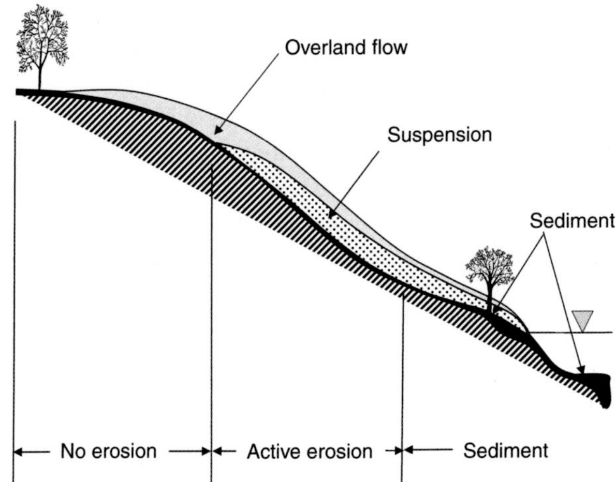

Figure 2.24 Conceptual illustration of Horton's theory of hillslope formation.

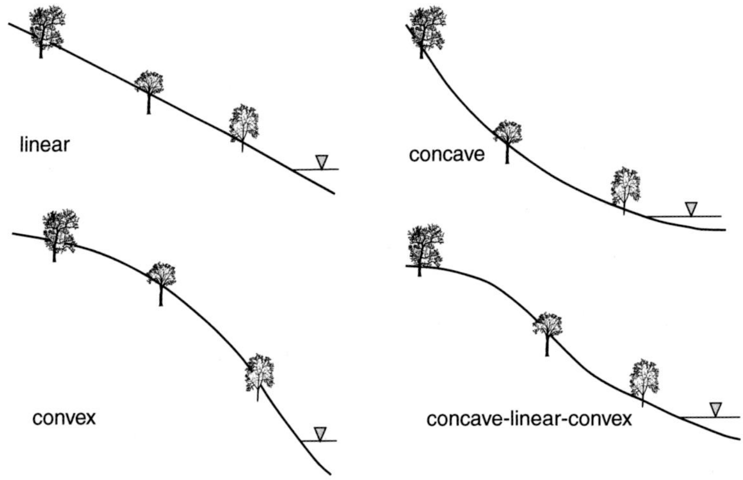

Figure 2.25 Illustration of common hillslope profiles.

In realty, the shape of hillslope profiles can be of many forms, and even a specific hillslope can evolve in morphology from one distinct form to another over time. Figure 2.25 shows some of the common shapes of hillslope profiles.

2.4.2 Subsurface flow processes and runoff generation

From the beginning, the simplifying assumptions of Horton's theory have been challenged. The main drawbacks are its emphasis on overland flow to the relative neglect of the roles of infiltration and groundwater flow, and the importance of soil hydrologic and mechanical

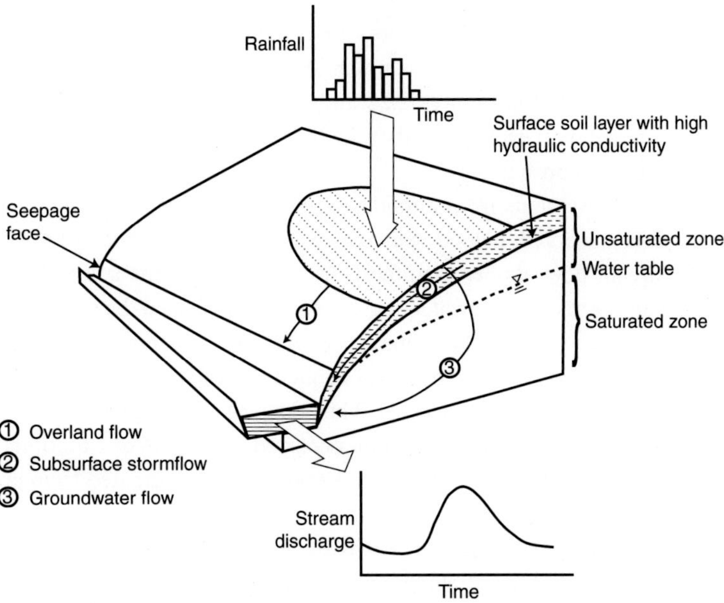

Figure 2.26 Illustration of interplay among overland flow, subsurface stormflow (through flow), and groundwater flow (after Freeze, 1972).

properties in shaping slope morphology and controlling the hydrologic cycle. The importance of subsurface hydrologic processes in stream flow is illustrated in Figure 2.26. As shown, while rainfall characteristics can be correlated to stream discharge, the timing and pattern of discharge is often significantly influenced by subsurface flow. The governing principles describing subsurface flow processes will be covered in Chapters 4 and 5.

It is generally accepted that overland flow is less than the Hortonian prediction, and infiltration and groundwater flow play major roles in generating stream flow. While the exact nature of flow distribution between surface and subsurface is site specific, the qualitative relationship is shown in Figure 2.22. In general, hillslope runoff and infiltration are controlled by two major factors: characteristics of precipitation and characteristics of hillslope. As mentioned previously, there are four precipitation characteristics that are associated with landslide occurrence: annual precipitation, storm duration, antecedent precipitation, and storm intensity. The hillslope characteristics important to landslides are slope gradient, vegetation form and coverage, soil profile and hydrologic and mechanical properties, and the water table location.

Depending on surface and subsurface conditions, runoff can be through several flow processes as depicted in Figure 2.22. If the slope gradient is relatively gentle, soils are permeable, and the water table location is shallow, return flow from infiltration and saturated overland flow can occur as shown on the slope on the right side of the stream. On the other hand, if hillslope materials are impermeable and slope is steep, as shown on the upper left side of the stream in Figure 2.22, Hortonian overland flow best describes the runoff situation. Pipe flow or return flow from pipe outlets can also occur if fractures, cracks, cavities and other macropores are present.

Infiltration generally is much slower than overland flow, depending on its flow paths, as illustrated in Figures 2.22 and 2.26. Because these so-called "preferential" flow paths interact with each other, their hydro-mechanical consequences are not straightforward and are often obscured.

Observations from humid, well-vegetated hillslopes, where overland flow was rarely, if ever observed, helped spur the investigation of other explanations for the generation of stream flow. Rather than the Hortonian view of the soil mantle as a boundary separating surface and groundwater, these studies emphasized the role of subsurface flow processes in generating stream flow. One aspect of the Horton model is that when the infiltration capacity is exceeded, overland flow is generated from all hillslopes in the catchment or drainage basin more or less simultaneously. This idea was challenged in the work by Betson (1964), who showed that storm runoff only occurred over parts of experimental watersheds in western North Carolina. Runoff during heavy rainfall originated from areas where the surface was saturated (swampy areas) and the relative amount of runoff increased during the winter season when soil moisture was greatest. Ragan (1968) introduced the term "partial area concept" to describe this mechanism for generating storm runoff. He presented soil moisture measurements that showed that very little rainwater entered the soil mantle near a stream channel, and that the opposite was the case at locations some distance from the stream where the water table was more than 2 m below the surface. Localized zones adjacent to the stream that expanded and contracted with rainfall intensity and duration were the source of runoff.

The recognition of the dynamic extent of the area of a watershed that contributes overland flow to streams and the recognition that subsurface flow is significant to stream flow generation gave rise to the "variable source-area concept" (e.g., Hewlett and Hibbert, 1967; Dunne and Black, 1970). This concept describes the temporal and spatial variation in the source of stream flow that is controlled by soil moisture conditions and available storage in the vadose zone (Chorley, 1978). Runoff, termed "saturated excess overland flow" is generated by rain falling on areas adjacent to stream channels where the ground surface is saturated (Figure 2.22). The spatial extent of the saturated zones expands during a storm as infiltrating rainfall and subsurface lateral flow towards stream channels increases soil moisture and decreases available storage capacity. As rainfall intensity decreases, the saturated zone decreases in areal extent and stream flow decreases.

Greater appreciation of the role of subsurface processes on stream flow generation followed the innovative findings of Sklash and Farvolden (1979). They used the relative presence of environmental isotopes in "old" groundwater and "new" rainfall as tracers, and determined that for wet initial conditions in a small (1 km^2), low relief drainage in Ontario, Canada, both overland flow and stream flow were dominated by groundwater. Conversely, rainwater dominated both overland and stream flow if intense rainfall fell on the basin under relatively dry initial conditions. The suggested mechanism, in which infiltration creates a total potential that drives groundwater to discharge to the stream is known as "groundwater ridging." The displacement of pre-storm water by new rainfall is often referred to as translatory flow (Hewlett and Hibbert, 1967).

2.4.3 Subsurface stormflow

Subsurface stormflow is the lateral downslope movement of water through soil layers or permeable bedrock that occurs during heavy rainfall or snowmelt (e.g., Hursh, 1936; Whipkey and Kirkby, 1978; Weiler *et al.*, 2005) and may dominate stream flow generation in some settings. Whipkey (1965) reported results from a hillslope in east-central Ohio, where the lateral subsurface stormflow was monitored during applied rainfall. A trough was dug at the base of the slope through a soil profile that consisted of an upper 90 cm layer of permeable sandy loam and a lower layer of more compact, less permeable loam and clay loam. Flow was collected from the free face at various depths and the pressure head conditions in the hillslope were monitored with tensiometers. Observations showed that infiltrating water moved vertically in the soil until it reached an impeding layer of lower permeability, where it formed a saturated layer that led to seepage from the upper layer. The quantitative findings of this study have been questioned based on the influence of the free face and the applied rainfall rates on the development of saturated zones, but qualitatively, the study provides insight into the significance of subsurface lateral flow in hillslope hydrology.

One of the earliest quantitative field investigation employing the total potential concept to subsurface unsaturated and saturated flows in hillslopes was by Weyman (1973). Troughs 60 cm deep were cut into a hillside through the soil layer and into the underlying impermeable bedrock. Lateral discharge and water table variations as functions of time were measured during several natural rainfall events. The study revealed the following seemingly contradictory findings regarding subsurface flow in hillslopes. Infiltration, instead of subsurface stormflow, dominated discharge to the stream as the discharge patterns follow Darcy's law at the hillslope scale. Through flow only occurred above breaks (impedance) in the vertical permeability profile of the soil. Other than at the wetting front, full saturation only occurred above some impediment to further vertical movement of water. Once saturated conditions were generated, lateral flow occurred as the equipotential lines within the saturated soil were nearly orthogonal to the slope. As a result of through flow in a soil layer, the water table rises some time after rainfall, but contracts during the drainage process. Weyman (1973) also provided both conceptual illustrations and measurement data using the total water potential concept to conclude that, during rainfall, vertical flow should dominate, after the cessation of rainfall, lateral downslope flow will dominate, and long after rainfall, upslope flow can occur due to evaporation.

This understanding and evidence of unsaturated flow regimes in hillslopes laid the foundation for other studies that explored flow regimes in other geologic and geomorphologic settings. Results of a field study of a small, gauged catchment by Harr (1977) show that although flow through a soil layer may be unsaturated during rainfall, saturated flow can develop and persist for some time after prolonged heavy rainfall. The study confirmed that, between rainfall events, the downslope component of lateral flow can be greater than the vertical flow component. However, during rainfall, the downslope and vertical components can be equal. Harr's study also confirmed that subsurface flows, both unsaturated flow and stormflow, can account for a considerable portion of the total precipitation. Unsaturated flow was a much greater proportion of stream flow than stormflow (97% vs. 3%) at the study site.

Anderson and Burt (1978) demonstrated the importance of topography in controlling lateral downslope flow in headwater catchments by showing that the lateral subsurface flow toward the hollow (Figure 2.15) can be directly correlated to stream discharge records. Topographic control on the dynamic moisture distribution at the catchment scale has been a classical problem in hillslope hydrology under different climate and geographic settings. Based on studies at catchments in Vermont and Ontario, Dunne *et al.* (1975) showed that the extent of the area of surface saturation is variable throughout the year. By late in the winter season a significant part of the surficial soil of the catchment was saturated, whereas by late in the spring season, as the soils dried, only a narrow strip along the axis of the catchment was saturated. Based on observations in temperate regions in Australia, Grayson *et al.* (1997) developed a two-state conceptual model that describes the spatial distribution of soil moisture at the catchment scale. Specifically, the soil moisture distribution switches between either a wet or a dry state. The wet state is associated with the rainy season when lateral flow dominates both on the surface and in the subsurface. The dry state is associated with the period when evapo-transpiration exceeds rainfall and vertical subsurface flow dominates the moisture distribution in the catchment. The flow directions in the two-state model of Grayson *et al.* (1997) are somewhat inconsistent with Weyman's (1973) conceptual model. Other recent investigations on the exact conditions that determine flow directions and regimes are covered in greater detail in Chapter 4.

The existence of a capillary fringe above the water table in close proximity to stream channels has been hypothesized by some researchers (e.g., Martinec 1975; Fritz *et al.*, 1976; Sklash and Farvolden, 1979; Gillham, 1984). Results from a laboratory sandbox experiment by Abdul and Gillham (1984) showed that the capillary fringe is a potentially important mechanism in providing quick overland and subsurface flow to stream discharge. However, the notion has been challenged ever since its introduction. In comments to Abdul and Gillham's (1984) article, Zaltsberg (1986) used physical evidence of the discrepancy between the observed capillary fringe height and the depths to the water table to discount the capillary fringe mechanism. Typically, the height of the capillary fringe is a fraction of the depth of the water table at the hillslope scale, so the laboratory sandbox observation and conclusions drawn from those observations are not applicable at the hillslope scale. In reply, Gillham and Abdul (1986) reiterated that the capillary fringe effect is one of the proposed mechanisms that can reconcile many observations and requirements for the stream discharge increase associated with rainfall. They also pointed out that the evidence Zaltsberg used to discount their argument was partially based on information from wells that were located far from stream channels. Further investigations supporting each side of the arguments (Jayatilaka and Gillham, 1996; McDonnell and Buttle, 1998) have been presented. In summary, the question of the importance of the capillary fringe mechanism to stream flow generation at the hillslope scale remains unresolved.

Although perhaps intuitive, recent studies (e.g., McDonnell *et al.*, 1996; Burns *et al.*, 1998; Freer *et al.*, 2002) provide compelling evidence that subsurface bedrock topography can greatly control downslope flow at the hillslope scale. The general concept is that depressions in the bedrock fill with water following rainfall and infiltration. During rainfall

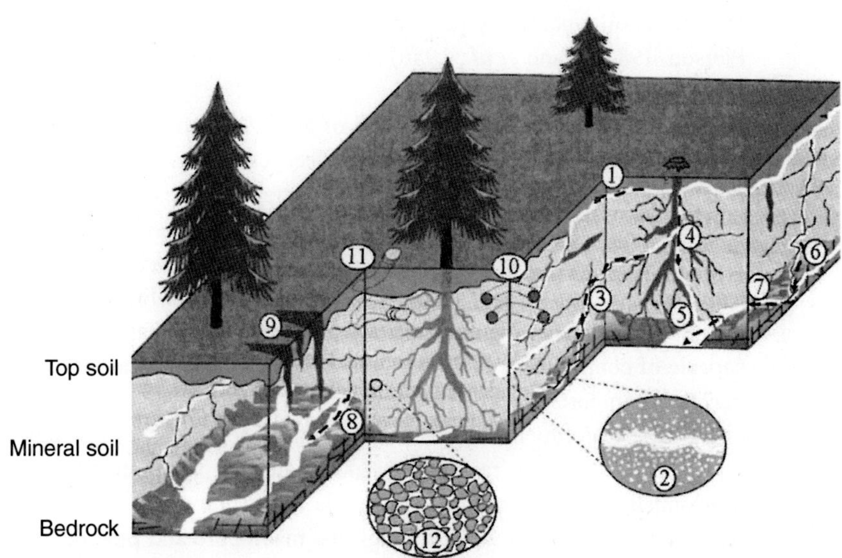

Figure 2.27 Illustration of subsurface flow paths (from Sidle and Ochiai, 2006).

and infiltration the thickness of the saturated zone above the bedrock increases and at some point the discontinuous saturated zones that are collected in the bedrock depressions become connected and downslope lateral saturated flow is enhanced. This lateral downslope flow continues until the saturated thickness above the bedrock diminishes and only the depressions are saturated.

2.4.4 Subsurface stormflow and landslide initiation

In steep terrain, rapid downslope movement of water is often attributed to flow through macropores or soil pipes or fractured bedrock (e.g., Mosely, 1979; Sidle *et al.*, 2000; see also Figure 2.22). Because stormflow can contribute to the generation of positive pore-water pressures it has been linked to slope instability and the generation of landslides (e.g., Montgomery *et al.*, 1997; Uchida *et al.*, 2001; Fox and Wilson, 2010). Macropores are soil pores of relatively large size compared to those in the soil matrix and are generally visible to the naked eye. They result from biological activity, such as tree root growth and decay, cracking of the soil that occurs under tension, and shrink–swell processes resulting from cyclic wetting and drying or freezing and thawing (see Figure 2.27). The term "soil pipes" generally refers to large macropores that are oriented roughly parallel to the slope. The development of soil pipes is often attributed to the rapid infiltration of water along cracks in the soil surface that then moves laterally along a low permeability horizon (e.g., Selby, 1993). Flow along this contrast may dislodge and erode soil particles creating a soil pipe. Soil pipes are commonly found in the head scarps of landslides (e.g., Pierson, 1983; Brand *et al.*, 1986) and may range from centimeters to meters in diameter (Uchida *et al.*, 2001).

In a field study in the South Island of New Zealand, Mosely (1979) demonstrated that stream flow generation from a small, steep catchment during rainfall is dominated by flow through macropores. Conservative dye tracers were applied to the soil surface several meters upslope of soil pits. The pits allowed access to the soil profile and collection of tracer fluid. Results showed that macropores with diameters greater than about 3 mm provide a pathway for water to essentially bypass the soil matrix to form a saturated zone perched on the relatively impermeable bedrock. Mosely concluded that flow through macropores was capable of contributing to stream flow during rainfall and that the translatory mechanism may not be important. However, work by Pearce *et al.* (1986), using environmental isotopes on the same experimental catchments, produced contradictory findings showing that stream flow generated during rainfall was dominated by water in storage prior to the storm. McDonnell (1990) was able to reconcile the two studies by showing that the flow of a large volume of stored water through the macropore and pipe network satisfied both the observations of stream flow volume and the isotopic composition of the stream waters.

Tsuboyama *et al.* (1994) presented a conceptual model of stormflow through macropores and soil pipes that describes the expansion of the hydrologically active portion of a soil pipe network with increasing moisture content or an increase in the thickness of a perched saturated zone. The role of soil pipe flow on the initiation of landslides is difficult to generalize. Soil pipes provide a route to rapidly drain saturated zones in steep hillslopes. Alternately, if soil pipes become blocked, they may provide a locus for pore-pressure increase and landslide initiation (e.g., Harp *et al.*, 1990). Similarly, flow out of fractured bedrock into the overlying soil has been identified as a mechanism capable of initiating landslides (e.g., Montgomery et al., 1997), but bedrock fractures are also effective pathways to drain overlying soil. Further quantitative insight into the role of subsurface stormflow and flow through macropores and pipes on landslide initiation awaits advances in geophysical methods that promise to provide spatial and temporal information on shallow groundwater flow (e.g., Kuras *et al.*, 2009).

2.5 Mechanical processes in hillslopes

Mechanical processes in hillslopes are described in terms of force variation and their distribution. At the hillslope scale, it is convenient to describe forces in terms of stresses. Stress distribution in a hillslope, whether due to water movement, external loading, or gravity, is a dynamic process. Based on the mechanical equilibrium principle of Newton's second law, any non-equilibrium state will cause instability or movement of hillslope materials and thus landsliding. Mechanical processes also alter the properties of hillslope materials by promoting physical transport and chemical reactions of materials. In nature, many mechanical processes can provide driving mechanisms for hillslope instability. Some

Table 2.1 Major mechanical processes contributing to shallow landslides

Mechanical processes	Controlling factors	Failure modes
Seismic shaking	Seismic characteristics, soil hydro-mechanical properties	Shear/compaction/ liquefaction
Infiltration-induced pore-water variation	Precipitation characteristics, saturated soil hydrologic properties	Shear
Infiltration-induced soil suction variation	Precipitation characteristics, unsaturated soil hydrologic properties	Shear
Change of soil weight	Precipitation characteristics, soil hydrologic properties, slope morphology	Shear
Weathering	Hillslope material composition, precipitation characteristics	Strength reduction
Mineral dissolution and precipitation	Precipitation characteristics, clay mineralogy	Strength reduction
Slickenside	Precipitation characteristics, clay mineralogy, and movement history	Strength reduction
Creep	Material composition, precipitation characteristics	Strength reduction
Erosion	Precipitation characteristics, soil mechanical properties	Stress change and strength reduction
Tensile cracking	Material composition	Stress change and strength reduction
Shrink–swell	Precipitation characteristics, soil mineralogy	Stress change and strength reduction

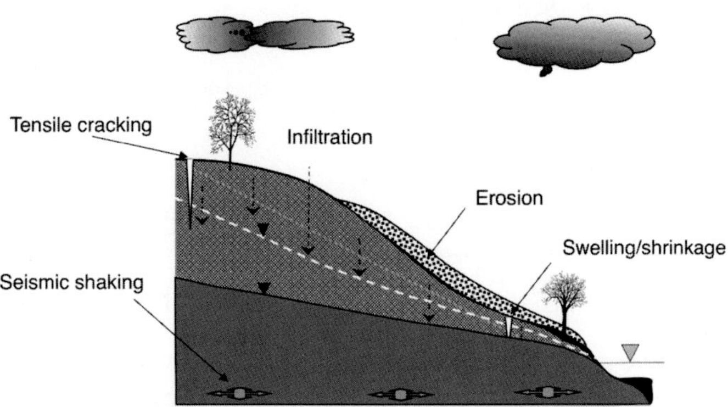

Figure 2.28 Illustration of major mechanical processes in a hillslope.

mechanical processes, such as seismic shaking, pore-water pressure diffusion (saturated), and soil suction stress variation (unsaturated), can cause pure stress variation in hillslope. Other processes change the strength of materials, such as weathering, slickenside, and creeping, and yet others cause changes in both stress and strength, such as tensile cracking, swelling, and shrinkage. The major processes, their controlling factors, and the resulting failure modes are summarized in Table 2.1 and illustrated in Figure 2.28.

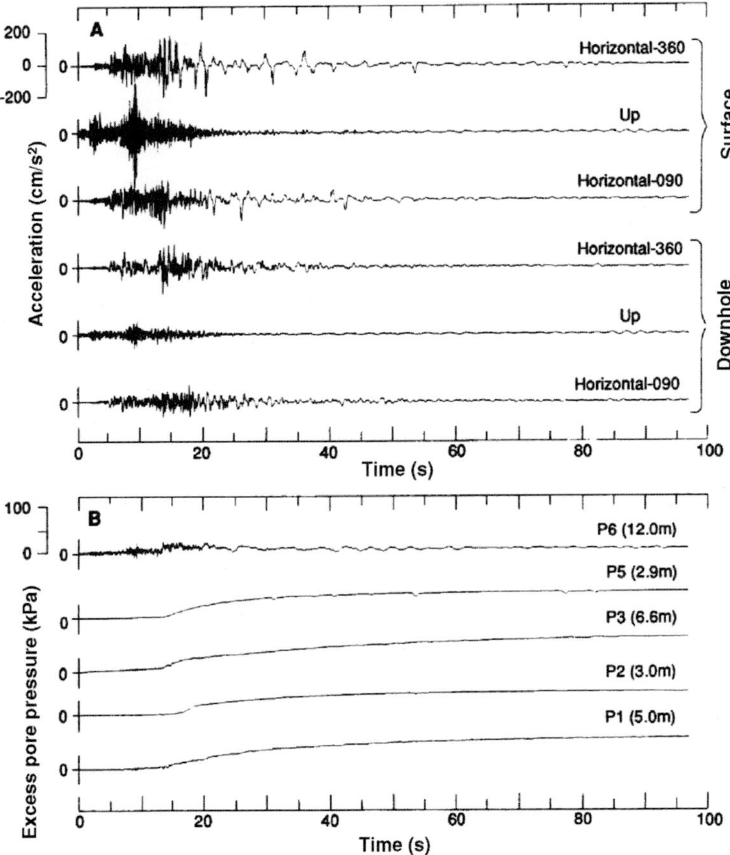

Figure 2.29 (A) Seismograms and (B) induced pore-water pressure response in soil during the 1987 Superstition Hill, California, earthquake (after Holzer *et al.*, 1989).

2.5.1 Stress variation mechanisms

Seismic shaking

Stress is a physical quantity that does not depend on material type or composition. In one-dimensional space in hillslopes, it describes the intensity of the force acting on soil. All major stress variation mechanisms involve water movement in hillslopes. As illustrated in Figure 2.28, earthquakes typically cause ground motion in both horizontal and vertical directions. Stress waves due to earthquake shaking can be in either compressive or shear form. Since earth materials are multi-phase porous media, wave propagation involves hydro-mechanical coupling, resulting in complex and dynamic stress variation patterns. A single strike of a large earthquake, of moderate or large magnitude, can trigger thousands of shallow landslides (Keefer, 1984). For example, the 1994 M6.7 Northridge earthquake triggered more than 10,000 shallow landslides near Los Angeles, California (Harp and Jibson, 1996).

For a specific site, the stress waves are often recorded in terms of acceleration in both the horizontal and vertical directions, as illustrated in Figure 2.29A. As shown, seismic waves in both horizontal directions typically arrive later and attenuate more slowly than vertical waves. The entire episode typically lasts for tens of seconds to several minutes depending on the source, magnitude, and depth of the earthquake. Due to the free-surface boundary condition and possible soil amplification effect, amplitudes of acceleration at or near the ground surface are generally much greater than in the deep subsurface. The measured acceleration can be directly used to calculate total stress changes in hillslope by Newton's law or equation of motion.

Diffusion of pore-water pressure due to seismic shaking depends on soil hydro-mechanical properties and hillslope configuration, as illustrated in Figure 2.29b. Pore pressure waves with high frequencies typically arrive earlier in deep locations, but are attenuated as they propagate toward the ground surface. However, pore-water pressure diffuses slowly as it approaches the ground surface. For example, in the first 10 seconds of the earthquake shaking, pore-water pressure fluctuates frequently at the depth of 12 m, but very little at shallower depths. After 30 seconds, pore-water pressure at the depth of 12 m quickly returns to the pre-earthquake condition, whereas at shallow depths, pore-water pressures still increase gradually.

Because the mechanical behavior of saturated soils is governed by Terzaghi's effective stress principle, records of acceleration (total stress) and pore water at different depths provide important and necessary information to calculate changes in effective stress, and thus the state of slope stability. Under saturated or nearly saturated conditions, it is well known that excess pore-water pressure reduces effective stress and can cause soil liquefaction and soil volume reduction. Soil liquefaction is a phenomenon whereby saturated soil loses significant strength and stiffness under dynamic loading. Under unsaturated soil conditions, which are the predominant conditions for many shallow landslide-prone regions, Terzaghi's effective stress principle can be expanded with a generalized framework proposed by Lu and Likos (2004a, 2006). The application of this generalized framework to landslide initiation under static unsaturated conditions will be illustrated throughout this book.

Weight of water

Stresses near the ground surface can significantly change simply due to the weight of water. Rapid infiltration near the surface can result in changes in soil weight by 30%. This could induce an increase in the predominantly vertical stresses, leading to an increase in shear stress in the slope and possible failure. Intense precipitation and rapid soil drainage will promote such conditions. Quantitative analysis of the effect of soil weight change due to wetting will be provided in Chapter 9.

Saturated pore-water diffusion

Pore-water pressure variations due to groundwater movement are by far the most commonly recognized mechanism for stress changes. Conceptually, two types of pore-water pressure variation exist: pore-water pressure in the saturated zone below the water table and soil suction in the vadose zone above the water table. Traditional approaches in geotechnical engineering design of earthworks, landslide science, and geologic engineering have focused on saturated pore-water pressure changes due to rainfall, as they are well described using

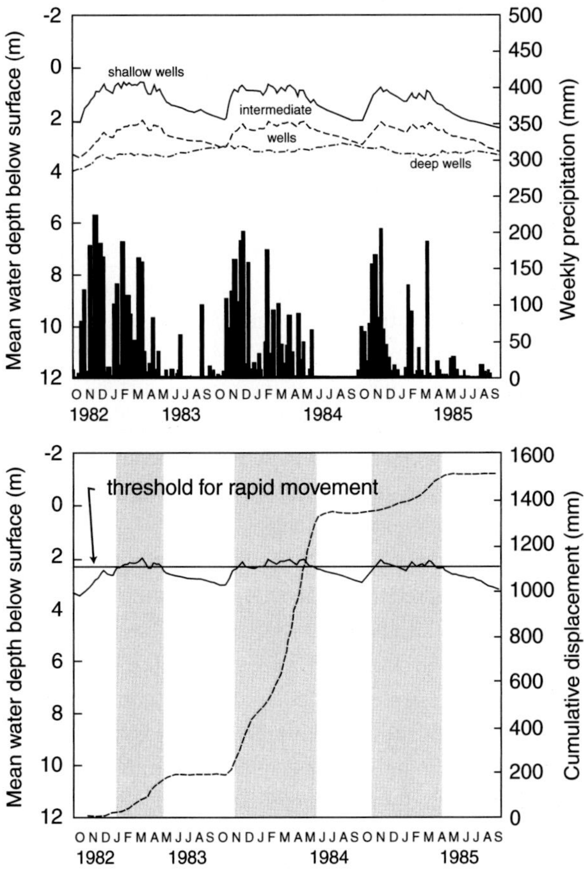

Figure 2.30 Temporal correspondence of precipitation, groundwater levels, and landslide displacement at the Minor Creek landslide monitoring site in northwestern California. The vertical grey bars in the lower panel indicate periods of rapid landslide motion (redraw from Iverson and Major, 1987).

Terzaghi's effective stress principle. Pore-water (or soil suction) changes in the vadose zone and the consequent changes in soil inter-particle stress are subjects that have drawn increasing attention recently. This book will systematically introduce principles of soil suction changes (Chapters 3 and 4), principles of soil suction stress changes (Chapter 6), and principles of how to use geomorphology, vadose zone hydrology, and suction stress for landslide initiation prediction (Chapters 9 and 10).

Pore-water pressure changes due to groundwater seepage and water table rise are generally well-documented phenomena in hillslope and landslide environments. This mechanism is commonly considered in the slope stability analysis. Figure 2.30 illustrates results from a monitoring study of the Minor Creek landslide in northwestern California (Iverson and Major, 1987) that show the temporal link between precipitation, pore-water response, and landslide motion. Sixty open-standpipe piezometers (wells) were installed in the landslide to allow for weekly water level measurements. These observations are grouped into three categories: shallow (<3 m depths), intermediate (3 to 6 m depths), and deep (>6 m depths). Pore-water response lags the onset of precipitation by several days, and the greater hydraulic head in the shallow wells compared to the deeper ones, indicate a general downward gradient. The overall record shows a strong seasonal cycle with a peak in groundwater levels in the late fall and winter months. Landslide motion generally coincides with increased water levels and rapid motion occurs once a critical groundwater level is exceeded in the intermediate depth wells. Movement of the landslide continues at a generally steady velocity as long as the water level remains high, but movement tends to cease once water levels fall below the threshold.

The impact of infiltration on water table rise is illustrated in Figure 2.28. If a soil layer of relatively low hydraulic conductivity is present at depth, the overlying water table can rise quickly under heavy rainfall. As the water table rises, some areas originally under partially saturated conditions become saturated (or the area between the original water table and the elevated water table). There are two hydrologic consequences that result from the water table rise: pore pressure increases below the water table and soil suction dissipates in the vadose zone. According to Terzaghi's effective stress principle, any point subject to an increase in pore-water pressure without a change in total stress will decrease its effective stress, thus increasing the likelihood that the shear strength of the soil will be exceeded. In contrast, until recently, the consequence of reduction in soil suction to slope instability has not been addressed. The basic concept of soil suction reduction leading to soil failure is illustrated below. Detailed treatment of this subject can be found in subsequent chapters.

Unsaturated soil suction diffusion

Soil suction variation occurs in the unsaturated or vadose zone due to infiltration. It can vary greatly from zero to several hundred thousand kPa, depending on type of soil and moisture content. In hillslopes, changes in soil suction can occur with or without changes in the water table configuration. Mechanically, soil suction controls and provides the mechanism for particle-scale forces. There are three particle-scale forces fully or partially controlling soil suction; namely, the electric double layer force, van der Waals attraction, and capillarity. The resulting stress is the suction stress theorized by Lu and Likos (2004a, 2006). This stress can be considered as part of effective stress and a generalization of pore-water pressure in Terzaghi's effective stress concept. Suction stress generally provides pulling forces among soil grains. Consequently, it can increase effective stress up to 600 kPa in most soils. Such changes in stress are particularly important for shallow soil (<1 m), as effective stress due to the self-weight of soil could be very low (several kPa). The theoretical and practical basis for the suction stress concept, as well as its experimental validation, will be provided in detail in Chapter 6.

In sandy soils, the range over which soil suction accompanies changes in water content is from 0 to as much as 200 kPa. The resulting suction stress (effective stress) reduction from dry (200 kPa) to wet (0 kPa) could be up to 5 kPa. Furthermore, the variation of suction stress with suction or soil water content is highly non-linear, being zero at both zero suction (saturated) and high suction (dry). Such an amount of stress reduction could cause soil failure in steeply sloping environments ($\sim >30^{\circ}$).

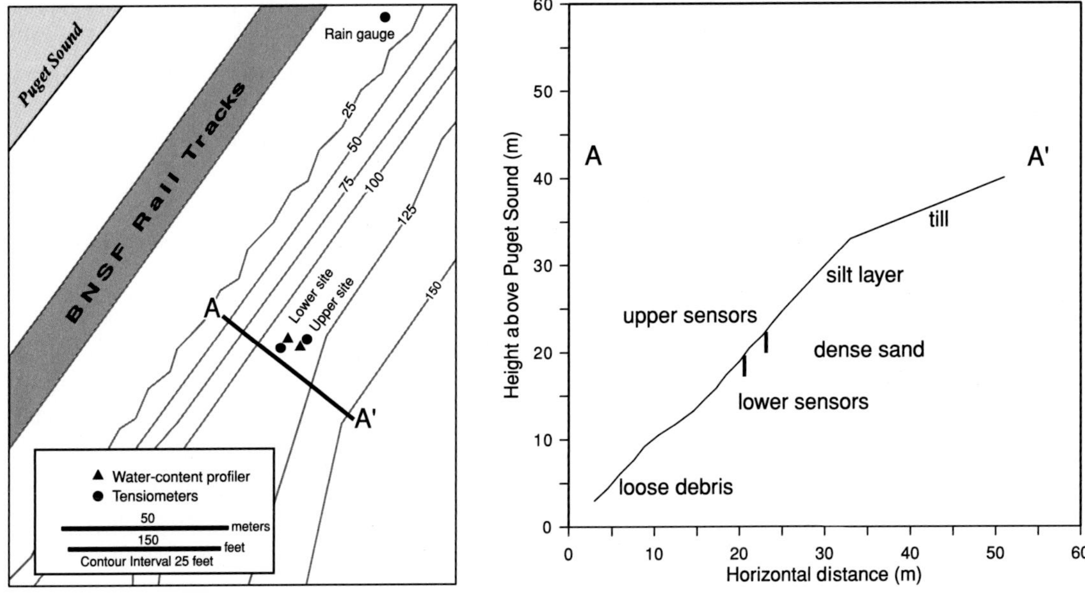

Figure 2.31 Map showing the location of soil moisture and pore-pressure monitoring instruments at the USGS site near Seattle, Washington (from Baum *et al.*, 2005).

In silty soils, the controlling range of soil suction for water content is up to 500 kPa. The resulting suction stress (effective) reduction from dry (500 kPa) to wet (0 kPa) could be up to 100 kPa. Here again, the variation of suction stress with suction or soil water content is highly non-linear, being zero at both zero suction (saturated) and high suction (dry). Such an amount of stress reduction could cause soil failure in steep to moderately sloping environments ($\sim 25^{\circ}$).

In clayey soils, the controlling range of soil suction for water content is 0 to 100,000 kPa. The resulting suction stress (effective) reduction from dry (100,000 kPa) to wet (0 kPa) could be up to 500 kPa. However, the variation of suction stress with soil suction or soil water content is a monotonic function, being zero at zero soil suction (saturated) and increasing monotonically up to 500 kPa at high soil suction (dry). Such an amount of stress reduction could cause soil failure in any slope environment.

As an example to illustrate the strong correlation between intense precipitation and the reduction in soil suction, let us examine field-monitoring data collected from the coastal bluff at the USGS site near Seattle, Washington, described previously. Rainfall, soil moisture, and soil suction data were collected during the period between October 2003 and May 2004 (Baum *et al.*, 2005). A plan-view map and slope cross section are shown in Figure 2.31. The slope is inclined at about 45° and the surficial materials are sandy colluvium with a poorly developed O horizon. The height of the bluff at this location is about 50 m.

Instruments to monitor soil moisture and suction were installed in two profiles at two different depths. The time series of moisture content and soil suction are plotted in Figure 2.32. In the first two weeks of November 2003, there was very little rainfall, leading to a decrease

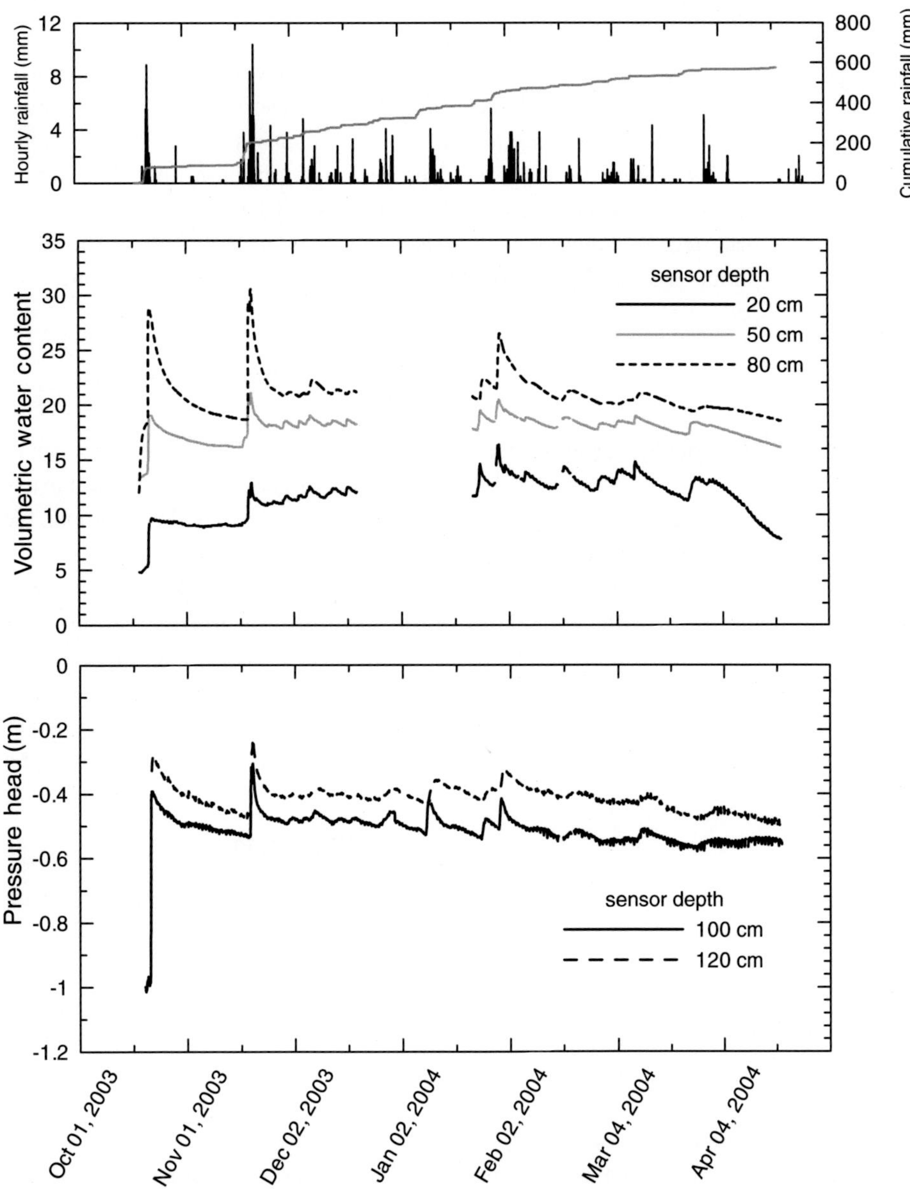

Figure 2.32 Illustration of correlation between precipitation, soil moisture content, and soil suction variation at the Edmonds, Washington, site (from Baum *et al.*, 2005).

in soil moisture contents (drying) and a decrease in pore-water pressures (a decrease in pore-water pressures indicates a similar increase in soil suction, which is defined as the difference between atmospheric pressure and soil water pressure). It is also evident that at this site, for the observation period, soil moisture content generally increases (soil suction decreases) with increasing depth from the ground surface.

Around 18 November 2003, there were several days of intense precipitation with a cumulative amount of 100 mm. Soil moisture content increased by about 4.5% in the several days after the initial precipitation, with the shallow station (180 cm from the ground surface) ahead of deep station (200 cm from the ground surface) in response time. Correspondingly, soil suction reduced by an amount of 25 cm in pressure head at both depths, with the shallow station ahead of the deep station in response time. The timing in precipitation, soil moisture content, and suction changes indicates that downward infiltration of rainfall occurred. Precipitation, soil moisture, and soil suction data for the rest of the observation period ending May 2004 all confirm the strong correlation among each of the observed processes.

2.5.2 Strength reduction mechanisms

Intrinsically, strength is a material variable, which means it is dependent on the composition and quality of the material. For earth materials, strength is often cast in a mathematical relation coupling stress components with some parameters. For example, the Mohr–Coulomb criterion is often used to describe a linear relationship between shear stress and normal stress. When this linear relation is plotted with shear stress on the vertical axis and normal stress on the horizontal axis, two parameters, namely cohesion (intercept) and friction angle (slope of the line) are used to represent shear behavior of earth materials. Thus, each material has its own cohesion and friction angle. Over time scales of several to hundreds or thousands of years, weathering processes could change these strength properties of hillslope materials. The strength of hillslope materials will be described in Chapter 7.

Weathering

Weathering is the process of breaking down earth materials under natural forces of air, water, other chemicals, and biological action. It generally occurs slowly due to the rates of physical, chemical, and biological reactions. As a result of these interactions, the strength of earth materials (parameters) changes over time. Weathering could either strengthen or weaken the shear strength characteristics of earth materials, although it generally weakens them. Because weathering is a slow process, its effect on landslides is often not seen or considered over short periods of time, say in days, but can be a factor over a long period of time, say months and years. The most powerful agent causing physical, chemical, and biological reactions is water. Therefore, humid environments usually promote deeper and more rapid weathering, i.e., thicker soils result from the generally more efficient breaking down of rock. In general, weathering has a more marked influence on soil cohesion than on its friction angle.

Mineral dissolution or precipitation

Mineral dissolution or precipitation provides another mechanism for strength change. It typically occurs in the vadose zone. When liquid water flows through surficial soils either by infiltration or capillary rise, the water will dissolve or precipitate salts or minerals and carry or remove them. In the case of infiltration, the dissolution process is also called leaching, and it will remove dissolvable matter from the original site, leading to weak bonds among soil grains and an increase in porosity. In expansive soils, leaching of salt can result in an increase in swelling pressure upon wetting. In the case of capillary rise, water can evaporate into the atmosphere, leaving behind dissolved matter among soil grains, leading to strong inter-particle bonds. Both chemical dissolution and precipitation usually take a long time. The cumulative effect is often physically evident as a thick layer of calcium carbonate-rich soils near the ground surface in regions where evaporation is dominant in dry regions and loose topsoil where infiltration is dominant in wet regions.

Slickenside

Slickensides provide another strength reduction mechanism for clayey soils that are sheared. Nearly saturated and low salt conditions promote strong double layer repulsive forces among clay particles in high plasticity clays. Under the soil's self-weight or external shear loading conditions, the repulsive force will cause the plate-like clay particles to realign parallel to the slip plane. This parallel structure along the slip plane leads to a very small friction (angle) in the slipping direction, causing strong anisotropy in the friction angle. A slickenside can be considered as a local-scale soil failure, so it reduces both cohesion and friction angle significantly. Slickenside zones are considerably weaker than non-slickenside zones where particles are randomly oriented. Since the slip surface has already reached the failure state, the friction angle along the slip plane is similar to the residual friction angle of the soil, often in the range of a few to ten degrees.

Soil creep

Soil creep is a phenomenon that can occur in wet clays or unconsolidated soils where soils deform continuously under a sustained loading. Creep generally decreases the shear strength of soils and can cause soil failure. It can be exacerbated under cyclic loading conditions such as wet–dry and freeze–thaw cycles. Thus, it is a common landslide mode in permafrost regions. Creep is a permanent displacement that accumulates over time and can result in sudden sliding on a continuous failure plane.

Root growth and decay

Root growth and decay result following the removal of vegetation by natural or human causes. In industrial forests, during the typical cycle of tree harvest and replanting, root tensile strength will be drastically reduced in the first few years following cutting. As new trees establish root systems, root tensile strength increases. Full recovery of root strength may take a decade or more. During the period when root tensile strength is at a minimum, rainfall-induced landslides may occur more frequently. In hillslope environments, root tensile strength is often mobilized as root shear strength. The decay of tree roots following harvest may result in a reduction of several tens of kPa of shear strength leading to an

Figure 2.33 Tensile cracking developed near the top of a hillslope in Jefferson County, Colorado. See also color plate section.

increase in landslide susceptibility. The quantitative treatment of root tensile strength and induced shear strength are described in Chapter 7.

2.5.3 Combined change in stress and strength

Processes such as soil erosion and tensile cracking involve changes in both the fields of total stress that result from gravity (either by slope geometry or by weight), and effective stress (suction stress) and the previously mentioned strength reduction mechanisms. Quantitative descriptions of the total stress field due to gravity are provided in Chapter 5 and quantitative treatments of changes in the effective stress field are described in Chapters 6, 8, 9, and 10.

Erosion

Runoff described in the previous section provides a driving mechanism for soil erosion. The downslope movement of surface soil driven by runoff is the combined result of stress increase (pore-water and mechanical drag) and strength reduction (material softening and loss of cohesion). Since a slope is reshaped after an episodic event, some portion of the slope (see Figures 2.24 and 2.25) becomes steeper, increasing the potential for further landslides in the future.

Tensile cracking

Tensile cracking can be considered a result of combined changes in soil stress and strength. As illustrated in Figure 2.33, tensile cracking is often found near the crest of hillslopes and is evident in many translational shallow landslides. However, the exact role of tensile cracking in the initiation of shallow landslides is not clear at the present time. There is little physical evidence indicating whether tensile cracks precede or follow the formation of a

Figure 2.34 Local-scale cracks or cavitations due to swelling and shrinking along the Qinghai-Tibet Highway. See also color plate section.

translational failure surface. Conceptually, tensile cracks result from the lateral movement of a slope soil under the free stress at the slope surface and developed shear stress within soils. Tensile cracking is typically more pronounced in clayey soil than in sandy soil. Because soil is a deformable material, changes in moisture content often produce nearly vertical cracks that develop progressively under cyclic changes in moisture content. The progressive development of small or micro-cracks not only weakens the strength of the soil mass as a whole, but also provides pathways for infiltration into the hillslope, causing further reduction in soil suction and suction stress in clayey soils.

Shrink–swell

Swelling or shrinkage, as shown in Figure 2.34, not only weakens soil strength but also causes internal stress (suction stress) changes. So, it may useful to consider swelling–shrinkage as a combined stress and strength failure mechanisms. Upon wetting or drying, the high content of smectitic soil minerals, such as montomorillonite, may hydrate or dehydrate, causing great variation in either void ratio or internal pressure, depending on the confining conditions. This may result in a decrease in suction stress (swelling due to increase in electric double layer repulsion), an increase in suction stress (development of shrinkage and cracks), or a reduction of shear strength characteristics (either due to crack development at local scale or swelling that causes irreversible pore volume enlargement).

The ultimate failure of hillslopes due to swelling and shrinkage takes a long time, often years. This type of failure often is associated with slickenside development over time. The failure surface is often translational and shallow. Engineered hillslopes made of clayey soil in cyclic wet–dry climatic environments often experience such failure years after the

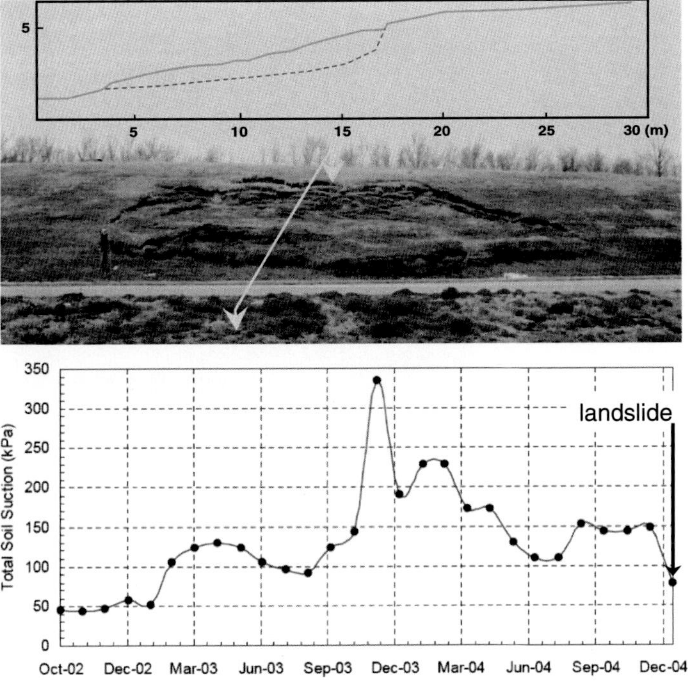

Figure 2.35 Case history of shallow landslide (Garvin landslide) due to swelling–shrinkage at a site along US-70 in McCurtain County, Oklahoma (redrawn from Clarke and Nevels, 2007).

completion of construction. The following case history provides an example illustrating the interplay among the degradation of hillslope materials, the variation in weather and precipitation, subsurface soil suction responses, and the final occurrence of a shallow landslide in unsaturated clayey soil.

The Garvin landslide in southeastern Oklahoma occurred in January 2005 in a naturally occurring stiff clay in a highway road cut, as shown in Figure 2.35 (Clarke and Nevels, 2007). The slope is inclined at about 17°, and the stiff clay is structured and has slickensides that form parallelepipeds and cracks that open up at various times of the year. The slope material can be characterized as a stiff, very high plasticity, mottled, structured, and expansive clay. The road cut was completed in September 2001. During the 42 months prior to the landslide, the overall weather was dryer than normal; however, numerous rainfall events occurred. The average soil suction variation is shown in Figure 2.35, indicating that the soil experienced large suction variation 24 months prior to the landslide.

The site investigation and slope stability analysis conducted by Clarke and Nevels (2007) indicated that this landslide resulted from cyclic stress variations and clay property degradation that occurred entirely in the vadose zone. The degradation of the internal angle of friction of the soil progressed to the extent that it was reduced to nearly its residual value. The cyclic effect of water infiltration into a cracked soil structure was evidently sufficient for the development of the large cumulative shear strains needed for the development of the residual shear strength failure condition.

2.6 Problems

1. What happens to moist airflow as it encounters mountains?
2. Describe "Pineapple Express" storms. How often do they impact the west coast of North America in a year?
3. What is a monsoon? Where do they occur?
4. In what months does Nepal typically suffer the most fatalities related to landslides? Why?
5. What does the Clausius–Clapeyron relation describe?
6. In general, how much additional water can the atmosphere hold with each degree Celsius increase in temperature?
7. Where have the most significant increases in daily precipitation in North America been observed?
8. What climate change process might result in a decrease in landslide frequency or activity?
9. Describe the three hillslope zones identified by Hack and Goodlett in their study in the Appalachians.
10. What processes tend to increase the potential for landslides in topographic hollows?
11. What fraction of debris flows originated in topographic hollows in Marin County following the storm in January 1982?
12. Describe the data structure most commonly used in digital elevation models.
13. What is the slope, as a percentage, between two points separated by a horizontal distance of 100 m and a vertical distance of 20 m?
14. Describe some of the assumptions that are made in applying distributed hydrologic models to assess landslide potential.
15. Why do we need to classify hillslope materials?
16. What is the major criterion used to classify hillslope materials in different systems?
17. What are the major categories used in naming hillslope materials?
18. If a hillslope material consists of soil particles with uniform size of 0.06 mm, what will be the name of the material in the systems of USDA, International, USCS, and AASHTO?
19. What are the major mechanical processes in hillslopes?
20. Will pore-water pressure (under saturated and unsaturated conditions) increase or decrease the stability of hillslopes? Why?
21. What are the major failure modes for hillslopes?
22. Will seismic loading affect total stress or pore-water pressure?
23. How much could the weight of water potentially change total stress in hillslopes?

24 Figure 2.30 shows the delay of slope movement after precipitation. What is the reason for such delay?

25 What is orographic precipitation? What are atmospheric rivers? What are monsoons? What are tropical cyclones?

26 What are the major climatologic and meteorologic facts triggering landslides? What are the effects of topography on precipitation? What are the major climatologic events affecting precipitation?

27 What is El Niño? What is the opposite of El Niño? During an El Niño, where would you expect higher or lower than normal precipitation?

28 What are the major mechanical failure mechanisms due to soil water changes?

29 Calculate the slope (as a percentage and in degrees) of a 10 m DEM cell that lies at the center of a 3 × 3 matrix of elevations where $a = 5.0$, $b = 6.0$, $c = 8.5$, $d = 4.5$, $e = 7.0$, $f = 8.0$, $g = 4.0$, $h = 5.0$, and $i = 8.5$. Elevations are in meters.

PART II

HILLSLOPE HYDROLOGY

3 Steady infiltration

3.1 Water movement mechanisms

3.1.1 Introduction

The mechanism that drives the movement of water, in either the liquid or vapor phase, or as mixture of the two, in hillslopes is the gradient of water potential. For unsaturated hillslope materials, water potential can be cast in either liquid or vapor (gas) form. Several distinct physical mechanisms can contribute to water potential in pore water, namely, gravity, pressure, kinetics, and osmosis. In the near-surface atmospheric and subsurface environments, water potential in the vapor phase directly reflects the number of water molecules in a unit volume of air. This is because the other major gas molecules, such as oxygen and nitrogen, are not involved in phase changes at temperatures typical in these environments, thus the composition of air is relatively constant.

From the second law of thermodynamics (matter moves from high energy places to low energy places), a change in phase from liquid water to vapor in soil pores will occur if a gradient in potential exists. If the total water potential of pore water is greater than the vapor (air) potential, evaporation will occur. By the same thermodynamic equilibrium concept, within each of the phases, if a gradient exists in the total potential, liquid or vapor flow will occur.

The total potential (in terms of head) is the energy stored in liquid pore water that is available to drive fluid motion in the absence of any chemical reaction. The total potential is often simplified using the superposition principle, so that it can be expressed as the sum of the head due to pore-water pressure h_m, the head due to gravity h_g, the head due to osmosis h_o, and the head due to kinetic energy h_v, i.e.,

$$h_t = h_m + h_g + h_o + h_v \tag{3.1}$$

Among the four possible mechanisms for energy or potential to be stored in pore water, kinetic energy h_v is typically negligible, as the velocity of most subsurface flow is less than 10^{-3} m/s and thus the corresponding kinetic energy ($v^2/2g$) is generally small, less than 10^{-7} m in terms of head. At the hillslope scale, the three other mechanisms are important; they can vary up to hundreds of meters of head both spatially and temporally. The importance of gravitational and pressure mechanisms in hillslopes is illustrated in Figure 3.1.

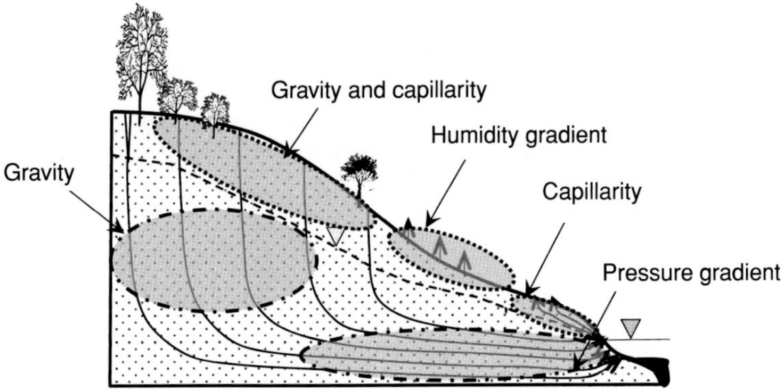

Figure 3.1 Illustration of some of the major water potential processes (gravitational and pressure) in a humid hillslope with a perennial effluent channel.

In a hillslope, gravity drives the flow of liquid water in the unsaturated zone predominantly in the downward direction (infiltration) invariant of time and location. Near the ground surface, both gravity and the gradient of pore-water pressure (also called capillarity in unsaturated soil) can cause significant water movement in both horizontal and vertical directions. In the saturated zone, the geometry of the hillslope often results in the horizontal movement of water due to the existence of a pressure gradient between a stream located at the toe of the slope and the hillslope above. Capillary rise can cause upward seepage in the unsaturated zone near a stream. Evaporation and transpiration can induce significant upward water vapor movement near the ground surface if the gradient of relative humidity between the atmosphere and subsurface is high.

The total potential is commonly described in three inter-changeable units: length (m), pressure (pascal or Pa), and chemical potential (joule per mole or J/mol). They are inter-changeable in that they are defined by the same physical concept; the ability of a unit of water to store energy. They differ in the way a unit of water is defined. In the case of head potential, it is defined as energy per unit weight of pore water, i.e.,

$$h_t = \frac{\text{force} \times \text{distance}}{\text{force}} = \text{distance} = [\text{m}]$$

In the case of pressure potential, it is defined as energy per unit volume of pore water, i.e.,

$$\psi_t = \frac{\text{force} \times \text{distance}}{\text{volume}} = \frac{\text{force}}{\text{distance}^2} = \frac{[\text{newton}]}{[\text{m}^2]} = [\text{Pa}]$$

In the case of chemical potential, it is defined as energy per unit mass either in moles or kilograms, i.e.,

$$\mu_t = \frac{\text{force} \times \text{distance}}{\text{mole}} = \frac{[\text{J}]}{[\text{mol}]}$$

or

$$\mu_t = \frac{\text{force} \times \text{distance}}{\text{mass}} = \frac{[\text{J}]}{[\text{kg}]}$$

Table 3.1 Conversion chart for pore-water potentials

	Chemical potential μ_t (J/mol)	Head h_t (m)	Pressure ψ_t (kPa)
Chemical potential (J/mol)	–	$\mu_t = h_t g \omega_w$	$\mu_t = \psi_t v_w$
Head (m)	$h_t = \mu_t / g\omega_w$	–	$h_t = \psi_t / \rho_w g = \psi_t v_w / g\omega_w$
Pressure (kPa)	$\psi_t = \mu_t / v_w$	$\psi_t = hg\omega_t / v_w = hg\rho_w = \gamma_w h$	–

Conversion among these three potentials is shown in Table 3.1. In the table, g is gravitational acceleration (m/s^2), ω_w is the molecular mass of water (~0.018 kg/mol), γ_w is the unit weight of water (9800 N/m^3), v_w is the partial molar volume of water (m^3/mol), and ρ_w is the density of water (~1,000 kg/m^3) equal to ω_w / v_w. So, the partial molar volume of water v_w is 0.018 m^3/kmol. The reader is encouraged to obtain a full grasp of the relations by deriving these conversions.

As an example to illustrate the conversion, if the total energy at a point in soil pore water is 10 m in terms of head, from the above table, the total energy in terms of pressure is 98 kPa, and 1.764 J/mol in terms of chemical potential.

3.1.2 Gravitational potential

Gravity provides a perpetual stress for all matter on earth, regardless of state, whether gas, liquid, or solid. The effect of gravity on matter can be accurately described by gravitational field theory. For the shallow subsurface, gravity is simply defined as the gravitational potential of matter with respect to some reference point (e.g., sea level, or the toe of the hillslope, or the water table). The gravitational potential of soil pore water is the vertical distance from the chosen reference point. Gravitational potential at each point has a single value that is invariant to direction. The gravitational potential field varies only in the vertical direction toward the center of the mass of the earth, so it only changes value with elevation. The gravitational potential of matter increases linearly and proportionally moving away from the center of the earth. As described above, the magnitude of the gravitational potential of soil pore water can be described in three equivalent ways: head potential in meters, pressure potential in Pa, or chemical potential in J/mol. The gravitational potential in terms of head is

$$h_g = z \tag{3.2}$$

where z is the distance from a reference point and increases upward (shown in Figure 3.2). The physical meaning of the value of z is the energy change per unit weight of pore water against the earth's gravitational pull from a reference point of interest.

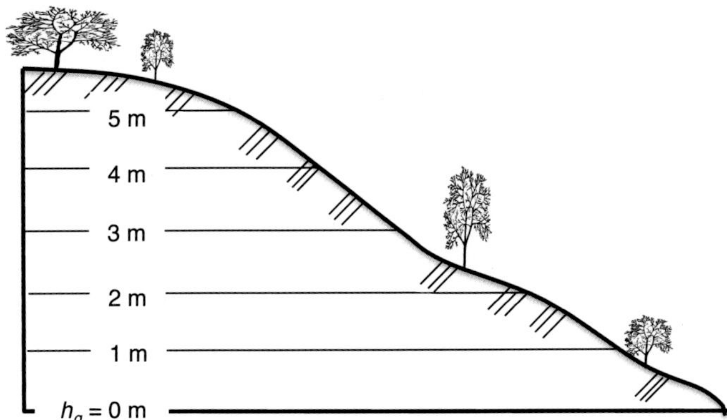

Figure 3.2 Field of gravitational potential in a hillslope, which is equal to the elevation above the reference point or datum.

3.1.3 Pressure potential

Pressure potential explicitly accounts for energy stored in pore water under pressure. For soil water below the water table, pore pressure is compressive, thus it stores more pressure energy than water at ambient atmospheric pressure conditions (typically about 100 kPa at sea level). On the other hand, pore-water pressure above the water table is tensile, thus water in the unsaturated zone stores less pressure energy than water at ambient atmospheric pressure. Because compressive and tensile pore-water pressures are mutually exclusive, only one term in the total potential, Equation (3.1), is necessary. When soil is under unsaturated conditions, pore-water pressure is often called capillary pressure, as the word "capillary" or "capillarity" indicates that the gas, liquid, or/and solid phase(s) of water co-exist in the soil.

Determination of the magnitude of compressive or tensile pore-water pressure (stored energy) under saturated conditions does not require soil properties as it is independent of the type of soil involved. However, determination of the magnitude of tensile pore-water pressure under unsaturated conditions does involve soil properties, as it is dependent upon the type of soil and water content. For example, a given sand with different water contents θ shown in Figure 3.3 will have different suction h_m (tensile pressure) values.

As another example, consider two soils, sand and clay, both with volumetric water contents of 0.2. At this water content, the sand will have a suction (tensile pore-water pressure) on the order of several to tens of kPa, whereas clay will have a suction of several thousands to tens of thousands of kPa.

This difference in suction for a given moisture content originates from differences in the soil composition, particle-size distribution, and pore size distribution. Soil suction, also called matric suction, is defined as the pressure difference between the ambient air and pore water. It is a combination of two quite different physical mechanisms; meniscus curvature or capillarity, which occurs across an air–water interface, and adsorption due to surface hydration, van der Waals attraction, and electrical double-layer interaction, which occurs at and near the solid–water interface. Since clay particles are much smaller than

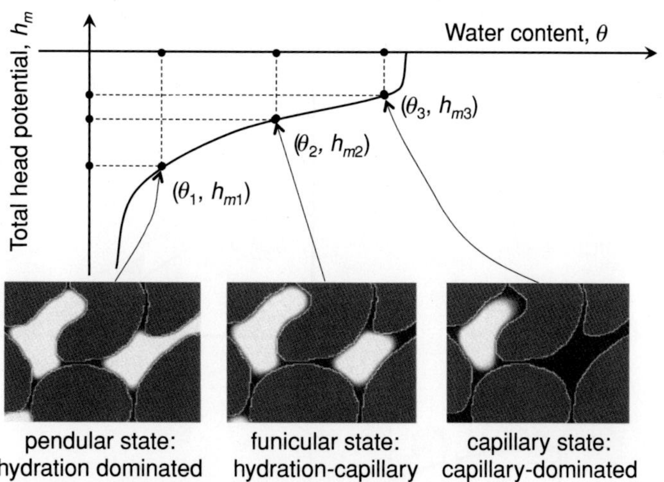

Figure 3.3 Illustration of the general relationship between water content and total head potential. When the water content is low, surface hydration (the ability of the solid surface to attract water molecules) dominates, leading to a much lower potential (negative) relative to free water. This state of water retention is also called the pendular regime. As the water content increases (middle panel), menisci or capillary water increases and both capillary and hydration effects are present. Eventually, as more water is drawn into the complex, the capillary effect is the dominant mode for suction.

sand and their surface area per unit volume, or specific surface, is much larger, clay has a much higher matric suction or lower water potential than sand under the same volumetric water content. Most soil materials have many oxygen and hydroxide ions near the particle surface. This provides very strong adsorptive forces along the solid surface that attract the hydrogen cations of water. Thus, when the water content of the soil is low, matric suction predominantly results from surface hydration forces. Hydration forces will quickly diminish as the water content increases beyond three water molecular layers and the soil water forms menisci. This yields a matric suction of several thousands of kPa. At higher water contents, capillarity starts to take over. As the soil becomes wetter and wetter towards full saturation, the area of the air–liquid interface diminishes and matric suction reduces to nearly zero.

The relationship between water content and matric suction in soil is controlled by the particle surface area and pore structure. Because particle surface area and pore structure vary from soil to soil, each soil has a unique constitutive relationship between matric suction and soil water content. This constitutive relationship is called the soil water characteristic curve (SWCC) or soil water retention curve (SWRC) as shown in Figure 3.3. Common methods for measuring SWCCs as well as a powerful technique for SWCC measurement will be introduced in Chapter 8.

In hillslope environments, the distribution of pressure potential is a transient and spatially varying process that is a function of the distribution of hillslope materials and results from the boundary conditions of precipitation, humidity of the atmosphere, the shape of the slope surface, and water table locations. A conceptual illustration (Figure 3.4) shows the pressure potential field under both drying and wetting scenarios.

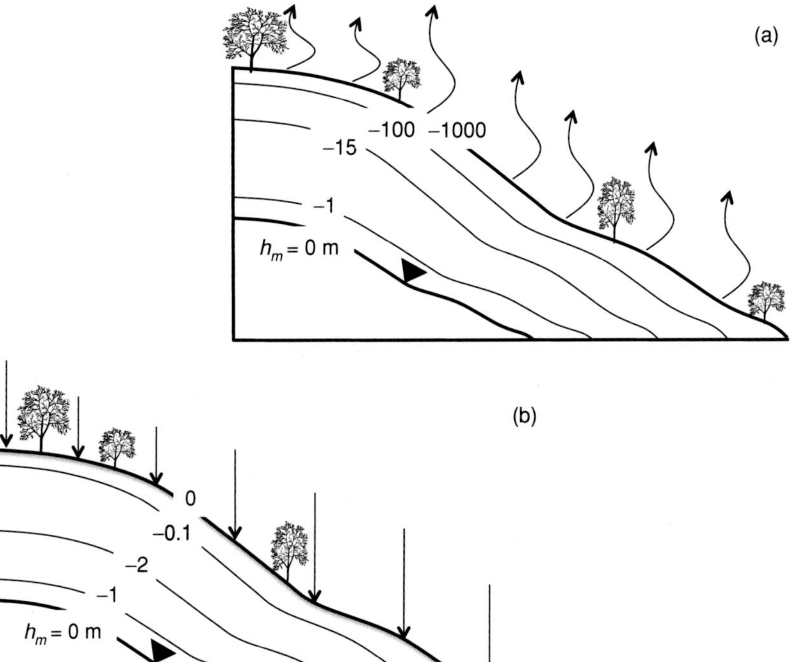

Figure 3.4 Illustration of pressure head potential field in a hillslope as a function of wetting or drying state (i.e., boundary conditions): (a) drying state and (b) wetting state.

3.1.4 Osmotic potential

Salts easily dissolve in water, resulting in a solution with lower water potential relative to pure water. For a solution with low salt concentration, the lowering of potential by salt is linearly proportional to the salt concentration and can be described by the van 't Hoff equation:

$$h_o = -\frac{cRT}{\rho_w g} \tag{3.3}$$

where c is solute concentration (mol m^{-3}), R is the universal gas constant (8.314 N m mol^{-1} K^{-1}), and T is temperature (K). From Table 3.1, the osmotic suction (negative sign for less than that of pure water) is

$$\psi_o = -cRT \tag{3.4}$$

The dependence of the osmotic potential on temperature and salt concentration is illustrated in Figure 3.5a. As described by Equation (3.4), increasing temperature or solution concentration will reduce osmotic potential. If a semi-permeable membrane, which allows water to move through freely but prevents salt ions from doing so, is used to separate free water from a solution, the thermodynamic equilibrium law states that the total potentials on

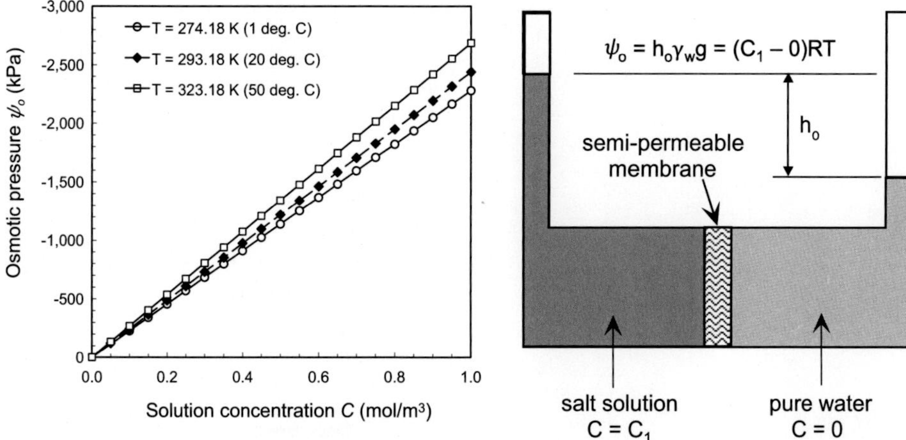

Figure 3.5 Illustration of (a) dependence of osmotic potential (suction) on salt concentration as a function temperature and (b) osmotic potential (suction) due to salt solution in system where a semi-permeable membrane permits water molecules to flow through but prevents dissolved salts from doing so.

both sides of the membrane will be the same. Since the solution side initially has a lower total potential, water will move toward the solution and an osmotic pressure will build up on the side with the solution. When equilibrium is reached, we have (setting the reference point at the top of the free water surface):

$$(h_t)_{\text{left}} = h_o + h_p = -\frac{cRT}{\rho_w g} + h_p = (h_t)_{\text{right}} = 0$$

which leads to the osmotic pressure at the reference elevation in the solution:

$$h_o = \frac{cRT}{\rho_w g} \tag{3.5}$$

Physically, Equation (3.5) states that a hydraulic pressure equivalent to h_o is needed to make the solution on the left side of Figure 3.5b have the same energy level as the pure water on the right side.

The reader is encouraged to calculate the total head and the pressure potentials at the either side of the semi-permeable membrane.

The osmotic potential in soil is often measured using the electrical conductivity of the pore-water solution free from the soil sample. Because the solute has a strong impact on the adsorption part of the pore-water potential (through interaction with the electrical double-layer formation and van der Waals forces), it is possible that a coupling exists between pressure potential and osmotic potential implying that Equation (3.1) may not be suitable for determining total potential. However, this topic is poorly studied and awaits clarification.

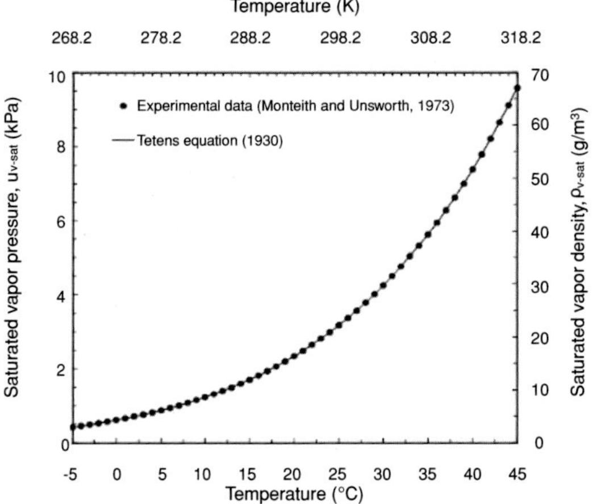

Figure 3.6 Saturated water vapor pressure and water vapor density as a function of temperature.

3.1.5 Water vapor potential

As mentioned previously, because changes in the concentration of water vapor dominate changes in the chemical composition of air near the ground surface, water vapor potential can be accurately described using the concept of relative humidity. Based on the ideal gas law, the maximum partial pressure of a particular gas is proportional to its concentration in equilibrium with other gases. Nitrogen and oxygen make up most of the atmosphere near the ground surface. Depending on the prevailing temperature and pressure, water vapor typically only makes up a small percentage of the total composition of air. The maximum vapor pressure for a given set of temperature and total pressure conditions in equilibrium with pure water, used as a reference point, is called the saturated vapor pressure. Experimental evidence shows that the saturated vapor pressure is very sensitive to temperature and insensitive to the total air pressure so that it can be accurately described by Tetens' equation (1930):

$$u_{vo} = 0.611 \exp\left(17.27\frac{T - 273.2}{T - 36}\right) \tag{3.6}$$

Like Equation (2.1), the above equation is another approximation of the complex Clausius–Clapeyron relation. Equation (3.6) is plotted in Figure 3.6 for the saturated vapor pressure as a function of temperature. For temperature variation between 268.2 K (−5 °C) to 318.2 K (45 °C), vapor pressure varies about by 10 kPa.

According to the ideal gas law, the gas density of a particular gas species follows the relationship

$$\rho_v = \frac{\omega_w u_v}{RT} \tag{3.7}$$

where u_v is the prevailing vapor pressure that is usually less than or equal to the saturated vapor pressure near the ground surface. When the vapor pressure is at the saturated vapor pressure, the vapor density is called absolute humidity and can be obtained by simply substituting Equation (3.6) into Equation (3.7):

$$\rho_{vo} = \frac{\omega_w u_{vo}}{RT} = \frac{0.611\omega_w}{RT}\exp\left(17.27\frac{T-273.2}{T-36}\right) \tag{3.8}$$

which is also plotted in Figure 3.6. For temperatures varying between 268.2 K (−5 °C) and 318.2 K (45 °C), the saturated vapor density varies within a range of 0 to 70 g/m^3.

The relative humidity is a measure of the prevailing vapor pressure relative to the saturated vapor pressure and is defined as the ratio of the prevailing vapor pressure to the saturated vapor pressure, or as the ratio of the prevailing vapor density to the absolute vapor density:

$$\mathrm{RH} = \frac{u_v}{u_{vo}} = \frac{\rho_v}{\rho_{vo}} \tag{3.9}$$

Thus, the relative humidity represents the degree of water vapor potential with respect to the saturated vapor state (or pure water). The total potential of water vapor with respect to pure water can be expressed as

$$h_{vap} = \frac{RT}{\omega_w g}\ln(\mathrm{RH}) \tag{3.10}$$

As the relative humidity is usually less than or equal to 1.0, the total potential of vapor (air) is given in negative values, representing potential lower than the reference point of that of distilled water. For example, if the relative humidity in pore air is 0.9 and the prevailing temperature is 300 K (26.8 °C), the head potential of the water vapor is

$$h_{vap} = \frac{RT}{\omega_w g}\ln(\mathrm{RH}) = \frac{(8.31)\frac{[\mathrm{N\,m}]}{[\mathrm{mol\,K}]}(300)\,[\mathrm{K}]}{(0.018)\frac{[\mathrm{kg}]}{[\mathrm{mol}]}(9.8)\frac{[\mathrm{m}]}{[\mathrm{s}^2]}}\ln(0.9) = -1310\,[\mathrm{m}]$$

The negative sign implies a deficit or suction relative to the reference point of distilled water, which has 1.0 relative humidity and zero vapor pressure head. The negative sign also implies that if pure water free from soil is placed in this environment, the water will evaporate, as the total potential of water (zero) is higher than the total potential of the water vapor, a reflection of the second law of thermodynamics.

3.1.6 Chemical potential equilibrium principle in multi-phase media

Because adsorption, osmosis, and the existence of the air–water interface (capillary mechanism) all reduce liquid water potential, the total potential of soil pore water is less than zero (relative to free distilled water). What will happen if a limited amount of soil with total soil water potential head $h_t = h_x$ is exposed to a constant relative humidity of RH_y environment as shown in Figure 3.7?

The total potential of water vapor can be first calculated by Equation (3.10):

$$h_y = \frac{RT}{\omega_w g}\ln(\mathrm{RH}_y)$$

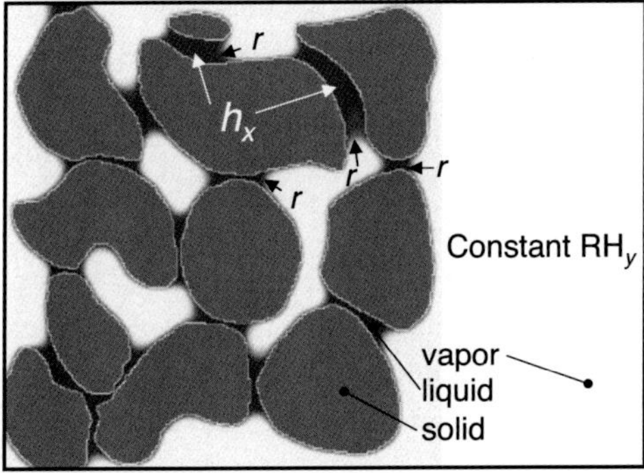

Figure 3.7 Illustration of thermodynamic equilibrium between gas and liquid phases in porous materials where the gas phase maintains a constant humidity while the liquid phase has a limited amount of water.

There are three possible initial states ($h_x = h_y$, $h_x > h_y$, $h_x < h_y$), but following the second law of thermodynamics, all three lead to the same equilibrium potential state, i.e., $h_t = h_y$ in both vapor and liquid phases. If $h_x = h_y$, there is no gradient in total potential between vapor and liquid phases, so no evaporation or condensation will occur. If $h_x > h_y$, evaporation will occur, and the soil potential becomes smaller until the new total potential of the soil water is equal to h_y of the total vapor potential. If $h_x < h_y$, condensation will occur as the pore water has less total potential and thus exhibits suction. The action will cease when the total pore-water potential increases to the value of h_y. At thermodynamic or potential equilibrium, a unique air–liquid interface with a radius of curvature r is reached and the total potentials of both phases are equal in value, i.e.,

$$h_t = h_m + h_g + h_o = \frac{RT}{\omega_w g} \ln(\mathrm{RH}) \tag{3.11}$$

According to Table 3.1, the above equation can also be written in terms of chemical potential or pressure potential, respectively:

$$\mu_t = \mu_m + \mu_g + \mu_o = RT \ln(\mathrm{RH}) \tag{3.12}$$

$$\psi_t = \psi_m + \psi_g + \psi_o = \frac{RT}{v_w} \ln(\mathrm{RH}) \tag{3.13}$$

In hillslope materials, because the multi-phase equilibrium described above occurs at the pore-size scale, the assumption of local equilibrium is sufficient compared to the time scale for equilibrium at the hillslope scale. One important practical implication is in the event that the total potential of pore water cannot be directly measured using a pressure probe (for suction greater than ~2,000 kPa); it can be inferred from measurements of the relative humidity of the soil gas nearby. For example, if the osmotic pressure and gravitational pressure potential can be ignored, and distilled water at atmospheric pressure ψ_a is used as a reference, Equation (3.13) leads to

$$\psi_a - \psi_t = \psi_a - \psi_m = \frac{RT}{v_w} [\ln(1) - \ln(\mathrm{RH})] = -\frac{RT}{v_w} \ln(\mathrm{RH}) \tag{3.14}$$

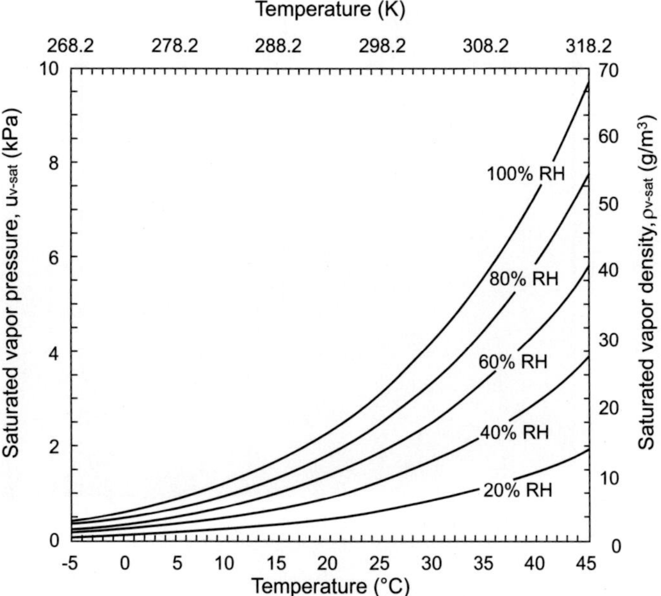

Figure 3.8 Vapor pressure and vapor density as functions of temperature and relative humidity.

The quantity $(\psi_a - \psi_m)$ is commonly written as $(u_a - u_w)$ and called matric suction or soil suction. By subtracting water pressure from a reference atmospheric pressure, soil suction is a positive quantity in most soils. The relative humidity or suction value at the total potential equilibrium can also be linked to the radius of the air–liquid interface r (Figure 3.7) by Kelvin's equation, with the knowledge of the interface surface tension T_s:

$$\psi_a - \psi_t = \psi_a - \psi_m = -\frac{RT}{v_w}\ln(\mathrm{RH}) = \frac{2T_s}{r} \tag{3.15}$$

A physical implication of Equation (3.14) or Equation (3.15) is that the value of equilibrium relative humidity in porous materials is less than 100% or unity. This phenomenon in porous materials is often called "vapor pressure lowering." In light of the definition of relative humidity by Equation (3.9), vapor pressure and vapor density at a given equilibrium relative humidity in porous materials can be plotted as functions of temperature, as shown in Figure 3.8.

3.1.7 Pressure profiles under hydrostatic conditions

As an application of the total water potential to field problems, the pressure profile across the saturated–unsaturated zone under hydrostatic conditions is examined here. Hydrostatic conditions are a special and rare case in reality at the field scale when thermodynamic

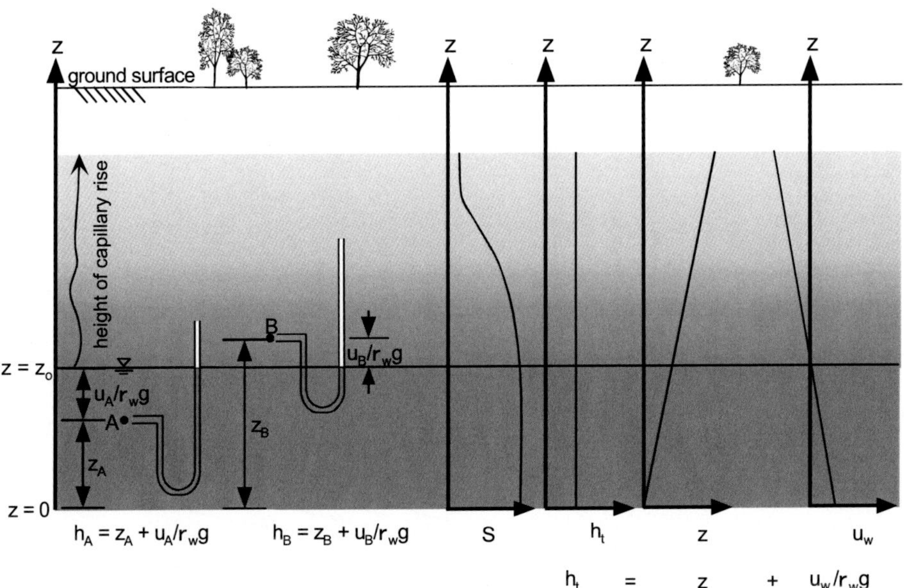

Figure 3.9 Schematic illustration of saturation S, pressure u, and total potential profiles h_t under hydrostatic conditions in variably saturated soil.

equilibrium is reached, so that there is no flow across the entire domain. Following the second law of thermodynamics, the total water potential head h_t is a constant equal to z_o. The elevation head increases linearly as $h_g = z$. Since the total head consists of the elevation and pressure heads, we have

$$h_t = z_o = h_m + h_g = z + \frac{u_w}{\rho_w g}$$

which leads to the linear pressure profile

$$u_w = (z_o - z)\rho_w g \tag{3.16}$$

with compressive pressure (positive) below the water table and tensile pressure (negative) above the water table shown in Figure 3.9. The equilibrium saturation profile is also shown in Figure 3.9, where it is constant below the air-entry pressure at the elevation around point B and decreases gradually to its residual value at the height of capillary rise. The air-entry pressure is the pressure above which the soil begins to desaturate.

If a piezometer (device to measure pressure head) is used to measure the total head in this soil, say at points A and B respectively, the total head at each of these points will be the same and equal to z_o, i.e.,

$$h_t = z_o = z_A + \frac{u_{wA}}{\rho_w g} = z_B + \frac{u_{wB}}{\rho_w g}$$

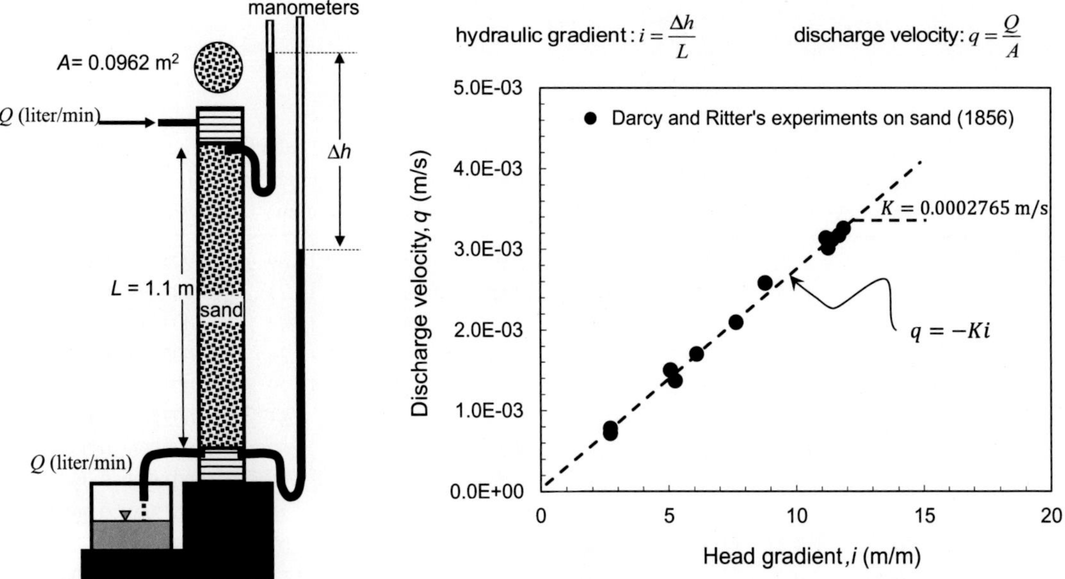

Figure 3.10 Illustration of Darcy and Ritter's original experimental setup and test results that form the foundation of Darcy's law.

which leads to

$$u_{wA} = (z_o - z_A)\,\rho_w g$$
$$u_{wB} = (z_o - z_B)\,\rho_w g$$

Pressure at point A is compressive (positive value), and pressure at point B is tensile (negative value). As illustrated, both points are under saturated conditions, but above the water table, suction is present.

3.2 Darcy's law

3.2.1 Darcy's experiments

Darcy's law is a reflection of the second law of thermodynamics: the total energy of the system stored in the form of the total water potential is always lost during flow in porous material and the amount of energy or total potential lost depends on the material. The illustration of this statement and its applications to quantify total head and water content distributions in simple subsurface settings is shown in this section.

In 1857, Henry Darcy published a report on the water supply of the city of Dijon, France. In this report, a series of experiments examining the flow of water through a vertical column filled with sand is described, as shown in Figure 3.10. The original results of one series of these experiments are shown in the first seven columns of Table 3.2. The height of the

Table 3.2 Darcy and Ritter's original experimental results (the first series on their phase 2 experiments) and interpretation of flow through a sand column

Experiment	Duration (minutes)	Average flow rate (l/min)	Average pressure over the filter (m)	Average pressure below the filter (m)	Pressure differential (m)	Ratio of flow rate to pressure (l/m-min)	Hydraulic gradient m/m	Darcy flux m/s	Hydraulic conductivity m/s
1	15	18.8	9.48	–3.60	13.08	1.44	11.89	0.003257	0.000274
2	15	18.3	12.88	0.00	12.88	1.42	11.71	0.003170	0.000271
3	10	18	9.80	–2.78	12.58	1.43	11.44	0.003118	0.000273
4	10	17.4	12.87	0.46	12.41	1.40	11.28	0.003014	0.000267
5	20	18.1	12.80	0.49	12.31	1.47	11.19	0.003135	0.000280
6	10	14.9	8.86	–0.83	9.69	1.54	8.81	0.002581	0.000293
7	15	12.1	12.84	4.40	8.44	1.43	7.67	0.002096	0.000273
8	15	9.8	6.71	0.00	6.71	1.46	6.10	0.001698	0.000278
9	20	7.9	12.81	7.03	5.78	1.37	5.25	0.001369	0.000260
10	20	8.65	5.58	0.00	5.58	1.55	5.07	0.001498	0.000295
11	20	4.5	2.98	0.00	2.98	1.51	2.71	0.000780	0.000288
12	20	4.15	12.86	9.88	2.98	1.39	2.71	0.000719	0.000265

sand-filled column and the cross-sectional area are also shown in Figure 3.10. The corresponding hydraulic gradient and the discharge velocity or "Darcy's flux" of the 12 experiments can be calculated and are shown in columns 8 and 9 of Table 3.2, respectively. Plotting the discharge velocity q vs. hydraulic gradient i, as shown in Figure 3.10, it becomes apparent that a linear relation between these two quantities exists. This relation is Darcy's law:

$$q = -Ki \tag{3.17}$$

where K is hydraulic conductivity, which can be determined by applying Equation (3.17) to the experimental data as shown in column 10 of Table 3.2. Inspecting the calculated hydraulic conductivities in Table 3.2 indicates that Darcy's law is valid within 10% of the difference in hydraulic conductivity measurements. Darcy's law has been confirmed in numerous experiments on porous media and has been shown to be valid for all directions in space.

3.2.2 Darcy's law in three-dimensional space

For fluid flow in multi-dimensional porous media, the magnitude and direction of liquid water flow in saturated porous media under the different driving mechanisms of pressure and gravity can be unified by the total water potential concept and Darcy's law:

$$\mathbf{q} = -\mathbf{K}\nabla h_t = -\mathbf{Ki} \tag{3.18}$$

where $\mathbf{q}$ is the specific discharge vector (m/s), $\mathbf{K}$ is the hydraulic conductivity tensor, and $\mathbf{i}$ is the gradient of the total head. For most practical problems related to the stability of hillslopes, the total soil water head potential can be considered using only the pressure head and gravity head as

$$h_t = h_m + h_g = h_m + z \tag{3.19}$$

The components the gradients of soil water potential are

$$i_x = \frac{\partial h_t}{\partial x} = \frac{\partial h_m}{\partial x} \tag{3.20}$$

$$i_y = \frac{\partial h_t}{\partial y} = \frac{\partial h_m}{\partial y} \tag{3.21}$$

$$i_z = \frac{\partial h_t}{\partial z} = \frac{\partial h_m}{\partial z} + 1 \tag{3.22}$$

The vector form of Darcy's law can be expressed as

$$\begin{Bmatrix} q_x \\ q_y \\ q_z \end{Bmatrix} = -\begin{bmatrix} K_x & 0 & 0 \\ 0 & K_y & 0 \\ 0 & 0 & K_z \end{bmatrix} \begin{Bmatrix} i_x \\ i_y \\ i_z \end{Bmatrix} = -\begin{bmatrix} K_x & 0 & 0 \\ 0 & K_y & 0 \\ 0 & 0 & K_z \end{bmatrix} \begin{Bmatrix} \dfrac{\partial h_m}{\partial x} \\ \dfrac{\partial h_m}{\partial y} \\ \dfrac{\partial h_m}{\partial z} + 1 \end{Bmatrix} \tag{3.23}$$

for the case in which the principal directions of the hydraulic conductivity tensor are aligned with the coordinate directions. For unsaturated soil, both hydraulic conductivity, K_i, in the $i = x$, y, and z directions and matric suction head are highly non-linear functions of soil water content, i.e.,

$$q_x = -K_x(\theta)\frac{\partial h_m(\theta)}{\partial x} \tag{3.24}$$

$$q_y = -K_y(\theta)\frac{\partial h_m(\theta)}{\partial y} \tag{3.25}$$

$$q_z = -K_z(\theta)\frac{\partial h_m(\theta)}{\partial z} - K_z(\theta) \tag{3.26}$$

where θ is the volumetric water content defined as the relative volume of water to volume of soil. For isotropic materials, the three hydraulic conductivity functions (HCF) reduce to one, and hydraulic conductivity can be considered a scalar quantity.

3.2.3 Hydraulic properties

The dependence of matric potential head in terms of matric suction ψ_m on soil water content is called the soil water characteristic curve (SWCC) and is illustrated qualitatively for various soils in Figure 3.11.

For sandy soils (e.g., dune sand), the volumetric water content at full saturation is typically about 0.4. As matric suction increases to around 1 kPa, air enters the soil or the soil begins to desaturate. The suction at which air starts to enter the soil is called the air-entry pressure ψ_b. As matric suction increases above 100 kPa, the soil nearly reaches its residual water content state. The residual water content θ_r is the water content beyond which large

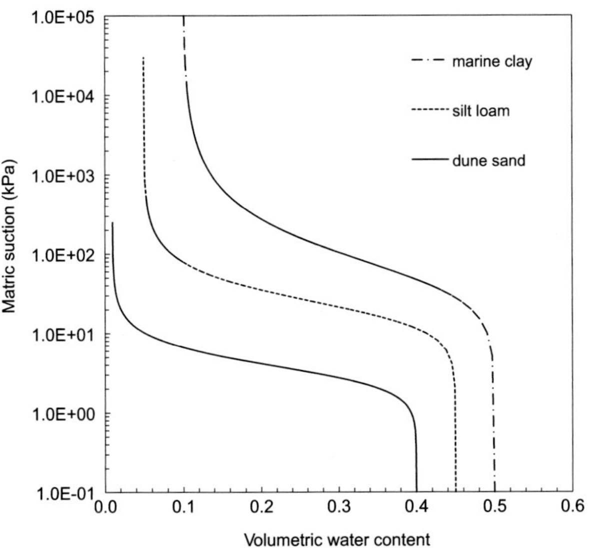

Figure 3.11 Soil water characteristic curves for three soils. Because larger specific surface area promotes higher surface hydration, clay soils generally exhibit higher water content than silty soils under the same matric suction. Similarly, silty soils have higher water content than sandy soils under the same matric suction.

increases in soil suction lead to very little change in water content. For uncompacted silty soils (e.g., silt loam), in which soil pores are dominantly tens to hundreds of μm in size, the volumetric water content at full saturation is about 0.45, air-entry pressure is about 10 kPa, and the residual water retention state will not be reached until matric suction is on the order of several hundreds to thousands of kPa. For clayey soils (e.g., marine clay), pore sizes are dominated by μm scales, leading to relatively large pore volumes at full saturation (about 0.5), high air-entry pressures up to tens of kPa, and very high matric suctions up to tens of thousands to hundreds of thousands of kPa before the residual state is reached.

Both air-entry pressures and residual water content are important parameters for describing the hydrologic and mechanical characteristics of soils. A more precise depiction of these properties is shown in Figure 3.12. The water content at air-entry is also a boundary that defines two soil water retention regimes; for water contents greater than this value air is occluded, or isolated, in bubbles. This is the capillary state (Figure 3.3). For water contents less than this value, air and water co-exist forming a continuous web that surrounds soil particles (funicular regime). As the water content decreases approaching the residual value, water exists in discontinuous lenses (pendular regime). Each of these regimes has different behavior in both water movement and inter-particle stress distribution.

One unique feature of the SWCC is that the drying path differs from the wetting path, meaning that prior wetting or drying conditions determine the relation between moisture content and matric suction. This feature is known as soil hysteresis. For a given water content, pore water resides at a higher energy state along a wetting path than it does along a drying path. Hysteresis generally results from tortuous soil pore structure and/or difference

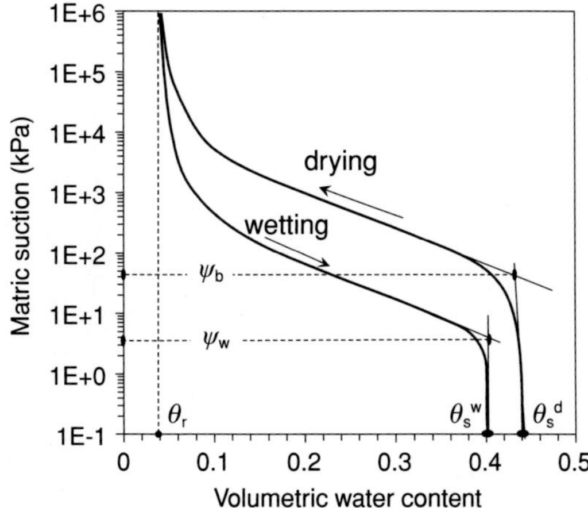

Figure 3.12 Commonly used definitions of air-entry pressure and residual water content. Solid line shows the drying (desorption) curve whereas dash line shows the wetting (adsorption) curve (from Lu and Likos, 2004a).

in affinity between soil grains and water under different wetting histories. Because landslides are often initiated under infiltration, the wetting path is generally more relevant in describing the physical process than the drying path. For example, as illustrated in Figure 3.12, the water-entry pressure ψ_w along the wetting path is a key hydrologic parameter indicating whether infiltrating water will enter the soil domain. Along the wetting path, soil will never become fully saturated under natural (*in-situ*) conditions. The difference between the saturated water content at the drying state θ_s^d and the saturated water content at the wetting state θ_s^w could be as much as 30%, as shown in Chapter 8. Nevertheless, almost all laboratory techniques are limited to measurements along the drying path. Thus there is great need for further research to develop better techniques to measure the wetting path of the SWCC. A technique capable of identifying both wetting and drying SWCCs will be introduced in Chapter 8.

The HCF of soil is also highly dependent on soil water content or soil suction, as conceptually illustrated in Figure 3.13a, and shown in Figure 3.13b for measurements of two soils: silty loam and sand.

In general, the HCF is at its highest value when the soil is fully saturated and decreases drastically as suction increases. An interesting feature shown in Figure 3.13b is that the sand has much higher saturated hydraulic conductivity than the silt, but the relative reduction of hydraulic conductivity with increasing matric suction is much greater for sand than for silt or clay. For many soils, a reduction of many orders of magnitude of hydraulic conductivity from the saturated state to the residual state is common. The highly non-linear feature of the HCF makes laboratory measurements difficult to perform. Only a few reliable measurement techniques have been developed so far, each with limitations in terms soil suction range, water content, and testing conditions. Often, indirect estimations such as using SWCC and HCF models are widely employed. A systematic review of the major techniques can be found in Chapters 10–12 in Lu and Likos (2004a).

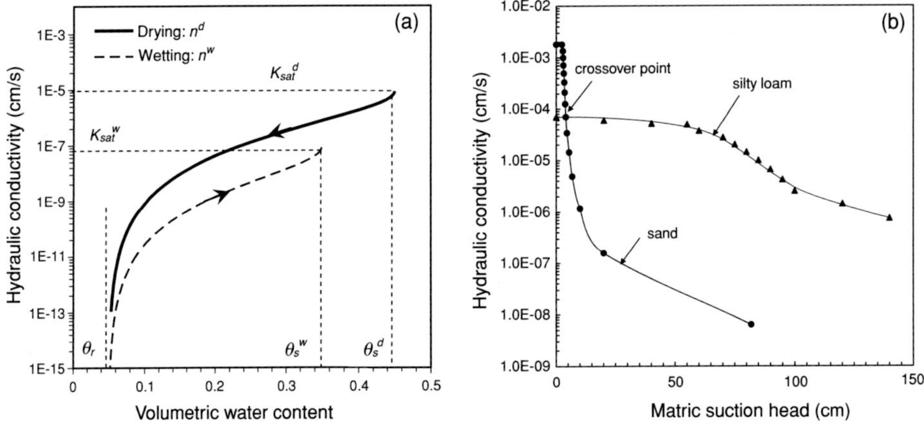

Figure 3.13 (a) Conceptual illustration of the dependence of hydraulic conductivity on wetting history, and (b) HCFs for silty loam and sand (data from Hillel, 1982).

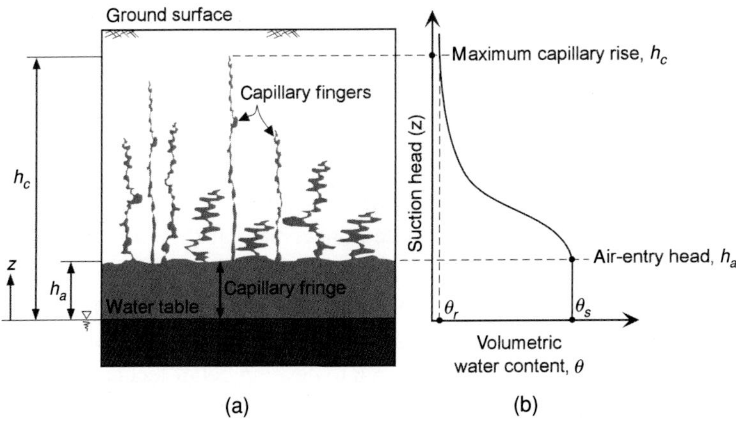

Figure 3.14 Schematic illustration of capillary rise and associated pore-water retention in an unsaturated soil profile where the moisture profile reveals the SWCC when hydrostatic conditions are reached (Lu and Likos, 2004a).

3.3 Capillary rise

3.3.1 Height of capillary rise in soils

An example of the second law of thermodynamics cast in the form of Darcy's law is the phenomenon called "capillary rise" above a water table. Capillary rise refers to the natural phenomenon of upward water movement above the water table under the driving force of capillarity. As illustrated in Figure 3.14, for a given soil, capillary rise will cease at a certain height h_c, above which the pore water exists in discontinuous or pendular form. When the action of capillary rise ceases, the total potential head everywhere reaches a constant equal to that at the water table. Two fundamental questions are often raised regarding the capillary rise: how high is the capillary rise? And how fast is the capillary rise?

Conceptually, the height of capillary rise is directly related to the suction at which a soil reaches the residual water content. Therefore, clay soils have the highest height of capillary rise, on the order of tens of meters; capillary rise in silty soils is in the middle of the range, from several to 10 meters; and sandy soil has the lowest capillary rise ranging from several centimeters to several decimeters. To better quantify the height of capillary rise, several empirical expressions have been proposed. Peck *et al.* (1974) correlated the height of capillary rise to the void ratio and the particle size at the 10% and finer fraction D_{10}:

$$h_c = \frac{C}{eD_{10}} \tag{3.27}$$

where e is the void ratio defined as the ratio of the volume of void to the volume of solid and C is an empirical parameter varying between 10 and 50.

Kumar and Malik (1990) defined the height (in cm) as a function of the air-entry head h_a (cm) and the mean pore size r (μm):

$$h_c = h_a + 134.84 - 5.16\sqrt{r} \tag{3.28}$$

Lu and Likos (2004a) defined the height (in mm) as a function of the particle size (mm) at the 10% and finer fraction:

$$h_c = -990\ln(D_{10}) - 1540 \tag{3.29}$$

Two conditions in which the rate of rise is of practical importance are: an initially dry soil profile subject to a sudden introduction of a water table, and initially moist soil profile subject to a sustained humidity gradient at the ground surface. The first condition is often encountered in laboratory tests using soil columns or in newly constructed soil structures. The second is more common in many field settings. The first condition is controlled by soil properties, whereas the second condition is controlled by the environment (gradient of relative humidity) near the ground surface.

3.3.2 Rate of capillary rise in soils

While there are a number of theories for capillary rise in ideal tubes, there are two theories representative of soil in laboratory or field conditions. Detailed derivations of these two theories can be found in Lu and Likos (2004a, Chapter 4), but the major results are presented and illustrated here. Terzaghi (1943) presented a theory predicting the rising wetting front z from the water table as a function of elapsed time t using the saturated permeability (m/s):

$$t = \frac{nh_c}{K_s}\left(\ln\frac{h_c}{h_c - z} - \frac{z}{h_c}\right) \tag{3.30}$$

In deriving the above equation, it is assumed that the driving head gradient at the wetting front z above the water table can be approximated by $i = (h_c - z)/z$ and Darcy's law can

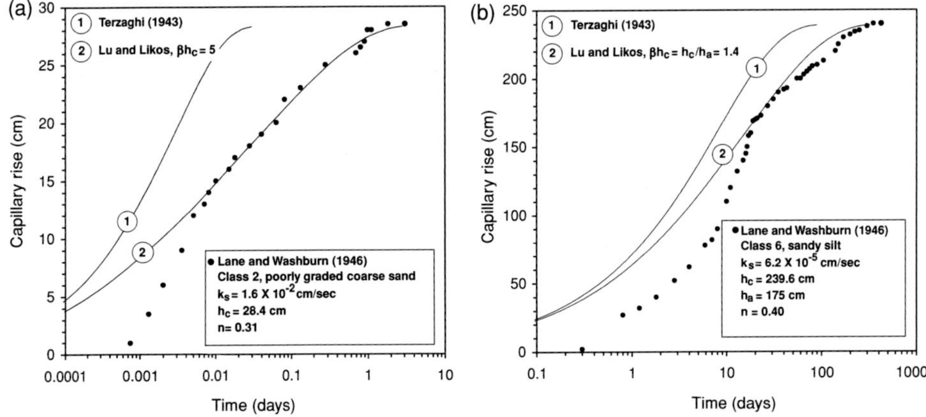

Figure 3.15 Comparison of two theoretical predictions of capillary rise by Terzaghi (1943) and Lu and Likos (2004) and experimental data in (a) coarse sand and (b) fine sand (Lu and Likos, 2004a).

be used to describe the upward seepage velocity v, i.e., $v = K_s i = n\, dz/dt$, where K_s is the saturated hydraulic conductivity, n is the porosity, and t is time. The solution is plotted against the capillary rise from tests for two soils by Lane and Washburn (1946) shown in Figure 3.15.

Lu and Likos (2004a) arrived at a theory by explicitly considering the unsaturated hydraulic conductivity function:

$$t = \frac{n}{K_s} \sum_{j=0}^{\infty} \frac{\beta^j}{j!} \left(h_c^{j+1} \ln \frac{h_c}{h_c - z} - \sum_{s=0}^{j} \frac{h_c^s z^{j+1-s}}{j+1-s} \right) \tag{3.31}$$

where β is the decay rate of the unsaturated hydraulic conductivity used in the following expression (Gardner, 1958):

$$K(h_m) = K_s \exp(\beta h_m) \tag{3.32}$$

The solution is also plotted against the test data in Figure 3.15. Because the highly non-linear reduction of hydraulic conductivity under increasing suction was recognized, the above solution significantly improves the predictability of the rate of capillary rise over the previous theory.

The steady rate of capillary rise controlled by the gradient of relative humidity at the land surface can be estimated by the vapor flow theory described below.

3.4 Vapor flow

Vapor diffusion is an important physical process at or near the boundary between the atmosphere and the subsurface where the humidity of air varies greatly. Fick's law can be used to describe vapor flow within the soil domain or at the atmosphere and subsurface interface:

$$q_v = -D_v \nabla \rho_v \tag{3.33}$$

where D_v is the vapor diffusion coefficient in unsaturated soil. It can be estimated by the air-filled porosity n_a and the free vapor diffusion coefficient D_o as

$$D_v = \frac{2}{3} n_a D_o \tag{3.34}$$

For pure air, D_o ranges approximately from 10^{-9} to 10^{-6} m^2/s.

Because vapor density can be described by the ideal gas law:

$$\rho_v = \frac{\omega_w}{RT} u_v \tag{3.35}$$

and vapor pressure can be accurately described by Tetens' (1930) equation:

$$u_{vo} = 0.611 \exp\left(17.27 \frac{T - 273.2}{T - 36}\right) \tag{3.6}$$

Lu and Likos (2004a) showed that the water flux through vapor near the land surface can be given as

$$q_v = D_v \rho_{v-sat} \left(-\frac{\nabla \mathrm{RH}}{\mathrm{RH}} + \frac{\nabla T}{T} - \frac{\lambda \omega_w \nabla T}{RT^2} \right) \tag{3.36}$$

where λ is the latent heat of vaporization, about 2.48 kJ/g at 10 °C. The saturated vapor density and vapor density for a given relative humidity can be assessed using Figure 3.8. The negative signs in front of the first and third terms on the right hand side of the above equation imply that vapor flows from regions of high relative humidity to regions of low relative humidity and from regions of high temperature to regions of low temperature. The positive sign for the second term implies a counteracting flux from low temperature to high temperature. This offsetting term arises from the fact that low temperature causes air to contract, resulting in a higher vapor density, and vice versa.

3.5 Vertical flow

3.5.1 One-layer system

Water movement in a typical hillslope can be multi-dimensional, as illustrated in Figure 3.16. However, in many climates for most of the year, soil is drier near the ground surface than it is at depth. Under these conditions both gravitational force and matric suction tend to pull water vertically downward so that one-dimensional vertical flow occurs in many circumstances, as illustrated in the circled area in Figure 3.16. Understanding one-dimensional vertical infiltration provides insight into the characteristics of unsaturated flow under gradients of matric suction and gravitational potential.

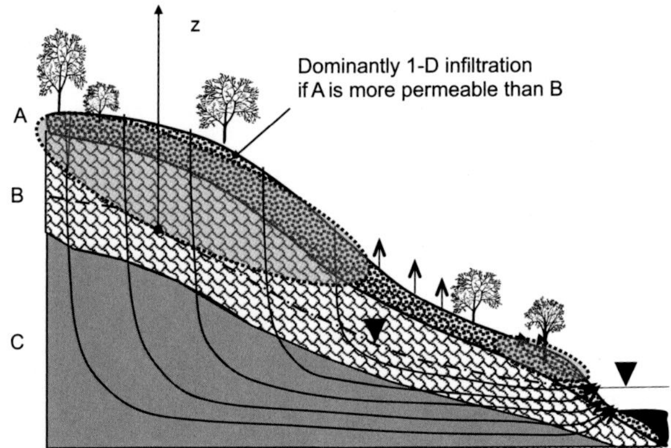

Figure 3.16 Illustration of possible flow patterns within a hillslope.

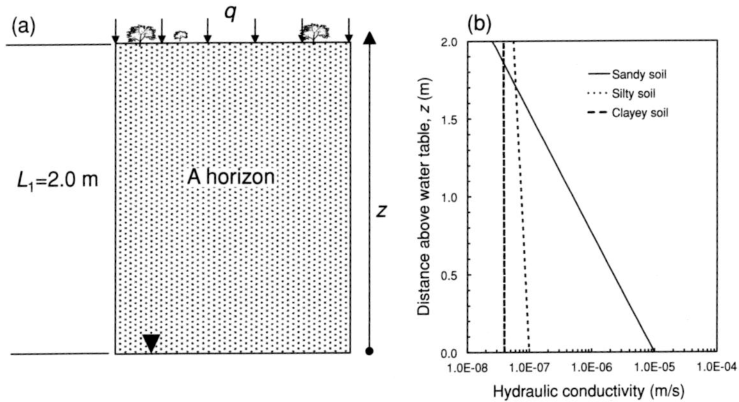

Figure 3.17 (a) A 2 m thick unsaturated soil layer bounded by a water table and (b) hydraulic conductivity profiles for three materials.

As discussed in the previous section, the unified quantity to describe flow in a variably saturated hillslope is the total potential. For ideal one-dimensional vertical conditions in isotropic materials shown in Figure 3.17, Darcy's law connects the gradients of the total potential to the flow rate q by

$$q = -K(\theta)\frac{\partial h_t}{\partial z} \tag{3.37}$$

The function $K(\theta)$ is the hydraulic conductivity function of the hillslope material, which can also be expressed as a function of matric suction head, i.e., $K(h_m)$. The total head in most hillslope environments can be considered as a sum of gravitational (z) and matric potentials (h_m), i.e.,

$$h_t = h_m + z = h_m(S) + z = h_m(\theta) + z \tag{3.38}$$

where $h_m(\theta)$ is determined from the SWCC of the hillslope material of interest.

Table 3.3 Range of unsaturated hydrologic parameters for different soils

Soil type	β (m^{-1})	n (unitless)	θ_s (unitless)	K_s (m/s)
Sand	1.0–5.0	4.0–8.5	0.3–0.4	10^{-2}–10^{-5}
Silt	0.1–1.0	3.0–4.0	0.4–0.5	10^{-6}–10^{-9}
Clay	0.01–0.1	1.1–2.5	0.4–0.6	10^{-8}–10^{-13}

The HCF and SWCC can be described by Gardner's (1958) and Brooks and Corey's (1964) models, respectively:

$$K = K_s \exp(\beta h_m) = K_s \left(\frac{\theta}{\theta_s}\right)^n \tag{3.39}$$

where K_s is the saturated hydraulic conductivity, θ_s is the saturated moisture content, and β and n are fitting parameters. In the above equation, four unsaturated hydrologic parameters n, K_s, θ_s, and β define the SWCC and HCF of the hillslope material. The ranges of these parameters for typical soils are listed in Table 3.3.

The so-called "Kirchhoff" integral transformation can be used for the HCF:

$$H = \int_{-\infty}^{h_m} K dh_m \tag{3.40}$$

Substituting the first part of Equation (3.39) into the above leads to

$$H = \frac{K}{\beta} \tag{3.41}$$

Substituting Equations (3.41) and (3.38) into Equation (3.37) and integrating leads to

$$H(z) = -\frac{q}{\beta} + c_1 \exp(-\beta z) \tag{3.42}$$

Imposing a boundary condition of $h_m = 0$ at $z = 0$ (the water table) on the above equation leads to

$$c_1 = \frac{q + K_s}{\beta} \tag{3.43}$$

and

$$K(z) = -q + (q + K_s)\exp(-\beta z) \tag{3.44}$$

It can be shown that

$$h_m(z) = \frac{1}{\beta} \ln\left(\frac{K(z)}{K_s}\right) \tag{3.45}$$

$$\theta(z) = \theta_s \left(\frac{K(z)}{K_s}\right)^{1/n} \tag{3.46}$$

Equations (3.43) to (3.45) define soil suction and water content profiles under a steady infiltration rate q. In what follows, a 2 m layer unsaturated system composed of three representative soils: sand, silt, and clay, with the profiles of hydraulic conductivity shown

Table 3.4 Unsaturated soil hydrologic parameters and depths

Soil type	β (m^{-1})	n (unitless)	θ_s (unitless)	K_s (m/s)	L (m)
Sand	3.0	4.0	0.35	10^{-5}	2.0
Silt	0.3	3.0	0.45	10^{-7}	2.0
Clay	0.03	2.0	0.55	4×10^{-8}	2.0

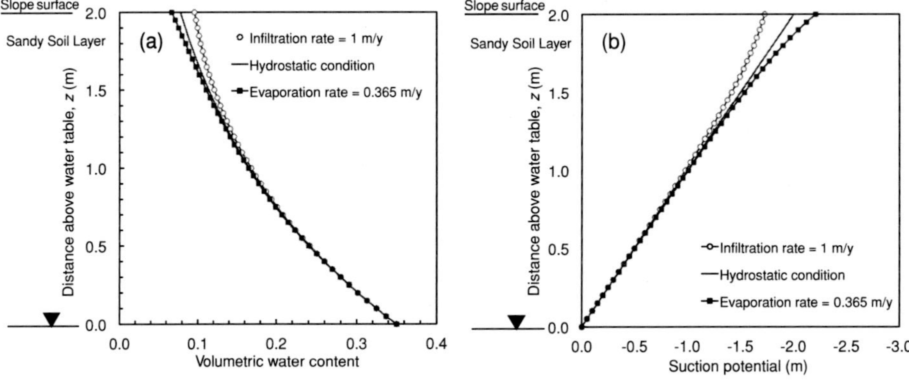

Figure 3.18 In a 2 m thick sandy layer: (a) soil suction head profiles and (b) soil water content profiles under various steady flux conditions.

in Figure 3.17, is used to examine the characteristics of profiles of soil suction and moisture. The hypothetical soils have unsaturated parameters shown in Table 3.4 and are subject to three hydrologic conditions: hydrostatic (zero flux), infiltration of 1 m/year, and evaporation of −0.365 m/year.

Case i: sandy soil layer

Profiles of soil suction and volumetric water content are plotted in Figure 3.18. For hydrostatic conditions (solid line), the only case where the system is under thermodynamic equilibrium, i.e., no energy lost and the total head is a constant everywhere in the soil layer, the suction potential is distributed linearly from zero at the water table to −2.0 m at the ground surface. The suction potential is invariant with respect to the material type, and is completely compensated by the elevation head. Volumetric water content decreases from 0.35 at the water table to 0.08 at the ground surface. The water content profile here (volumetric water content vs. minus elevation) is the SWCC of this soil. Notice that because of simplifications in the SWCC model (Equation (3.39)), air enters the soil immediately above the water table so that there is no capillary fringe.

Steady infiltration of 1 m/year into the soil domain moves the soil water to a higher energy state, thus increasing suction potential relative to that of the hydrostatic conditions. Because zero pressure head is maintained at the water table, the cumulative effect causes the soil suction head to increase from −2.0 m under the hydrostatic conditions to about

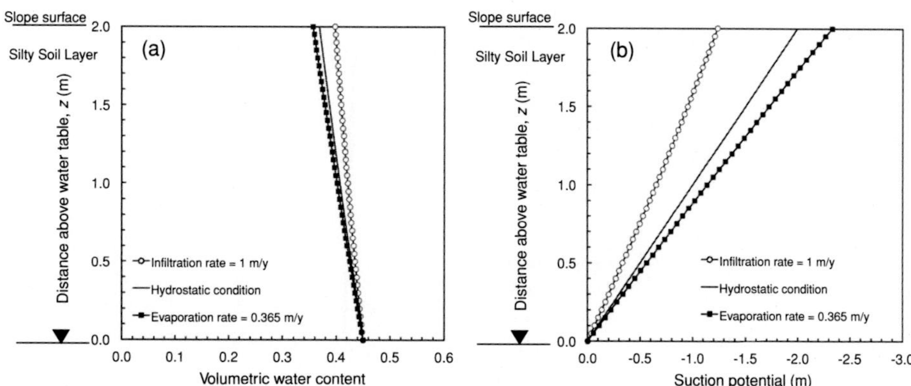

Figure 3.19 In a 2 m thick silt layer: (a) steady soil suction head profiles and (b) soil water content profiles under various flux conditions.

−1.75 m at the ground surface. Corresponding increases in the water content profile are shown in Figure 3.18b.

For the same soil, steady evaporation of −0.365 m/year moves the soil water to a lower energy state, thus decreasing the soil suction potential relative to that of the hydrostatic conditions. The suction potential reduces from zero at the water table to −2.22 m at the ground surface, which is 0.22 m less than that under the hydrostatic conditions. Correspondingly, the soil is drier than it is under the hydrostatic conditions (Figure 3.18b). Overall, steady infiltration or evaporation at the rates illustrated here have a quite limited effect on the distribution of soil suction or water content, with most of the change concentrated near the ground surface relative to the hydrostatic conditions.

For thicker (e.g., 5 m) sand layers under steady evaporation, Equation (3.45) predicts that the soil suction head will be more than several tens of meters at the ground surface. Correspondingly, the water content will be only a small percentage near the ground surface.

Case ii: silty soil layer

The distribution of suction potential is nearly linear throughout the entire soil layer under hydrostatic, infiltration, and evaporation conditions. Infiltration of 1 m/year has a significant effect on the suction potential distribution, as shown in Figure 3.19. The suction head is about −1.25 m at the ground surface; a 0.75 m increase over hydrostatic conditions. The increase in water content due to infiltration is quite limited, about 0.03 at the ground surface. Evaporation of 0.365 m/year causes a significant reduction in suction potential. The soil suction potential is −2.4 m at the ground surface; a 0.4 m decrease over hydrostatic conditions. Water content reduction under evaporation in this 2 m silty soil is insignificant. The water content profile generally varies nearly linearly with depth under the two flux conditions. Compared to sandy soil, the variation in the water content profile is relatively small across the entire 2 m layer.

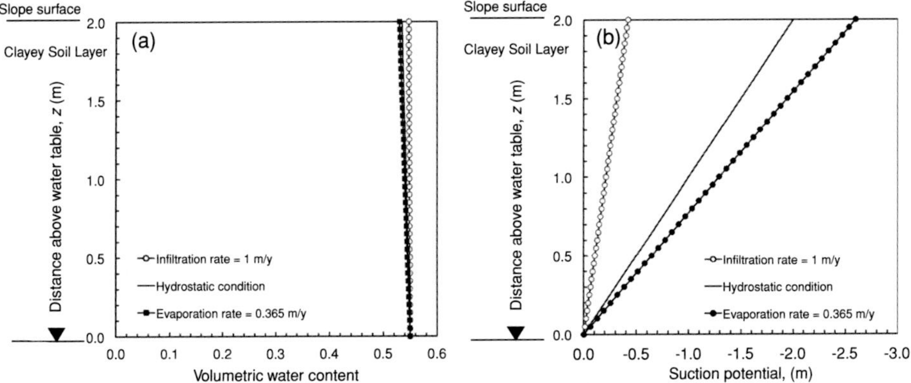

Figure 3.20 In a 2 m thick clay layer: (a) steady soil suction head profiles and (b) soil water content profiles under various flux conditions.

Case iii: clayey soil layer

The suction potential is distributed nearly linearly throughout the entire soil layer. Infiltration of 1 m/year has a significant effect on the suction potential distribution, as shown in Figure 3.20. The suction potential is only −0.4 m at the ground surface; a 1.60 m increase over hydrostatic conditions. There is very little increase in water content due to infiltration throughout the entire layer. Evaporation of 0.365 m/year causes a significant reduction of soil suction potential. The soil suction potential is −2.65 m at the ground surface; a 0.65 m decrease from that under the hydrostatic condition. The variation in the water content profile in this 2 m clayey soil is insignificant. The water content profile, for all examples, generally varies nearly linearly with depth under the various flux conditions at the upper boundary, fluctuating less than 0.02 beyond the hydrostatic case.

3.5.2 Two-layer system

Because weathering and other geologic processes lead to stratigraphy in hillslope soils, it is also useful to examine the hydrologic response in a layered system. In this example each soil horizon is 1 m thick and the system is illustrated in Figure 3.21.

To arrive at an analytical solution for the two-layer system, the same procedure as for the one-layer system can be followed. For the bottom layer bounded by the water table ($0 \leq z \leq L_1$):

$$H(z) = -\frac{q}{\beta_1} + c_1 \exp(-\beta_1 z) \tag{3.47}$$

$$c_1 = \frac{q + K_{s1}}{\beta_1} \tag{3.48}$$

$$h_m(z) = \frac{1}{\beta_1} \ln\left(\frac{K(z)}{K_{s1}}\right) \tag{3.49}$$

$$\theta(z) = \theta_{s1}\left(\frac{K(z)}{K_{s1}}\right)^{1/n_1} \tag{3.50}$$

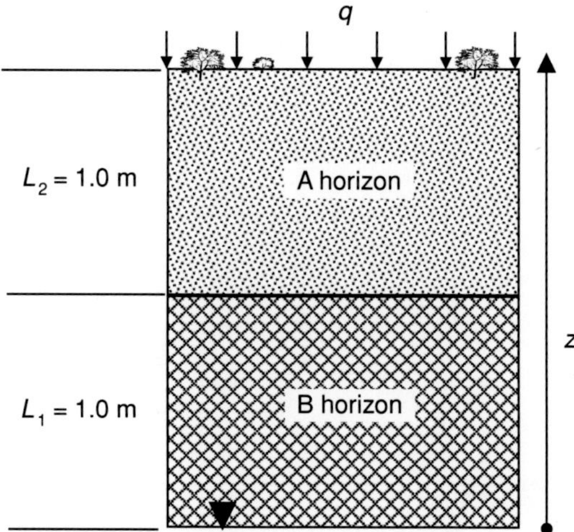

Figure 3.21 A 2 m thick, two-horizon unsaturated soil with a water table at the lower boundary.

For the top layer bounded by the atmosphere ($L_1 \leq z \leq L_1 + L_2$):

$$H(z) = -\frac{q}{\beta_2} + c_2 \exp(-\beta_2 z) \tag{3.51}$$

$$h_m(z) = \frac{1}{\beta_2} \ln\left(\frac{K(z)}{K_{s2}}\right) \tag{3.52}$$

$$\theta(z) = \theta_{s2}\left(\frac{K(z)}{K_{s2}}\right)^{1/n_z} \tag{3.53}$$

The relations between the integration coefficients c_1 and c_2 can be arrived at by imposing a suction head continuity condition at the interface of the two layers:

$$c_2 = \exp(\beta_2 L_1)\left\{\frac{q}{\beta_2} + \frac{q + K_{s2}}{\beta_2}\left[\frac{\beta_1}{K_{s1}}\left(-\frac{q}{\beta_1} + c_1 \exp(-\beta_1 L_1)\right)\right]^{\beta_2/\beta_1}\right\} \tag{3.54}$$

For the two-layer system with three different types of soil (sand, silt, and clay), there are six possible cases, but the profiles of suction head and water content can all be defined by Equations (3.47)–(3.54). The hydrologic parameters for all these three hypothetical soils are listed in Table 3.4.

Case iv: sand–silt system

Suction potential and water content profiles under three different flux boundary conditions are plotted in Figure 3.22. Again, the suction head profile under hydrostatic conditions is invariant to material type and is linear with a slope of 1:1. However, the water content profile is discontinuous across the contact between the overlying sand and underlying silt. Here, volumetric water content decreases abruptly from 0.4 to 0.16. This discontinuity is expected as sand and silt under the same suction conditions would have very different equilibrium water contents, which are controlled by their SWCCs.

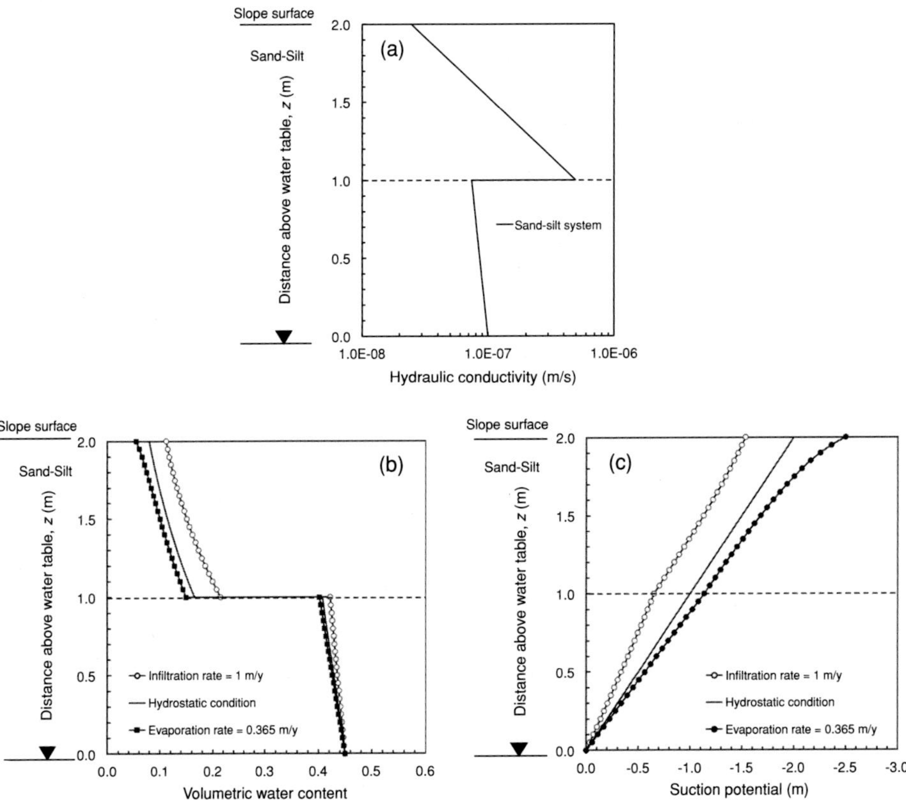

Figure 3.22 In a two-layer sand–silt system: (a) profile of hydraulic conductivity under hydrostatic conditions, (b) steady soil water content, and (c) suction potential profiles under various flux conditions.

Under infiltration conditions, suction head increases nearly linearly both above and below the contact between the sand and silt. The variation in suction head in the lower silt layer is very similar to that in the single-layer silty soil (Case ii), and the variation in the overlying sand layer is likewise very similar to that in the single-layer sandy soil (Case i). The difference between infiltration and hydrostatic conditions is significant. The water content difference at the contact of the silt and sand layer is reduced by about 0.02 compared to the profile for hydrostatic conditions.

Under evaporation conditions, suction varies non-linearly with depth. Towards the ground surface, suction head is increasingly reduced compared to hydrostatic conditions. However, change in the slope of the suction profile at the contact is not obvious. The water content abruptly decreases at the boundary between the silt and sand layers. Here, water content in the silt layer is slightly greater than that under hydrostatic conditions (by about 0.03) but in the sand layer is much greater (by about 0.06) than that under hydrostatic conditions.

For all flux conditions, the water content reduction in the sand layer is greater than that in the silt layer, indicating that water content in sandy soils is sensitive to variation in suction head. As predicted by Equations (3.47)–(3.50), suction potential and water content values

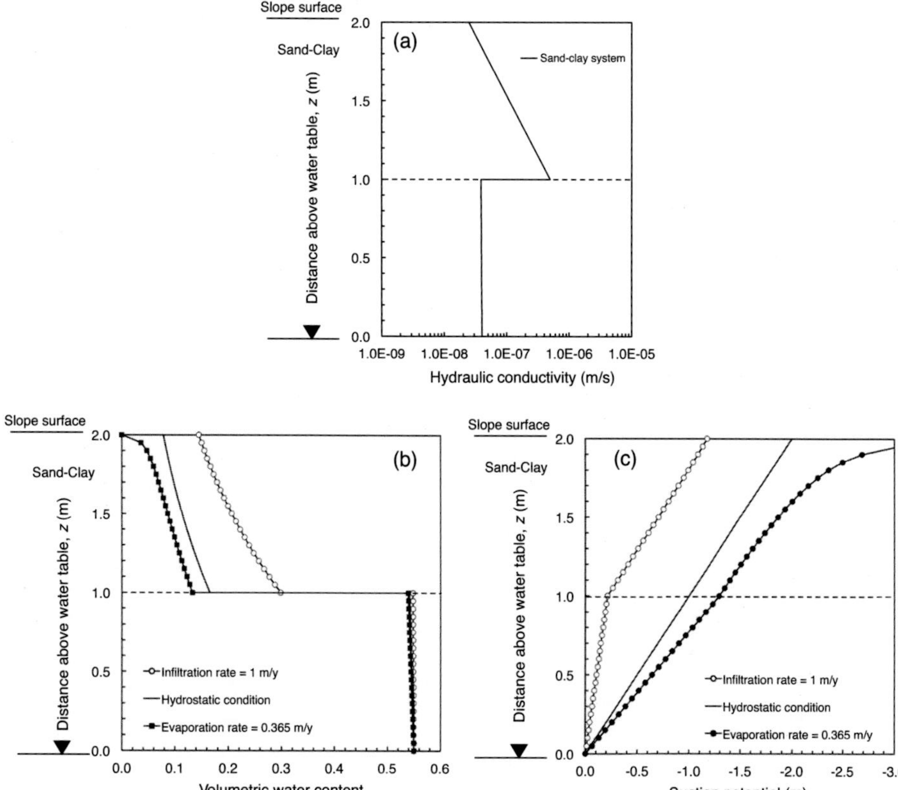

Figure 3.23 In a two-layer sand–clay system: (a) profile of hydraulic conductivity under hydrostatic conditions, (b) steady soil water content, and (c) suction potential profiles under various flux conditions.

at the interface in the silt layer are completely determined by the hydrologic properties and flux conditions and independent of the overlying sand layer.

Case v: sand–clay system

Water flux, either infiltration or evaporation, has substantial impact on both suction head and water content profiles in the sand–clay system, as shown in Figure 3.23. The abrupt reduction in water content at the soil interface under hydrostatic conditions is more pronounced than that in the sand–silt system (Case iv), changing from 0.53 in the clay to 0.16 in the sand. The significant difference between suction potential profiles under hydrostatic and infiltration conditions is largely controlled by the clay layer, as the suction potential profile in the sand layer is nearly parallel to that under hydrostatic conditions.

The water content profile under infiltration changes little from hydrostatic conditions in the clay layer, but the profile in the sand is greatly influenced by the infiltration. For an infiltration rate of 1 m/year, the water content at the contact between the sand and the clay increases from 0.16 under hydrostatic conditions to 0.31 in the sand. The water

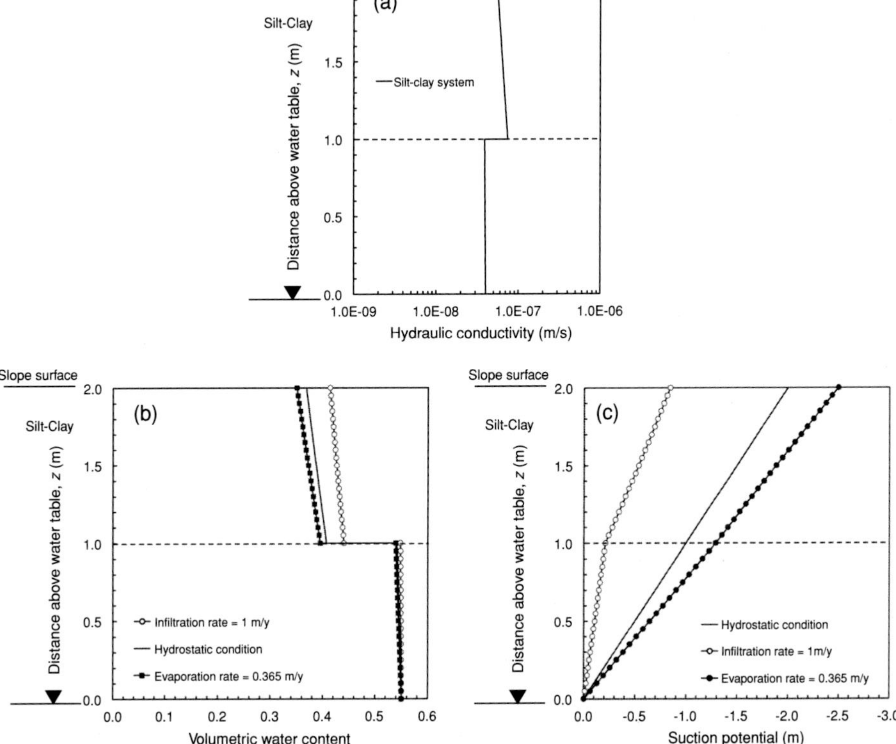

Figure 3.24 In a two-layer silt–clay system: (a) profile of hydraulic conductivity under hydrostatic conditions, (b) steady soil water content, and (c) suction potential profiles under various flux conditions.

content profile is also obviously non-linear in the sand. At the ground surface, the suction potential increases from −2.0 m under hydrostatic conditions to −1.2 m under infiltration of 1.0 m/year, and the corresponding water content increases from 0.08 to 0.14.

Under evaporation of 0.365 m/year, the suction potential changes significantly compared to the hydrostatic case. The profile is nearly linear in the clay layer, but is non-linear in the sand. At the contact, the suction potential reduces from −1.0 m under hydrostatic conditions to −1.3 m. The water content profile in the clay layer is not affected by evaporation, but evaporation has a significant effect on the profile in the sand layer. At the soil contact, water content decreases from 0.16 under hydrostatic conditions to 0.12 under evaporation of 0.365 m/year.

Case vi: silt–clay system

Suction potential and water content profiles are shown in Figure 3.24. For this system, suction potential and water content profiles are approximately linear under both infiltration and evaporation conditions. Under hydrostatic conditions, the water content profile changes abruptly from 0.54 to 0.40 at the contact between the overlying silt and underlying clay.

Both infiltration and evaporation have a significant effect on the distribution of suction head. The difference across the soil interface is much more pronounced under infiltration than that under evaporation at the rates shown here. Under infiltration, the reduction of suction potential in the silt layer is greater than that in the clay layer. Under the infiltration rate of 1.0 m/year, the suction potential reduces with depth at a rate of 0.8 m/m in the silt layer and by 0.7 m/m in the clay layer. The change in water content in the clay layer is insignificant, but could be significant in the silt layer. Under infiltration of 1.0 m/year, the increase in the water content in the silt layer varies between 0.04 at the soil interface to 0.06 at the ground surface.

Under evaporation, suction potential is affected significantly compared to hydrostatic conditions; the water content much less so. This is due to the fact that the water content of clay within this suction range is relatively insensitive to suction change. Compared to the sand–clay system (Case v), the suction potential varies linearly within the upper silt layer and is much less at the ground surface. The corresponding water content is much higher in this case than that in sand–clay system.

Case vii: silt–sand system

Suction potential and water content profiles are shown in Figure 3.25. Although this system shares the same silt with the previous case (silt–clay system), the suction potential and water content profiles are quite different. Flux under both infiltration and evaporation conditions has little effect on either the distribution of suction potential or water content in the underlying sand layer, but does have an effect on the suction profile in the silt layer. The suction potential profile is linear in the sand layer, but slightly non-linear in the silt layer. The water content profile is non-linear in the sand layer, but nearly linear in the silt layer. The water content varies little from the hydrostatic case in the sand layer.

Under infiltration, the abrupt change in the slope of the suction potential profile at the soil contact can be clearly seen. Under hydrostatic conditions the suction potential increases nearly linearly with depth over the entire profile. However, under infiltration of 1 m/year the suction head increases by about 0.45 m at the ground surface over that obtained for hydrostatic conditions. The corresponding water content increases from 0.35 to 0.38. The profile of water content abruptly increases at the contact between the silt and the sand from 0.17 to 0.41.

Subject to an evaporation rate of 0.365 m/year at the ground surface, suction potential varies nearly linearly in both layers. The effect of evaporation is small in the sand layer, but substantial in the overlying silt. At the ground surface, suction potential decreases from −2.0 m under hydrostatic conditions to −2.4 m under evaporation. However, the corresponding water content profile shows very little change from hydrostatic conditions throughout the entire two-layer system.

Case viii: clay–sand system

The profiles of suction potential and water content for this system are very similar to those described previously for the silt–sand system (Case vii). Both suction potential

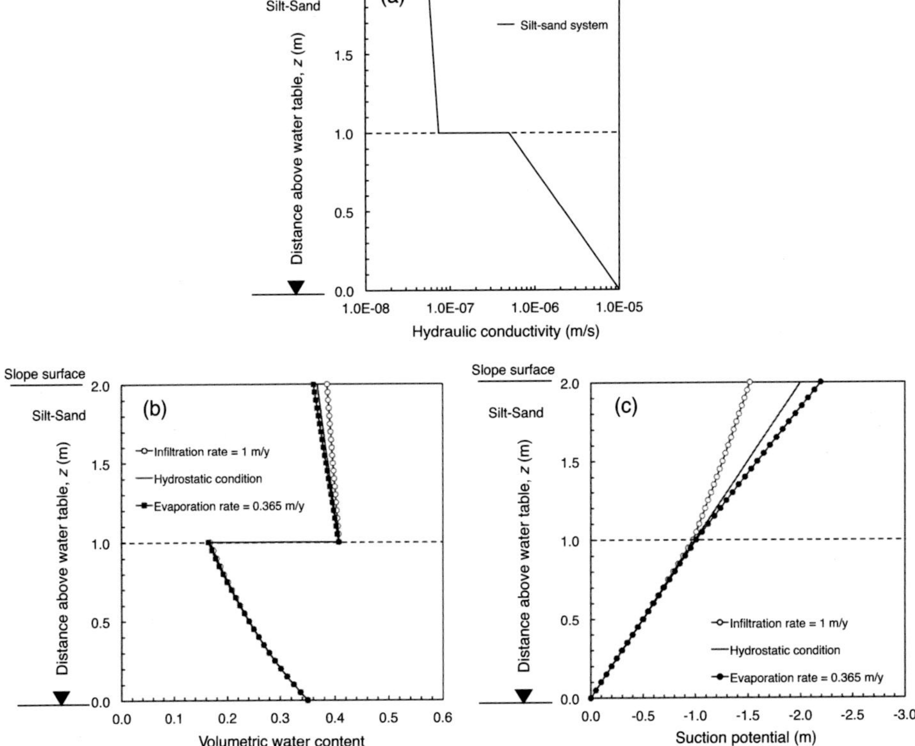

Figure 3.25 In a two-layer silt–sand system: (a) profile of hydraulic conductivity, (b) steady soil water content, and (c) suction potential profiles under various flux conditions.

and water content profiles under various flux conditions are shown in Figure 3.26. For the suction potential profile under either infiltration or evaporation, the change from hydrostatic conditions is generally greater than that in the silt–sand system (compared to Figure 3.25a). However, the deviation of the water content profile from hydrostatic conditions is generally smaller (compared to Figure 3.25b). The sudden increase in water content at the soil contact is larger than that in the silt–sand system; the water content increases from 0.16 in the sand to 0.54 in the clay.

Case ix: clay–silt system

Suction potential and water content profiles show some unique patterns in this system (Figure 3.27). Both suction potential and water content profiles are nearly linear under the various flux conditions in both layers. Infiltration generally has a substantial influence on the suction potential distribution in both layers. The deflection of the suction potential profile at the soil contact is substantial. Suction is reduced with depth in the 1 m thick silt layer by about 0.6 m/m and in the 1 m clay layer by a much smaller amount

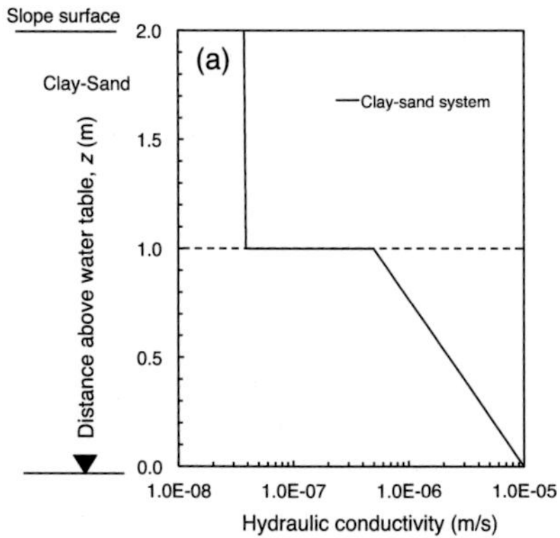

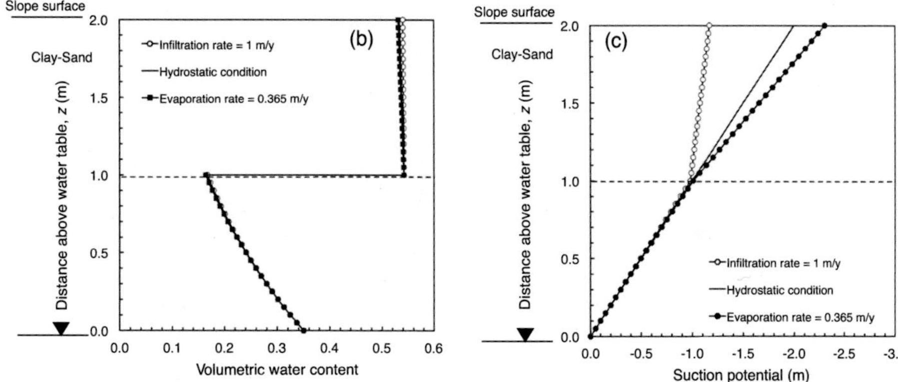

Figure 3.26 In a two-layer clay–sand system: (a) profile of hydraulic conductivity, (b) steady soil water content, and (c) suction potential profiles under various flux conditions.

(about 0.25 m/m). At the soil contact, the suction potential increases from -1.0 m under the hydrostatic conditions to -0.6 m, and at the ground surface, the suction potential increases from -2.0 m to -0.9 m under infiltration of 1 m/year. The water content profile under infiltration appears very similar to that under hydrostatic conditions, except at the soil interface and ground surface where the water content increases by about 0.02.

Under evaporation, the deflection of suction potential at the soil contact is not obvious. For an evaporation rate of 0.365 m/year, the maximum deviation of the suction potential profile from that under hydrostatic conditions occurs at the ground surface where it is reduced by 0.5 m. The water content profile essentially follows the same distribution as that for hydrostatic conditions with a sudden increase of about 0.14 at the soil contact.

In summary, suction potential and water content profiles under steady-state conditions are greatly influenced by the unsaturated soil properties, which control the profiles of hydraulic conductivity, and flux conditions at the upper boundary. The variation in water content and suction potential with depth can be calculated using analytical solutions (3.43)–(3.46) for a single-layer system and analytical solutions (3.47)–(3.54) for a two-layer system. Insight regarding the suction head and water content profiles in both systems is gained by

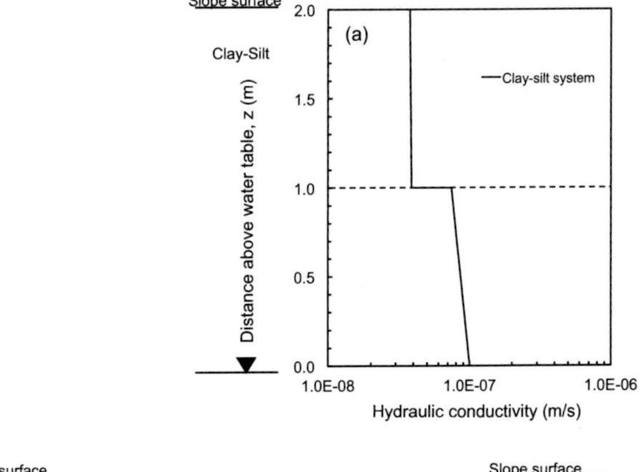

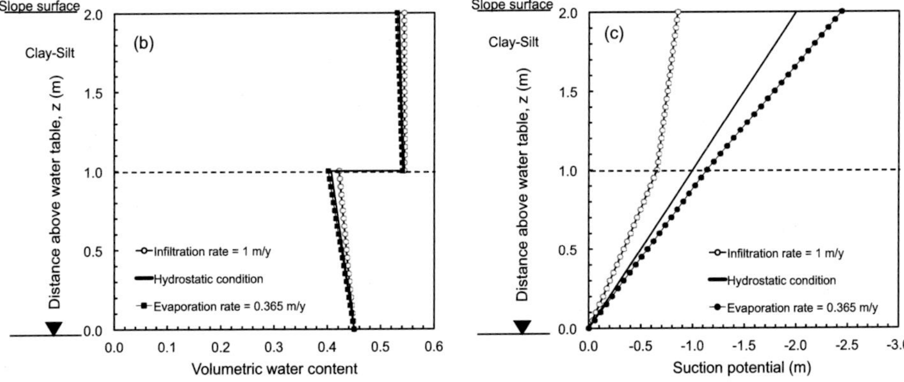

Figure 3.27 In a two-layer clay–silt system: (a) profile of hydraulic conductivity, (b) steady soil water content, and (c) suction potential profiles under various flux conditions.

examining results for a number of 2 m thick systems. For example, water content changes abruptly across the contact between soils of different textures, whereas the flux conditions at the upper boundary have little effect for the cases where fine-grained materials underlie coarse materials. While these analyses are useful for illustrating the complex patterns in these profiles, the variation can be further complicated and controlled by the thickness of each of the layers and the variations in unsaturated soil properties, which can be assessed effectively by Equations (3.47)–(3.54). Readers are encouraged to try these equations for other configurations of layered systems.

3.6 Hydrologic barriers

3.6.1 Flat capillary barriers

Because hillslope materials are often anisotropic and heterogeneous, water movement is multi-dimensional with different flow rates. Material anisotropy refers to the directional

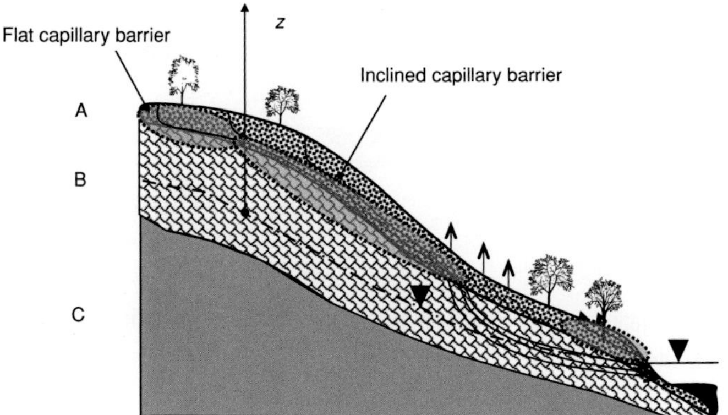

Figure 3.28 Illustration of a capillary barrier within a hillslope in which horizon A is less permeable than horizon B.

dependence of material properties. In an anisotropic material the value of a property such as hydraulic conductivity or elastic modulus varies with direction. The degree of anisotropy in material properties at a point is controlled by geologic, geomorphologic, and hydrologic processes such as sedimentation, consolidation, and weathering. Material heterogeneity commonly refers to location dependency of material properties such that the value of any given property varies from one point to another. The degree of heterogeneity is similarly controlled by geologic, geomorphologic, and hydrologic processes. Heterogeneity occurs both within a single material type, and from one material to another. For example, as shown in Figure 3.28, if the hydrologic properties between horizon A and B are very different such that horizon A is composed of much finer particles than horizon B, the flow of infiltrating water may be slowed dramatically near the contact between the materials, depending on the location. This is called the "capillary barrier effect." In the relatively flat area near the top of the hillslope, infiltration could cease and a lens of water could build up above the interface, whereas in the sloping area, water could be diverted laterally and move along the interface for some distance before it enters horizon B. On the other hand, if horizon A is much more permeable than horizon B, a perched water body could form above the interface. The underlying physics for both capillary barriers and perched water will be discussed in detail in this section.

If the interface between two materials with sharply different hydrologic properties is horizontal, water movement, if any, is due to a gradient in matric potential. Gravitational potential across the contact is absent in this case. Though counterintuitive, the conceptual illustration of why water might move from coarse to fine soil is illustrated using a capillary pore model in Figure 3.29. For an initial position of water shown in Figure 3.29a, the total head potential in the fine pore can be described by the Laplace equation:

$$(h_t)_{fine} = -\frac{2T_s}{\gamma_w r_{fine}} \tag{3.55}$$

where T_s is surface tension of water, γ_w is the unit weight of water, and r_{fine} is the radius of an idealized soil pore for fine soil. The negative sign indicates that the total potential is less

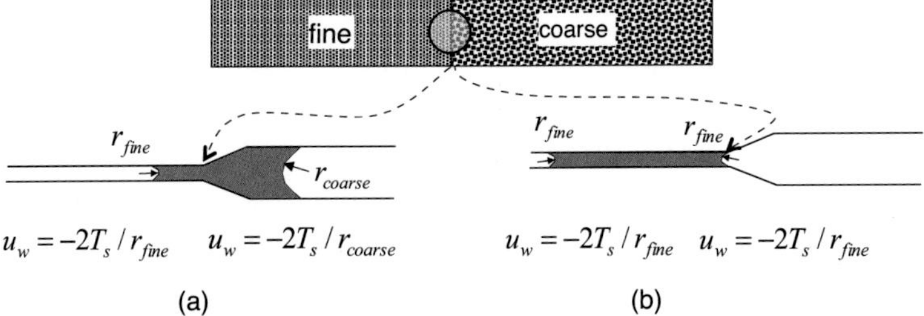

Figure 3.29 Illustration of why water moves horizontally from coarse soil to fine soil: (a) non-equilibrium state and (b) equilibrium state.

than the free water outside the soil pore. The corresponding tensile pore-water pressure or soil suction can be written as

$$(u)_{fine} = -\frac{2T_s}{r_{fine}} \tag{3.56}$$

By the same token, the total head potential and pore-water pressure in the coarse pore are, respectively,

$$(h_t)_{coarse} = -\frac{2T_s}{\gamma_w r_{coarse}} \tag{3.57}$$

$$(u)_{coarse} = -\frac{2T_s}{r_{coarse}} \tag{3.58}$$

Because the radius of the fine pore is less than the radius of a coarse pore, the total head potential in the coarse soil is higher than that in the fine soil, i.e.,

$$(h_t)_{coarse} = -\frac{2T_s}{\gamma_w r_{coarse}} > (h_t)_{fine} = -\frac{2T_s}{\gamma_w r_{fine}} \tag{3.59}$$

According to the second law of thermodynamics, fluid flows from a place of higher energy to where energy is lower, i.e., from the coarse to the fine pore. In the absence of gravitational potential, fluid flows from a place of higher pressure (coarse pore) to where pressure is lower (fine pore) as well. A state of thermodynamic potential equilibrium will eventually be reached when all the water moves into the fine pore, as shown in Figure 3.29b.

If the interface between fine and coarse soils is horizontal, a gravitational potential exists and the final equilibrium state can also be determined by the total potential concept even though the pressure across the interface is different, as illustrated in Figure 3.30. Initially, when the water layer at the interface is thin (Figure 3.30a), all the water is retained in the fine pore as the difference in the gravitational potential between the top and bottom of the water layer is small in comparison with the capillary potential, i.e., the total head potential at the bottom h_b is close to the total head potential at the top h_t when the elevation head between the bottom z_b and top z_t is negligible ($z_b \approx z_t \approx z_o$):

$$h_b = -\frac{2T_s}{\gamma_w r_{fine}} \approx h_t = \frac{u_{wb}}{\gamma_w} + z_b = \frac{u_{wt}}{\gamma_w} + z_t u_{wb} = u_{wt}$$

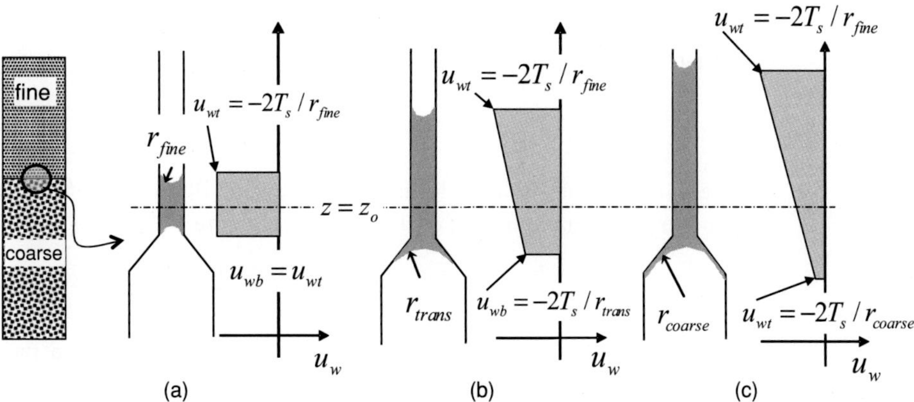

Figure 3.30 Illustration of why water does not move from fine soil to coarse soil vertically: (a) thin suspended water layer, (b) intermediate suspended water layer, and (c) water layer at the threshold of breakthrough.

The above equilibrium condition leads to

$$u_{wb} \approx u_{wt}$$

As more water accumulates near the interface, the gravitational potential increases, and water starts to enter the coarse pore shown in Figure 3.30b. The equilibrium condition for the total head potential leads to the linear pressure increase profile in the downward direction across the interface as follows:

$$h_b = -\frac{2T_s}{\gamma_w r_{trans}} = \frac{u_{wb}}{\gamma_w} = h_t = \frac{u_{wt}}{\gamma_w} + (z - z_0) = \frac{-T_s}{\gamma_w r_{fine}} + (z - z_o)$$
$$u_{wb} = u_{wt} + \gamma_w (z - z_o)$$

The elevation head difference is completely compensated by the pressure head difference, which is provided by the pore size difference in terms of pore radii r_{fine} and r_{tran}. This difference reaches a maximum when r_{tran} reaches r_{coarse}, as shown in Figure 3.30c. Further accumulation of the elevation head will not be possible, as the water will break through into the coarse layer. The pressures at the critical state before the breakthrough of the water layer can be established by equilibrating the total head between the top and bottom of the water lens and setting the corresponding pore radii:

$$u_{wb} = u_{wt} + \gamma_w (z - z_o)$$
$$(z - z_o) = h_{critical} = \frac{u_{wb} - u_{wt}}{\gamma_w} = \frac{2T_s}{\gamma_w}\left(\frac{1}{r_{fine}} - \frac{1}{r_{coarse}}\right) \quad (3.60)$$

The above equation identifies that the controlling factor for the thickness of the water layer at the interface is the pore size contrast between the two adjacent soils. For real soil with various pore sizes, the pore radius r_{fine} can be related to the air-entry head of the fine soil $h_{air\text{-}fine}$ due to the possibility of air entering from the top (Lu and Likos, 2004a):

$$r_{fine} = \frac{2T_s}{u_b} = \frac{2T_s}{\gamma_w h_{air-fine}} = 2T_s \alpha_{fine} \quad (3.61)$$

where $\gamma_w h_{air\text{-}fine} = 1/\alpha_{fine}$ is also called air-entry pressure. The pore parameter r_{coarse} can be related to the water-entry head of the coarse soil $h_{w\text{-}coarse}$ due to the possibility of water entering the coarse soil (Bouwer, 1966):

$$r_{coarse} = \frac{2T_s}{u_w} = \frac{2T_s}{\gamma_w h_{w-coarse}} \tag{3.62}$$

where $\gamma_w h_{w\text{-}coarse}$ is also called water-entry pressure. Bouwer (1966) further suggested the water-entry pressure to be half of the air-entry pressure $1/\alpha_{coarse}$:

$$\gamma_w h_{w-coarse} = \frac{1}{2\alpha_{coarse}} \tag{3.63}$$

Substituting Equations (3.61)–(3.63) into Equation (3.60) leads to the critical head or thickness of the water layer:

$$h_{critical} = \frac{1}{\gamma_w}\left(\frac{1}{\alpha_{fine}} - \frac{2}{\alpha_{coarse}}\right) \tag{3.64}$$

For example, a two-layer interface consisting of a clayey silt with $\alpha_{fine} = 0.05\ \text{kPa}^{-1}$, and a sandy soil with $\alpha_{coarse} = 0.3\ \text{kPa}^{-1}$, will be able to hold a water column of

$$h_{critical} = \frac{1}{\gamma_w}\left(\frac{1}{\alpha_{fine}} - \frac{2}{\alpha_{coarse}}\right) = \frac{1}{(9.8)\left(\frac{\text{kN}}{\text{m}^3}\right)}\left(\frac{1}{0.05}\left(\frac{\text{kN}}{\text{m}^2}\right) - \frac{2}{0.3}\left(\frac{\text{kN}}{\text{m}^2}\right)\right)$$
$$= 2.04\,(\text{m}) - 0.62\,(\text{m}) = 1.42\,(\text{m})$$

before water can break through into the underlying sandy soil.

3.6.2 Dipping capillary barriers

In hillslopes any sharp interface between two materials is likely to be inclined. Under such conditions a great deal of water can be held above the interface. Of equal importance is that infiltrating flow can change direction from predominantly vertical to preferentially parallel to the interface, as illustrated in Figure 3.31.

Under the influence of gravitational potential, water will infiltrate predominantly vertically until it approaches a contact where the capillary barrier effect will begin to divert flow. As more and more water diverts laterally along and above the interface, pore-water pressures above the interface will increase. In the same manner as the horizontal capillary barrier illustrated in Figure 3.29, if the pressure head difference overcomes the capillary barrier, water will enter the underlying coarse layer. The maximum horizontal distance along the interface where no water can cross the interface is called the diversion distance (Figure 3.31), and the corresponding flux equal to the infiltration times the diversion distance is called the diversion capacity.

By employing the total potential concept and considering the interface geometry and soil hydrologic properties, the diversion distance and capacity can be determined. Following the approach of Ross (1990), Steenhuis *et al.* (1994) provided analytical expressions for

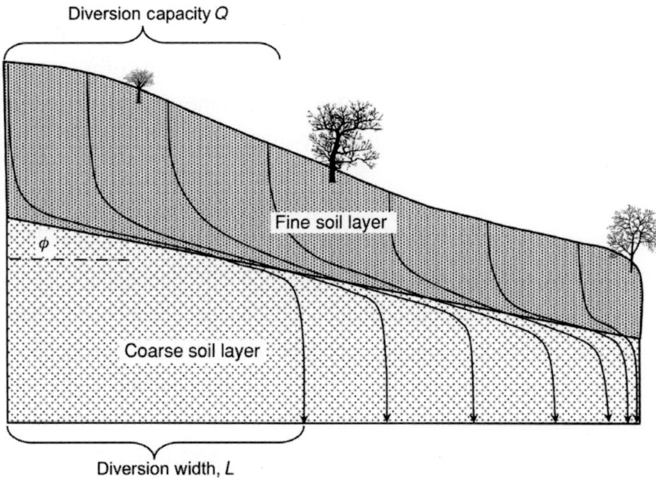

Figure 3.31 Steady flow field in dipping two-layer capillary barrier system.

the diversion distance L and capacity Q_{max} of the capillary barrier with saturated hydraulic conductivity k_s for the fine soil layer and a dip angle of ϕ:

$$L \leq \frac{K_s \tan\phi}{q}\left[\left(2h_{air\text{–}fine} - h_{w\text{–}coarse}\right)\right] \tag{3.65}$$

$$Q_{\max} \leq K_s \tan\phi\left[\left(2h_{air\text{–}fine} - h_{w\text{–}coarse}\right)\right] \tag{3.66}$$

For example, for a two-layer system, with an upper silty soil with $h_{air\text{-}fine} = 2.0$ m, and $K_s = 10^{-7}$ m/s, and a lower sandy soil with $h_{w\text{-}coarse} = 0.33$ m, a steady flux $q = 10^{-8}$ m/s, and a dip angle $\phi = 20^{\circ}$, the diversion distance and capacity can be calculated as

$$L \leq \frac{K_s \tan\phi}{q}\left[2h_{air\text{–}fine} - h_{w\text{–}coarse}\right] = \frac{\left(10^{-7}\ \text{m/s}\right)\tan 20^{\circ}}{\left(10^{-8}\ \text{m/s}\right)}(4\ \text{m} - 0.33\ \text{m}) = 13.3\,(\text{m})$$

$$Q_{\max} \leq K_s \tan\phi\left[h_{air\text{–}fine} - h_{w\text{–}coarse}\right] = Lq = (13.3\ \text{m})(10^{-8}\ \text{m/s}) = 1.33 \times 10^{-7}\,(\text{m}^2/\text{s})$$

For an ideal capillary system, the water-entry pressure is small and can be ignored so that

$$L_{\max} = \frac{2K_s \tan\phi}{q} h_{a\text{–}fine} \tag{3.67}$$

The efficiency of a capillary barrier ω then can be defined from Equations (3.65) and (3.67):

$$\frac{L}{L_{\max}} = \omega = 1 - \frac{h_{w\text{–}coarse}}{2h_{air\text{–}fine}} \tag{3.68}$$

For example, if the air-entry head of the fine layer is 10 times that of the water-entry head of the coarse layer, the efficiency of the capillary barrier is

$$\omega = 1 - \frac{1}{2(10)} = 0.95 \text{ or } 95\%$$

Recognize that for real-world cases, the infiltrating water starts to enter the lower soil layer over a much shorter distance than the diversion capacity suggests, as both the air-entry and water-entry heads are indicators of large amounts of water or air starting to enter the soil. A more accurate analysis of water flow and moisture distribution can be done using numerical solutions of the governing flow equations described in Chapter 4.

3.6.3 Hydraulic barriers due to heterogeneity

Small variations in hydraulic conductivity can cause significant changes in flow pattern and slope stability. For flow in a hillslope composed of two materials, it is the contrast in the hydraulic conductivity between the two materials that controls the variation in pore-pressure distribution, rather than the absolute magnitude of hydraulic conductivities. Variation in hydraulic conductivity arises from a contrast in geologic materials, such as a contact between permeable sandstone and less permeable silt or mudstone. Variation in hydraulic conductivity can also be present in a single geologic material that has undergone different processes or degrees of the same process, such as consolidation or weathering. In hillslopes, such contrasts are common at the contact between the upper layer of soil or regolith and the better consolidated and less weathered underlying bedrock.

In a two-dimensional cross section of a hillslope, under gravity, the steady-state seepage flow field and pore-pressure potential can be obtained using the mathematical solution of the following equation (see derivation of Equation (4.9) in Chapter 4) (e.g., Freeze and Cherry, 1979):

$$\frac{\partial}{\partial x}\left(K\frac{\partial u}{\partial x}\right) + \frac{\partial}{\partial z}\left(K\frac{\partial u}{\partial z} + K\rho_w g\right) = 0 \tag{3.69}$$

where u is the pore-water pressure, ρ_w is the density of water, g is the gravitational acceleration, and K is the hydraulic conductivity.

Reid (1997) solved Equation (3.69) using a numerical model to examine the effects of contrasts in hydraulic conductivity. In a saturated hillslope, groundwater flow is driven exclusively by gravity. The model domain was established assuming periodic topography, yielding a symmetric hillslope that has no flow boundaries along the lateral margins and the bottom. Figure 3.32 shows results for a 34° hillslope with material properties of varying geometry marking a five-fold contrast in hydraulic conductivity. The slope shown in Figure 3.32a is composed of homogeneous material. Figure 3.32b is a hillslope in which a surficial slope-parallel layer overlies a less permeable material, and Figure 3.32c is for a horizontal contact between an upper, relatively permeable material that is underlain by a less permeable material. Finally, Figure 3.32d is for a hillslope with a vertical contact between the more permeable material on the left and less permeable on the right. Hydraulic head is related to pore-water pressure u by

$$h = \frac{u}{\rho_w g} + z \tag{3.70}$$

The head gradient determines the seepage body force $-\rho_w g \nabla h$ associated with groundwater flow. Results shown in Figure 3.32 are displayed using normalized seepage-force vectors

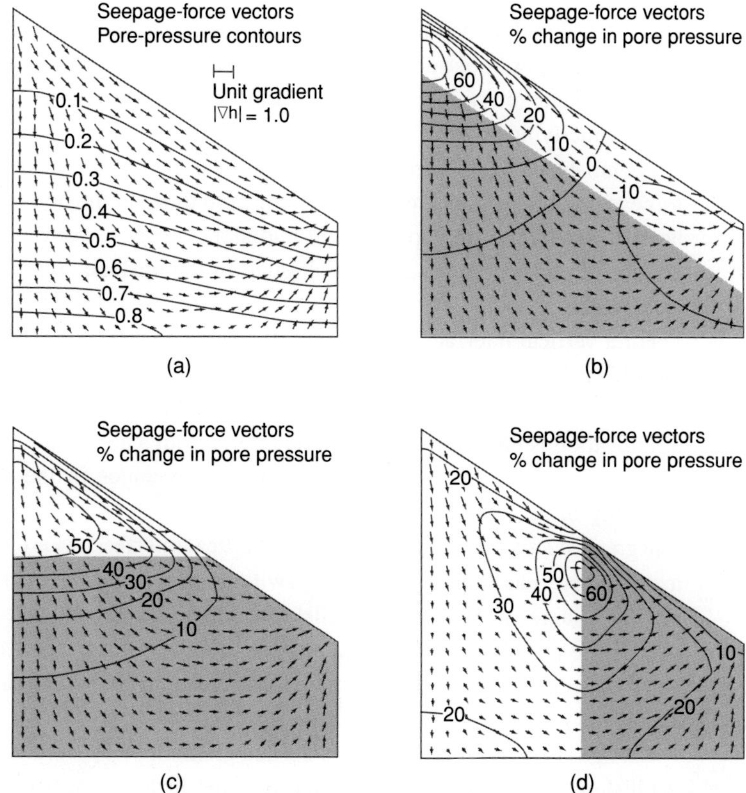

Figure 3.32 Groundwater flow. Gravity-driven groundwater in saturated hillslopes with (a) homogeneous material, (b) slope-parallel layer interface, (c) horizontal layer interface, (d) vertical layer interface. The slope has ratio of 1:1.5 or 34° and the hydraulic conductivity contrast is 5, with the shaded regions having a relatively smaller hydraulic conductivity (from Reid, 1997).

$-\rho_w g \nabla h / \rho_w g$ and normalized pore pressures $u/\rho_w g H$, where H is the height of the sloping part of the domain to remove any scale dependence from the results (Reid, 1997).

The general pattern of the simulated flow field for a saturated hillslope under gravity is as follows. The direction of the hydraulic gradient of the groundwater varies with slope position and distance from the slope face. Near the left-hand boundary in the upslope area the gradient ranges from nearly vertically downward to nearly parallel to the slope face; in the mid-slope region, the gradient ranges from slope parallel to horizontal; and in the downslope discharge region at the right hand boundary, the dominant hydraulic gradient is in the laterally upward direction. This pattern is also evident in the analytical solution of Equation (3.69) for regional-scale gravitationally driven groundwater flow by Tóth (1963).

For hillslopes in which there is a contrast in hydraulic conductivity between different layers, seepage forces (hydraulic gradient) and pore pressure distributions will be modified. In some areas elevated seepage forces and pore-water pressures will result. For hillslopes with a slope-parallel interface between the two layers (Figure 3.32b), seepage within the permeable layer near the ground surface is dominantly sub-parallel to the slope surface, and seepage within the underlying, less permeable layer resembles the homogeneous case. However, the pore-pressure distribution has been modified significantly. In the upslope area above the layer interface, pore pressures increase by as much as 80% due to the impedance of the underlying layer.

For hillslopes with a horizontal layer interface (Figure 3.32c), the patterns in both seepage force and pore pressure are very different from those in the homogeneous or slope-parallel layer cases. In both the upslope and mid-slope areas, seepage becomes more dominantly horizontal and pore pressures are elevated in the upslope area both above and below the layer interface.

For a vertical interface (Figure 3.32d), seepage forces are generally smaller in the permeable layer than those simulated in the homogeneous case. Seepage forces become larger and rotated upward near the interface and within the less permeable layer. Pore pressure modification is pronounced around the interface area and near the ground surface.

The study by Reid (1997) also shows that changes in the maximum pore pressure generally increase as the contrast in hydraulic conductivity between the layers increases. The greatest effect is for hillslopes with a vertical interface, where increases in pore pressure may be as large as 250%. For hillslopes with slope-parallel layers the pore pressure increases may be as large as 150%, and for hillslopes with the horizontal layer interface increases may be as large as 80%. However, the increase in pore pressure as a function of permeability contrast declines exponentially and becomes insignificant when the hydraulic conductivity contrast is greater than 100.

Changes in both pore-water pressure and gradient of pore-water pressure (seepage force) in hillslopes have also been shown to cause significant changes in slope stability. This point will be elaborated upon in Chapters 9 and 10.

3.7 Problems

1. What are the driving mechanisms for water movement in hillslopes?
2. In a hillslope, where is the capillary head important for flow, and why? Where is the gravity head important for flow and why?
3. If two adjacent points in a hillslope have the same total head, but different pressure heads, would you expect flow between these two points?
4. If a point near the slope surface is drying due to evaporation, would you expect the total head to be increasing or decreasing with depth, and why?
5. In Figure 3.5b, is the total head in the salt solution the same as the total head in the pure water? What is the difference in the osmotic head between the salt solution and the pure water shown in the figure?
6. In a closed chamber consisting of air with 100% relative humidity, if the temperature increases, do you expect vapor pressure to change? Do you expect relative humidity to change?

7 What is the meaning of the negative sign used for vapor pressure head?

8 In soils, can we measure the relative humidity of the pore air to determine the pore-water pressure? Assuming the pore water is pure water, how do we calculate pore-water pressure from the measurement of the pore air's relative humidity?

9 If we know the pore pressure by the above (Problem 8), how can we estimate the curvature of the water meniscus in the soils?

10 A soil layer is under hydrostatic conditions. Do you expect that the total heads above and below the water table are the same or different, and why? Do you expect that the pressure values are different above and below the water table?

11 Will Darcy's law apply if soil is under hydrostatic conditions?

12 Why is there a negative sign in front of the right-hand side of Darcy's law?

13 According to Equation (3.25), what will happen if soil moisture content is constant within a horizontal soil layer?

14 What are the characteristic functions used to quantify flow in unsaturated isotropic soils?

15 In which state, drying or wetting, does soil water have greater energy?

16 Will the hydraulic conductivity of unsaturated sand always be higher than the hydraulic conductivity of unsaturated silt?

17 What is capillary rise?

18 In what types of soil would you expect the ultimate capillary rise to be the greatest? Why?

19 What is the total potential head when capillary rise ceases?

20 Describe the two conditions where capillary rise is of practical importance. What controls these two conditions?

21 Figure 3.15 shows two theoretical predictions of experimental capillary rise data. Explain why one model fits the data better than the other.

22 In a field setting, what environmental conditions control the rate of capillary rise?

23 In which regions of the world or physiographic setting would you expect capillary rise and vapor flow to be an important process? Why?

24 In determining vertical flow in unsaturated soils, what are the two characteristic functions of hillslope materials?

25 In a 2 m layer of unsaturated sand, do you expect the matric suction head profile to be linear under hydrostatic conditions? Why?

26 In a 2 m layer of unsaturated sand, do you expect the soil moisture profile to be linear under hydrostatic conditions? Why?

27 Which type of soil, sand, silt, or clay, would you expect to have the greatest suction variation in the region above the water table?

28 Which type of soil, sand, silt, or clay, would you expect to have the greatest moisture content variation in the region above the water table?

29 When rainfall rate increases, will soil suction in the region near the ground surface increase or decrease?

30 Given a vadose zone consisting of two layers of materials, sand and clay, would you expect the suction profile to be continuous at the interface between the two layers? Why?

31 Given a vadose zone consisting of two layers of materials, sand and clay, would you expect the moisture content profile to be continuous at the interface between the two layers? Why?

32 What is a capillary barrier? How does it work?

33 The water distribution in a two-material system is shown in Figure 3.29. Will the water move towards the left or right? Why?

34 How would you describe the diversion capacity of a dipping capillary barrier?

35 If an unsaturated soil has a water potential of 1,000 J/kg, what is the equivalent soil suction value? If the soil at air-dry conditions has a matric suction of 100 MPa, what is the soil water potential in joules per kilogram and in meters?

36 Three soils, clay, silt, and sand, are all equilibrated at the same matric suction; which soil will have the highest water content and why?

37 The relative humidity at equilibrium in an unsaturated soil is measured to be 80% at 22 °C. (a) What is the vapor pressure in the soil? (b) What is the vapor density in the soil? (c) What is the absolute humidity if the temperature in the soil is maintained constant but vaporization is allowed to occur? And (d) what is the free energy per unit mass of the pore water?

38 At a prevailing temperature of 25 °C and pressure of 95 kPa, how much does the density of air change from a completely dry state to a 100% relative humidity state?

39 Derive Terzaghi's (1943) solution for the rate of capillary rise:

$$t = \frac{nh_c}{K_s}\left(\ln\frac{h_c}{h_c - z} - \frac{z}{h_c}\right)$$

where t is the wetting front arrival time, n is porosity, h_c is the ultimate height of capillary rise, K_s is saturated hydraulic conductivity, and z is the wetting front location above the water table.

Show that Lu and Likos' (2004a) solution below for the rate of capillary rise can be reduced to Terzaghi's if the summation index m is zero.

$$t = \frac{n}{K_s}\sum_{j=0}^{m=\infty}\frac{\beta^j}{j!}\left(h_c^{j+1}\ln\frac{h_c}{h_c - z} - \sum_{s=0}^{j}\frac{h_c^s z^{j+1-s}}{j+1-s}\right)$$

40 Consider a horizontal capillary barrier consisting of a silt layer overlying a coarse sand layer. If the silt has an air-entry pressure of 10 kPa and the sand has an air-entry pressure of 2 kPa, what is the maximum height of water that can be suspended in the overlying silt? What is the efficiency of the capillary barrier if it has a tilted interface?

41 Consider a flat slope with a two-layer system of sand overlying silt. The thickness of the sandy layer is 1 m, and of the silty layer is 1 m. The unsaturated hydraulic properties are as follows: $\beta_{sand} = 3\ \text{m}^{-1}$, $\beta_{silt} = 0.3\ \text{m}^{-1}$, $n_{sand} = 4$, $n_{silt} = 3$, $\theta_{sand} = 0.4 = \theta_{silt}$, $K_{s\text{-}sand} = 10^{-5}$ m/s, and $K_{s\text{-}silt} = 10^{-7}$ m/s. Calculate and plot steady suction profiles and water content profiles under the following three infiltration conditions: $q = -10^{-8}$ m/s, 0.0 m/s, and 10^{-8} m/s. Draw your major conclusions regarding patterns of soil moisture and suction distributions in the two-layer system from this exercise.

4 Transient infiltration

4.1 Governing equation for transient water flow

4.1.1 Principle of mass conservation

Fluid flow and moisture content in hillslope environments vary spatially and temporally mainly due to time-dependent environmental changes and the storage capacity of soils. The former is often cast into boundary conditions and the latter into the governing flow equations or laws.

When flow occurs in hillslopes, the total water potential is in disequilibrium; it varies from one point to another point within the hillslope and from time to time. The general governing equation for liquid flow in variably saturated soils can be derived by applying the principle of mass conservation. This principle states that for a given elemental volume of soil, the rate of water loss or gain is conservative and is equal to the net flux in and out of the element. The mass conservation principle is also called the continuity principle. For a representative elementary volume (REV) of soil shown in Figure 4.1 with porosity n and volumetric water content θ, the rate of total inflow of water flux (kg/s) along the positive coordinate direction is

$$q_{in} = \rho(q_x \Delta y \Delta z + q_y \Delta x \Delta z + q_z \Delta x \Delta y) \tag{4.1}$$

and the total outflow is

$$q_{out} = \rho\left(\left(q_x + \frac{\partial q_x}{\partial x}\Delta x\right)\Delta y \Delta z + \left(q_y + \frac{\partial q_y}{\partial y}\Delta y\right)\Delta x \Delta z + \left(q_z + \frac{\partial q_z}{\partial z}\Delta z\right)\Delta x \Delta y\right) \tag{4.2}$$

where ρ is the density of water (kg/m^3), and q_x, q_y, and q_z are fluxes in the x, y, and z directions respectively (m/s).

When a steady-state condition has not been reached, the rate of loss or gain of water mass is

$$\frac{\partial(\rho\theta)}{\partial t}\Delta x \Delta y \Delta z \tag{4.3}$$

and is equal to the net inflow and outflow:

$$-\rho\left(\frac{\partial q_x}{\partial x}\Delta x \Delta y \Delta z + \frac{\partial q_y}{\partial y}\Delta x \Delta y \Delta z + \frac{\partial q_z}{\partial z}\Delta x \Delta y \Delta z\right) = \frac{\partial(\rho\theta)}{\partial t}\Delta x \Delta y \Delta z \tag{4.4}$$

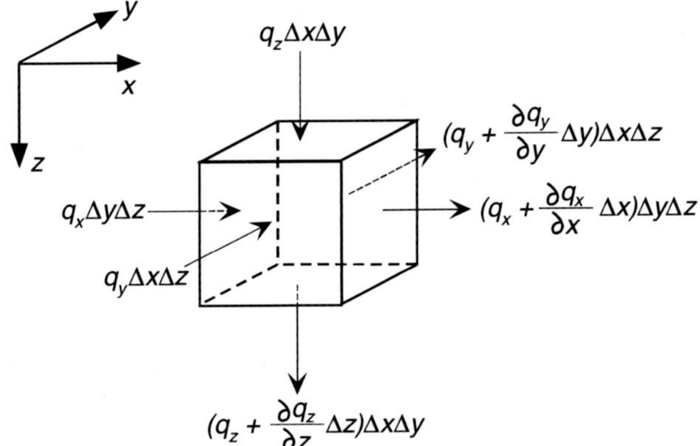

Figure 4.1 Elemental soil volume showing the continuity requirements for fluid flow.

or

$$-\rho\left(\frac{\partial q_x}{\partial x}+\frac{\partial q_y}{\partial y}+\frac{\partial q_z}{\partial z}\right)=\frac{\partial(\rho\theta)}{\partial t} \tag{4.5}$$

The above equation is the governing unsteady or transient flow equation and is applicable for both saturated and unsaturated soils.

The energy stored or released by an element of soil due to change in the fluid volume $\rho\theta$ can be related to the hydraulic (pressure) head. For both saturated and unsaturated soils, following the same approach illustrated in Freeze and Cherry (p.51, 1979) for saturated porous media ($\theta = n$), the right hand side of Equation (4.5) can be defined as

$$\frac{\partial(\rho\theta)}{\partial t}=\rho S_s\frac{\partial h_m}{\partial t} \tag{4.6}$$

$$S_s=\rho g\left(\frac{1}{v_s}\frac{\partial v_s}{\partial u}+\frac{n}{v_v}\frac{\partial v_w}{\partial u}\right)=\rho g(\alpha_s+n\beta_w) \tag{4.7}$$

where S_s is the specific storage of the soil, u is pore-water pressure ($\rho g h_m$), v_s is the volume of the solid in the REV shown in Figure 4.1, v_v is the volume of void in the REV, v_w is the volume of water in the REV, α_s is the compressibility of bulk soil (m^2/N) and β_w is the compressibility of water (m^2/N). The compressibility of soil ranges from approximately 10^{-6} to 10^{-8} m^2/N for clay, 10^{-7} to 10^{-9} m^2/N for sand, and 10^{-8} to 10^{-10} m^2/N for gravel. The compressibility of water is approximately constant with a value of about 4.4×10^{-10} m^2/N.

Darcy's law (Equation (3.18)) allows us to write fluid flow in terms of principal hydraulic conductivity and hydraulic gradient for each principal coordinate direction as

$$q_x=-K_x\frac{\partial h}{\partial x} \tag{4.8a}$$

$$q_y=-K_y\frac{\partial h}{\partial y} \tag{4.8b}$$

$$q_z=-K_z\frac{\partial h}{\partial z} \tag{4.8c}$$

Accordingly, the governing flow Equation (4.5) for transient variably saturated flow in anisotropic soils becomes

$$\frac{\partial}{\partial x}\left(K_x\frac{\partial h}{\partial x}\right)+\frac{\partial}{\partial y}\left(K_y\frac{\partial h}{\partial y}\right)+\frac{\partial}{\partial z}\left(K_z\frac{\partial h}{\partial z}\right)=S_s\frac{\partial h_m}{\partial t} \tag{4.9}$$

4.1.2 Transient saturated flow

The governing flow equation (4.9) for transient saturated flow in isotropic and homogeneous (i.e., $K = K_x = K_y = K_z$) soils becomes

$$\left(\frac{\partial^2 h}{\partial x^2}+\frac{\partial^2 h}{\partial y^2}+\frac{\partial^2 h}{\partial z^2}\right)=\frac{S_s}{K}\frac{\partial h}{\partial t} \tag{4.10}$$

which may also be written in the form of a diffusion equation:

$$\frac{\partial h}{\partial t}=\frac{K}{S_s}\left(\frac{\partial^2 h}{\partial x^2}+\frac{\partial^2 h}{\partial y^2}+\frac{\partial^2 h}{\partial z^2}\right)=D\left(\frac{\partial^2 h}{\partial x^2}+\frac{\partial^2 h}{\partial y^2}+\frac{\partial^2 h}{\partial z^2}\right) \tag{4.11}$$

where D is called the water or hydraulic diffusivity (m^2/s) and is equal to K/S_s.

4.1.3 Richards equation for unsaturated flow

Under unsaturated conditions, the compressibility of the bulk soil usually is smaller than the compressibility of the fluid, i.e., displacement of pore fluid is much easier than displacement of solid under the same pore pressure variation:

$$\frac{1}{v_s}\frac{\partial v_s}{\partial u} \ll \frac{n}{v_v}\frac{\partial v_w}{\partial u} \tag{4.12}$$

Thus the left hand side in the above inequality can be ignored and the specific storage term (Equation (4.7)) becomes specific moisture capacity:

$$S_s=\rho g\left(\frac{n}{v_v}\frac{\partial v_w}{\partial u}\right)=\rho g\frac{n}{v_v}\frac{v}{v}\frac{\partial v_w}{\partial \rho g h_m}=\frac{\partial \theta}{\partial h_m} \tag{4.13}$$

where θ is the volumetric water content (v_w/v) and h_m is the pore-water pressure head. The Richards (1931) equation for unsaturated soil flow problems can be arrived at by substituting Equation (4.13) into Equation (4.9):

$$\frac{\partial}{\partial x}\left(K_x\frac{\partial h}{\partial x}\right)+\frac{\partial}{\partial y}\left(K_y\frac{\partial h}{\partial y}\right)+\frac{\partial}{\partial z}\left(K_z\frac{\partial h}{\partial z}\right)=\frac{\partial \theta}{\partial t} \tag{4.14}$$

Noting that total head is the sum of the matric suction head and the elevation head in the absence of the osmotic head ($h_t = h_m + z$), we have

$$\frac{\partial}{\partial x}\left(K_x\frac{\partial h_m}{\partial x}\right)+\frac{\partial}{\partial y}\left(K_y\frac{\partial h_m}{\partial y}\right)+\frac{\partial}{\partial z}\left(K_z\left(\frac{\partial h_m}{\partial z}+1\right)\right)=\frac{\partial \theta}{\partial t} \tag{4.15}$$

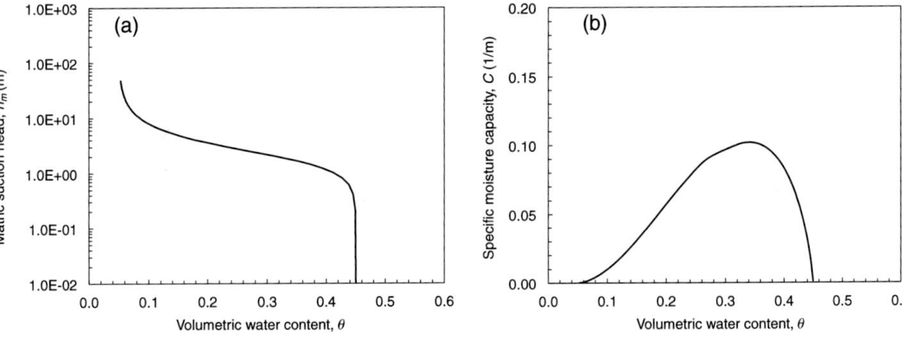

Figure 4.2 Conceptual illustration of the relationship between SWCC and the specific moisture capacity for a silty soil.

The rate of water loss or gain expressed on the right side of Equation (4.15) can be rewritten in terms of the matric suction by applying the chain rule:

$$\frac{\partial \theta}{\partial t} = \frac{\partial \theta}{\partial h_m}\frac{\partial h_m}{\partial t} \tag{4.16}$$

The specific moisture capacity $\partial\theta/\partial h_m$ is the slope of a curve that defines the relation between volumetric water content and suction head described by the SWCC and is commonly referred to as the specific moisture capacity C:

$$S_s = C(h_m) = \frac{\partial \theta}{\partial h_m} \tag{4.17}$$

Figure 4.2a illustrates an example of the volumetric water content as a function of suction head for a typical silty soil. The corresponding specific moisture capacity as a function of water content (the derivative of the water content with respect to suction head is shown in Figure 4.2b. The magnitude of the specific moisture capacity is largest where the slope of the SWCC is flattest at $\theta \approx 0.32$. For this soil, a change in each centimeter of suction head at $\theta \approx 0.32$ will result in a change in soil water content of about 0.10.

Substituting Equations (4.16) and (4.17) into Equation (4.15), we have

$$\frac{\partial}{\partial x}\left(K_x\frac{\partial h_m}{\partial x}\right) + \frac{\partial}{\partial y}\left(K_y\frac{\partial h_m}{\partial y}\right) + \frac{\partial}{\partial z}\left(K_z\left(\frac{\partial h_m}{\partial z}+1\right)\right) = C\left(h_m\right)\frac{\partial h_m}{\partial t} \tag{4.18}$$

The above equation is the Richards equation. The solution of the Richards equation with appropriate boundary and initial conditions will give a field of soil suction or pressure head in space and time. As implied in Equation (4.18), two characteristic functions of the soil are required for its solution: the hydraulic conductivity function, and the SWCC.

The Richards equation may also be written in terms of soil water content, as is often done in soil physics. Following the chain rule, Darcy's law can be expressed as

$$q_x = -K_x(\theta)\frac{\partial h_m}{\partial x} = -K_x(\theta)\frac{\partial h_m}{\partial \theta}\frac{\partial \theta}{\partial x} \tag{4.19a}$$

$$q_x = -D_x(\theta)\frac{\partial \theta}{\partial x}; \quad D_x(\theta) = \frac{K_x\left(\theta\right)}{\dfrac{\partial \theta}{\partial h_m}} \tag{4.19b}$$

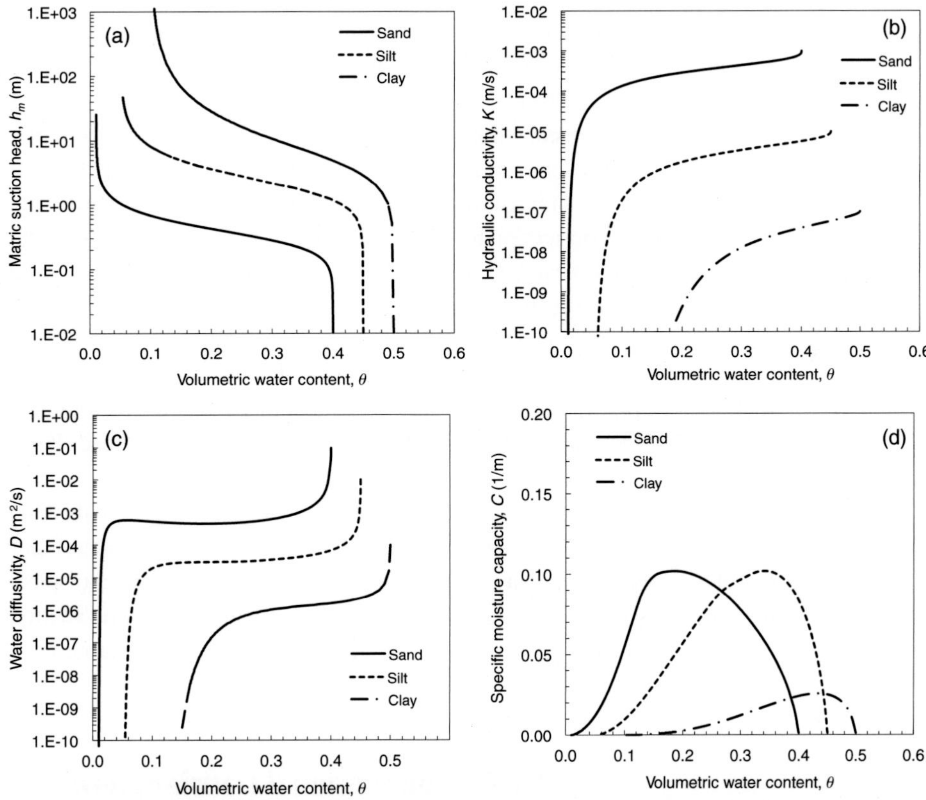

Figure 4.3 Conceptual illustration of the relationship among the soil water characteristic curve (SWCC), hydraulic conductivity function (HCF), water diffusivity function (WDF), and the specific moisture capacity function (SMCF) defined by Equation (4.19): (a) SWCC, (b) HCF obtained independently from test or theory, (c) WDF obtained by HCF divided by SMCF, and (d) SMCF obtained from the inverse of the slope of SWCC.

where D_x is defined as the ratio of the hydraulic conductivity to the specific moisture capacity, and is called the water diffusivity for unsaturated soil. The relationship among the water diffusivity, the hydraulic conductivity function, SWCC, and the specific moisture capacity is illustrated in Figure 4.3 for sand, silt, and clay.

Similarly, we can define fluxes in the y and z (gravity) directions as

$$q_y = -D_y(\theta)\frac{\partial \theta}{\partial y}; \quad D_y(\theta) = \frac{K_y(\theta)}{\dfrac{\partial \theta}{\partial h_m}} \tag{4.20a}$$

$$q_z = -K_z(\theta)\left(\frac{\partial h_m}{\partial \theta}\frac{\partial \theta}{\partial z} + 1\right) = -D_z(\theta)\frac{\partial \theta}{\partial z} - K_z(\theta) \tag{4.20b}$$

Substituting Equations (4.19b), (4.20a), and (4.20b) into Equation (4.14), we have

$$\frac{\partial \theta}{\partial t} = \frac{\partial}{\partial x}\left(D_x(\theta)\frac{\partial \theta}{\partial x}\right) + \frac{\partial}{\partial y}\left(D_y(\theta)\frac{\partial \theta}{\partial y}\right) + \frac{\partial}{\partial z}\left(D_z(\theta)\frac{\partial \theta}{\partial z}\right) + \frac{\partial K_z(\theta)}{\partial z} \tag{4.21}$$

Analytical solutions of Equation (4.21) under various initial and boundary conditions constitute a rich body of classical problems in soil physics and groundwater hydrology and are sampled in further reading (e.g., Philip, 1957; Srivastava and Yeh, 1991; Warrick *et al.*, 1997; Smith *et al.*, 2002).

If two of the three constitutive functions (C, D, K) are known, the remaining one can be inferred from the others. For example, if the SWCC is known (Figure 4.3a for three typical soils), $C(\theta)$ can be obtained by taking the derivative of the volumetric water content θ with respect to the matric suction head h_m and is shown in Figure 4.3d. Further, if $D(\theta)$ is known independently, $K(\theta)$ can be inferred by Equation (4.19) and is shown in Figure 4.3c.

Figure 4.3 also illustrates the ranges of magnitude in h_m, K, and C for typical soils. Beginning from saturated conditions, the volumetric water content of sandy soil is typically reduced to residual conditions with an increase of two or three orders of magnitude of matric suction head ($h_m < 100$ m). For silty soil, most of the change in volumetric water content occurs within a range of 1,000 to 10,000 m of matric suction head. For clayey soil, most of the change in moisture content could be in the range of up to 60,000 m (not shown in Figure 4.3a). Correspondingly, the unsaturated hydraulic conductivity and water diffusivity could decrease 6 to 10 orders of magnitude from saturated to residual moisture conditions for all soil types, as shown in Figures 4.3b and 4.3c. The specific moisture capacity reflects the sensitivity of the volumetric water content to changes in matric suction head. The peak values of C for sandy and silty soils (Figure 4.3d) are relatively large with respect to those for clayey soil.

4.2 One-dimensional transient flow

4.2.1 Richards equation in hillslope settings

In the solution of transient hillslope hydrology problems it is common to consider a long planar slope with a constant slope angle. The Richards equation (4.21) is often written in a rotated Cartesian coordinate system where x_* and z_* are aligned with the slope angle γ, as shown in Figure 4.4a. This is done mainly to seek analytical solutions for infiltration in hillslopes. The Richards equation can be rewritten in terms of x_* and z_* with the following coordinate transformations:

$$x = x_* \cos\gamma - z_* \sin\gamma \tag{4.22a}$$

$$z = x_* \sin\gamma - x_* \cos\gamma \tag{4.22b}$$

Substituting the above equation into Equation (4.21) leads to the Richards equation in terms of x_* and z_*:

$$\frac{\partial\theta}{\partial t} = \nabla\cdot(D\nabla\theta) - \frac{dK}{d\theta}\left(\frac{\partial\theta}{\partial x_*}\sin\gamma + \frac{\partial\theta}{\partial z_*}\cos\gamma\right) \tag{4.23}$$

To date, few analytical solutions for the above equation have been established. However, the above equation can be further simplified using Philip's (1991) assumption that moisture

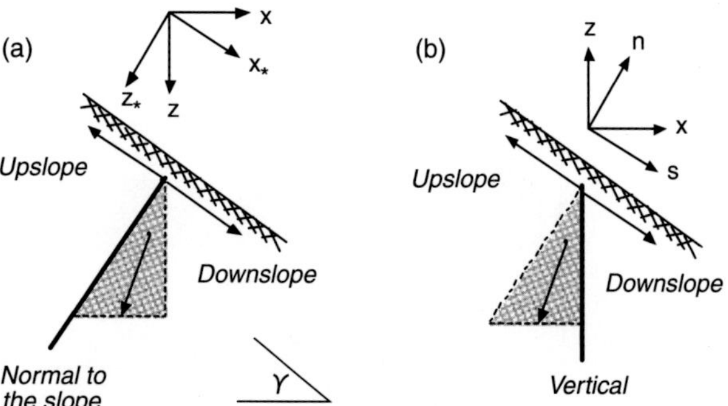

Figure 4.4 Two commonly used definitions of downslope and upslope flow: (a) using the direction normal to the slope surface as the divide (e.g., Philip, 1991), and (b) using the vertical direction as the divide (e.g., Harr, 1977; Zaslavsky and Sinai, 1981b; and Jackson, 1992). According to (a), flux shown in the shaded area would be called downslope flow, whereas according to (b) it would be called upslope flow. In this section, we adopt the definition presented in (b) (from Lu *et al.*, 2011).

distribution is independent of x_* in hillslopes except near the slope crest. Imposing this assumption, Equation (4.23) can be further simplified as a one-dimensional problem in terms of two independent variables z_* and t:

$$\frac{\partial \theta}{\partial t} = \nabla \cdot (D \nabla \theta) - \frac{dK}{d\theta} \frac{\partial \theta}{\partial z_*} \cos \gamma \tag{4.24}$$

The above equation is amenable for many analytical solutions (e.g., Philip, 1991; Chen and Young, 2006). However, several investigations from both theoretical and experimental perspectives over the past two decades (e.g., Jackson, 1992, 1993; Philip, 1993; Sinai and Dirksen, 2006; Lu *et al.*, 2012) show that Philip's (1991) assumption is not valid. Section 4.4 provides a more detailed and critical analysis of this assumption. The following sections describe two analytical solutions that provide instructive insight into the transient nature of infiltration in partially saturated hillslopes.

4.2.2 The Green–Ampt infiltration model

One of the first successful attempts at describing the transient process of water infiltration into unsaturated soil was Green and Ampt's (1911) analytical solution for flow into an initially dry, uniform column of soil. The approach is not a direct solution of the Richards equation, but rather an analytical solution of Darcy's law under two assumptions that arise from observations of a sharp wetting front during infiltration: (1) the soil suction head beyond the wetting front (in the dry portion of the column) is constant in both space and time, and (2) the water content and the corresponding hydraulic conductivity of the soil behind the wetting front (in the wet portion of the column) are also constant in both space and time. Green and Ampt's analysis described the wetting front progression for three flow conditions: vertical downward flow, vertical upward flow, and horizontal flow. Their solution for horizontal infiltration is derived below to illustrate the diffusive process of water

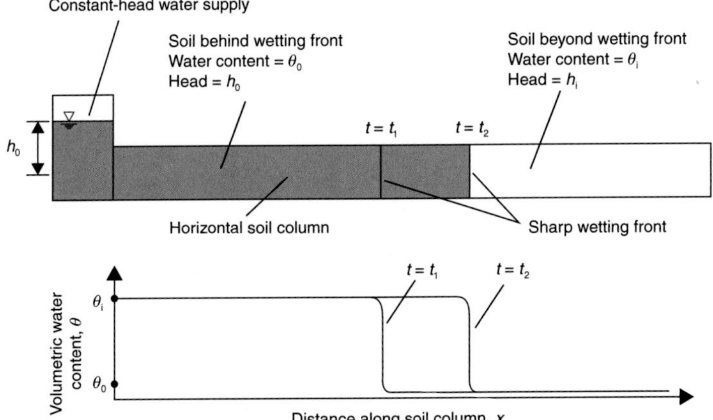

Figure 4.5 Transient infiltration of a sharp wetting front in a horizontal soil column and the corresponding water content distributions with time under the Green–Ampt assumptions (from Lu and Likos, 2004a).

infiltration under the gradient of soil suction head, and the vertical downward solution is described below to illustrate the relative importance of the gradients of soil suction head and gravity.

Horizontal infiltration is shown in Figure 4.5. Under the two previously mentioned assumptions and a constant head h_o, the total volumetric infiltration Q, at any time t, is equal to the change in water content of the soil column from the initial value θ_i to the wetting front water content θ_o times the distance x that the wetting front has traveled, i.e.,

$$Q = (\theta_o - \theta_i)x \tag{4.25}$$

The rate of total infiltration is equal to the infiltration rate at the influent boundary (left end of column in Figure 4.5) and can be approximated by Darcy's law, i.e.,

$$q = \frac{dQ}{dt} = \frac{(\theta_o - \theta_i)\,dx}{dt} = -K_o\frac{dh}{dx} = -K_o\frac{h_i - h_o}{x} \tag{4.26}$$

where K_o is assumed to be the saturated hydraulic conductivity, although in general this may not be the case. Integrating with respect to the two variables x and t and imposing the initial condition of $x = 0$ at $t = 0$ yields

$$x^2 = \left(2K_o\frac{h_o - h_i}{\theta_o - \theta_i}\right)t \tag{4.27a}$$

or

$$\frac{x}{\sqrt{t}} = \sqrt{2K_o\frac{h_o - h_i}{\theta_o - \theta_i}} = \lambda = \text{constant} \tag{4.27b}$$

The quantity in parentheses in front of time t in Equation (4.27a) has units of length squared over time and can be considered as an effective water diffusivity D so that

$$x = \sqrt{2K_o \frac{h_o - h_i}{\theta_o - \theta_i} t} = \sqrt{Dt} \tag{4.28}$$

The above equation predicts that the wetting front advances at a rate proportional to the square root of time. The infiltration rate q at the influent boundary and the cumulative infiltration Q according to Equations (4.26) and (4.28) are

$$q = K_o \frac{h_o - h_i}{\sqrt{Dt}} \tag{4.29}$$

$$Q = (\theta_o - \theta_i)\sqrt{Dt} = s\sqrt{t} \tag{4.30}$$

where the parameter s was first termed "sorptivity" by Philip (1969). Equation (4.27b) provides a way to physically justify the use of the so-called Boltzmann transformation to convert the more rigorous governing fluid flow Equation (4.21) from a partial differential equation to an ordinary differential equation such that it becomes amenable to analytical solution under various initial and boundary conditions.

For a varying moisture content profile during infiltration, i.e., a "non-sharp" wetting front, the sorptivity under the Boltzmann transformation according to Philip (1969) is

$$s = \int_{\theta_i}^{\theta_o} \lambda(\theta) d\theta \tag{4.31}$$

The sharp wetting front assumptions (1) and (2) imply that the new variable λ is a constant or $h(x) = h_i$. Inherent in Equations (4.29) and (4.30) is ambiguity in the appropriate value for the constant h_i. Thus, results based on the assumptions (1) and (2) should be considered semi-analytical in nature. According to Green and Ampt, and numerous other experimental works, the soil suction head constant h_i ranges between -0.5 and -1.5 m. In general, Equations (4.29) and (4.30) cannot be used to accurately predict the rate of wetting front progression and the infiltration rate unless h_i is determined experimentally. If the wetting front is not sharp, which may be the case for a fine-grained soil or any soil with moist initial conditions, D is not a constant and explicit solution of Equation (4.21) is required. For example, Philip's (1957) work is a classic treatment of such a problem.

When constant head is applied at the upper surface of a vertical column as shown in Figure 4.6, the wetting front advances downward under the gradients of soil suction head and gravity. The total head at the wetting front is $h = -z + h_i$ such that Equation (4.26) becomes

$$q = \frac{dQ}{dt} = \frac{(\theta_o - \theta_i)dz}{dt} = -K_o \frac{dh}{dz} = -K_o \frac{-z + h_i - h_o}{z} = K_o \left(1 + \frac{h_o - h_i}{z}\right) \tag{4.32a}$$

or

$$q = K_o + \frac{K_o(h_o - h_i)(\theta_o - \theta_i)}{Q} \tag{4.32b}$$

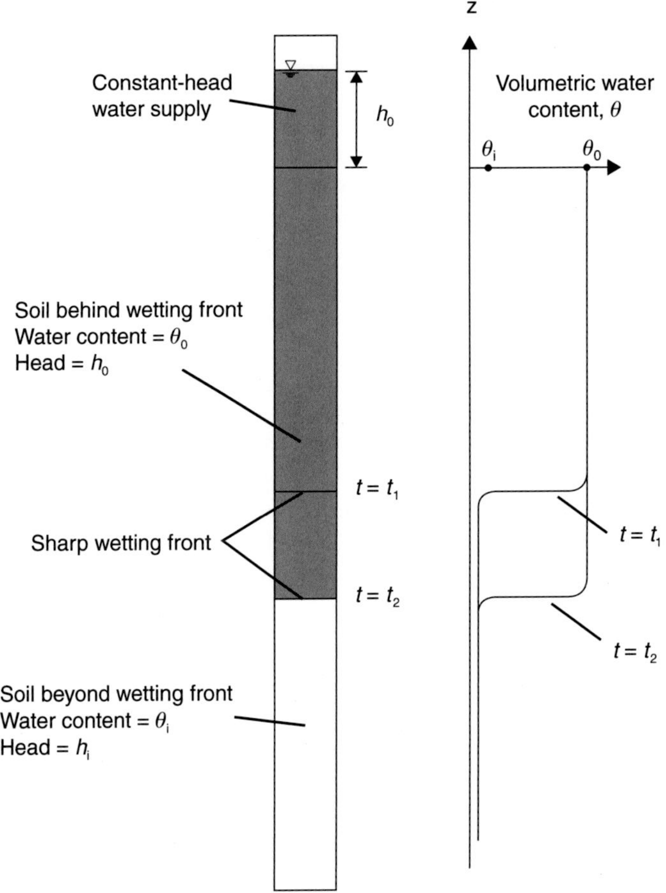

Figure 4.6 Transient infiltration of a sharp wetting front in a vertical soil column and the corresponding water content distribution with time under the Green–Ampt assumptions (from Lu and Likos, 2004a).

or

$$\frac{dz}{dt} = \frac{K_o}{\theta_o - \theta_i}\left(1 + \frac{h_o - h_i}{z}\right) \tag{4.33}$$

One implication of Equations (4.32a) or (4.32b) is that as the time, t, of infiltration becomes sufficiently large, the depth z of infiltration is also large. As such, the infiltration rate q becomes a constant value that is equal to K_o, the hydraulic conductivity that corresponds to the water content θ_o.

Integrating the above equation and imposing the initial condition of $z = 0$ at $t = 0$ yields

$$\frac{K_o}{\theta_o - \theta_i} t = z - (h_o - h_i) \ln\left(1 + \frac{z}{h_o - h_i}\right) \tag{4.34}$$

This equation predicts the wetting front arrival time as a function of the wetting front distance from the influent boundary (top of column). The cumulative infiltration Q, a quantity easily

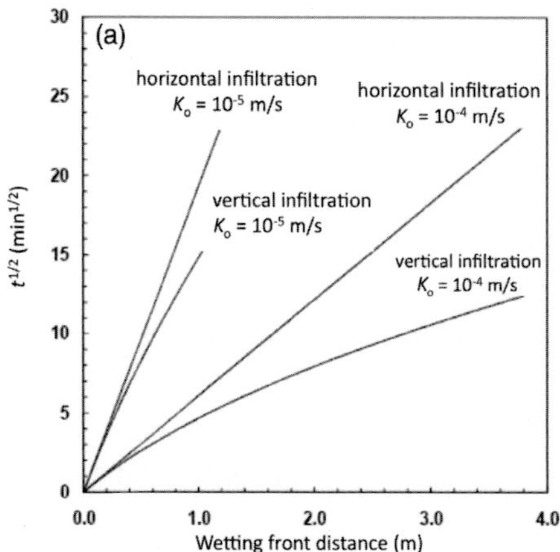

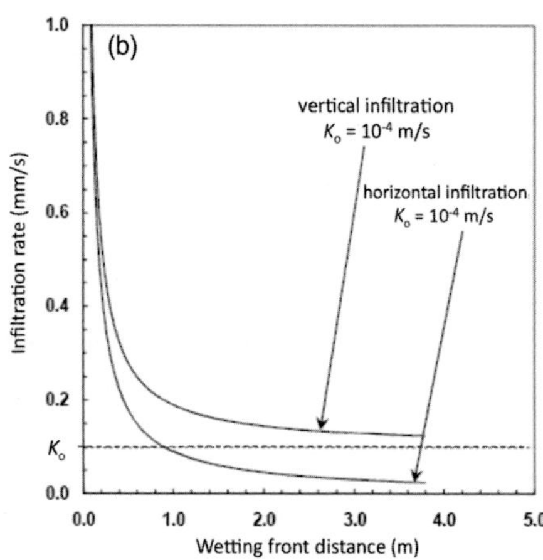

Figure 4.7 (a) Square root of wetting front arrival time and (b) infiltration rate as function of wetting front distance for horizontal and vertical infiltration for $h_o - h_i = 0.9$ m and $\theta_o - \theta_i = 0.4$.

measured or controlled during testing, can also be obtained by imposing Equation (4.25) on Equation (4.32a) to eliminate the variable z, and integrating and imposing the initial condition of $Q = 0$ at $t = 0$, or can be obtained directly from substituting Equation (4.25) into Equation (4.32a):

$$t = \frac{1}{K_o}\left\{Q - (h_o - h_i)(\theta_o - \theta_i)\ln\left[1 + \frac{Q}{(h_o - h_i)(\theta_o - \theta_i)}\right]\right\} \tag{4.35}$$

The general behavior of both horizontal and vertical infiltrations can now be examined by using Equations (4.28), (4.29), (4.32), and (4.34), as depicted in Figure 4.7. For horizontal infiltration, where gravity is absent and flow is driven solely by the gradient in the soil suction head, the wetting front advances linearly with respect to the square root of time, $\sqrt{t}$. The wetting front for vertically downward infiltration, on the other hand, advances non-linearly (Figure 4.7a) with respect to the square root of time. Due to the presence of gravity, vertical infiltration advances faster than horizontal infiltration. As shown in Figure 4.7b, the infiltration rate for both horizontal and vertical infiltration decreases exponentially with respect to the wetting front distance, but asymptotically approaches zero for horizontal infiltration and approaches K_o for vertical infiltration. The asymptotic feature of vertical infiltration can be used to estimate K_o from infiltration or ponding experiments. However, as illustrated in Figure 4.7b, the test could be costly in amount of water and time due to the fact that the infiltration rate decays slowly in many soils.

The Green–Ampt infiltration model was developed for constant-head or ponded infiltration boundary conditions. It has been expanded to constant (Mein and Larson, 1973) and unsteady rainfall flux conditions (Chu, 1978).

Chen and Young (2006) expanded the Green–Ampt constant-head model shown in Equations (4.34) and (4.35) to hillslope conditions by employing Philip's (1991) assumption of no transient moisture variation in the downslope x_* direction (Figure 4.4) and that the gravity head acts along the direction perpendicular to the slope $z_*\cos\gamma$

$$\frac{K_o \cos\gamma}{\theta_o - \theta_i} t = z_* - (h_o - h_i)\ln\left(1 + \frac{z_* \cos\gamma}{h_o - h_i}\right) \tag{4.36}$$

$$t = \frac{1}{K_o \cos\gamma}\left\{Q - (h_o - h_i)(\theta_o - \theta_i)\ln\left[1 + \frac{Q\cos\gamma}{(h_o - h_i)(\theta_o - \theta_i)}\right]\right\} \tag{4.37}$$

The above equations can be used to examine the impact of the slope angle on the wetting front arrival time in hillslope environments.

4.2.3 The Srivastava and Yeh infiltration model

Setting the origin of the z coordinate at the water table and upward positive, the one-dimensional form of the Richards equation (Equation (4.15)) used to described vertical flow of water in the unsaturated zone is

$$\frac{\partial\theta}{\partial t} = \frac{\partial}{\partial z}\left[K(h_m)\frac{\partial(h_m - z)}{\partial z}\right] \tag{4.38}$$

Equation (4.38) is a non-linear partial differential equation, which can be linearized using the exponential model of Gardner (1958) to provide an analytical solution for transient infiltration above a water table (Srivastava and Yeh, 1991; Baum *et al.*, 2008). The Gardner model provides the constitutive relations between suction head and hydraulic conductivity and volumetric moisture content,

$$K(h_m) = K_s \exp(\beta h_m) \tag{4.39a}$$

$$\theta = \theta_r + (\theta_s - \theta_r)\exp(\beta h_m) \tag{4.39b}$$

respectively, where K_s is the saturated hydraulic conductivity, θ_s is the saturated volumetric water content, and θ_r is the residual moisture content. The reciprocal of the vertical height of the capillary fringe β describes the reduction in hydraulic conductivity and water content with increasing suction head. Coarse-grained soils such as sand have relatively large β values compared to finer-grained soils such as silt or clay.

Substitution of (4.39a) and (4.39b) into the one-dimensional form of the Richards equation (4.38) yields a linear partial differential equation in terms of $K(z, t)$

$$\frac{\beta(\theta_s - \theta_r)}{K_s}\frac{\partial K}{\partial t} = \frac{\partial^2 K}{\partial z^2} - \beta\frac{\partial K}{\partial z} \tag{4.40}$$

Introducing the following dimensionless variables for distance, Z, hydraulic conductivity, K^*, flux at the ground surface (Q_A and Q_B), and time, T, Equation (4.40) can be further simplified:

$$Z = \beta z \qquad K^* = \frac{K}{K_s} \qquad Q_A = \frac{q_A}{K_s} \qquad Q_B = \frac{q_B}{K_s} \qquad T = \frac{\beta K_s t}{\theta_s - \theta_r}$$

where q_A is the steady initial flux into the ground surface at time zero and q_B is the flux at times greater than zero. Substituting the above dimensionless variables into Equation (4.40) leads to the expression

$$\frac{\partial^2 K^*}{\partial Z^2} + \frac{\partial K^*}{\partial Z^2} = \frac{\partial K^*}{\partial T} \tag{4.41}$$

For the problem of vertical infiltration from the ground surface towards a water table of fixed depth, the initial and boundary conditions in terms of the dimensionless variables are

$$K^*(Z,0) = Q_A - \left(Q_A - e^{-\beta h_o}\right)e^{-Z} = K_o(Z) \tag{4.42a}$$

$$K^*(0,T) = e^{-\beta h_o} \tag{4.42b}$$

$$\left[\frac{\partial K^*}{\partial Z} + K^*\right]_{Z=L} = Q_B \tag{4.42c}$$

where h_o is the prescribed suction head at the water table, which is typically zero, and L is the depth of the water table below the ground surface. Equation (4.42a) describes the initial steady suction head profile above a water table, Equation (4.42b) describes the time-dependent suction head at the water table, and Equation (4.42c) describes constant flux at the ground surface.

Applying the Laplace transform to Equation (4.41) and considering the initial and boundary conditions above, Srivastava and Yeh (1991) arrived at the following analytical solution in terms of dimensionless hydraulic conductivity K^*:

$$\begin{aligned} K^* &= Q_B - \left(Q_B - e^{-\beta h_o}\right)e^{-Z} \\ &\quad - 4\left(Q_B - Q_A\right)e^{(L-Z)/2}e^{-T/4}\sum_{n=1}^{\infty}\frac{\sin(\Lambda_n Z)\sin(\Lambda_n L)e^{-\lambda_n^2 T}}{1 + L/2 + 2\Lambda_n^2 L} \end{aligned} \tag{4.43a}$$

where the values of Λ_n are the positive roots of the pseudoperiodic characteristic equation

$$\tan(\Lambda L) + 2\Lambda = 0 \tag{4.43b}$$

The flux at the water table ($Z = 0$) for any time is

$$\begin{aligned} Q(T) &= K_s\left[\frac{\partial K^*}{\partial Z} + K^*\right]_{Z=0} \\ &= K_s Q_B - 4K_s\left(Q_B - Q_A\right)e^{L/2}e^{-T/4}\sum_{n=1}^{\infty}\frac{\Lambda_n \sin(\Lambda_n L)e^{-\Lambda_n^2 T}}{1 + L/2 + 2\Lambda_n^2 L} \end{aligned} \tag{4.44}$$

The figures below (4.8–4.10) show suction head and soil moisture profiles for two soils, coarse sand and silt, calculated using the analytical solutions in Equation (4.43). To assist in comparing results for the two soils, the soil moisture profiles are given in terms of effective saturation S_e:

$$S_e = \frac{\theta - \theta_r}{\theta_s - \theta_r} \tag{4.45}$$

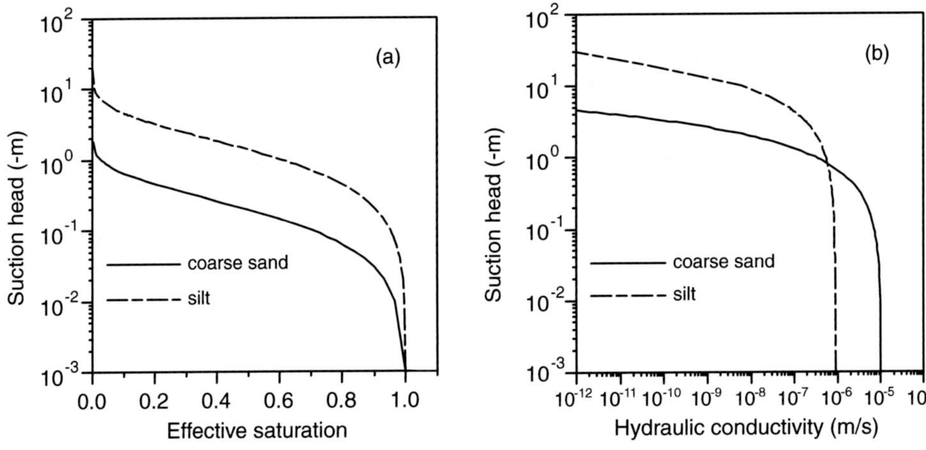

Figure 4.8 (a) Soil water characteristics (b) and hydraulic conductivity functions for the two example soils described with the exponential models (Equation 4.39) of Gardner (1958).

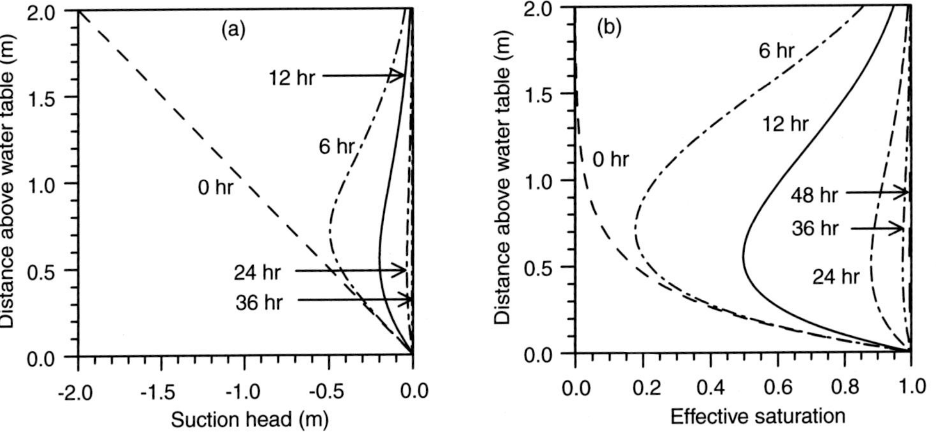

Figure 4.9 Simulated (a) Suction head (b) and effective saturation profiles for homogeneous, 2 m thick coarse sand layer with $\beta = 3.5\ \mathrm{m}^{-1}$.

The soil water characteristics for the two soils are described using Equation (4.39) with $\beta = 3.5\ \mathrm{m}^{-1}$ for the coarse sand and $\beta = 0.5\ \mathrm{m}^{-1}$ for the silt. The saturated volumetric moisture contents θ_s are 0.41 and 0.45, and residual moisture contents θ_r are 0.05 and 0.1 for the coarse sand and silt, respectively. The saturated hydraulic conductivity for the coarse sand is 1×10^{-5} m/s and 9×10^{-7} m/s for the silt.

Figure 4.9 shows simulated (a) suction head and (b) effective saturation profiles for the coarse sand. The model domain is a homogeneous, 2 m thick unsaturated zone above a fixed water table. Initial conditions ($t = 0$) are prescribed as hydrostatic with a steady surface infiltration rate, q_A, of zero. For times greater than zero, an infiltration rate q_B equal to the saturated hydraulic conductivity, $K_s = 1 \times 10^{-5}$ m/s is applied. Note that at early

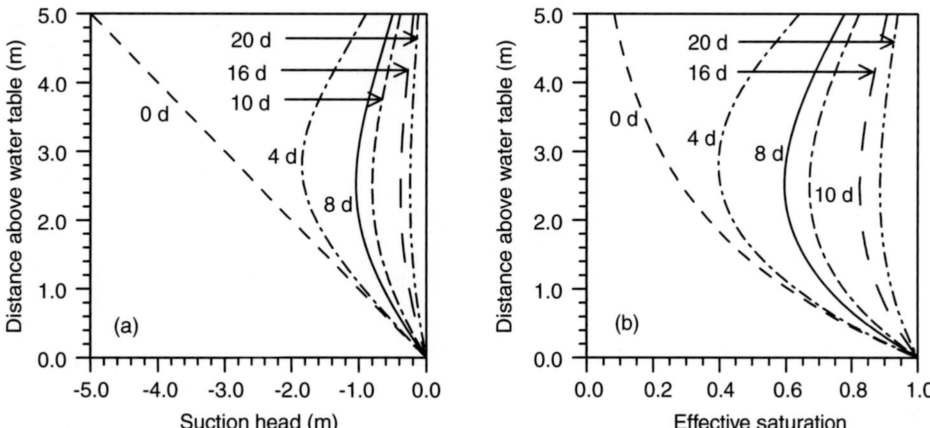

Figure 4.10 Simulated (a) suction head (b) and effective saturation profiles for homogeneous, 5 m thick silt layer with $\beta = 0.5\ \mathrm{m}^{-1}$

times the suction head increases (becomes less negative) near the ground surface. This trend then progresses downward toward the water table as time goes on and the wetting front progresses and infiltration continues. After about 36 hours, the suction head profile approaches a steady condition near to zero. Similar patterns are shown in Figure 4.9b for the simulated effective saturation profile.

Figure 4.10 shows (a) simulated suction head and (b) effective saturation profiles for the silt example. The model domain for this example is a homogeneous, 5 m thick unsaturated zone above a fixed water table. Initial conditions ($t = 0$) are prescribed as hydrostatic with a steady surface infiltration rate, q_A, of zero. For times greater than zero, an infiltration rate q_B equal to the saturated hydraulic conductivity, $K_s = 9 \times 10^{-7}$ m/s is applied. The wetting front for the silt example (Figure 4.10b) is generally more diffuse or spread out over the model domain compared to the results for the coarse sand example (Figure 4.9b). Because the saturated hydraulic conductivity is smaller and the model domain is thicker in the silt example, the time to an approximate steady state near zero suction head takes tens of days rather than the tens of hours for the coarse sand example. Figure 4.11 shows the normalized flow rate Q/K_s at the water table as a function of time in response to rainfall flux applied at the ground surface. The effect of the unsaturated zone results in attenuation and smoothing of the basal flux at the water table at the bottom of the domain. Extension of the analytical solution described in these sections has been implemented in a publicly available computer program that can be used to assess landslide potential over broad geographic areas (Baum *et al.*, 2008; 2010).

4.3 Numerical solutions for multi-dimensional problems

Numerical solutions of the Richards equation (4.21) to simulate flow in the vadose zone have been widely applied to agricultural, water supply, and contaminant transport problems over

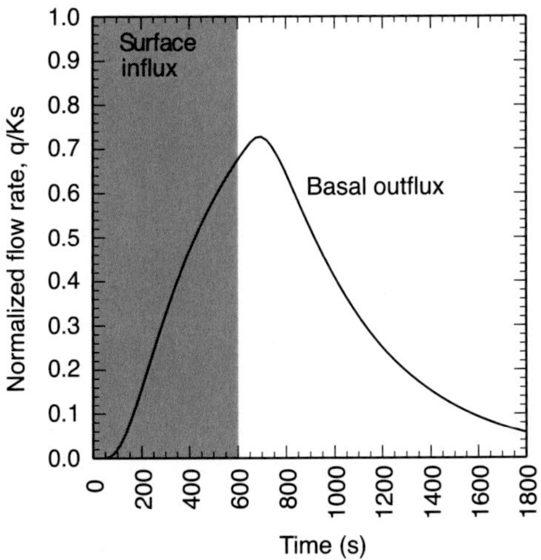

Figure 4.11 Input and output from the analytical solution of Equation (4.44) (from Baum *et al.*, 2008).

the past three decades, beginning with the work of Rubin and Steinhardt (1963) and Freeze (1969). The advantages of numerical models over the analytical solutions such as those described in sub-sections 4.2.2 and 4.2.3 are the ability to simulate water flow in multiple dimensions, incorporate heterogeneity in material properties and initial conditions, accommodate complex boundary geometries, and include a variety of time-varying boundary conditions. For example, simulating the effects of rainfall pattern on the direction and magnitude of water flow in the vadose zone (Figures 4.19 and 4.21) described in subsequent sections is not possible with currently available analytical solutions and thus requires numerical models. The main disadvantage of numerical solutions is the need for greater computational power over what is typically required for solving the analytical equations. This need has been lessened by the availability of increasingly powerful computers, although it remains an issue for multi-dimensional applications over broad regions with fine spatial and temporal discretization.

Numerical models of variably saturated flow have also been combined with slope stability models to examine a wide variety of problems in slope engineering and geomorphology. This coupled approach has been used to assess the conditions that lead to perched water table formation (Reid *et al.*, 1988), examine the effect of pedogenesis and climate change on landslide frequency (Brooks and Richards, 1994; Brooks *et al.*, 1995), quantify the role of antecedent rainfall on landslide occurrence (Rahardjo *et al.*, 2001), assess the impacts of hysteresis of soil water characteristics on slope failure (Ebel *et al.*, 2010), and to evaluate the suitability of vegetation (Collison *et al.*, 1995) and drainage (Cai *et al.*, 1998) for landslide mitigation. Table 4.1 lists several of the freely or commercially available models, among many others, for simulating variably saturated flow.

Table 4.1 Examples of numerical codes for simulation of water flow and other processes in the vadose zone

Model	Dimensions	Licensing	Description	Reference
VS2DI	2	Freely available	Finite-difference solution of the Richards equation, solute and heat transport, graphical user interface	Hsieh *et al.*, 1999; Healy, 2008
HYDRUS 1-D	1	Freely available	Finite-element solution of the Richards equation, graphical user interface, inverse modeling of material properties from observations	Šimůnek *et al.*, 2005, 2008
HYDRUS[1] 2D/3D	3	Commercial	3-D finite-element solution to the Richards equation, hysteresis of soil hydraulic properties, non-linear solute transport	Šimůnek *et al.*, 1999, 2008
TOUGH2	3	Freely available	3-D integrated finite-difference solution to the Richards equation, hysteresis of soil hydraulic properties, non-linear heat transfer	Pruess *et al.*, 2011

[1] Any use of trade on firm names does not imply endorsement by the U.S. Government.

4.4 Transient flow patterns in hillslopes

4.4.1 Controlling factors for flow direction

The temporal and spatial distribution of moisture content in the shallow subsurface, or flow regime, is vital to biological, chemical, and geomorphologic processes. For example, if rainfall-induced unsaturated flow has a significant lateral component parallel to the slope surface under wetting conditions, it could mechanically destabilize the slope leading to surface erosion or landsliding. On the other hand, if the rainfall-induced transient unsaturated flow is in the vertical direction under drying conditions, it could stabilize the slope (Iverson and Reid, 1992; Reid and Iverson, 1992).

In general, the flow regime in a homogeneous and isotropic hillslope is governed by rainfall characteristics, hillslope geometry, and the hydrologic properties of the hillslope materials (e.g., Hewlett and Hibbert, 1963). As discussed in detail in Chapter 3, the major driving mechanisms for moisture movement are gravity and gradients in soil suction head or moisture content (e.g., Sinai *et al.*, 1981; McCord and Stephens, 1987). Flow in a saturated homogeneous and isotropic hillslope under the driving force of gravity and a constant pressure boundary at the slope surface is, although not uniform, always in the laterally downslope direction under both transient and steady-state conditions. (e.g., Tóth, 1963; Freeze and Cherry, 1979; Reid, 1997).

Under variably saturated conditions, both gravity and moisture content gradients (or soil suction head gradients as suction head and moisture content are constitutively related) drive fluid motion leading to complex flow patterns (Sinai *et al.*, 1981; Torres *et al.*, 1998; Silliman *et al.*, 2002; Thorenz *et al.*, 2002). In general, the flow field near the ground surface is variably saturated and transient, and the direction of flow could be laterally downslope, laterally upslope, or vertical. The division between downslope and upslope described in Section 4.2.1 can be defined as either the direction normal to the slope (Figure 4.4a), or the vertical (Figure 4.4b).

The exact conditions under which fluid flow is upslope, vertical, or downslope have been a subject of research over the past three decades. It is well known that layering and heterogeneity can promote strong lateral flow (e.g., Zaslavsky and Sinai; 1981a; Miyazaki, 1988; Ross, 1990; Reid, 1997; Warrick *et al.*, 1997). The highly heterogeneous and sometimes fractured nature of shallow subsurface environments, patterns of rainfall, and interactions with vegetation (e.g., Redding and Devito, 2007) can also induce non-Darcy and non-equilibrium unsaturated water flow (e.g., Swartzendruber, 1963; Hassanizadeh and Gray, 1987; Ritsema *et al.*, 1993; Sinai and Dirksen, 2006). For a homogeneous and isotropic hillslope with an initially uniform moisture distribution, the possible existence of lateral flow has been studied by Zaslavsky and Sinai (1981b) and McCord and Stephens (1987) who found that shallow unsaturated flow moves nearly parallel to the slope surface and converges in topographically concave areas in natural slopes.

For a hillslope of initially constant moisture content that is subsequently subject to a constant but greater moisture content at the slope surface, Philip (1991) arrived at a seminal analytical solution for the transient moisture content field as a function of two-dimensional space (planar) and time. He found that there is a time-independent component of horizontal flow into the slope and a time-dependent downslope (parallel to the slope surface) flow component. However, Philip's (1991) solution does not apply to flow in hillslopes following the cessation of rainfall. Jackson (1992) solved the Richards equation (4.18) numerically to simulate flow both during and following precipitation in a homogeneous and isotropic hillslope. He concluded that the lateral downslope flow is largely a drainage phenomenon driven by the change from flux to no-flow boundary conditions at the surface. He showed that infiltration is nearly vertical during rainfall, but when rainfall ceases unsaturated flow near the surface becomes predominantly parallel to the slope surface.

Because of the difference in defining the flow components in a hillslope compared to those in a flat environment, confusion about the meaning of "downslope" and "upslope" has persisted over the years. Based on the two definitions shown in Figure 4.4, Jackson (1992) noted that Philip's (1991) analytical solution under infiltration conditions can produce downslope flow using the definition in Figure 4.4a, but can never predict downslope flow using the definition in Figure 4.4b because the resultant of gravity (vertical) and gradient of moisture content (inward normal to the slope surface) is in the upslope direction depicted in Figure 4.4b. Philip (1993) could not accept the definition of upslope and downslope defined in Figure 4.4b and was unable to reconcile the contention of no downslope flow under infiltration made by Jackson (1992). In the reply to Philip's (1993) comments, Jackson (1993) correctly pointed out, "This controversy boils down to an issue of semantics." Based

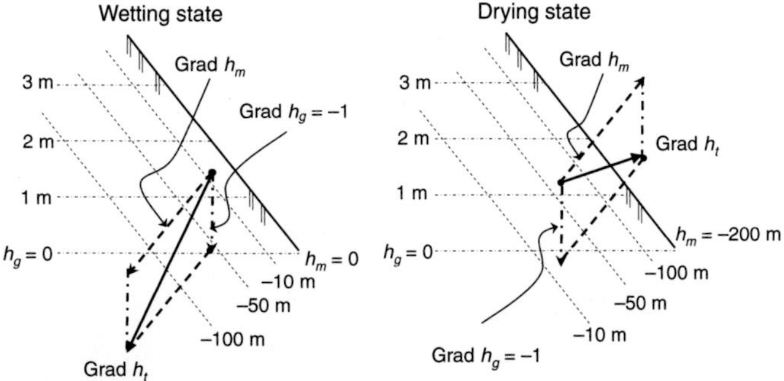

Figure 4.12 Conceptual illustration of gradient of total potential (or flow direction) in regions near a hillslope surface; a sudden wetting after a long drought would promote strong upslope lateral flow whereas rapid drying following a long period of rainfall could promote strong downslope lateral flow.

on the definition by Philip (1991), shown in the shaded area in Figure 4.4a, downward flow in the region between vertical (z) and the slope normal directions (z_*) is called downslope flow. This is impractical and inconvenient in studying hillslope hydrology, as unsaturated flow under such constraints will always reach the water table before ever moving toward the toe of the slope. The definition shown in Figure 4.4b provides a clear way to identify the "drainage" mechanism (i.e., downslope flow upon cessation of rainfall) identified numerically by Jackson (1992) and confirmed experimentally by Sinai and Dirksen (2006). Throughout this section we use upslope and downslope flow relative to the vertical direction as shown in Figure 4.4b. Upslope flow is flow in the direction from vertically downward toward the direction normal to the slope and downslope flow is flow in the direction from vertically downward toward the slope surface.

Sinai and Dirksen (2006) performed laboratory sandbox infiltration experiments and showed that cessation of rainfall is not a general condition for the occurrence of downslope unsaturated lateral flow. Based on their experiments, they concluded, "The necessary condition for downslope lateral flow to occur is not zero-flow at the slope surface, but decreasing rain intensity" (Sinai and Dirksen, 2006, p. 11). In general, the movement of soil moisture in a homogeneous and isotropic hillslope is driven by both gravity and the gradient of moisture content (or matric potential). As illustrated in Figure 4.12, the gradient of total head potential is the resultant of the gradients of soil suction potential and gravity. The gradient of gravity is a constant of -1 anywhere in a hillslope, but the gradient of soil suction potential head is highly dependent on the state of wetting or drying and its absolute value could be very high, easily exceeding several hundred (e.g., suction change of 500 kPa crossing the top 1 m near the ground surface).

Because moisture redistribution in a hillslope is a transient process governed by Equation (4.21), the flow direction may not be directly and concurrently correlated to changes in the boundary conditions. While all previous work regarding unsaturated lateral flow defines flow patterns as a function of boundary conditions, i.e., constant rainfall, increasing rainfall intensity, decreasing rainfall intensity, or cessation of rainfall (e.g. Jackson, 1992; Sinai and Dirksen, 2006), more general conditions that clearly delineate unsaturated lateral flow can be established by identifying appropriate hydrologic conditions within a hillslope.

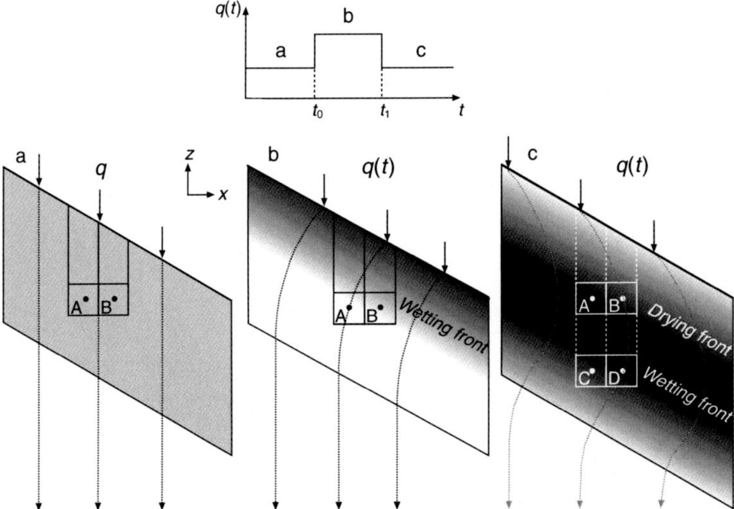

Figure 4.13 Conceptual illustration of flow regimes in a hillslope: (a) steady state, (b) wetting state, and (c) drying and wetting state. Dashed lines with arrows show the hypothetical path of a particle in the flow field. The water table is assumed to be far below the slope surface such that it has no effect on flow direction near the slope surface (from Lu *et al.*, 2011).

The states of wetting, steady moisture conditions, and drying can be uniquely determined by the moisture content, θ, at a point with respect to time, t. Specifically:

$$\begin{aligned} \text{wetting:} &\quad \frac{\partial \theta}{\partial t} > 0 \\ \text{neutral:} &\quad \frac{\partial \theta}{\partial t} = 0 \\ \text{drying:} &\quad \frac{\partial \theta}{\partial t} < 0 \end{aligned}$$

A practical implication is that the dynamic lateral flow direction in a hillslope can be inferred from time-series data of moisture content collected in the field. This criterion also becomes instructional in any numerical modeling and experimental testing of hillslope hydrology.

Lu *et al.* (2011) present the hypothesis that at any point within a homogeneous and isotropic hillslope, downslope lateral unsaturated flow will occur if that point is in the state of drying (Figure 4.13c) while upslope lateral unsaturated flow will occur if that point is in the state of wetting (Figure 4.13b). This hypothesis provides a quantitative criterion in the time domain. Mathematically the hypothesis postulates:

$$\begin{aligned} \text{upslope lateral flow:} &\quad \frac{\partial \theta}{\partial t} > 0 \quad \text{or wetting} \\ \text{vertical downward flow:} &\quad \frac{\partial \theta}{\partial t} = 0 \quad \text{or neutral} \\ \text{downslope lateral flow:} &\quad \frac{\partial \theta}{\partial t} < 0 \quad \text{or drying} \end{aligned}$$

Thus, together with the spatial distribution of the state of wetting or drying, flow regimes in a hillslope can be completely defined.

In the following section, the above hypothesis is examined by using a well-calibrated two-dimensional numerical model. The extensive experimental data collected by Sinai and Dirksen (2006) under increasing and decreasing rainfall intensity are used to calibrate the model, reconfirm their conditions necessary for unsaturated lateral flow, and test the above hypothesis on flow regimes. The numerical model is then used to quantify a more complex rainfall scenario qualitatively studied by Sinai and Dirksen (2006) to further validate the general conceptual model.

4.4.2 General conceptual model for "wetting" and "drying" states

All previous conceptualizations of flow regimes in hillslopes consider drying or wetting at the slope surface and overlook a simple fact that wetting and drying is a dynamic process, such that regions of wetting and drying can concurrently occur and change even when the boundary condition can be defined by a single state of wetting or drying. Strictly speaking, the conditions identified in previous work for upslope lateral flow under wetting and downslope lateral flow under decreasing rainfall intensity (e.g., Jackson, 1992) are only correct in the region immediately below the slope surface. Regions away from the slope surface could be in any state, i.e., wetting, neutral, and drying, that is governed by the unsaturated flow process under the competing driving mechanisms of gravity and the gradient in soil suction head or moisture content. This leads to the concept of time-and-space regimes of flow in hillslope environments and a conceptual model that, at any point within a homogeneous and isotropic hillslope, downslope lateral unsaturated flow will occur at a point if that point is in the state of drying, and upslope lateral unsaturated flow will occur at the point if that point is in the state of wetting. The physical reasoning for such a model is illustrated in Figure 4.13 and is described in the following paragraph.

Consider two adjacent points A and B in a homogeneous and isotropic hillslope with the same elevation near the surface of the hillslope (see Figure 4.13). If the two points reach a steady state (i.e., the moisture content becomes invariant with time) and assuming the rainfall intensity is less than the saturated hydraulic conductivity of the slope materials, then the water flux at these two points should be equal to the constant vertical infiltration rate at the surface, leading to an absence of a gradient in soil suction potential or moisture content between the two points. Thus, the resulting unsaturated flow is predominantly vertical due to gravity, as shown in Figure 4.13a. If the rainfall intensity increases to another value (still less than the saturated hydraulic conductivity) at some time t_o, the soil at the surface becomes wetter leading to an additional gradient of soil suction in the direction normal to and inward from the slope surface. This will result immediately in upslope lateral flow in the soil adjacent to the surface but not in the region of points A and B. Flow at points A and B will remain vertical under gravity at time t_o. With time, the region of upslope lateral flow will propagate into the slope directly normal and inward from the surface as shown in Figure 4.13b. The additional wetting front will arrive earlier at point B than at point A as the slope-normal distance from the surface to point B is shorter than that to point A, yielding a gradient in soil suction potential between points A and B. When this happens, upslope lateral flow occurs. Using the same logic, if the rainfall intensity decreases or the rainfall ceases, the opposite will occur, as point B will drain earlier than point A, leading to downslope lateral flow as shown in Figure 4.12c. At the same time, on the slope-parallel horizon between points C and D shown in Figure 4.13c, the wetting front has just arrived, leading to lateral upslope flow by the same mechanism shown in Figure 4.13b under the wetting state. Because rainfall intensity varies and moisture movement in a hillslope is a dynamic process, definition of upslope or downslope lateral flow is better accomplished at the local scale rather than by the simple hydrologic conditions at the slope surface.

4.4.3 Flow patterns under constant rainfall intensity

To test the conceptual model that the state of wetting and drying of any point in homogeneous and isotropic hillslopes completely defines the upslope or downslope lateral flow at the point described in the previous section, Lu *et al.* (2011) used HYDRUS-2D, a finite-element-based software that solves the Richards equation for flow in variably saturated porous media (Šimůnek *et al.*, 1999). The numerical model was calibrated using results from a series of sandbox experiments conducted by Sinai and Dirksen (2006). The Richards equation (4.21), in terms of volumetric moisture content θ, can be expressed as (e.g., Lu and Likos, 2004a)

$$\frac{\partial}{\partial x_j}\left(\frac{K(\theta)}{C(\theta)}\frac{\partial \theta}{\partial x_j} + K(\theta)\delta_{3j}\right) = \frac{\partial \theta}{\partial t} \tag{4.21}$$

where x_j are coordinate directions with $j = 1$ and 2 being horizontal and $j = 3$ being vertical, h_m is matric suction head, $h_m(\theta)$ is the soil water retention curve, $K(\theta)$ is the hydraulic conductivity function, $C(\theta)$ is the specific moisture capacity function, and δ_{3j} is the Kronecker delta with non-zero value when $j = 3$. The first term in the bracket of the left hand side of Equation (4.21) represents the flux driven by the gradient of moisture content and the second term represents the flux driven by the gravity.

Sinai and Dirksen (2006) performed a series of laboratory sandbox experiments to capture unsaturated lateral flow in homogeneous isotropic sloping soils. In their experiments V-shaped trenches were carved into originally horizontal sand surfaces to represent hillslopes. The dimensions of the sandbox were 0.81 m in height, 1.12 m in length and 0.05 m in width (Figure 4.14). Special effort was taken to pack the sandbox with fine sand (i.e., particle sizes $<$ 0.25 mm) to create as homogeneous conditions as possible. They controlled the infiltration rate using a rainfall simulator composed of one hundred hypodermic needles on the top of the sandbox. To prevent water accumulation at the bottom of the sandbox and in order to conduct series of experiments without repacking the sand, ceramic porous filter tubes were placed on the bottom of the sandbox and a constant suction head of 2 m was maintained. To analyze flow regimes visually, food dye was used as a tracer. The dye was injected into the soil pack using hypodermic needles through small holes drilled through the sides of the sandbox. The locations of injection points for the experiment are roughly shown in Figure 4.14a and those used in the current modeling analyses are clearly shown in Figure 4.14b.

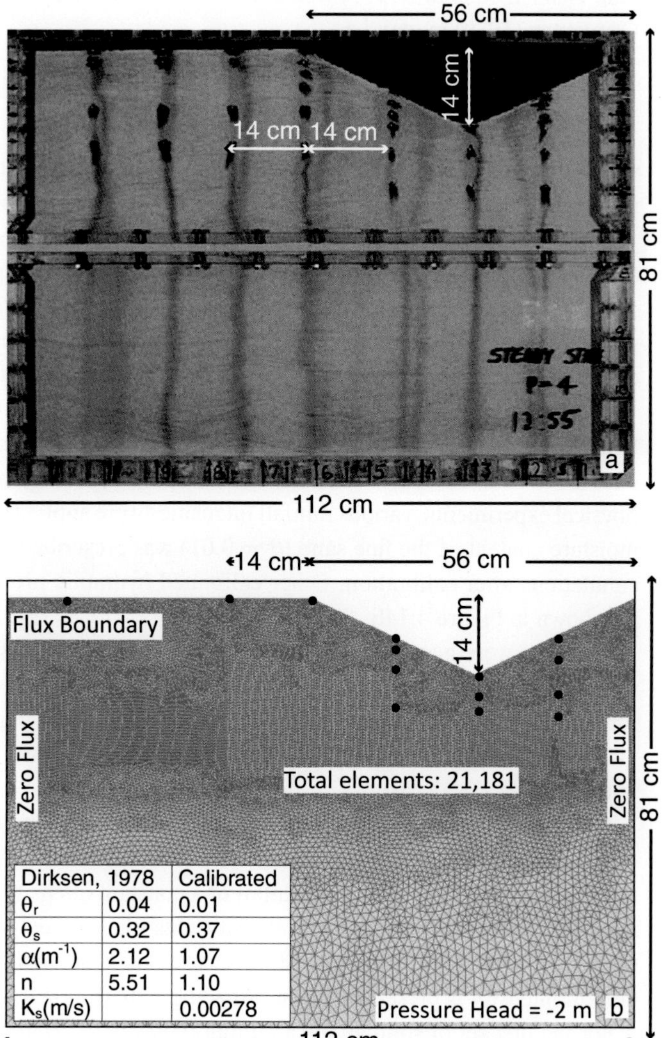

Figure 4.14 (a) Experimental setup of Sinai and Dirksen (2006), and (b) the simulation domain, boundary conditions, locations of dye (particle) injection used in numerical modeling (black dots), and hydrologic properties with the original sand reported by Dirksen (1978) and with the calibrated parameters (from Lu *et al.*, 2011).

The dimensions of the model domain shown in Figure 4.14b used by Lu *et al.* (2011) were the same as that of the sandbox used in Sinai and Dirksen's (2006) experiments. The geometry was a single 56 cm wide, 14 cm deep V-trench (26.6° slope) with a flat surface at the left of the sandbox as a reference. The model domain contained 21,181 elements with finer mesh at the top and coarsening towards the bottom boundary to increase computational accuracy and to focus on the upper part of the domain near the sloping surface (Figure 4.14b). The SWCC and HCF reported by Dirksen (1978) for another similar set of sandbox experiments were used for material hydrologic properties. The RETC code (van Genuchten *et al.*, 1991) was used to calculate the hydrologic parameters for the van Genuchten (1980) model, i.e.,

$$\frac{\theta - \theta_r}{\theta_s - \theta_r} = \left[\frac{1}{1 + (\alpha \gamma_w h)^n} \right]^{1-\frac{1}{n}} \tag{4.46}$$

where θ_s is the porosity, γ_w is the unit weight of water, θ_r is the residual water content, α is the inverse of the air-entry pressure for soil, and n is the pore-size parameter.

Some adjustments were made to the hydrologic parameters to reproduce the timing of the wetting front movement and the path lines in the experiment. The original (Dirksen, 1978) and the final calibrated hydrologic parameters used in the numerical model are also shown in Figure 4.14b. The effect of hysteresis was not accounted for in the analysis, as it does not appear to have a dominant effect on the direction of flow (Scott *et al.*, 1983).

A flux boundary condition was prescribed at the top of the model domain, no flow boundaries on the sides, and constant pressure head boundary of −2 m at the bottom consistent with that used in the sandbox experiments (Figure 4.14b). As in the original physical experiments, various rainfall intensities were applied to the upper surface. Residual moisture content of the fine sand ($\theta = 0.01$) was prescribed as the initial condition for all simulations after calibration. Other calibrated hydraulic properties used in the modeling are shown in Figure 4.14b.

Pathlines were generated during the simulations to visually compare the traces of dyes in the experiments with the numerical results. To capture the trends of the pathlines formed by the dye, the starting locations of the water "particles" tracked by the simulation were selected as close as possible to the reported locations and times of dye injection.

Four rainfall conditions were examined using the numerical simulations and compared with the experimental results quantitatively and qualitatively: (1) initial change to a constant rainfall intensity, (2) prolonged rainfall of a constant intensity, (3) decrease to zero rainfall intensity, and (4) varying rainfall intensity. Quantitative comparisons are presented in Sub-sections 4.4.4 and 4.4.5 for the first three cases and a qualitative comparison under varying rainfall intensity is presented in the following Sub-section 4.4.6.

For the first set of simulations a rainfall intensity of 50 mm/hr was applied to the upper boundary and the movement of the wetting front and water particles as function of time was simulated. In the original experiments, both pathlines and streak lines were used to track the flow field. A pathline is the trajectory of dye particles carried by the flow from an initial injection point, whereas a streak line is the trajectory of a pulse of dye particles injected at a given point in the flow domain during the experiment. Pathlines were used to visualize fluid motion from the top boundary and streak lines were used to visualize flow direction over a shorter period of time near the point of interest. Because no information was available on the release times for the dye injected for the streak lines, only pathlines were used in the numerical simulations for comparisons and for visualization. The traced particles were released at the beginning of the simulation from the top surface. At early times, less than 0.5 hour after a 50 mm/hr rainfall was introduced, the wetting front was approximately parallel to the slope surface (Figure 4.15a). As the wetting front proceeded, the effect

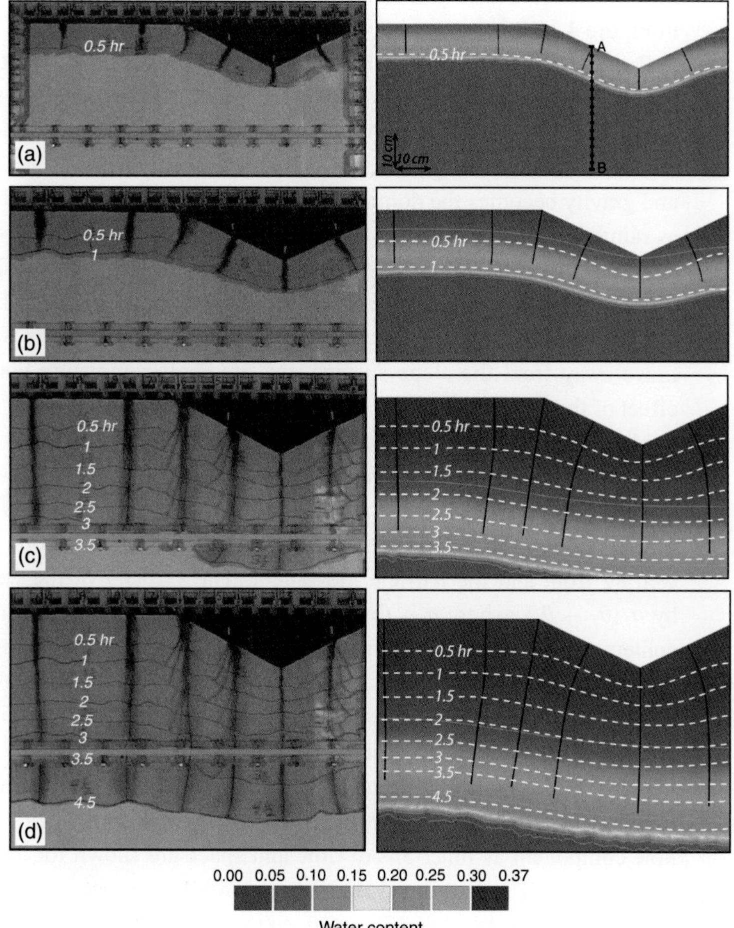

Figure 4.15 Comparison of flow field subject to a constant rainfall intensity of 50 mm/hr between experimental results (left column) and numerical simulations (right column) at (a) 30 minutes after rainfall, (b) 1 hour after rainfall, (c) 3.5 hours after rainfall, and (d) 4.5 hours after rainfall. Black lines in experimental results (left column) are streak lines. Black lines in numerical simulations (right column) are pathlines. Contour lines are wetting front locations at labeled times. See text for explanation of the differences between streak lines and pathlines. The saturated moisture content, θ_s, for the simulations was 0.37 (from Lu *et al.*, 2011). See also color plate section.

of the slope gradually diminished (Figure 4.15b, c), and the wetting front approached horizontal (Figure 4.15d). Sinai and Dirksen (2006) determined the position of the wetting front visually and traced its location on the glass wall at 0.5-hour intervals. The simulated movement of the wetting fronts accord well with the experimental observations. Note that the comparisons of pathlines in the later times (Figure 4.15c, d) do not appear to be as good as those in the earlier times (Figure 4.15a, b). This is attributed to the use of pulse injections of dye for the streak lines, which would tend to smear and alter the pathlines (see the difference in pathlines near the slope surface in the middle of the slope shown in Figure 4.15d).

Pathlines near the wetting fronts are almost perpendicular to them at all times in both experiments and simulations, and as the wetting fronts proceed they become nearly vertical through time. This can be explained with the conceptual model presented earlier in Figure 4.13, where the gradient in moisture content is diminished as infiltration progresses and gravity becomes the dominant driving mechanism for water flow. During earlier times, as rainfall infiltrates through the slope, the moisture content of the soil near point B in the downslope region increases before the moisture content in the upslope region near point A. Therefore moisture content will be lower in the upslope area such that water moves in the upslope direction under the gradient in moisture content. As the wetting front and particles move away from the slope surface, the gradient in moisture content is reduced and the effect of the slope diminishes. Under this condition, gravitational force dominates over the gradient in moisture content, and the particles tend to move vertically.

As can be seen from Figure 4.15, the particle movement lags behind the wetting front and this lag becomes more obvious in time. This phenomenon is pertinent to unsaturated flow. The reason for this lag is the difference in the calculation of the velocities for the wetting front and the particle movement. The velocity of the wetting front is calculated by $q/(\theta_f - \theta_i)$, where q is the Darcian flux, and θ_i and θ_f are the initial and final water contents, respectively, meaning that the water partly or completely replaces the initial air phase. However, the velocity of the particle is calculated by q/θ_f, which means that the particle moves in all of the water in the soil regardless of whether it is initial or added water (J. Šimůnek, personal communication, 2009).

A more quantitative illustration of upslope flow during the wetting state can be seen in Figures 4.16a, b, where the deviation of the flow direction from vertical and the horizontal flow component as functions of time and space are shown for the profile in the middle of the slope (A–B in Figure 4.15a). When rainfall commences, the flow direction is normal to the slope surface, about 26.6° on average from vertical (Figure 4.16a) and the upslope component of flux (negative in this coordinate system) is the strongest (Figure 4.16b). In time, the upslope deviations from vertical diminish (Figure 4.16a). For example, after 10 minutes of rainfall, the flow in the upper 5 cm deviates upslope from vertical by between 23° and 32°. At 1 hour, the deviation from vertical of the flow directions in the upper 15 cm have all decreased to around 15°. At 4.5 hours when the sandbox experiment was terminated (Sinai and Dirksen, 2006), the deviations from vertical in the upper 48 cm have all decreased to around 6°. Initially, the upslope flux (or the magnitude of horizontal flux) at the slope surface is at its highest and diminishes over time, whereas the vertical flux at the slope surface increases slightly in time and approaches the infiltration rate. The region of change in both horizontal and vertical fluxes, as well as changes in total flux direction, coincide in time with the downward propagation of the wetting fronts as shown in Figures 4.16a–d. After 7 days a steady state (vertical flux equal to the infiltration rate, and zero horizontal component) in the numerical simulations was reached. Three distinct features can be observed. First, compared to the arrival times of the wetting fronts shown in Figure 4.16 flow only changes direction in the domain behind the wetting fronts. Second, at a given time the angle of the upslope flow direction remains relatively constant with depth except near the wetting front, where it is drastically reduced to zero (i.e., no change in flow

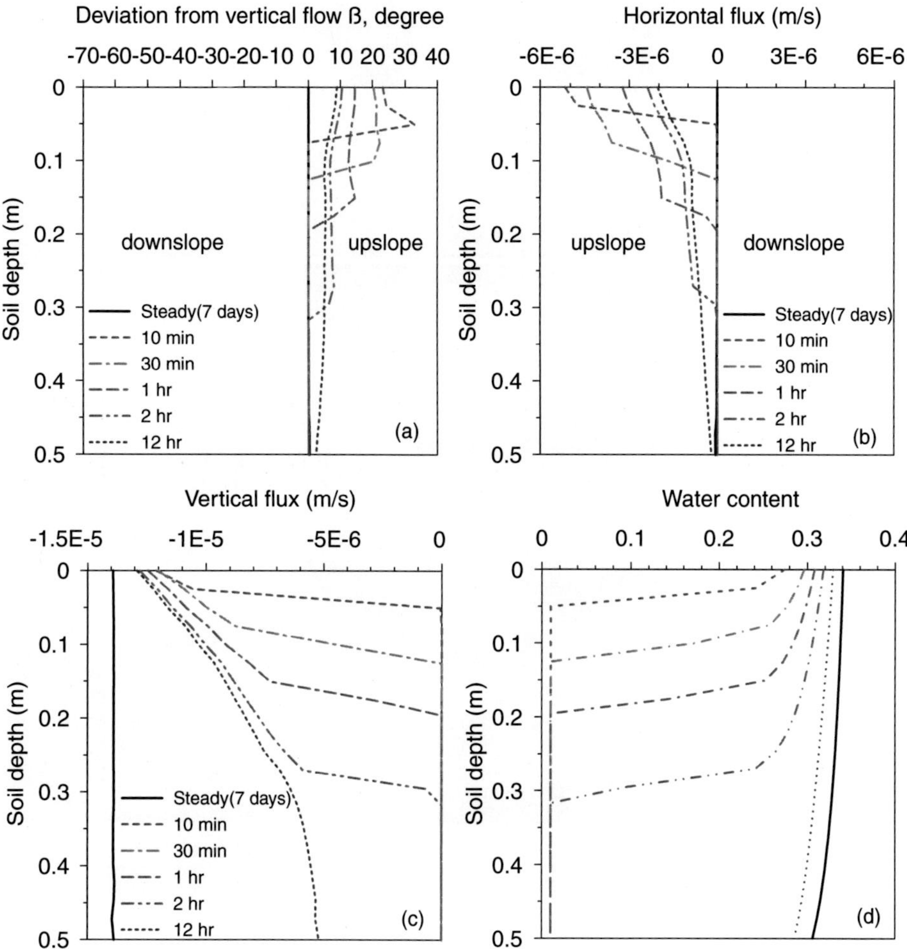

Figure 4.16 Profiles of simulated quantities in the middle of the slope (14 cm horizontally from the toe of the slope shown as A–B cross section in Figure 4.15a) at different times under constant rainfall intensity: (a) deviation of flux from the vertical direction, (b) horizontal flux, (c) vertical flux, and (d) moisture content (from Lu *et al.*, 2011).

direction from vertical). Finally, the magnitude of the upslope flow direction is related to the moisture content. These three features indicate that it is the gradient of moisture content at the point of interest that controls the flow direction at that point. This is most pronounced and best illustrated at the moving wetting front, where the gradient of moisture content (shown in Figure 4.16d) is the highest.

4.4.4 Flow patterns following the cessation of rainfall

For the third set of simulations the rainfall was stopped after an initial steady-state condition was established by applying continuous rainfall of 100 mm/hr for 7 days. The movement of particles through time was then observed.

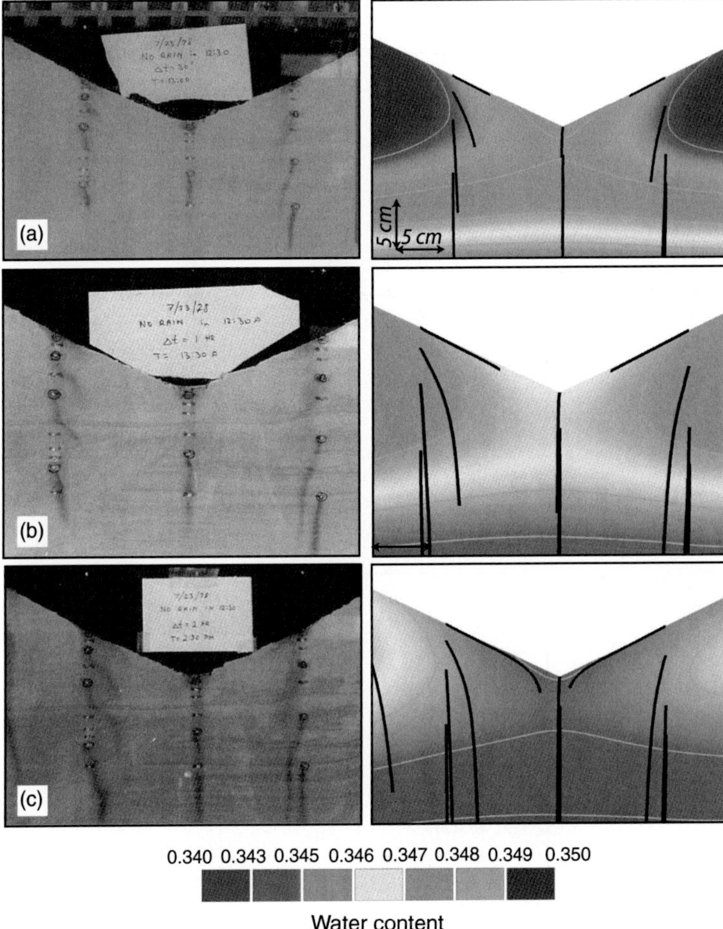

Figure 4.17 Comparison of flow fields after cessation of steady rainfall intensity of 100 mm/hr between experimental results (left column) and numerical simulations (right column) at (a) 30 minutes after rainfall ended, (b) 1 hour after rainfall ended, and (c) 2 hours after rainfall ended (from Lu *et al.*, 2011). See also color plate section.

After the rainfall ends the pathlines simulated numerically change direction indicating downslope lateral flow (shown on the right column in Figure 4.17) and compare favorably with the pathlines from the sandbox experiments (shown on the left column in Figure 4.17). Near the slope surface the pathlines are almost parallel to the sloping surface. This can be observed in both experiments and simulations. For the particles released further away from the slope surface the pathlines become more vertical. Because the effect of hydrodynamic dispersion is evident in the experimental results, pathlines are difficult to quantify, making precise one-to-one comparison between the experiments and simulations impossible. However, the phenomena of downslope flow and the overall flow patterns can be confidently observed in both the experiments and simulations. The conceptual model illustrated in Figure 4.13 can be used to understand such a phenomenon. Drainage in the downslope region near point B will be faster than that in the upslope region near point A due to the smaller volume of soil above point *B* than that above point A. This difference in the drainage rate leads to a downslope lateral gradient in moisture content and downslope lateral flow when the drying front arrives. As a particle moves away from the slope, the volume difference between that above point A and point B becomes less significant and the lateral gradient in moisture content diminishes, leading to vertical water flow (Figure 4.17c).

The drying front is not as visually prominent as the wetting front in the sandbox experiments but can be clearly identified in the simulations by plotting contours of water content shown on the right column in Figure 4.17 (see color plate section). It is also clear from the moisture content contours, the model scale, model materials (sand), and duration of infiltration, that boundary effects strongly influence results in the area below the bottom of the valley.

Further quantitative understanding of downslope lateral flow during the drying state can be gained by examining the total flux direction, the horizontal and vertical flow components, and moisture content as functions of time and space along a vertical profile located in the middle of the slope shown as A–B in Figure 4.15a. Semi-quantitative comparison with the sandbox experiments is also possible as Sinai and Dirksen (2006) recorded the ultimate change in the direction of the infiltration vector with soil depth during the drying experiments, as shown in Figure 4.18a. As in the wetting case, although the downslope horizontal flux is strongest at the slope surface and diminishes with time, the vertical flux is strongest at the slope surface and increases with time. The largest change in flow direction occurs at the slope surface and remains at about 62° down from vertical, indicating that the lateral downslope flow near the surface is nearly parallel to the slope surface (28°). The horizontal flow component (downslope is positive in the coordinate system shown in Figure 4.13) is the strongest at the cessation of rainfall and diminishes as time elapses. For example, 10 minutes after the rainfall stops, the deviation of the flow direction from vertical in the upper 10 cm varies from 0° at 10 cm depth to ~60° at the slope surface (Figure 4.18a). Over time, the zone of downslope horizontal flow progresses deeper into the slope (Figure 4.18b). From the simulation, we can see that the drying process is much slower than that of wetting and that the sandbox experiment was terminated long before a new steady state was reached, as shown in Figure 4.18c, d. Even 4 months after the cessation of rainfall, water contents (Figures 4.18d) are well above the residual moisture content, θ_r, of 0.02.

Three distinct features of the drying state, in contrast to the wetting state, can be identified. First, the downslope lateral flow direction is always most pronounced at the slope surface, invariant with time, and approximately parallel to the slope surface (about 62° downslope from vertical in this case), whereas in the wetting state, the lateral flow direction is initially constant behind the wetting front but decreases with time (Figure 4.16a). The occurrence of the maximum deviation of the flow direction from vertical at the slope surface and the diminishing deviation away from the slope surface compares favorably with the sandbox experiment results, as shown in Figure 4.18a. Second, the downward progress of the drying front is much slower than that of wetting front. For the slope with the same fine sand, 7 days is sufficient to reach a steady state under constant rainfall, whereas a strong drying front does not occur after the rainfall ceases. This is probably due to the fact that the gradient

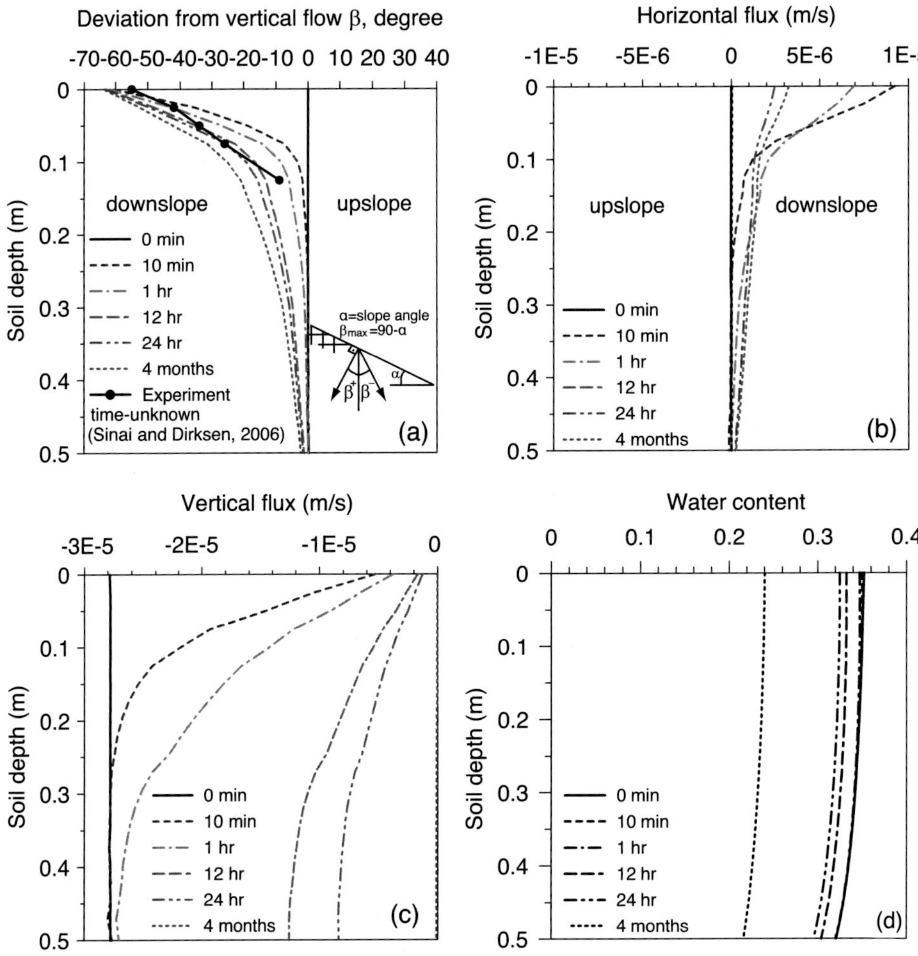

Figure 4.18 Profiles of simulated quantities in the middle of the slope (14 cm from the toe of the slope shown as A–B cross section in Figure 4.15a) at different times after rainfall ended: (a) deviation of flux from vertical direction, (b) horizontal flux, (c) vertical flux, and (d) water content (from Lu *et al.*, 2011).

of moisture content during drying generally acts against gravity and hydraulic conductivity decreases as drying occurs. As illustrated in Figures 4.18a and 4.18b, 4 months after the rainfall stops, the direction of flow becomes more downslope, although the magnitude of the flux is small. Finally, the angle of the flow direction is related to the gradient in moisture content, which is always at its maximum at the slope surface. These distinct features again strongly indicate that it is the gradient of moisture content at the point of interest that controls the flow direction at that point.

4.4.5 Flow patterns resulting from a step-function in rainfall

Sinai and Dirksen (2006) reported only qualitative results on four sandbox experiments in which they varied the rainfall intensity. The four experiments involved (1) periodic alternation of the rainfall rate from 100 mm/hr to 0 mm/hr every 15 minutes, (2) periodic alternation of the rainfall rate from 100 mm/hr to 0 mm/hr every 30 minutes, (3) periodic alternation of the rainfall rate from 150 mm/hr to 50 mm/hr with no specified period, and (4) fluctuations around a rainfall rate of 100 mm/hr with maximum and minimum intensities of 250 mm/hr and 0 mm/h, respectively. From these experiments they found that upslope lateral flow always occurred after rainfall rate increases and downslope lateral flow always occurred when the rainfall rate decreased. The latter finding supersedes Jackson's (1992) previous zero-rainfall condition for downslope lateral flow. Based on a series of experiments, Sinai and Dirksen (2006) constructed a hypothetical 6-hour variable rainfall rate experiment and qualitatively predicted the expected temporal changes in the direction of unsaturated lateral flow. Lu *et al.* (2011) used a numerical model of a field-scale sandy hillslope subject to two different rainfall conditions; constant rainfall of 40 mm/hr for 1 hour, and the same 6-hour varying rainfall intensity used by Sinai and Dirksen (2006) to examine the effects of these boundary conditions on flow patterns.

Results from these simulations illustrate quantitatively that the general conditions determining upslope, vertical, or downslope flow are not the rainfall conditions at the slope surface, but rather the state of wetting or drying within the hillslope. The conceptual model that the state of wetting or drying is the sole controlling factor determining lateral flow direction can be quantitatively confirmed by comparing time series of horizontal flux and moisture content variation. The changes of flow direction completely coincide with the changes in moisture content.

The simulation domain, boundary conditions, and material hydrologic properties are shown in Figure 4.19a. One hour of constant rainfall of 40 mm/hr is imposed on the slope surface with initially unsaturated hydrostatic condition throughout the entire hillslope domain. The evolution of flow patterns in this sandy hillslope is illustrated in the combined water content contour and flow direction vector plots for the upper-middle region of the hillslope as shown in Figures 4.19b–f at different times after the rainfall ceases.

Qualitatively, the behavior of the flow regimes is consistent with the conceptual model (Figure 4.13), i.e., the wetting or drying state at a particular point in the hillslope controls the flow direction (Figure 4.19). When rainfall begins, upslope flow is most pronounced near the slope surface (see lateral flow direction at 10 minutes in Figure 4.19b). The upslope lateral flow propagates into the hillslope with the wetting front (Figures 4.19c–e). Behind the wetting front a quasi-steady zone develops (Figures 4.18c–e). This zone is characterized by dominantly vertical flow. Upon the cessation of rainfall, strong downslope lateral flow develops near the slope surface, as shown in Figure 4.18d. The downslope flow direction near the slope surface persists after the rainfall ceases (Figures 4.19d, e). Based on these observations Lu *et al.* (2011) concluded that both upslope lateral flow and vertical flow can occur during rainfall and all three flow regimes – upslope lateral flow, vertical flow, and downslope lateral flow – can occur after the rainfall ends (Figure 4.19f).

The dynamic variation of the lateral flow directions and moisture content at different depths are quantitatively illustrated in Figure 4.20a, b and profiles of lateral flow direction and flow magnitude in the middle of the hillslope are shown in Figures 4.20c, d. For given points in the hillslope at three different vertical depths (0.1, 0.5, and 1.0 m) from the slope surface, upslope lateral flow occurs when the water content at those

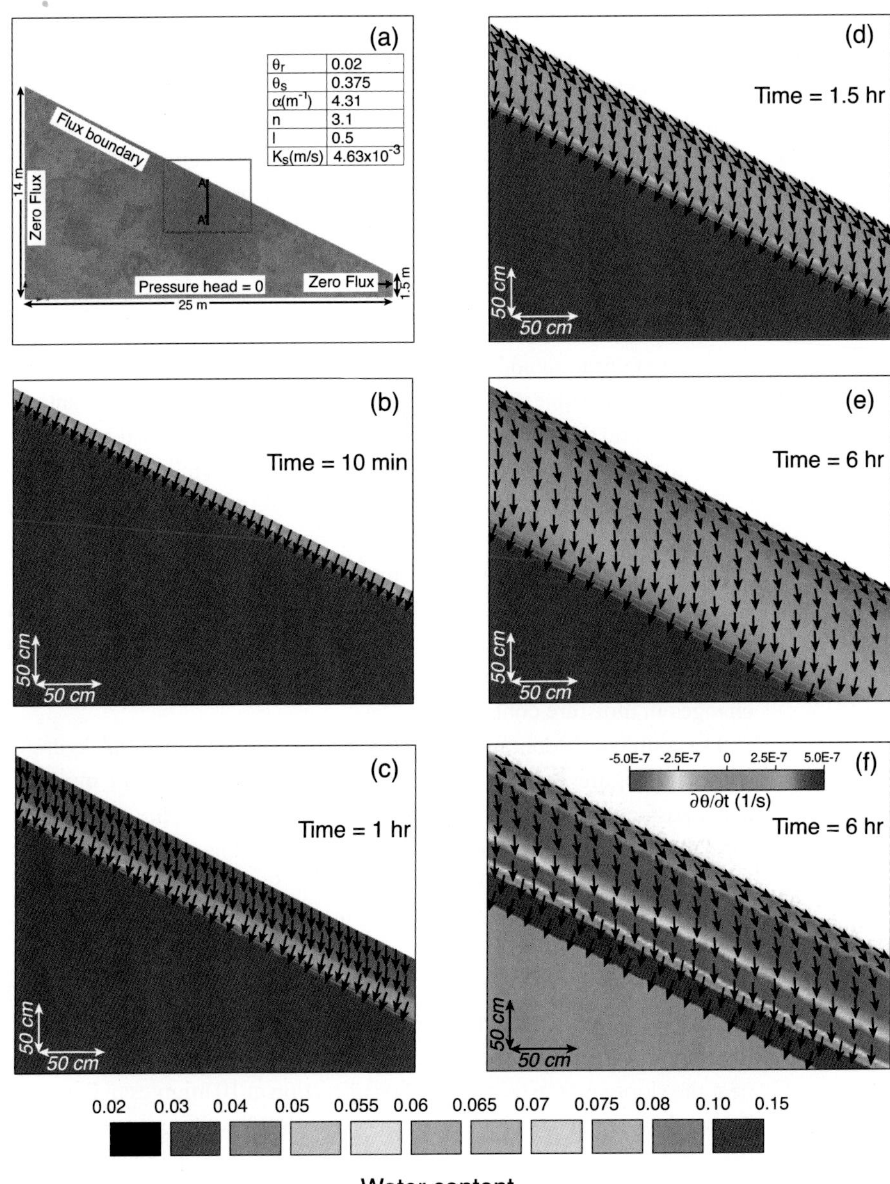

Figure 4.19 Simulation results from a 40 mm/hr step-increase-then-cease rainfall intensity boundary condition: (a) simulated domain, (b) velocity direction and moisture content at 10 minutes, (c) at 1 hour, (d) at 1.5 hours, (e) at 6 hours and (f) velocity direction and $\partial\theta/\partial t$ at 6 hours (from Lu *et al.*, 2011). See also color plate section.

points increases, nearly vertical flow occurs when water content is quasi-steady, and downslope lateral flow occurs when water content decreases. The onset of upslope lateral flow occurs with the change to a wetting state at the depths of 0.1, 0.5, and 1.0 m in Figure 4.20b. At the depth of 0.1 m the onset time is 3 minutes, at the depth of 0.5 m

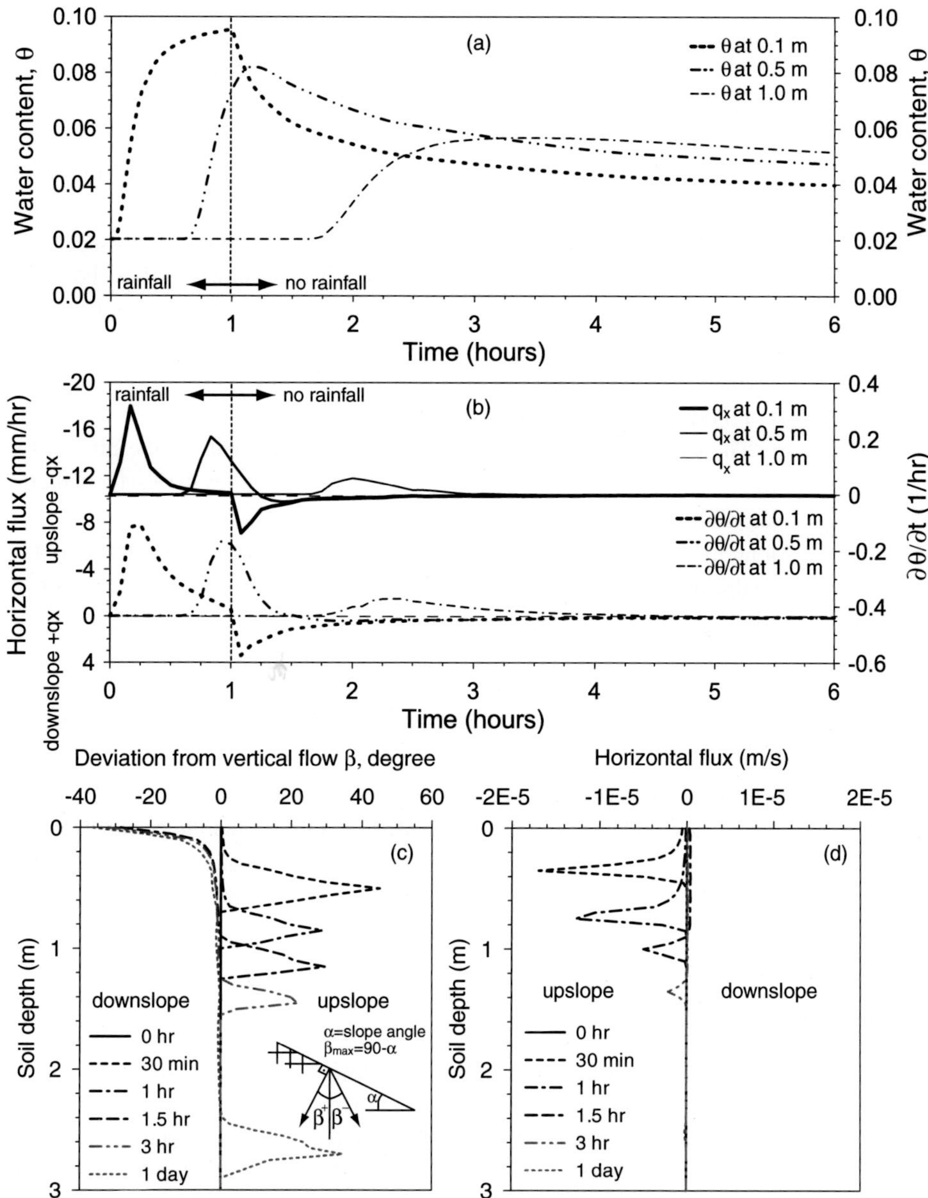

Figure 4.20 Simulation results from a step-increase-then-cease rainfall intensity boundary condition: (a) moisture content as a function of time at different depths, (b) horizontal flux and rate of moisture content change as a function of time at different depths, (c) profiles of deviation of flow direction from vertical at the middle of the slope at different times, and (d) profiles of horizontal flux at the middle of the slope at different times (from Lu *et al.*, 2011).

the onset time is about 40 minutes, and at the depth of 1.0 m the onset time is about 1 hour and 45 minutes. A zone of nearly vertical flow behind the wetting front develops as it reaches a steady state (e.g., depth between 0.0 and 0.6 m at $t = 1$ hour in Figure 4.20c). Upon the cessation of rainfall the onset of downslope lateral flow coincides with the change to a drying state as shown in Figure 4.20b for the 0.1, 0.5, and 1.0 m depths. At the depth of 0.1 m the onset of downslope flow is almost immediate, at the depth of 0.5 m the onset of downslope flow occurs after about 1.5 hours, and at the depth of 1.0 m the onset of downslope flow is at about 4.5 hours. While strong downslope lateral flow develops near the slope surface and advances downward as shown in Figure 4.19d–f, concurrently strong upslope flux continues near the wetting front (Figure 4.20b–d). As the wetting front continues advancing into the slope, three different flow regimes – upslope, vertical, and downslope – occur simultaneously and evolve dynamically, as is clearly shown in Figures 4.19c–f and 4.20b–d.

The simulation also confirms the observations made in the previous section. Upslope flow is most pronounced near the wetting front (Figures 4.19b–f and 4.20c, d). Downslope lateral flow is mostly restricted to the region near the slope surface and can persist for a long time (Figures 4.19f and 4.20c). The general condition determining upslope, vertical, or downslope flow is not the rainfall condition at the slope surface; rather it is the state of wetting or drying within the hillslope.

4.4.6 Flow patterns resulting from transient rainfall

The final numerical test of the conceptual model presented by Lu *et al.* (2011) was a simulation under varying rainfall intensity rather than under constant rainfall conditions. For comparison and discussion, the rainfall history suggested by Sinai and Dirksen (Figure 11, 2006) was chosen. Here they suggested that the necessary condition predicted by their hypothesis for changes in the direction of downslope lateral flow was decreasing rainfall intensity. The same sandy slope defined in the previous sub-section (Figure 4.19a) was used with varying rainfall intensity shown in Figure 4.21a and the anticipated temporal directions of lateral flow by Sinai and Dirksen (2006) shown in Figure 4.21b.

The 6-hour varying rainfall history consists of five periods (Figure 4.21a): (I) increasing rainfall intensity (from 0.5 to 1.5 hours), (II) decreasing rainfall intensity (1.5 to 3.5 hours), (III) no-rainfall (from 3.5 to 4.75 hours), (IV) increasing rainfall (4.75 to 5.5 hours), and (V) constant rainfall (time > 5.5 hours). Sinai and Dirksen's (2006) hypothesis predicts that three flow regimes will occur, as shown in Figure 4.21b: upslope lateral flow from 0.5 to 1.5 hours, downslope lateral flow from 1.5 to about 4.75 hours, and upslope lateral flow for times greater than 4.75 hours. Results using the Lu *et al.* (2011) conceptual model are shown in Figure 4.21b, where the lateral flow component at the slope surface follows a similar pattern as that proposed by Sinai and Dirksen (2006), but the arrival times of changes in lateral flow direction are determined by the wetting and drying states, not by the rainfall history at the slope boundary.

As shown in Figure 4.21b, at the surface of the slope the upslope lateral flux begins once the rainfall starts. However, when rainfall intensity starts to decline at 1.5 hours, the lateral

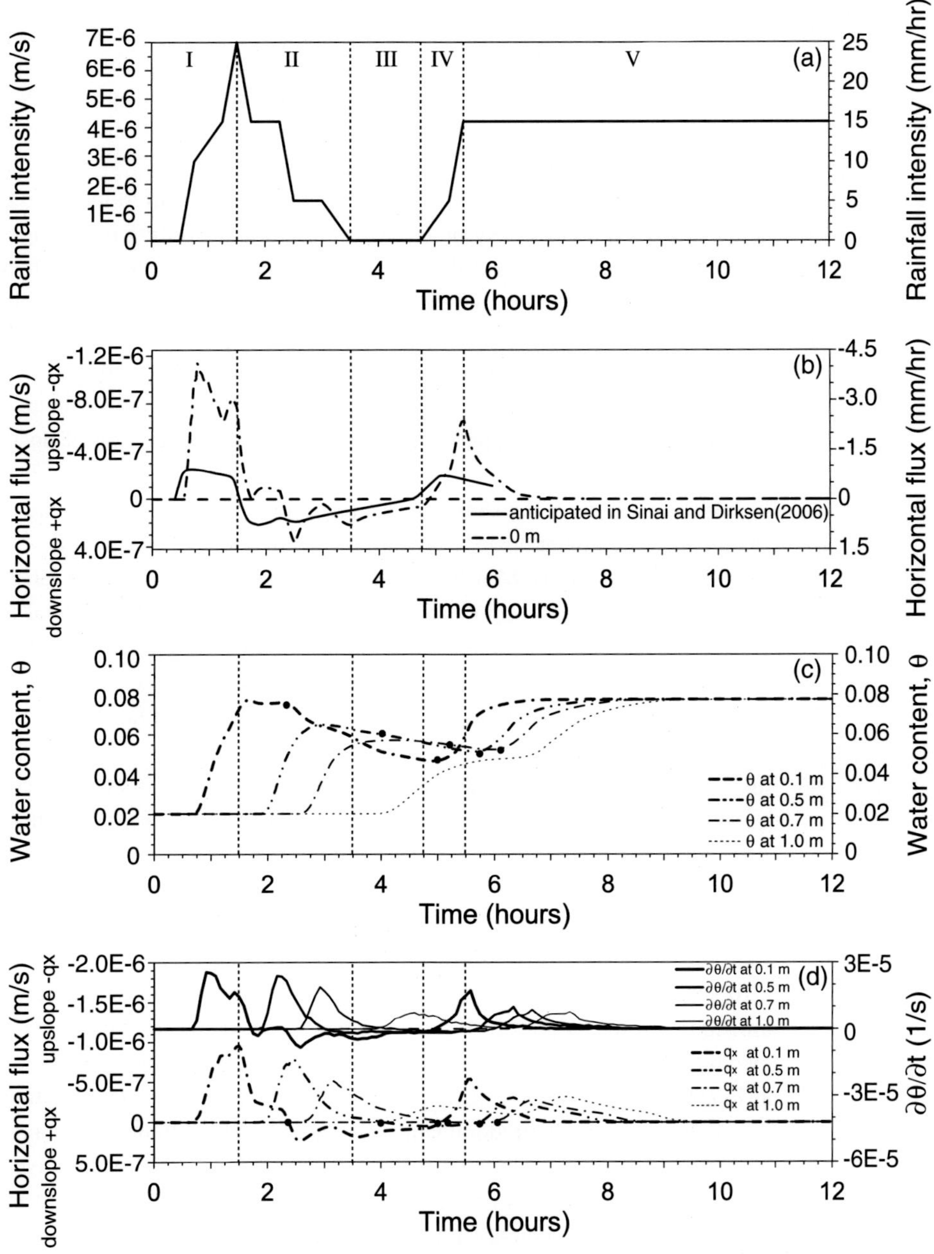

Figure 4.21 Simulation results from varying rainfall intensity: (a) infiltration intensity vs. time at the slope surface, (b) simulated lateral flux in the hillslope at the slope surface and anticipated flow direction from Sinai and Dirksen (2006), (c) water content at different depths as a function of time, and (d) horizontal flux and moisture content change at different depths as a function of time (from Lu *et al.*, 2011).

flow direction does not switch immediately to the downslope direction, as predicted by Sinai and Dirksen (2006); rather the change in direction occurs at around 2 hours 20 minutes and persists until the end of the period III, when it switches to the upslope direction again in response to increasing rainfall intensity at the beginning of period IV. The simulation results shown in Figure 4.21b, however, confirm that vertically downward flow occurs when a state-steady (i.e., for times > 7 hours) is reached.

The flow regime in regions away from the slope surface will follow far more complicated patterns, but can be precisely predicted by the state of wetting or drying at each point in the hillslope. The strong correlation between variations in the lateral flow direction and moisture content change with time is illustrated in Figure 4.21c, d at different depths from the slope surface. For example, at the depth of 0.1 m the duration of the upslope component of horizontal flux (from 0.75 to 2.3 hours) corresponds to the period when moisture content is increasing; the duration of downslope lateral flow (from 2.3 to 5.0 hours) coincides with the period when moisture content is decreasing; and the duration of upslope lateral flow (from 5.0 to 7.0 hours) coincides with the period when moisture content is increasing. For regions away from the slope surface (e.g., at the depth of 0.5 m), increase in moisture content and consequent upslope lateral flow begin at about 2 hours, i.e., 1 hour 30 minutes after the rainfall begins. Moisture content decreases and downslope lateral flow starts at around 4 hours, i.e., 30 minutes after rainfall ceases. Downslope lateral flow persists for another 1 hour and 15 minutes until about 5 hours and 45 minutes into the simulation, when moisture content increases again and upslope lateral flow occurs. For a region deeper beneath the slope surface (e.g., at the depth of 0.7 m), lateral flow is mostly upslope. For a brief period (5.25 to 6.0 hours) flow is laterally downslope and then the magnitude of the flux is small. This brief period of downslope lateral flow occurs concurrently with a small decrease in moisture content (Figure 4.21c, d). For a region even deeper from the slope surface (e.g., at the depth of 1.0 m) flow never changes direction and is always upslope during the entire simulation period of 12 hours. This is again consistent with the change in moisture content at this point in that it only increases and over time reaches a constant value.

To summarize the observations above; the state of wetting or drying controls the flow direction, and the rainfall history is inadequate to define the flow regime in a hillslope. Specifically, when a point in a hillslope is subject to rapid increases in moisture content, upslope lateral flow occurs; when moisture content at the point stays relatively steady or the rate of change is small, predominantly downward vertical flow occurs; and when moisture content decreases, downslope lateral flow occurs. This can be quantitatively confirmed by comparing the time series of horizontal flux and the rate of change in moisture content (Figure 4.21d). At all points within the hillslope the changes of flow direction completely coincide with the rate of changes in the moisture content.

4.5 Summary of flow regimes in hillslopes

Until recently, the direction of flow in unsaturated hillslopes has not been well understood. The direction of saturated groundwater flow in isotropic and homogeneous hillslopes is generally laterally downslope under the driving force of gravity and constant pressure boundary conditions along the slope surface, regardless of transient or steady rainfall conditions. However, under unsaturated conditions an additional mechanism, namely the gradient of moisture content or suction potential, plays a significant role. Furthermore, the boundary condition along the slope surface is better described by time-dependent flux rather than time-dependent moisture content. Such a transient flux boundary condition and concurrent flow driving mechanisms of gravity and gradient in moisture content yield spatially and temporally complex flow patterns in hillslopes, which include upslope lateral flow, downward vertical flow, and downslope lateral flow. In general, all three patterns can simultaneously exist within the variably saturated flow regime in an isotropic and homogeneous hillslope subject to time-varying flux boundary conditions.

Previous studies considered flow patterns to be solely controlled by rainfall conditions at the slope surface (Jackson, 1992; Sinai and Dirksen, 2006). They found that the condition for upslope lateral flow is increasing rainfall intensity, the condition for vertical flow is steady-state infiltration, and the condition for downslope lateral flow is decreasing or zero rainfall intensity. These conditions were confirmed using laboratory-scale hillslope infiltration tests (Sinai and Dirksen, 2006). Using a two-dimensional numerical model calibrated to those laboratory tests, Lu *et al.* (2011) demonstrate that the previous conclusions for flow regimes in hillslopes are only valid in the region immediately adjacent to the slope surface. It is shown that flow directions in isotropic and homogeneous unsaturated hillslopes could be simultaneously upslope, vertical, and downslope throughout the soil profile. The hypothesis is that, at any point within a homogeneous and isotropic hillslope, downslope lateral unsaturated flow will occur at a point if that point is in a drying state, and upslope lateral unsaturated flow will occur at that point if that point is in a wetting state. It is found that a quasi-steady zone behind the wetting front will develop and is characterized by predominantly downward vertical flow. Numerical simulations of the previous laboratory experiments under constant rainfall conditions and of a field-scale sand hillslope under varying rainfall intensity conditions confirm our conceptual model. Thus, if the history of moisture content within a homogeneous and isotropic hillslope is known (e.g., from field measurements), the pattern of unsaturated flow in the hillslope can be unambiguously defined. For hillslopes with significant heterogeneity and anisotropy, the pattern of unsaturated flow could be complex, thus the role of the moisture content gradient vs. matric potential gradient in controlling flow direction needs to be further explored.

4.6 Problems

1. What are the laws and principles involved in describing transient flow processes?
2. What is the physical meaning of the time-dependent term in the transient governing flow equation?
3. How many unknowns (dependent variables) and independent variables are there in the head governing Equation (4.15)? How many characteristic functions?
4. If we know the soil water characteristic function and hydraulic conductivity function for a soil, how do we estimate the water diffusivity function for this soil?

5. In deriving the Richards equation, there are four characteristic functions of unsaturated soil that are commonly encountered; what are they? For a given soil, how many independent characteristic functions are needed in order to describe the total head distribution in space and time?
6. Flow direction in a hillslope is shown by the arrow in the shaded areas in Figures 4.4a, b. Will the flow be called upslope or downslope by the definition of Figure 4.4a? Will the flow be called upslope or downslope by the definition of Figure 4.4b?
7. According to the Green–Ampt model for horizontal infiltration, will the total infiltration be linear or non-linear with respect to time? What is (are) the driving mechanism(s) for horizontal infiltration? What will be the infiltration rate when the infiltration time becomes infinite?
8. What is (are) the driving mechanism(s) for vertical infiltration? What will be the infiltration rate when the infiltration time becomes infinite? What are the driving mechanisms when the infiltration time becomes infinite?
9. What are the major differences between the Green–Ampt model and the Srivastava–Yeh model in terms of type of equations solved, moisture profile, and governing parameters?
10. What is the fundamental difference between steady and transient unsaturated infiltration into an unsaturated layer?
11. What are the major advantages and disadvantages of numerical solutions of unsaturated flow problems?
12. What are the governing factors in controlling flow regimes in hillslopes?
13. What are the two common ways to define flow direction in hillslopes?
14. What are the main findings from Philip's 1991 work regarding flow direction in hillslopes?
15. What are the main findings of Jackson's 1992 work regarding flow direction in hillslopes?
16. What are the major disagreements between Philip and Jackson regarding flow directions in hillslopes ?
17. What are the main findings in Sinai and Dirksen's (2006) work regarding flow direction in hillslopes?
18. According to Lu *et al.*'s (2011) conceptual model shown in Figure 4.12, why does upslope lateral flow occur when the soil is in the wetting state? Why does downslope lateral flow occur when the soil is in the drying state?
19. What is the general hypothesis presented by Lu *et al.* (2011) regarding flow direction in hillslopes?
20. What are the major findings in Sinai and Dirksen's (2006) experiments under constant rainfall intensity conditions?
21. Do Sinai and Dirksen's (2006) experiments support the general hypothesis proposed by Lu *et al.* (2011) regarding flow direction in hillslopes?
22. Do the numerical simulations shown in Section 4.4 support the general hypothesis proposed by Lu *et al.* (2011) regarding flow direction in hillslopes?
23. What is the physical meaning of the specific moisture capacity? For a soil subject to variation of soil moisture content, when will the specific moisture capacity be zero?

Using van Genuchten's (1980) model to represent the SWCC, derive the moisture content when the maximum specific moisture capacity occurs.

24 A silty soil layer with a 20 m thick unsaturated zone (from the surface to the water table) has the following hydraulic properties: $\theta_s = 0.4$, $\theta_r = 0.05$, and $K_o = 10^{-7}$ m/s. Estimate and plot the arrival time of a downward infiltration front (Figure 4.7a) toward the water table after a heavy-rainfall storm. Assume that the initial suction head is $h_i = -1.5$ m and the heavy rain causes the surface suction head to be equal to the atmospheric pressure for a sustainable period of time.

25 A hillslope consists of an isotropic and homogeneous silty soil identical to that in Problem 24. The slope surface is tilted 20° from the horizontal plane and is subjected to the same hydrologic conditions as that in Problem 24. Estimate and plot the arrival time of the inward infiltration wetting front z_*. Draw conclusion on the impact of the slope angle on infiltration from comparisons with Problem 24.

26 The SWCC for a silty loam can be represented by van Genuchten's (1980) model with the following hydraulic parameters: $\alpha = 0.0028$ kPa^{-1}, $n = 1.3$, $\theta_r = 0.030$, and $\theta_s = 0.322$. Calculate and plot volumetric water content as a function of matric suction head. Calculate and plot specific moisture capacity as a function of water content θ.

27 In hillslopes with isotropic and homogeneous materials, what is the determining variable for flow direction? When will the upslope lateral flow occur? When will the downslope lateral flow occur? When will the vertical flow occur? Why?

Part III

TOTAL AND EFFECTIVE STRESS IN HILLSLOPES

LECTURE TODAY: *defining stress in slopes*

5 Total stresses in hillslopes

5.1 Definitions of stress and strain

5.1.1 Definition of total stress

The objective of this chapter is to provide the total stress distributions in hillslopes using linear elasticity theory. The actual stresses in hillslopes are typically unknown and they are not easily measured. The stress distribution in hillslopes is the product of geologic history – tectonic stress, erosion, and sedimentation as well as other processes. Thus, models that compute the distribution of stress in hillslopes are difficult to test and validate. The driving force for the distribution of total stress is the gravitational force, which varies spatially and temporally in hillslope environments. Stress is an average quantity of force over an area of interest. The analysis of stresses in solids, liquids, and gases, and the consequent deformation or flow of these materials belongs to a branch of mechanics known as continuum mechanics (Malvern, 1969). In continuum mechanics stresses are described by continuous mathematical functions, and the external forces acting on any free body causing total stresses can be classified in two types: body forces and surface forces. The main body force responsible for the total stress distribution in hillslopes is gravity **g**. It is commonly defined as force per unit volume and is an "action-at-a-distance" force in the direction toward the center of the earth, as shown in Figure 5.1. For a chosen unit volume in hillslopes, the body force **b** due to gravity is

$$\mathbf{b} = \rho \mathbf{g} \tag{5.1}$$

The bold notation in Equation (5.1) indicates that body force is a vector or directional quantity, and ρ is the density of material (in kg/m^3), which is typically a function of location and time in hillslopes. The density distribution in hillslopes varies from material to material and varies with the distribution of moisture. Because the moisture distribution depends on time, the density distribution is also dynamic or time-dependent. However, the moisture distribution usually varies within 10% of its annual average value within hillslopes, except in the upper meter or so near the ground surface, where changes in density of soil due to moisture can be as much as 20%. In indicial and matrices notations, the body force can be expressed as

$$b_i = \rho g_i \quad \text{or} \quad \begin{bmatrix} b_1 \\ b_2 \\ b_3 \end{bmatrix} = \rho \begin{bmatrix} 0 \\ 0 \\ g \end{bmatrix} \tag{5.2}$$

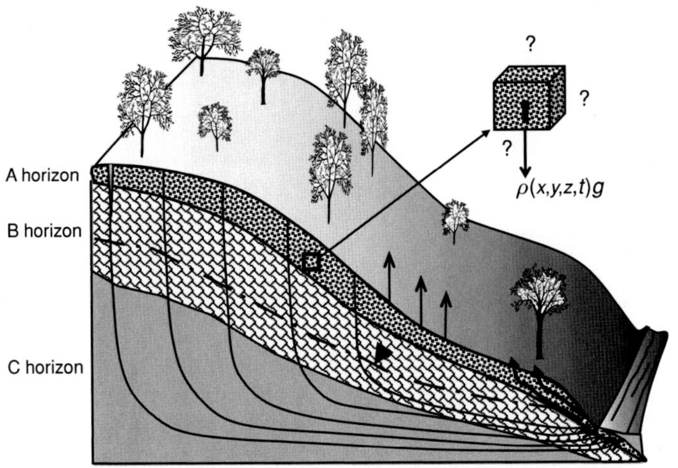

Figure 5.1 Illustration of hillslope geometry and various flow processes that affect body force distributions. Illustration not to scale.

with $i = 1$ and 2 for the horizontal directions (denoted as x and y) and $i = 3$ for the vertical direction (denoted as z), and g is the gravitational acceleration of the earth, equal to 9.81 m/s^2 for all practical landslide problems. The quantity ρg (in N/m^3) is called the unit weight. Another body force is the fictitious force due to the acceleration of materials and it is called the inertial force. The motion of hillslope materials under dynamic conditions such as earthquake shaking or mass flow after failure creates a fictitious state of equilibrium in a dynamic state. The body force under such conditions can be accounted for under the same principle described by Equation (5.1) with explicit knowledge of accelerations in each direction. Because the focus here is confined to static failure of hillslope materials, the inertia force is ignored in the subsequent discussion of total stress distribution.

Surface forces are contact forces acting on any volume (element) of soil at its bounding surfaces. Under static conditions, some of these forces are due to the gravitational forces and their values can vary over many orders of magnitude depending on the geometry and dimensions of a hillslope. The magnitude of these forces is commonly quantified by their area intensity or force per unit area over the bounding surfaces of the unit volume. The force intensities defined in such manner are called total stresses, as illustrated in Figure 5.2. Assuming a surface cut through the material perpendicular to the x direction, by Newton's third law, an equivalent force vector $\mathbf{F}$ can be placed on the surface with area S (Figure 5.2a). The stress vector (also called traction) on that surface can be obtained by taking the average of the force vector $\mathbf{F}$ over the surface area S as

$$\sigma_{xi} = \lim|_{S\to 0} \frac{F_{xi}}{S} \quad \text{or} \quad \begin{bmatrix} \sigma_{xx} \\ \sigma_{xy} \\ \sigma_{xz} \end{bmatrix} = \lim_{S\to 0} \begin{bmatrix} \dfrac{F_{xx}}{S} \\ \dfrac{F_{xy}}{S} \\ \dfrac{F_{xz}}{S} \end{bmatrix} \tag{5.3}$$

The first component of the stress vector is in the direction perpendicular to the surface and

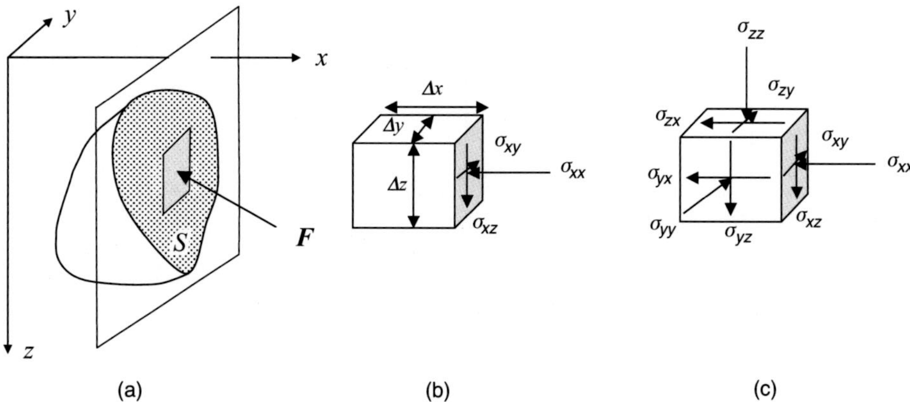

Figure 5.2 Illustration of surface forces and total stress for an element in Cartesian coordinates: (a) force vector **F** on a cross section of interest S, (b) convention in sign of stress components, and (c) stress tensor at a point in Cartesian coordinates.

is called normal stress σ_{xx}. This stress is defined as positive (compressive) if it results in a shortening effect (shown in Figure 5.2b) and negative (tensile) if it results in an elongating effect. This is conventionally done in soil and rock mechanics. The second component is parallel to the surface and in the y direction, and is called shear stress, as it creates distortion of the material. The third component also is a shear stress but in the z direction. The sign convention for shear stresses can be defined as positive, as shown in Figure 5.2c, and negative if the shear stresses point in the directions opposite of those shown in Figure 5.2c.

In order to define the state of stress at a point, three orthogonal planes bounding the element are used in the Cartesian coordinate system. The total stresses obtained in such manner are illustrated in Figure 5.2c, and expressed as

$$\sigma_{ji} = \lim|_{S\to 0}\frac{F_{ji}}{S} \quad \text{or} \quad \begin{bmatrix} \sigma_{xx} & \sigma_{xy} & \sigma_{xz} \\ \sigma_{yx} & \sigma_{yy} & \sigma_{yz} \\ \sigma_{zx} & \sigma_{zy} & \sigma_{zz} \end{bmatrix} = \lim_{\substack{S_{xz}=\Delta x\Delta z\to 0\\ S_{yx}=\Delta y\Delta x\to 0\\ S_{yz}=\Delta z\Delta y\to 0}} \begin{bmatrix} \frac{F_{xx}}{S_{yz}} & \frac{F_{xy}}{S_{yz}} & \frac{F_{xz}}{S_{yz}} \\ \frac{F_{yx}}{S_{zx}} & \frac{F_{yy}}{S_{zx}} & \frac{F_{yz}}{S_{zx}} \\ \frac{F_{zx}}{S_{xy}} & \frac{F_{zy}}{S_{xy}} & \frac{F_{zz}}{S_{xy}} \end{bmatrix} \tag{5.4}$$

Therefore, in general, the state of stress at a point cannot be defined completely with a single stress vector on one of the three orthogonal planes, and three stress vectors with nine components (three normal stresses and six shear stresses) are necessary.

From the theorem of conjugate shear stresses, or application of moment balance with respect to coordinate axes, it can be shown that the component of the shear stress on one of the planes (Figure 5.2c) that is perpendicular to the line of intersection is equal to the similar shear component on the other plane, i.e., $\sigma_{xy} = \sigma_{yx}$, $\sigma_{xz} = \sigma_{zx}$, and $\sigma_{zy} = \sigma_{yz}$. In other words, the total stress tensor is symmetric and the symmetrically placed off-diagonal elements of the stress matrix (Equation (5.4)) are equal:

$$\sigma_{ji} = \begin{bmatrix} \sigma_{xx} & \sigma_{xy} & \sigma_{xz} \\ \sigma_{yx} & \sigma_{yy} & \sigma_{yz} \\ \sigma_{zx} & \sigma_{zy} & \sigma_{zz} \end{bmatrix} = \begin{bmatrix} \sigma_{xx} & \sigma_{xy} & \sigma_{xz} \\ \sigma_{xy} & \sigma_{yy} & \sigma_{yz} \\ \sigma_{xz} & \sigma_{yz} & \sigma_{zz} \end{bmatrix} \tag{5.5}$$

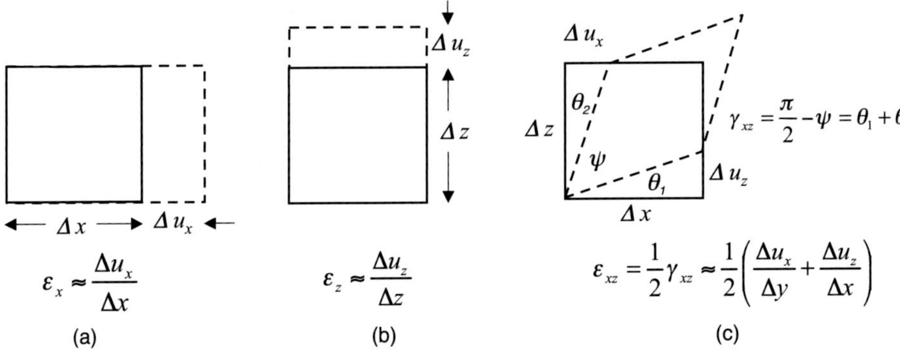

Figure 5.3 Illustration of small strain for an element in Cartesian coordinates: (a) uniaxial extension in the x direction, (b) uniaxial extension in the z direction, and (c) pure shear distortion without rotation. For large strains, the approximations shown are no longer valid and non-linear relationships (more terms) between strain and displacement may be needed.

Thus, six components are considered to be independent and can define the state of total stress at a point in three-dimensional space.

5.1.2 Definition of strain

Strain is the average quantity of displacement over a length or an angle of interest. Strain is generally a consequence of stress field distributions. Pure rotational movement can cause rotational strain that is not discussed here, as it is a consequence of rigid body movements. At any point, strain can be decomposed into two pure modes; extensional or compressional strain (also called normal strain) and shear strain, as illustrated in a two-dimensional x-z space in Figure 5.3.

The initial positions (solid lines) of the element are commonly used to normalize deformed positions (dashed lines). The normal strains, ε, (Figure 5.3a, b) can be defined by changes in lengths per unit initial lengths, i.e.,

$$\varepsilon_{xx} = \lim_{\Delta x \to 0} \frac{\Delta u_x}{\Delta x} = \frac{\partial u_x}{\partial x} \tag{5.6a}$$

$$\varepsilon_{zz} = \lim_{\Delta x \to 0} \frac{\Delta u_z}{\Delta z} = \frac{\partial u_z}{\partial z} \tag{5.6b}$$

where u_x, and u_z are the displacement in the x, and z directions, respectively.

The shear strain reflects a change in angle of the originally orthogonal element (Figure 5.3c) and is defined as half the decrease γ_{xz} in the right angle initially formed by the sides parallel to the x and z axes:

$$\varepsilon_{xz} = \frac{1}{2}\gamma_{xz} = \frac{1}{2}(\theta_1 + \theta_2) \approx \lim_{\Delta x = \Delta z \to 0} \frac{1}{2}\left(\frac{\Delta u_x}{\Delta z} + \frac{\Delta u_z}{\Delta x}\right) = \frac{1}{2}\left(\frac{\partial u_x}{\partial z} + \frac{\partial u_z}{\partial x}\right) \tag{5.7}$$

where θ_1 and θ_2 are angles of distortion for elements Δx and Δz, respectively, and γ_{xz} is the average angle of distortion between Δx and Δz.

As shown, strains are dimensionless quantities. To define the state of strain at any point, nine strain components on the three orthogonal planes of an element are necessary, i.e.,

$$\varepsilon_{ji} = \begin{bmatrix} \varepsilon_{xx} & \varepsilon_{xy} & \varepsilon_{xz} \\ \varepsilon_{yx} & \varepsilon_{yy} & \varepsilon_{yz} \\ \varepsilon_{zx} & \varepsilon_{zy} & \varepsilon_{zz} \end{bmatrix} \tag{5.8}$$

The symmetry of the conjugate shear strain is automatically satisfied by switching the order of the terms on the right hand side in Equation (5.7) and does not mathematically change the definition, i.e.,

$$\begin{aligned} \varepsilon_{xz} = \frac{1}{2}\gamma_{xz} &\approx \lim_{\Delta x=\Delta z\to 0}\frac{1}{2}\left(\frac{\Delta u_x}{\Delta z}+\frac{\Delta u_z}{\Delta x}\right) = \varepsilon_{zx} = \frac{1}{2}\gamma_{zx} \\ &\approx \lim_{\Delta x=\Delta z\to 0}\frac{1}{2}\left(\frac{\Delta u_z}{\Delta x}+\frac{\Delta u_x}{\Delta z}\right) = \frac{1}{2}\left(\frac{\partial u_z}{\partial x}+\frac{\partial u_x}{\partial z}\right) \end{aligned} \tag{5.9}$$

Therefore, six independent components of strain can fully define the state of strain at any point in three-dimensional space:

$$\varepsilon_{ji} = \begin{bmatrix} \varepsilon_{xx} & \varepsilon_{xy} & \varepsilon_{xz} \\ \varepsilon_{xy} & \varepsilon_{yy} & \varepsilon_{yz} \\ \varepsilon_{xz} & \varepsilon_{yz} & \varepsilon_{zz} \end{bmatrix} = \begin{bmatrix} \dfrac{\partial u_x}{\partial x} & \dfrac{1}{2}\left(\dfrac{\partial u_x}{\partial y}+\dfrac{\partial u_y}{\partial x}\right) & \dfrac{1}{2}\left(\dfrac{\partial u_x}{\partial z}+\dfrac{\partial u_z}{\partial x}\right) \\ \dfrac{1}{2}\left(\dfrac{\partial u_x}{\partial y}+\dfrac{\partial u_y}{\partial x}\right) & \dfrac{\partial u_y}{\partial y} & \dfrac{1}{2}\left(\dfrac{\partial u_y}{\partial z}+\dfrac{\partial u_z}{\partial y}\right) \\ \dfrac{1}{2}\left(\dfrac{\partial u_x}{\partial z}+\dfrac{\partial u_z}{\partial x}\right) & \dfrac{1}{2}\left(\dfrac{\partial u_y}{\partial z}+\dfrac{\partial u_z}{\partial y}\right) & \dfrac{\partial u_z}{\partial z} \end{bmatrix} \tag{5.10}$$

5.1.3 Stress–strain relationship

One fundamental hypothesis in continuum mechanics is that there exists an intrinsic relationship between stress and strain such that application of stress on an elastic body will result in strain or strain will result in stress. Such a relationship is called a constitutive law, as it is strongly dependent on the mechanical properties of a material. As stated at the beginning of this chapter, linear elasticity theory is introduced here to provide a mathematical description of a first approximation for hillslope materials. Ideal, isotropic, linear-elastic materials follow Hooke's law describing the link between stress and strain. Under the uniaxial stress and strain conditions shown in Figures 5.2 and 5.3, uniaxial stresses can be described by their conjugate uniaxial strains as

$$\sigma_{xx} = E\varepsilon_{\mathrm{xx}} \tag{5.11a}$$

$$\sigma_{yy} = E\varepsilon_{\mathrm{yy}} \tag{5.11b}$$

$$\sigma_{zz} = E\varepsilon_{\mathrm{zz}} \tag{5.11c}$$

where E is the elastic modulus or Young's modulus with units of stress. Thus, a physical interpretation of the elastic modulus is the change in the normal stress due to a unit change in normal strain (see Section 7.1). Similarly, under pure shear conditions, Hooke's law describes shear stresses in terms of shear strain with the shear modulus G:

$$\sigma_{xy} = 2G\varepsilon_{\mathrm{xy}} \tag{5.12a}$$

$$\sigma_{yz} = 2G\varepsilon_{\mathrm{yz}} \tag{5.12b}$$

$$\sigma_{zx} = 2G\varepsilon_{\mathrm{zx}} \tag{5.12c}$$

Thus, a physical interpretation of the elastic shear modulus is the change in the shear stress due to a unit change of double the shear strain. Under general stress and strain conditions, the isotropic Hooke's law links the stress tensor with the strain tensor in the following tensor indicial form:

$$\sigma_{ij} = \lambda\varepsilon_{kk}\delta_{ij} + 2G\varepsilon_{ij} \tag{5.13}$$

where δ_{ij} is the identity tensor; or in the following matrix form:

$$\begin{bmatrix} \sigma_{xx} & \sigma_{xy} & \sigma_{xz} \\ \sigma_{xy} & \sigma_{yy} & \sigma_{yz} \\ \sigma_{xz} & \sigma_{yz} & \sigma_{zz} \end{bmatrix} = \lambda\left(\varepsilon_{xx} + \varepsilon_{yy} + \varepsilon_{zz}\right)\begin{bmatrix} 1 & 0 & 0 \\ 0 & 1 & 0 \\ 0 & 0 & 1 \end{bmatrix} + 2G\begin{bmatrix} \varepsilon_{xx} & \varepsilon_{xy} & \varepsilon_{xz} \\ \varepsilon_{xy} & \varepsilon_{yy} & \varepsilon_{yz} \\ \varepsilon_{xz} & \varepsilon_{yz} & \varepsilon_{zz} \end{bmatrix} \tag{5.13}$$

where λ and G are the two independent Lamé elastic constants. Thus, from Equation (5.13), a physical interpretation of the Lamé elastic constant λ is the change in the volumetric stress due to a unit change in volumetric strain. According to linear elasticity theory, the two Lamé elastic constants can be expressed by Young's modulus E and Poisson's ratio ν:

$$\lambda = \frac{\nu E}{(1+\nu)(1-2\nu)} \tag{5.14}$$

$$G = \frac{E}{2(1+\nu)} \tag{5.15}$$

A physical interpretation of Poisson's ratio is the changes in the normal strain in a specific direction due to the change in a unit strain orthogonal to it. Equation (5.13) can also be inverted to write strain in terms of stress in tensor indicial form:

$$\varepsilon_{ij} = -\frac{\lambda\delta_{ij}}{2G(3\lambda+2G)}\sigma_{kk} + \frac{1}{2G}\sigma_{ij} \tag{5.16}$$

or in the following matrix form:

$$\begin{bmatrix} \varepsilon_{xx} & \varepsilon_{xy} & \varepsilon_{xz} \\ \varepsilon_{xy} & \varepsilon_{yy} & \varepsilon_{yz} \\ \varepsilon_{xz} & \varepsilon_{yz} & \varepsilon_{zz} \end{bmatrix} = \frac{-\lambda}{2G(3\lambda+2G)}\left(\sigma_{xx} + \sigma_{yy} + \sigma_{zz}\right)\begin{bmatrix} 1 & 0 & 0 \\ 0 & 1 & 0 \\ 0 & 0 & 1 \end{bmatrix}$$
$$+ \frac{1}{2G}\begin{bmatrix} \sigma_{xx} & \sigma_{xy} & \sigma_{xz} \\ \sigma_{xy} & \sigma_{yy} & \sigma_{yz} \\ \sigma_{xz} & \sigma_{yz} & \sigma_{zz} \end{bmatrix} \tag{5.16}$$

The above matrix form can be expanded and written in the six equations for the isotropic

Hooke's law that link the six total stresses to the six strains at any point:

$$\varepsilon_{xx} = \frac{1}{E}[\sigma_{xx} - \nu(\sigma_{yy} + \sigma_{zz})] \tag{5.16a}$$

$$\varepsilon_{yy} = \frac{1}{E}[\sigma_{yy} - \nu(\sigma_{xx} + \sigma_{zz})] \tag{5.16b}$$

$$\varepsilon_{zz} = \frac{1}{E}[\sigma_{zz} - \nu(\sigma_{xx} + \sigma_{yy})] \tag{5.16c}$$

$$\varepsilon_{xy} = \frac{\gamma_{xy}}{2} = \frac{1}{2G}\sigma_{xy} \tag{5.16d}$$

$$\varepsilon_{yz} = \frac{\gamma_{yz}}{2} = \frac{1}{2G}\sigma_{yz} \tag{5.16e}$$

$$\varepsilon_{zx} = \frac{\gamma_{zx}}{2} = \frac{1}{2G}\sigma_{zx} \tag{5.16f}$$

The bulk modulus K can be defined as the ratio of volumetric stress to volumetric strain and using Equation (5.16):

$$K = \frac{\sigma_{kk}}{3\varepsilon_{kk}} = \lambda + \frac{2}{3}G = \frac{E}{3(1-2\nu)} \tag{5.17}$$

In summary, for isotropic, linear-elastic materials, any two of the four parameters, namely G, E, λ, and ν are sufficient to quantify the stress–strain relationship.

5.2 Analysis and graphical representation of the state of stress

5.2.1 Mohr circle concept

As described in the previous section, there are six stress components for a point in three-dimensional space. Furthermore, depending on how the coordinate system is oriented, the values of the six components on the bounding faces generally vary. To interpret the state of stress from these components, it is instructional to present them in a two-dimensional plane with the aid of a Mohr circle.

If a bounding face perpendicular to the z-axis is chosen from Figure 5.2a, the state of stress on that face is determined by three stress components of the stress vector:

$$\begin{bmatrix} \sigma_{xx} \\ \sigma_{xy} \\ \sigma_{yy} \end{bmatrix} \tag{5.18}$$

In the literature, the off-diagonal stress components or shear stresses are commonly denoted by the symbol τ, and this will be used hereafter. The stress components for an element in the x-y plane are shown in Figure 5.4a.

The Mohr circle provides a convenient graphical way to determine the components of a stress vector on any arbitrarily oriented element in terms of the known stress vector

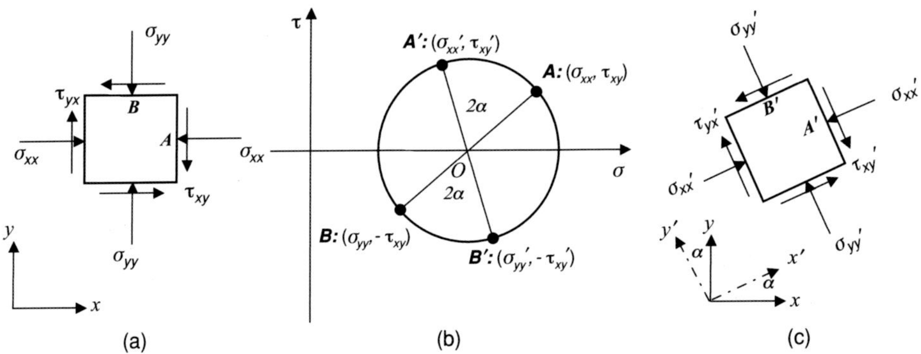

Figure 5.4 Illustration of the state of stress in 2-D space and the Mohr circle representation: (a) 2-D state of stress in Cartesian coordinates, (b) Mohr circle representation, and (c) 2-D state of stress after a counter-clockwise rotation of angle α.

$\{\sigma_{xx}, \tau_{xy}, \sigma_{yy}\}$, as illustrated in Figure 5.4b. To define a Mohr circle or a state of stress at a point on the normal stress versus shear stress plane, stress components on the two orthogonal bounding faces of an element must be known (e.g., Figure 5.4a). For the face containing point A in Figure 5.4a, the coordinates are (σ_{xx}, τ_{xy}) shown as dot A in Figure 5.4b. For the face containing point B in Figure 5.4a, the coordinates are $(\sigma_{yy}, -\tau_{xy})$ shown as dot B in Figure 5.4b. These two dots uniquely define the Mohr circle whose center is at the intercept of the line connecting the two dots and the normal stress axis (point O). An important property of the Mohr circle or the state of stress at a point is that the normal and shear stress components of the stress vector on any face passing through the point will reside on the circle. The exact location of the coordinates of the stress vector can be determined from the orientation angle α (Figure 5.4c) on which the stress vector is acting. As shown, it will require rotating line OA 180° counter-clockwise around the center of the circle to become OB (Figure 5.4b) where in a real physical setting (Figure 5.4a) it requires the face with point A to rotate 90° counter-clockwise to become the face containing point B. Thus, rotation of an axis counter-clockwise by angle α (Figure 5.4c) will be equivalent to a rotation of 2α counter-clockwise on the Mohr circle (Figure 5.4b). The normal and shear components of the stress vector on the rotated plane can be expressed in terms of the components of the stress vector on the initial plane:

$$\sigma'_{xx} = \sigma_{xx}\cos^2\alpha + \sigma_{yy}\sin^2\alpha + 2\tau_{xy}\sin\alpha\cos\alpha \tag{5.19}$$

$$\tau'_{xy} = \tau_{xy}(\cos^2\alpha - \sin^2\alpha) - (-\sigma_{xx} - \sigma_{yy})\sin\alpha\cos\alpha \tag{5.20}$$

Graphically, the above equations represent the coordinates of A' in Figure 5.4b.

By trigonometric relationships, it can be shown that Equations (5.19) and (5.20) can be written in the following forms:

$$\sigma'_{xx} = \frac{1}{2}(\sigma_{xx} + \sigma_{yy}) + \frac{1}{2}(\sigma_{xx} - \sigma_{yy})\cos 2\alpha + \tau_{xy}\sin 2\alpha \tag{5.21}$$

$$\tau'_{xy} = \tau_{xy}\cos 2\alpha - \frac{1}{2}(\sigma_{xx} - \sigma_{yy})\sin 2\alpha \tag{5.22}$$

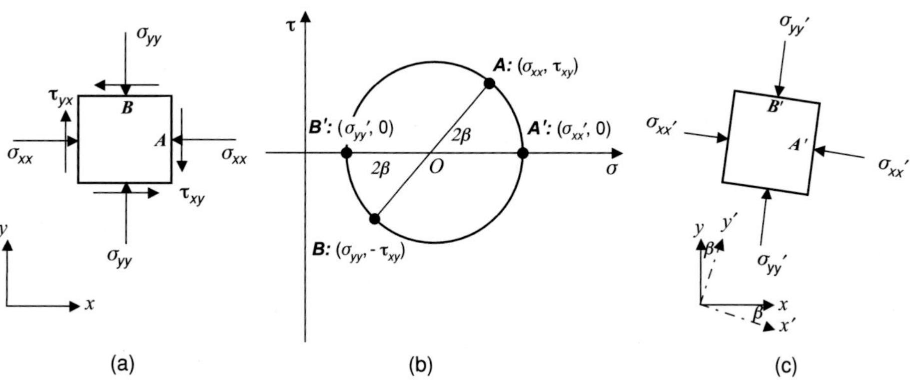

Figure 5.5 Illustration of the state of stress in 2-D space and its Mohr circle representation: (a) 2-D state of stress in Cartesian coordinates, (b) Mohr circle representation and principal stress locations, and (c) 2-D state of stress in the directions of principal stresses.

5.2.2 Principal stresses

From the known stress components on the two orthogonal faces shown in Figure 5.5a, a unique Mohr circle (Figure 5.5b) can be constructed. The radius of the circle R can also be determined from the equation of a circle in terms of the stress components as

$$R = \sqrt{\left[\frac{(\sigma_{xx} - \sigma_{yy})}{2}\right]^2 + \tau_{xy}^2} \tag{5.23}$$

To evaluate the magnitude of the stress components at a point, it is useful to determine maximum and minimum normal stresses and shear stresses. This can be done with the aid of a Mohr circle. From Figure 5.5b, it is evident that the maximum and minimum stresses occur at the intercepts of the circle (points A' and B') with the normal stress axis where shear stress is zero. If the angle between the vertical face (point A) and the plane where the principal stress locates is β (shown in Figure 5.5c), it is shown as 2β (angle AOA') on the Mohr circle representation. Substituting angle 2β into Equation (5.22) and recognizing zero shear stress ($\tau'_{xy} = 0$) leads to

$$\tan 2\beta = \frac{2\tau_{xy}}{\sigma_{xx} - \sigma_{yy}} \tag{5.24}$$

The magnitudes of the principal stresses can also be found using the Mohr circle. The coordinates of the center of the circle O are $(\sigma_{xx} + \sigma_{yy})/2$, 0. The principal maximum stress σ'_{xx} at point A' is the center plus the radius R, i.e.,

$$\sigma'_{xx} = \sigma_{\max} = \sigma_1 = \frac{1}{2}(\sigma_{xx} + \sigma_{yy}) + R \tag{5.25}$$

And the principal minimum stress σ'_{yy} at point B' is the center minus the radius R, i.e.,

$$\sigma'_{yy} = \sigma_{\min} = \sigma_3 = \frac{1}{2}(\sigma_{xx} + \sigma_{yy}) - R \tag{5.26}$$

Substituting Equation (5.23) into Equations (5.25) and (5.26) leads to

$$\sigma_1 = \frac{1}{2}(\sigma_{xx} + \sigma_{yy}) + \sqrt{\left[\frac{(\sigma_{xx} - \sigma_{yy})}{2}\right]^2 + \tau_{xy}^2} \tag{5.27}$$

$$\sigma_3 = \frac{1}{2}(\sigma_{xx} + \sigma_{yy}) - \sqrt{\left[\frac{(\sigma_{xx} - \sigma_{yy})}{2}\right]^2 + \tau_{xy}^2} \tag{5.28}$$

Adding Equations (5.27) and (5.28) yields

$$I_{1\sigma} = \sigma_1 + \sigma_3 = \sigma_{xx} + \sigma_{yy} \tag{5.29}$$

which implies that the sum of the normal stresses on any two orthogonal planes is a constant. Thus $I_{1\sigma}$ is called the first invariant of stress in two-dimensional space. Graphically, it determines the location of the center of the Mohr circle as being twice the value from the origin of the coordinate system to the center of the Mohr circle. Physically, it provides an indicator for the magnitude of normal stress at a point. In hillslope environments, this stress mainly depends on the weight of the overburden above a point of interest, or the depth from the slope surface.

From Figure 5.5b, it can also be seen that the maximum shear stress occurs at the top of the circle and its magnitude is the radius of the circle. With the knowledge of the principal stresses, the maximum shear stress is

$$\tau_{\max} = \frac{\sigma_1 - \sigma_3}{2} \tag{5.30}$$

Because it takes an exactly 90° (2α counter-clockwise) rotation from the principal maximum stress point (OA' in Figure 5.5) to the top of the circle, the maximum shear stress always occurs at 45° (α counter-clockwise) from the principal maximum stress. By the same token, the minimum shear stress occurs at the bottom of the circle at 45° (α clockwise) from the principal maximum stress. Its value is

$$\tau_{\min} = -\frac{\sigma_1 - \sigma_3}{2} \tag{5.31}$$

The maximum and minimum shear stresses are an important indicator for the stability of materials at a point, as earth materials are generally vulnerable to shear stress. In hillslope environments, this stress mainly depends on slope angle and depth. Steep and tall slopes generally promote higher shear stresses in the region near the toe of the slope.

5.3 Force equilibrium equations

5.3.1 Equations of motion

To describe a stress field in an isotropic and linear elastic body, continuum mechanics seeks a continuous form of governing partial differential equations for the stress field.

The momentum balance principle provides a theoretical basis to develop the governing equations for the stress field. It states that the time rate of change of the total momentum of a given volume is equal to the vector sum of all the external forces acting on the element. The momentum balance in each of the principal three directions can lead to the following set of three partial differential equations relating total stresses, gravitational forces, and inertial forces:

$$\frac{\partial \sigma_{ij}}{\partial x_j} + b_i = \rho \frac{d^2 u_i}{dt^2} \tag{5.32}$$

The first term represents the spatial change in total stress, the second term the body force due to gravity, and the third term inertial forces due to the acceleration of the element. For hillslope stability under static conditions, the acceleration of materials can be ignored, and the three Equations (5.32) can be simplified to the force equilibrium equations as:

$$\frac{\partial \sigma_{ij}}{\partial x_j} + b_i = 0 \tag{5.33}$$

5.3.2 Theory of linear elastostatics

If materials can be idealized as linearly elastic, a set of field equations with the two independent Lamé elastic constants can be developed. In three-dimensional space, there are six dependent stress variables, six dependent strain variables, and three dependent displacement variables. Thus, at least 15 independent equations are needed to complete the mathematical description. This can be accomplished using force equilibrium equations, Hooke's stress–strain laws, and strain–displacement relationships.

In Equation (5.33), the body forces can be accounted for by Equation (5.2), and Equation (5.33) can be further simplified in Cartesian coordinates:

$$\frac{\partial \sigma_{xx}}{\partial x} + \frac{\partial \tau_{xy}}{\partial y} + \frac{\partial \tau_{xz}}{\partial z} = 0 \tag{5.34}$$

$$\frac{\partial \sigma_{yy}}{\partial y} + \frac{\partial \tau_{xy}}{\partial x} + \frac{\partial \tau_{yz}}{\partial z} = 0 \tag{5.35}$$

$$\frac{\partial \sigma_{zz}}{\partial z} + \frac{\partial \tau_{xz}}{\partial x} + \frac{\partial \tau_{yz}}{\partial y} + \rho g = 0 \tag{5.36}$$

There are three equations and six dependent stress variables in the above equations. Hooke's law described by Equation (5.16), provides six equations linking stresses to strains, but introduces six additional dependent strain variables:

$$\varepsilon_{xx} = \frac{1}{E}[\sigma_{xx} - \nu(\sigma_{yy} + \sigma_{zz})] \tag{5.37}$$

$$\varepsilon_{yy} = \frac{1}{E}[\sigma_{yy} - \nu(\sigma_{xx} + \sigma_{zz})] \tag{5.38}$$

$$\varepsilon_{zz} = \frac{1}{E}[\sigma_{zz} - \nu(\sigma_{xx} + \sigma_{yy})] \tag{5.39}$$

$$\varepsilon_{xy} = \frac{\gamma_{xy}}{2} = \frac{1}{2G}\sigma_{xy} \tag{5.40}$$

$$\varepsilon_{yz} = \frac{\gamma_{yz}}{2} = \frac{1}{2G}\sigma_{yz} \tag{5.41}$$

$$\varepsilon_{zx} = \frac{\gamma_{zx}}{2} = \frac{1}{2G}\sigma_{zx} \tag{5.42}$$

The six strain components can be linked to three displacement variables through the six geometric equations (Equations (5.10)):

$$\varepsilon_{xx} = \frac{\partial u_x}{\partial x} \tag{5.43}$$

$$\varepsilon_{yy} = \frac{\partial u_y}{\partial y} \tag{5.44}$$

$$\varepsilon_{zz} = \frac{\partial u_z}{\partial z} \tag{5.45}$$

$$\varepsilon_{xy} = \frac{1}{2}\left(\frac{\partial u_x}{\partial y} + \frac{\partial u_y}{\partial x}\right) \tag{5.46}$$

$$\varepsilon_{xz} = \frac{1}{2}\left(\frac{\partial u_x}{\partial z} + \frac{\partial u_z}{\partial x}\right) \tag{5.47}$$

$$\varepsilon_{yz} = \frac{1}{2}\left(\frac{\partial u_y}{\partial z} + \frac{\partial u_z}{\partial x}\right) \tag{5.48}$$

The above 15 equations provide the basis of linear elasticity for solving fields of stress, strain, and displacement in three-dimensional isotropic materials. To define solvable static fields for specific problems, boundary conditions should be fully defined. There are three common types of boundary conditions for solution of the 15-variable fields: displacement boundary conditions, traction boundary conditions, and mixed boundary conditions.

Displacement boundary conditions involve three prescribed displacement values at the boundaries:

$$u_x = \overline{u_x} \tag{5.49}$$

$$u_y = \overline{u_y} \tag{5.50}$$

$$u_z = \overline{u_z} \tag{5.51}$$

Traction or stress boundary conditions prescribe three components of stress vectors on the boundaries of the problem. If **n** is the unit vector at a boundary point with its components

$$n_i = \begin{bmatrix} n_x \\ n_y \\ n_z \end{bmatrix} \quad \text{and} \quad n_x^2 + n_y^2 + n_z^2 = 1 \tag{5.52}$$

then traction at a point then can be expressed as

$$t_x = n_x\sigma_{xx} + n_y\tau_{xy} + n_z\tau_{xz} \tag{5.53}$$

$$t_y = n_y\sigma_{yy} + n_x\tau_{xy} + n_z\tau_{yz} \tag{5.54}$$

$$t_z = n_z\sigma_{zz} + n_x\tau_{xz} + n_y\tau_{yz} \tag{5.55}$$

Mixed boundary conditions can only be the combinations of the following two cases: (1) displacement boundary conditions on part of the bounding surface where the rest are traction boundary conditions, and (2) at each point either displacement or traction, but not both.

Solving 15 dependent variables in three-dimensional space is very challenging, even with today's high-speed supercomputers, not to mention complex geometries and heterogeneity of hillslopes. Often, simplifications are made in terms of dimensionality and the variables of interest.

5.4 Two-dimensional elastostatics

5.4.1 Navier's field equations in terms of displacement

If the primary interest is to formulate displacement-based numerical solutions, the 15 equations for 15 unknowns can be reduced to three equations for three displacements. Substituting the six geometric equations (5.43)–(5.48) into the six Hooke's laws (Equations (5.37)–(5.42)) to obtain stresses in terms of displacements, and substituting the result into the three force equilibrium equations (Equations (5.34)–(5.36)) to eliminate stresses, three second-order differential equations in terms of three displacements can be obtained:

$$(\lambda + G)\left(\frac{\partial^2 u_x}{\partial^2 x} + \frac{\partial^2 u_y}{\partial x \partial y} + \frac{\partial^2 u_z}{\partial x \partial z}\right) + G\left(\frac{\partial^2 u_x}{\partial^2 x} + \frac{\partial^2 u_x}{\partial^2 y} + \frac{\partial^2 u_x}{\partial^2 z}\right) = 0 \quad (5.56)$$

$$(\lambda + G)\left(\frac{\partial^2 u_x}{\partial x \partial y} + \frac{\partial^2 u_y}{\partial^2 y} + \frac{\partial^2 u_z}{\partial y \partial z}\right) + G\left(\frac{\partial^2 u_y}{\partial^2 x} + \frac{\partial^2 u_y}{\partial^2 y} + \frac{\partial^2 u_y}{\partial^2 z}\right) = 0 \quad (5.57)$$

$$(\lambda + G)\left(\frac{\partial^2 u_x}{\partial x \partial z} + \frac{\partial^2 u_y}{\partial y \partial z} + \frac{\partial^2 u_z}{\partial^2 z}\right) + G\left(\frac{\partial^2 u_z}{\partial^2 x} + \frac{\partial^2 u_z}{\partial^2 y} + \frac{\partial^2 u_z}{\partial^2 z}\right) + \rho g = 0 \quad (5.58)$$

With proper boundary conditions defined in Equations (5.49)–(5.51) and/or Equations (5.53)–(5.55), Equations (5.56)–(5.58) can be solved for three-dimensional fields of displacements. Once the displacement fields are obtained, strain and stress components can be estimated as secondary unknowns (derivatives of displacements) from Equations (5.43)–(5.48) and Equations (5.37)–(5.42).

Conceptualizations of two-dimensional problems are used quite often in hillslope hydrology and stability analysis as gravity in the vertical direction is the major driving force for landslides and fluid flow. The two-dimensional linear elastostatic field equations in terms of displacement are also called Navier's equations. There are two common types of two-dimensional simplifications; plane stress and plane strain. For plane strain, displacement perpendicular to the cross section in the x-z plane is zero ($u_y = 0$). Displacements u_x and u_z are independent of y, and Equations (5.56)–(5.58) reduce to two equations for two unknown displacements u_x and u_z:

$$(\lambda + G)\left(\frac{\partial^2 u_x}{\partial^2 x} + \frac{\partial^2 u_z}{\partial x \partial z}\right) + G\left(\frac{\partial^2 u_x}{\partial^2 x} + \frac{\partial^2 u_x}{\partial^2 z}\right) = 0 \quad (5.59)$$

$$(\lambda + G)\left(\frac{\partial^2 u_x}{\partial x \partial z} + \frac{\partial^2 u_z}{\partial^2 z}\right) + G\left(\frac{\partial^2 u_z}{\partial^2 x} + \frac{\partial^2 u_z}{\partial^2 z}\right) + \rho g = 0 \quad (5.60)$$

Plane strain is a good approximation for a long hillslope if the cross section goes through the direction perpendicular to its long axis. The traction boundary conditions in terms of u_x and u_z become

$$t_x = \frac{2G}{1-2\nu}\left((1-\nu)\frac{\partial u_x}{\partial x} + \nu\frac{\partial u_z}{\partial z}\right)n_x + G\left(\frac{\partial u_x}{\partial z} + \frac{\partial u_z}{\partial x}\right)n_y \tag{5.61}$$

$$t_z = \frac{2G}{1-2\nu}\left((1-\nu)\frac{\partial u_z}{\partial z} + \nu\frac{\partial u_x}{\partial x}\right)n_z + G\left(\frac{\partial u_x}{\partial z} + \frac{\partial u_z}{\partial x}\right)n_x \tag{5.62}$$

$$t_y = \frac{2G}{1-2\nu}\left(\frac{\partial u_x}{\partial x} + \frac{\partial u_z}{\partial z}\right)n_y \tag{5.63}$$

The second common simplification for the two-dimensional problem is the plane stress problem where stress in the direction perpendicular to the x-z plane is zero ($\sigma_{yy} = 0$), leading to special forms of Hooke's laws (Equations (5.37), (5.39), and (5.42)):

$$\sigma_{xx} = \frac{E}{1-\nu^2}\left(\varepsilon_{xx} + \nu\varepsilon_{zz}\right) \tag{5.64}$$

$$\sigma_{zz} = \frac{E}{1-\nu^2}\left(\varepsilon_{zz} + \nu\varepsilon_{xx}\right) \tag{5.65}$$

$$\tau_{xz} = 2G\varepsilon_{xz} \tag{5.66}$$

Substituting the above equations and the geometric equations into the equations of motion leads to

$$G\left(\frac{\partial^2 u_x}{\partial^2 x} + \frac{\partial^2 u_x}{\partial^2 z}\right) + G\frac{1+\nu}{1-\nu}\left(\frac{\partial^2 u_x}{\partial^2 x} + \frac{\partial^2 u_z}{\partial x \partial z}\right) = 0 \tag{5.67}$$

$$G\left(\frac{\partial^2 u_z}{\partial^2 x} + \frac{\partial^2 u_z}{\partial^2 z}\right) + G\frac{1+\nu}{1-\nu}\left(\frac{\partial^2 u_x}{\partial x \partial z} + \frac{\partial^2 u_z}{\partial^2 z}\right) + \rho g = 0 \tag{5.68}$$

The traction boundary conditions in terms of u_x and u_z become

$$t_x = \frac{2G}{1-2\nu}\left(\frac{\partial u_x}{\partial x} + \nu\frac{\partial u_z}{\partial z}\right)n_x + G\left(\frac{\partial u_x}{\partial z} + \frac{\partial u_z}{\partial x}\right)n_y \tag{5.69}$$

$$t_z = \frac{2G}{1-2\nu}\left(\frac{\partial u_z}{\partial z} + \nu\frac{\partial u_x}{\partial x}\right)n_z + G\left(\frac{\partial u_x}{\partial z} + \frac{\partial u_z}{\partial x}\right)n_x \tag{5.70}$$

It is noted that for both plane stress and plane strain problems, only two elastic material properties (Young's modulus and Poisson's ratio) are needed in order to fully define displacement fields. Analytical solutions of Equations (5.67)–(5.70) for slope problems are difficult if not impossible. Often numerical techniques such as finite elements and finite differences are used to compute displacement fields.

5.4.2 Beltrami–Michell's field equations in terms of stress

If the stress fields in hillslopes are the primary interest, the 15 field equations can be reduced to fewer equations in terms of stresses. In principle, one can do so by eliminating and substituting displacements, and strains in terms of stresses in the 15 equations. However, for single-valued displacement fields to exist, second-order differentiations of strains would be involved, which lead to implicit coupled stress equations called compatibility equations. These equations must be satisfied in order to ensure the existence of a displacement field. Beltrami (1892) and Michell (1899) showed that the 15 equations can be reduced to six compatibility equations and three equilibrium equations for six stress variables (e.g., Malvern, 1969; Lai *et al.*, 1978). However, of the six compatibility equations, only three are independent. It is difficult to solve analytically the six stress fields (Equation (5.5)) with nine high-order partial differential equations, and few cases have been done for three-dimensional problems. In two-dimensional cases, it is possible to further reduce to three equations with three unknown stress variables (e.g., $\sigma_{xx}, \sigma_{zz}, \tau_{xz}$).

For the plane strain problem, $u_y = 0$, and u_x and u_z are independent of y, leading to three independent strain components: ε_{xx}, ε_{zz}, and ε_{xz}. Hooke's laws (Equations (5.37), (5.39), and (5.42)) under plane strain conditions can be written as

$$\varepsilon_{xx} = \frac{1+\nu}{E}\left[(1-\nu)\sigma_{xx} - \nu\sigma_{zz}\right] \tag{5.71}$$

$$\varepsilon_{zz} = \frac{1+\nu}{E}\left[(1-\nu)\sigma_{zz} - \nu\sigma_{xx}\right] \tag{5.72}$$

$$\varepsilon_{xz} = \frac{1}{2G}\tau_{xz} \tag{5.73}$$

The equilibrium equations (Equations (5.34)–(5.36)) reduce to three stress components and two equations:

$$\frac{\partial \sigma_{xx}}{\partial x} + \frac{\partial \tau_{xz}}{\partial z} = 0 \tag{5.74}$$

$$\frac{\partial \sigma_{zz}}{\partial z} + \frac{\partial \tau_{xz}}{\partial x} + \rho g = 0 \tag{5.75}$$

One compatibility equation applied to the two-dimensional strain fields is

$$\frac{\partial^2 \varepsilon_{xx}}{\partial x^2} + \frac{\partial^2 \varepsilon_{zz}}{\partial z^2} = 2\frac{\partial^2 \varepsilon_{xz}}{\partial x \partial z} \tag{5.76}$$

Substituting the three Hooke's laws ((5.71)–(5.73)) into Equation (5.75) yields the compatibility equation in terms of the three independent stress components, two normal and one shear:

$$2\frac{\partial^2 \tau_{xz}}{\partial x \partial z} = -\frac{\partial^2 \sigma_{xx}}{\partial x^2} - \frac{\partial^2 \sigma_{zz}}{\partial z^2} - \frac{\partial(\rho g)}{\partial z} \tag{5.77}$$

Further eliminating the shear stress component τ_{xy}, by differentiating Equation (5.73) with respect to x and Equation (5.74) with respect to z and adding the two equations, yields the compatibility equation in terms of the two independent normal stress components:

$$\frac{\partial^2 \sigma_{xx}}{\partial x^2} + \frac{\partial^2 \sigma_{xx}}{\partial z^2} + \frac{\partial^2 \sigma_{zz}}{\partial x^2} + \frac{\partial^2 \sigma_{zz}}{\partial z^2} = -\frac{1}{1-\nu}\frac{\partial(\rho g)}{\partial z} \tag{5.78}$$

Thus three Equations (5.74), (5.75), and (5.78) fully define the plane strain problem for the stress field. These three equations can be used as the theoretical basis for solving total stress fields analytically or numerically in two-dimensional hillslopes. It is interesting to note that for two-dimensional plane strain slopes with linear isotropic elastic material, only one elastic property, namely Poisson's ratio, ν, is involved in the solution of stress fields.

For plane stress problems, $\sigma_{yy} = 0$, $\tau_{xy} = 0$, and $\tau_{zy} = 0$, lead to three independent stress components: σ_{xx}, σ_{zz}, and τ_{xz}. Hooke's laws (Equations (5.37), (5.39), and (5.42)) under plane stress conditions can be written as

$$\varepsilon_{xx} = \frac{1}{E}\left[\sigma_{xx} - \nu\sigma_{zz}\right] \tag{5.79}$$

$$\varepsilon_{zz} = \frac{1}{E}\left[\sigma_{zz} - \nu\sigma_{xx}\right] \tag{5.80}$$

$$\varepsilon_{xz} = \frac{1}{2G}\tau_{xz} \tag{5.81}$$

For plane stress problems, strain in the zero stress direction ($\sigma_{yy} = 0$) exists, but can be quantified by the two normal stress components in the x and z directions through Hooke's law (Equation (5.38)):

$$\varepsilon_{yy} = -\frac{\nu}{E}\left[\sigma_{zz} + \sigma_{xx}\right] \tag{5.82}$$

The equilibrium equations (Equations (5.34)–(5.36)) reduce to three stress components and two equations as in plane strain:

$$\frac{\partial\sigma_{xx}}{\partial x} + \frac{\partial\tau_{xz}}{\partial z} = 0 \tag{5.74}$$

$$\frac{\partial\sigma_{zz}}{\partial z} + \frac{\partial\tau_{xz}}{\partial x} + \rho g = 0 \tag{5.75}$$

One compatibility equation applied to the two-dimensional problem in terms of the two independent normal stress components can be arrived at:

$$\frac{\partial^2\sigma_{xx}}{\partial x^2} + \frac{\partial^2\sigma_{xx}}{\partial z^2} + \frac{\partial^2\sigma_{zz}}{\partial x^2} + \frac{\partial^2\sigma_{zz}}{\partial z^2} = -(1+\nu)\frac{\partial(\rho g)}{\partial z} \tag{5.83}$$

Thus three Equations (5.74), (5.75), and (5.83) can fully define the plane stress problem for the stress field. Again, for two-dimensional plane strain hillslopes with linear isotropic elastic material, only one elastic property, namely Poisson's ratio ν, is involved in the solution of the stress fields. Equivalency between plane stress and plane strain solutions can be inferred by comparison of the governing equations for each situation. For example, the solution of the stress field in the x-z plane under plane stress conditions can be obtained by the solution of the stress field of plane strain times $(1-\nu)/(1+\nu)$.

5.5 Total stress distribution in hillslopes

5.5.1 Savage's two-dimensional analytical solution

The linear elastostatic theory covered in the previous section provides a basis for computing total stress fields. Many complex theories considering non-linear or elasto-plastic behavior or even visco-elasto-plastic theories are built on linear elastostatic theory. Although stress fields in hillslopes are subject to potential uncertainty due to complex geologic history, linear elastostatics provides first-order quantitative solutions of total stress fields for non-linear and complex materials under various geometric and boundary conditions.

Typically, analytical and numerical techniques are employed to compute stress fields. The analytical techniques are classical and were common before the computer age. The major advantage of analytical solutions is that they are exact solutions, so that controlling parameters of geometry (e.g., slope angle and dimensions) and material properties (e.g., elastic modulus and Poisson's ratio) can be mathematically isolated to the solutions of stress fields. The major disadvantage of analytical solutions is their limitation in dealing with complex geometries and heterogeneity of material properties. In contrast, numerical solutions fully take advantage of the power of computers and provide greater flexibility in handling complex geometries and material heterogeneity. Nonetheless, specific models have to be constrained for each geometry and set of material properties. Thus, numerical solutions of three-dimensional stress fields provide a modern dilemma; on the one hand, they may provide accurate stress distributions for complex geometries, on the other hand, the computational demands and model parameters required to do so may be impractical and prohibitive even with today's computer power and techniques for characterizing material properties.

Solutions of total stress fields in two-dimensional space from each of these two methods are provided here to illustrate the major characteristics of stress variation in a variety of hillslope settings, i.e., slope angle and dimensions. The two-dimensional numerical solutions are presented in both dimensionless stress contours and table form so that one can use them as a quantitative basis for computing total stress distributions in finite slopes. The usefulness of such contours and tables for general slopes is fully illustrated in the stability analyses presented in Chapter 9.

Only a handful of analytical solutions for the state of stress in finite slopes have been developed (e.g., Silvestri and Tabib, 1983a, b; Savage, 1994), and few of them are adequately tested. Savage's solution is shown here as it has been compared with other analytical and numerical solutions.

In two-dimensional stress problems, instead of solving for stress fields directly, stress functions are often introduced (e.g., Muskhelishvili, 1953; Timoshenko and Goodier, 1987). Savage (1994) extended an original solution by Muskhelishvili (1953) to predict gravity-induced stresses beneath finite slopes of 15, 30, 45, and 90°. The mathematical development and solutions for Equations (5.74), (5.75), and (5.78) are described in detail in Savage (1994), but the results for the three independent stress fields (σ_1, σ_2, and τ_{max}) are presented here to illustrate the total stress distributions in finite slopes. The boundary conditions were set such that the top boundary is a free-stress boundary, and the lower and left and right sides are infinitely far from the slope. Because it is a plane strain problem, only one elastic property (Poisson's ratio) is needed. A Poisson's ratio of 1/3 is assumed in all cases shown here. All stresses are normalized by the quantity ρgh, where h is the height of the finite slope. Both the horizontal coordinate x and the vertical coordinate z are also normalized by the height of the finite slope h, making the results useful for finite slopes of any dimension. With these normalizations, the analytical solutions of the total stress distributions presented here could be used for hillslopes with any dimensions.

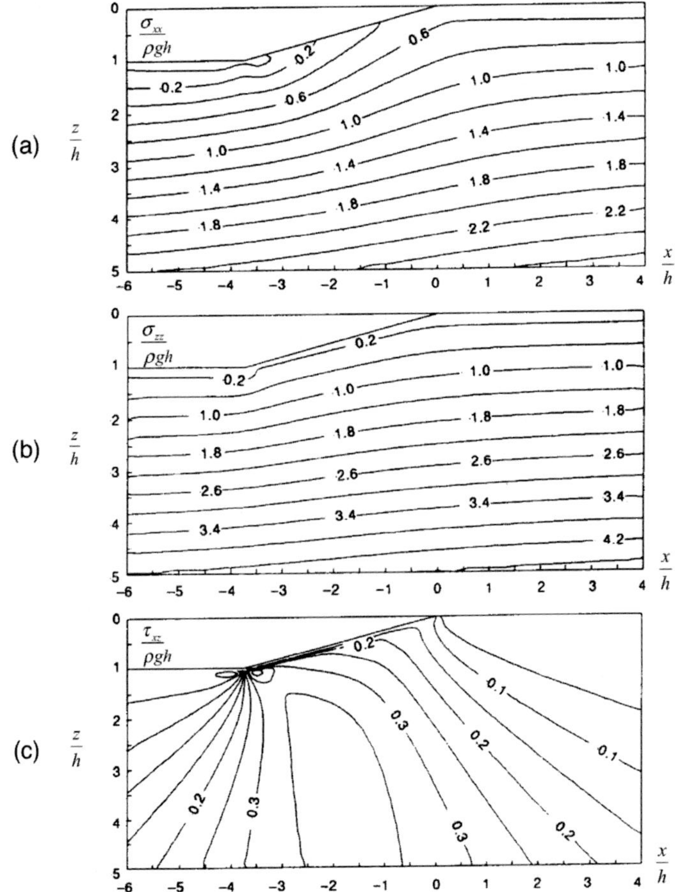

Figure 5.6 Stress contours for a 15° finite slope of height h for a Poisson's ratio of 1/3 (after Savage, 1994).

The total stress fields in dimensionless quantities for a 15° finite slope are shown in Figure 5.6. Both horizontal and vertical stresses are generally compressive or positive valued except for the minimum normal stress in the region in front of the toe of the slope. Stress concentrations occur around the toe region for both the horizontal normal stress and shear stress. A zone where the shear stress is nearly parallel to the slope surface can be observed near the lower portion of the slope, indicating that it is vulnerable to soil failure.

The total stress fields for a 30° finite slope are shown in Figure 5.7. Here the tensile stress zone expands further in front of the toe compared to the 15° slope (Figure 5.6). The zone of stress concentration for the horizontal normal and shear stresses expands upward from the toe. Contours of horizontal and vertical stresses (Savage, 1994) indicate that the minimum normal stresses near the toe area are mostly horizontal, in contrast to the vertical direction for the region away from the slope face. It is clear that, for such a slope, high

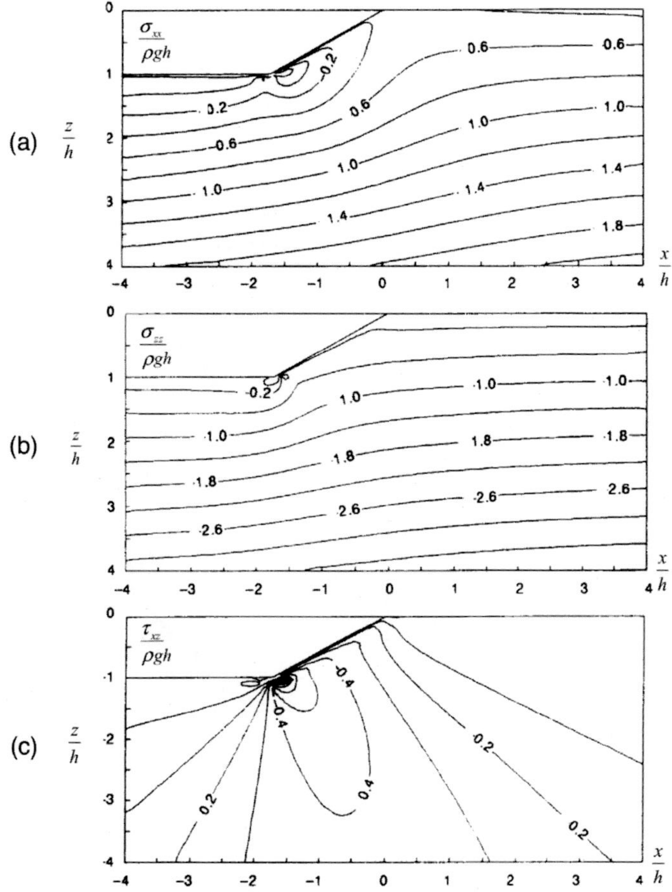

Figure 5.7 Stress contours for a 30° finite slope of height *h* for a Poisson's ratio of 1/3 (after Savage, 1994).

shear stress around the toe and tensile stress in front of the toe are two potential loci for soil failure.

For a 45° finite slope, the zone of shear stress concentration becomes wider and is distributed beneath the entire slope surface. Under such conditions, the slope can fail under shear failure modes (Figure 5.8c).

For a vertical cut slope, the zone of tensile stress expands to encompass the entire region immediately behind the vertical face (Figure 5.9a). The zero stress contours extend nearly the full distance *h* horizontally behind the cut face. Shear stress (Figure 5.9c) concentrates around the toe region. Slopes under such conditions are vulnerable to development of vertical tensile cracks.

In summary, hillslopes are the expression of topographic relief and gradient. Under gravity, topographic gradients promote shear stress; the steeper a slope, the higher the topographic gradient and resulting shear stress. For slopes inclined at the same angle, the magnitude of total stresses, both normal and shear stress, depends mainly on the unit weight

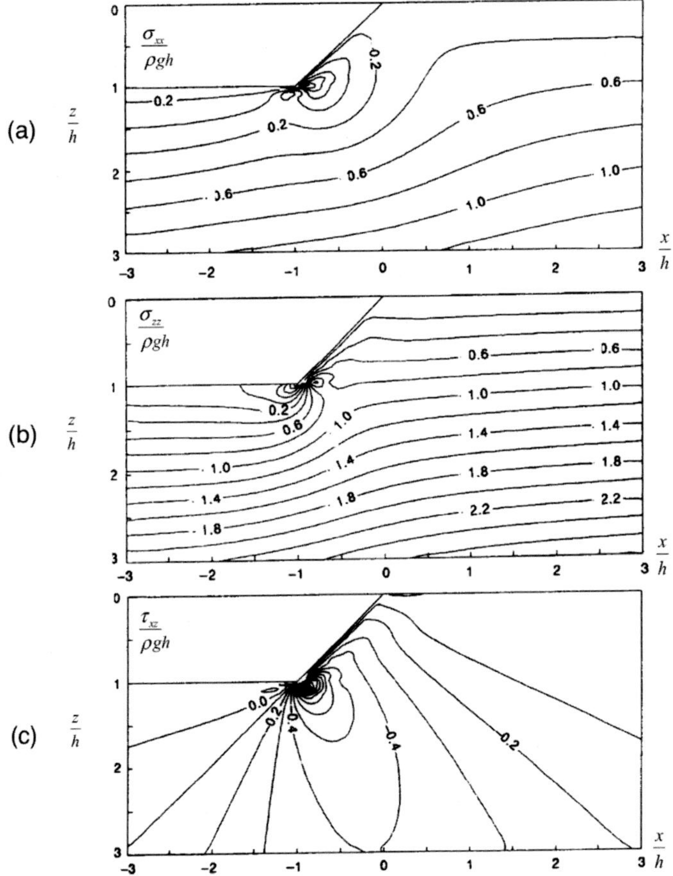

Figure 5.8 Stress contours for a 45° finite slope of height h for a Poisson's ratio of 1/3 (after Savage, 1994).

ρg of the material and the height of the slope h, and to a certain extent Poisson's ratio. The total stress distribution in hillslopes can be characterized by two modes that are critical to the stability of slopes: shear stress concentration, and tensile stress development. In the previously described examples, shear stress concentrates mainly near the toe of hillslopes and regions immediately beneath the slope face. Shear stress is zero for a flat surface and increases as a finite slope becomes steeper. Shear stress can reach several times ρgh beneath the toe in a vertical cut slope. When shear stress reaches the shear strength of hillslope materials, local failure occurs. According to the analytical solutions presented here, tensile stress can develop in front of the slope toe as a finite slope becomes steeper, and it is most pronounced in a vertical cut slope. Tensile stress on the order of ρgh can develop both in front and behind the slope face in a vertical cut slope. Tensile stress may also develop near the crest of steep slopes. Because hillslope materials are generally weak in tensile strength, tensile cracks or failures can occur in those regions of finite slopes. The strength of hillslope materials will be systematically described in Chapter 7 and stability analysis of hillslopes will be covered in Chapters 9 and 10.

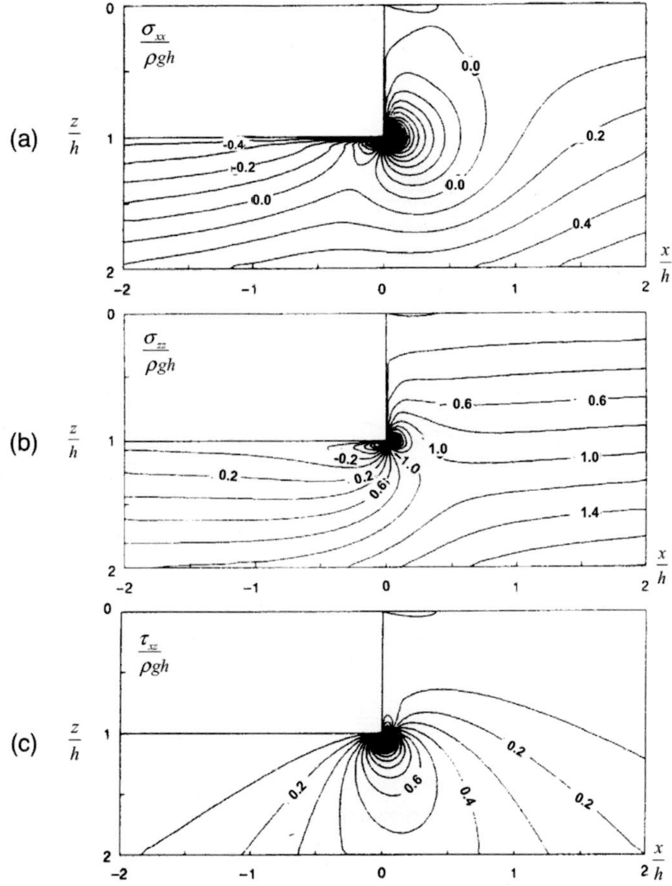

Figure 5.9 Stress contours for a 90° finite slope of height h for a Poisson's ratio of 1/3 (after Savage, 1994).

Although the analytical solutions presented here are useful for identifying weak zones and describing the characteristics of total stress distribution in hillslopes, verification of their accuracy and correctness remains incomplete. Savage (1994) conducted some comparisons with other independent analytical solutions for the same set of boundary value problems and finite-element-based solutions. The results were mixed; some are quantitatively similar, some are qualitatively similar, and still others are quite different. Reconciliation of these differences, both quantitatively and qualitatively, remains to be completed.

5.5.2 Finite-element solutions

Finite-element (FE) models have become widely available over the last several decades. They are numerical solution tools that solve governing partial differential equations under specific geometry and initial and boundary conditions. They offer great flexibility in computing the total stress distributions in hillslopes. Nevertheless, most FE formulations are

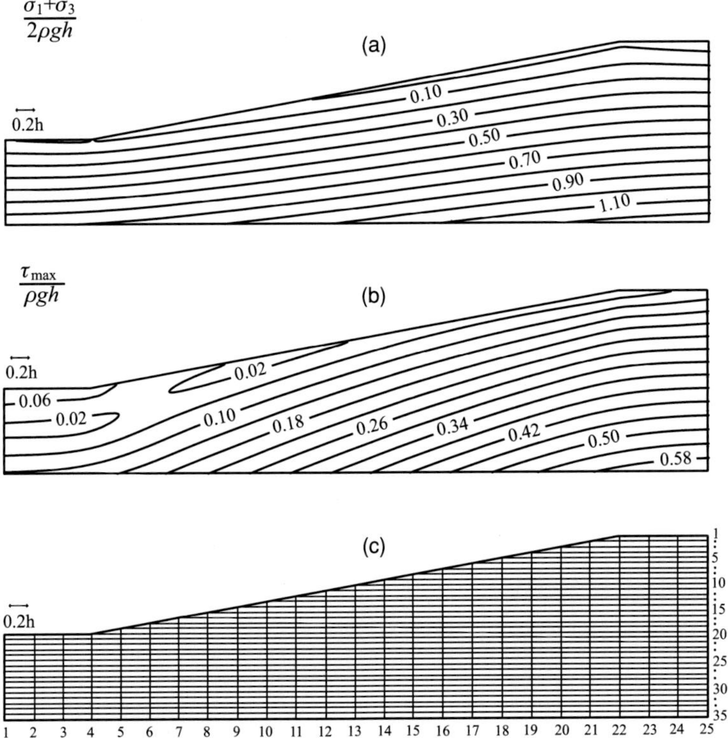

Figure 5.10 Stress contours for a 10° finite slope of height *h* for a Poisson's ratio of 1/3: (a) normalized mean stress, (b) normalized shear stress, and (c) grid for stress distribution Table 5.1.

limited by underestimation and "smoothing" of stress concentration patterns in regions near sharp changes in boundary geometries.

In this section, numerical solutions (GeoStudio, 2007) of the total stress distribution in homogeneous and isotropic hillslopes inclined at varying angles are provided. The lower, left, and right boundaries are no-displacement boundaries set far enough away from the slope such that the effect on the stress distribution within the slope is minimized. To make the solutions applicable for any scales, they are presented in dimensionless quantities in both contour and table forms so that estimates of total stress distributions can be obtained without conducting FE modeling. The solutions of the plane strain field Equations (5.74), (5.75), and (5.78) for total stresses in finite slopes under gravity are shown. All total stresses, namely mean normal stress and maximum shear stress, necessary to define the local factor of safety at every point in a finite slope described in Chapter 10, are normalized with respect to ρgh. Coordinates in both x (horizontal) and z (vertical) are also normalized by h so that the stress contours and tables can be used for finite slopes of any dimensions. For convenience, the increments of slope angles are as follows:

10° (Figure 5.10 and Table 5.1)
15° (Figure 5.11 and Table 5.2)

Table 5.1 Stress distributions for a 10° finite slope of height h for a Poisson's ratio of 1/3

$z(h)$ \ $x(h)$		1	2	3	4	5	6	7	8	9	10	11	12	13	14	15	16	17	18	19	20	21	22	23	24	25
				Toe																				Crest		
		0.00	0.32	0.63	0.95	1.26	1.58	1.89	2.21	2.52	2.84	3.15	3.47	3.78	4.10	4.41	4.73	5.04	5.36	5.67	5.99	6.30	6.62	6.93	7.25	7.56
1	0.00																						−0.01	−0.03	−0.05	−0.06
																							0.03	0.05	0.07	0.08
2	0.06																					−0.02	0.00	−0.01	−0.02	−0.03
																						0.04	0.05	0.07	0.08	0.09
3	0.11																				−0.03	0.00	0.03	0.02	0.01	0.01
																					0.05	0.06	0.08	0.09	0.10	0.11
4	0.17																			−0.03	0.00	0.04	0.06	0.06	0.05	0.04
																				0.05	0.06	0.08	0.10	0.11	0.12	0.12
5	0.22																		−0.03	0.00	0.04	0.07	0.10	0.10	0.09	0.08
																			0.05	0.06	0.08	0.10	0.12	0.13	0.13	0.14
6	0.28																	−0.03	0.00	0.04	0.07	0.11	0.13	0.13	0.13	0.12
																		0.05	0.06	0.08	0.10	0.12	0.14	0.15	0.15	0.16
7	0.33																−0.03	0.00	0.04	0.08	0.11	0.14	0.16	0.17	0.17	0.16
																	0.04	0.06	0.08	0.09	0.11	0.14	0.16	0.17	0.17	0.18
8	0.39															−0.02	0.00	0.04	0.08	0.11	0.14	0.17	0.20	0.20	0.20	0.20
																0.04	0.05	0.07	0.09	0.11	0.13	0.15	0.18	0.18	0.19	0.19
9	0.44														−0.02	0.01	0.04	0.08	0.12	0.15	0.18	0.21	0.23	0.24	0.24	0.23
															0.03	0.05	0.07	0.09	0.11	0.13	0.15	0.17	0.19	0.20	0.21	0.21
10	0.50													−0.01	0.01	0.05	0.09	0.12	0.15	0.19	0.22	0.25	0.27	0.28	0.28	0.28
														0.02	0.04	0.06	0.08	0.10	0.12	0.14	0.17	0.19	0.21	0.22	0.22	0.23
11	0.56												0.00	0.03	0.06	0.09	0.13	0.16	0.20	0.23	0.26	0.29	0.31	0.32	0.32	0.32
													0.02	0.03	0.06	0.08	0.10	0.12	0.14	0.16	0.19	0.21	0.23	0.24	0.24	0.25
12	0.61											0.01	0.04	0.07	0.10	0.13	0.16	0.20	0.24	0.27	0.30	0.32	0.34	0.35	0.36	0.36
												0.01	0.03	0.05	0.07	0.09	0.11	0.14	0.16	0.18	0.20	0.23	0.25	0.26	0.26	0.26
13	0.67										0.02	0.04	0.08	0.11	0.14	0.17	0.21	0.24	0.27	0.30	0.34	0.36	0.38	0.39	0.40	0.39
											0.00	0.02	0.04	0.06	0.08	0.10	0.13	0.15	0.17	0.20	0.22	0.24	0.26	0.27	0.28	0.28
14	0.72									0.03	0.05	0.09	0.12	0.15	0.18	0.21	0.25	0.28	0.31	0.34	0.37	0.40	0.41	0.43	0.43	0.43
										0.01	0.01	0.03	0.05	0.07	0.10	0.12	0.14	0.17	0.19	0.21	0.24	0.26	0.28	0.29	0.29	0.29
15	0.78								0.04	0.06	0.10	0.13	0.16	0.19	0.22	0.26	0.29	0.32	0.35	0.38	0.41	0.44	0.45	0.47	0.47	0.47
									0.02	0.01	0.02	0.04	0.06	0.09	0.11	0.13	0.16	0.18	0.21	0.23	0.25	0.28	0.29	0.30	0.31	0.31
16	0.83							0.05	0.07	0.11	0.14	0.17	0.20	0.23	0.27	0.30	0.33	0.36	0.39	0.42	0.45	0.47	0.49	0.50	0.51	0.51
								0.03	0.02	0.01	0.03	0.06	0.08	0.10	0.13	0.15	0.17	0.20	0.22	0.25	0.27	0.29	0.31	0.32	0.32	0.33

(cont.)

Table 5.1 (cont.)

z(h)	x(h)	1	2	3	4	5	6	7	8	9	10	11	12	13	14	15	16	17	18	19	20	21	22	23	24	25
				Toe																				Crest		
		0.00	0.32	0.63	0.95	1.26	1.58	1.89	2.21	2.52	2.84	3.15	3.47	3.78	4.10	4.41	4.73	5.04	5.36	5.67	5.99	6.30	6.62	6.93	7.25	7.56
17	0.89						0.06	0.08	0.11	0.15	0.18	0.21	0.24	0.28	0.31	0.34	0.37	0.40	0.43	0.46	0.48	0.51	0.53	0.54	0.55	0.55
							0.05	0.03	0.01	0.02	0.05	0.07	0.09	0.12	0.14	0.16	0.19	0.21	0.24	0.26	0.29	0.31	0.33	0.34	0.34	0.34
18	0.94					0.07	0.09	0.13	0.16	0.19	0.22	0.25	0.28	0.31	0.34	0.37	0.41	0.44	0.47	0.50	0.52	0.55	0.56	0.58	0.58	0.59
						0.06	0.04	0.02	0.02	0.04	0.06	0.08	0.10	0.13	0.15	0.18	0.20	0.23	0.25	0.28	0.30	0.32	0.34	0.35	0.36	0.36
19	1.00	0.11	0.10	0.10	0.10	0.10	0.14	0.17	0.20	0.23	0.26	0.29	0.32	0.35	0.38	0.42	0.45	0.48	0.51	0.54	0.56	0.59	0.60	0.61	0.62	0.63
		0.09	0.08	0.08	0.08	0.05	0.03	0.02	0.03	0.05	0.07	0.09	0.12	0.14	0.17	0.19	0.22	0.24	0.27	0.29	0.32	0.34	0.36	0.37	0.37	0.37
20	1.06	0.14	0.13	0.13	0.14	0.16	0.18	0.21	0.24	0.28	0.31	0.34	0.36	0.40	0.43	0.46	0.49	0.52	0.55	0.58	0.61	0.63	0.64	0.66	0.67	0.67
		0.08	0.08	0.07	0.07	0.04	0.03	0.02	0.04	0.06	0.08	0.11	0.13	0.16	0.18	0.21	0.23	0.26	0.29	0.31	0.34	0.36	0.37	0.38	0.39	0.39
21	1.11	0.19	0.18	0.17	0.18	0.20	0.23	0.25	0.29	0.32	0.35	0.38	0.41	0.44	0.47	0.50	0.53	0.57	0.59	0.62	0.65	0.67	0.68	0.70	0.70	0.71
		0.07	0.07	0.06	0.06	0.04	0.02	0.03	0.05	0.07	0.10	0.12	0.15	0.17	0.20	0.22	0.25	0.28	0.30	0.33	0.35	0.37	0.39	0.40	0.40	0.41
22	1.17	0.22	0.22	0.22	0.23	0.25	0.27	0.30	0.33	0.36	0.39	0.42	0.45	0.48	0.51	0.54	0.58	0.60	0.63	0.66	0.68	0.71	0.72	0.73	0.74	0.75
		0.06	0.05	0.05	0.04	0.03	0.02	0.04	0.06	0.09	0.11	0.13	0.16	0.18	0.21	0.24	0.27	0.29	0.32	0.34	0.37	0.39	0.40	0.41	0.42	0.42
23	1.22	0.27	0.26	0.26	0.27	0.29	0.32	0.35	0.37	0.40	0.43	0.46	0.49	0.52	0.55	0.58	0.61	0.64	0.67	0.70	0.72	0.74	0.76	0.77	0.78	0.78
		0.05	0.04	0.04	0.03	0.02	0.03	0.05	0.07	0.10	0.12	0.15	0.17	0.20	0.22	0.25	0.28	0.30	0.33	0.36	0.38	0.40	0.42	0.43	0.43	0.44
24	1.28	0.31	0.31	0.31	0.32	0.34	0.36	0.39	0.41	0.44	0.47	0.50	0.53	0.56	0.59	0.62	0.65	0.68	0.71	0.73	0.76	0.78	0.80	0.81	0.82	0.83
		0.03	0.03	0.03	0.02	0.02	0.04	0.06	0.08	0.11	0.14	0.16	0.19	0.21	0.24	0.27	0.29	0.32	0.35	0.37	0.40	0.42	0.43	0.44	0.45	0.45
25	1.33	0.36	0.36	0.35	0.37	0.38	0.41	0.43	0.46	0.49	0.51	0.55	0.58	0.61	0.64	0.67	0.70	0.73	0.75	0.78	0.80	0.82	0.84	0.85	0.86	0.87
		0.02	0.02	0.02	0.01	0.03	0.05	0.07	0.10	0.12	0.15	0.17	0.20	0.23	0.25	0.28	0.31	0.34	0.36	0.39	0.41	0.43	0.45	0.46	0.47	0.47
26	1.39	0.40	0.40	0.40	0.41	0.43	0.45	0.47	0.50	0.53	0.56	0.59	0.62	0.65	0.68	0.71	0.73	0.76	0.79	0.81	0.84	0.86	0.88	0.89	0.90	0.90
		0.01	0.01	0.01	0.01	0.03	0.06	0.08	0.11	0.13	0.16	0.19	0.21	0.24	0.27	0.30	0.32	0.35	0.38	0.40	0.43	0.45	0.46	0.47	0.48	0.48
27	1.44	0.44	0.44	0.45	0.45	0.47	0.49	0.52	0.55	0.57	0.60	0.63	0.66	0.69	0.71	0.75	0.77	0.80	0.83	0.86	0.88	0.90	0.92	0.93	0.94	0.94
		0.01	0.01	0.01	0.02	0.04	0.07	0.09	0.12	0.15	0.17	0.20	0.23	0.26	0.28	0.31	0.34	0.36	0.39	0.42	0.44	0.46	0.48	0.49	0.49	0.50
28	1.50	0.48	0.48	0.48	0.50	0.51	0.54	0.56	0.59	0.61	0.64	0.67	0.70	0.73	0.76	0.79	0.81	0.84	0.87	0.90	0.92	0.94	0.96	0.97	0.98	0.99
		0.02	0.02	0.02	0.03	0.05	0.08	0.11	0.13	0.16	0.19	0.21	0.24	0.27	0.30	0.32	0.35	0.38	0.41	0.43	0.46	0.48	0.49	0.50	0.51	0.51
29	1.56	0.53	0.53	0.54	0.55	0.56	0.58	0.61	0.63	0.66	0.69	0.72	0.75	0.77	0.81	0.83	0.86	0.89	0.92	0.94	0.96	0.98	1.00	1.01	1.02	1.03
		0.03	0.03	0.04	0.05	0.07	0.09	0.12	0.15	0.17	0.20	0.23	0.26	0.28	0.31	0.34	0.37	0.40	0.42	0.45	0.47	0.49	0.51	0.52	0.53	0.53
30	1.61	0.57	0.58	0.58	0.59	0.61	0.63	0.65	0.67	0.70	0.73	0.76	0.79	0.82	0.84	0.87	0.90	0.93	0.95	0.98	1.00	1.02	1.04	1.05	1.06	1.07
		0.04	0.04	0.05	0.06	0.08	0.10	0.13	0.16	0.19	0.21	0.24	0.27	0.30	0.32	0.35	0.38	0.41	0.44	0.46	0.49	0.51	0.52	0.53	0.54	0.54
31	1.67	0.62	0.62	0.62	0.63	0.65	0.67	0.69	0.72	0.74	0.77	0.80	0.83	0.86	0.88	0.92	0.94	0.97	1.00	1.02	1.04	1.06	1.08	1.09	1.10	1.11
		0.06	0.06	0.06	0.07	0.09	0.12	0.14	0.17	0.20	0.22	0.25	0.28	0.31	0.34	0.37	0.40	0.43	0.45	0.48	0.50	0.52	0.54	0.55	0.56	0.56
32	1.72	0.66	0.66	0.67	0.68	0.69	0.72	0.73	0.76	0.79	0.81	0.84	0.87	0.90	0.93	0.95	0.98	1.01	1.03	1.06	1.08	1.10	1.11	1.13	1.14	1.15
		0.07	0.07	0.07	0.09	0.10	0.13	0.16	0.18	0.21	0.24	0.27	0.30	0.32	0.35	0.38	0.41	0.44	0.47	0.49	0.51	0.54	0.55	0.56	0.57	0.57
33	1.78	0.70	0.70	0.71	0.72	0.74	0.75	0.77	0.80	0.83	0.86	0.88	0.91	0.94	0.97	1.00	1.02	1.05	1.07	1.10	1.12	1.14	1.15	1.17	1.18	1.19
		0.08	0.08	0.09	0.10	0.12	0.14	0.17	0.19	0.22	0.25	0.28	0.31	0.34	0.37	0.40	0.42	0.45	0.48	0.51	0.53	0.55	0.57	0.58	0.58	0.59
34	1.83	0.74	0.74	0.75	0.76	0.78	0.80	0.82	0.85	0.87	0.90	0.92	0.95	0.98	1.01	1.04	1.06	1.09	1.12	1.14	1.16	1.18	1.19	1.21	1.22	1.23
		0.09	0.09	0.10	0.11	0.13	0.15	0.18	0.21	0.24	0.27	0.29	0.32	0.35	0.38	0.41	0.44	0.47	0.50	0.52	0.54	0.56	0.58	0.59	0.60	0.60
35	1.89	0.78	0.79	0.79	0.81	0.82	0.84	0.86	0.89	0.91	0.94	0.97	0.99	1.02	1.05	1.07	1.10	1.13	1.16	1.18	1.20	1.22	1.23	1.25	1.26	1.27
		0.10	0.11	0.11	0.12	0.14	0.16	0.19	0.22	0.25	0.28	0.31	0.34	0.37	0.39	0.42	0.45	0.48	0.51	0.54	0.56	0.58	0.59	0.61	0.61	0.62

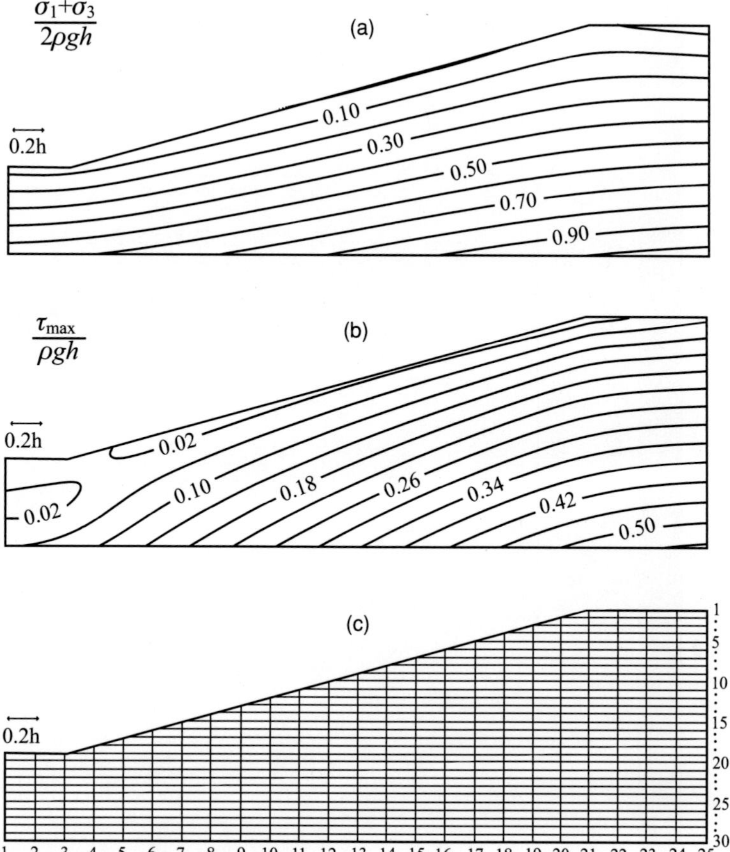

Figure 5.11 Stress contours for a 15° finite slope of height h for a Poisson's ratio of 1/3: (a) normalized mean stress, (b) normalized shear stress, and (c) grid for stress distribution Table 5.2.

20° (Figure 5.12 and Table 5.3)
30° (Figure 5.13 and Table 5.4)
40° (Figure 5.14 and Table 5.5)
45° (Figure 5.15 and Table 5.6)
50° (Figure 5.16 and Table 5.7)
60° (Figure 5.17 and Table 5.8)
70° (Figure 5.18 and Table 5.9)
80° (Figure 5.19 and Table 5.10)
90° (Figure 5.20 and Table 5.11).

The normalized figures and tables are applicable for a finite slope of any dimensions of similar slope geometry. For example, if one is interested in the total stress distributions of a 30° finite slope with a height $h = 15$ m, they can be found in Figures 5.13a and 5.13b and Table 5.4. With the knowledge of unit weight ρg and the height h, the magnitude of stresses

Table 5.2 Stress distributions for a 15° finite slope of height h for a Poisson's ratio of 1/3

$z(h)$ \ $x(h)$		1	2	3	4	5	6	7	8	9	10	11	12	13	14	15	16	17	18	19	20	21	22	23	24	25
					Toe																	Crest				
		0.00	0.21	0.41	0.62	0.83	1.04	1.24	1.45	1.66	1.87	2.07	2.28	2.49	2.70	2.90	3.11	3.32	3.52	3.73	3.94	4.15	4.35	4.56	4.77	4.98
1	0.00																					0.01	0.00	−0.01	−0.03	−0.03
																						0.00	0.02	0.03	0.04	0.05
2	0.06																				0.01	0.03	0.02	0.01	0.00	−0.01
																					0.01	0.02	0.04	0.05	0.06	0.07
3	0.11																			0.01	0.03	0.05	0.05	0.04	0.04	0.03
																				0.01	0.03	0.05	0.06	0.07	0.08	0.09
4	0.17																		0.00	0.03	0.06	0.08	0.08	0.08	0.07	0.06
																			0.01	0.03	0.05	0.07	0.08	0.09	0.10	0.11
5	0.22																	0.00	0.03	0.06	0.09	0.11	0.11	0.11	0.10	0.10
																		0.02	0.03	0.05	0.08	0.10	0.11	0.11	0.12	0.13
6	0.28																0.00	0.02	0.06	0.10	0.12	0.14	0.15	0.15	0.14	0.14
																	0.02	0.03	0.05	0.08	0.10	0.12	0.13	0.13	0.14	0.15
7	0.33															0.00	0.02	0.06	0.10	0.13	0.15	0.17	0.18	0.18	0.18	0.17
																0.02	0.03	0.05	0.07	0.10	0.12	0.14	0.15	0.15	0.16	0.16
8	0.39														0.00	0.02	0.06	0.10	0.13	0.16	0.19	0.20	0.21	0.21	0.21	0.21
															0.02	0.03	0.05	0.07	0.10	0.12	0.14	0.16	0.17	0.17	0.18	0.18
9	0.44													0.00	0.03	0.06	0.10	0.13	0.17	0.20	0.22	0.24	0.25	0.25	0.25	0.25
														0.02	0.03	0.05	0.07	0.09	0.12	0.14	0.16	0.18	0.19	0.19	0.20	0.20
10	0.50												0.00	0.03	0.07	0.11	0.14	0.17	0.20	0.23	0.25	0.27	0.28	0.29	0.29	0.28
													0.02	0.03	0.05	0.07	0.09	0.11	0.14	0.16	0.18	0.20	0.21	0.21	0.22	0.22
11	0.56											0.00	0.03	0.07	0.10	0.14	0.18	0.21	0.24	0.26	0.29	0.30	0.32	0.32	0.32	0.32
												0.01	0.03	0.05	0.07	0.09	0.11	0.13	0.16	0.18	0.20	0.22	0.23	0.23	0.24	0.24
12	0.61										0.00	0.04	0.07	0.11	0.15	0.18	0.21	0.24	0.27	0.30	0.32	0.34	0.35	0.36	0.36	0.36
											0.01	0.02	0.04	0.06	0.08	0.11	0.13	0.15	0.17	0.20	0.22	0.23	0.24	0.25	0.25	0.26
13	0.67									0.01	0.03	0.07	0.11	0.15	0.18	0.21	0.25	0.28	0.31	0.34	0.35	0.37	0.39	0.39	0.40	0.40
										0.01	0.02	0.04	0.06	0.08	0.10	0.12	0.15	0.17	0.19	0.22	0.24	0.25	0.26	0.27	0.27	0.27
14	0.72								0.02	0.04	0.08	0.12	0.15	0.19	0.22	0.25	0.29	0.32	0.35	0.37	0.39	0.40	0.42	0.43	0.43	0.43
									0.00	0.02	0.04	0.06	0.07	0.10	0.12	0.14	0.16	0.19	0.21	0.23	0.25	0.27	0.28	0.29	0.29	0.29

15	0.78							0.02	0.04	0.08	0.12	0.16	0.19	0.23	0.26	0.29	0.32	0.35	0.38	0.41	0.43	0.44	0.45	0.47	0.47	0.47
								0.01	0.01	0.03	0.05	0.07	0.09	0.11	0.13	0.16	0.18	0.20	0.23	0.25	0.27	0.29	0.30	0.30	0.31	0.31
16	0.83						0.02	0.05	0.09	0.13	0.16	0.20	0.23	0.27	0.30	0.33	0.36	0.39	0.42	0.44	0.46	0.48	0.49	0.50	0.50	0.51
							0.01	0.01	0.03	0.05	0.06	0.09	0.11	0.13	0.15	0.18	0.20	0.22	0.25	0.27	0.29	0.30	0.31	0.32	0.32	0.33
17	0.89					0.03	0.06	0.10	0.13	0.17	0.20	0.24	0.27	0.30	0.34	0.37	0.40	0.43	0.46	0.48	0.50	0.52	0.53	0.54	0.54	0.55
						0.02	0.01	0.02	0.04	0.06	0.08	0.10	0.12	0.14	0.17	0.19	0.22	0.24	0.26	0.29	0.30	0.32	0.33	0.34	0.34	0.34
18	0.94				0.04	0.07	0.11	0.14	0.18	0.21	0.25	0.28	0.31	0.35	0.38	0.41	0.44	0.47	0.49	0.52	0.54	0.55	0.57	0.58	0.58	0.59
					0.03	0.02	0.02	0.04	0.05	0.07	0.09	0.12	0.14	0.16	0.19	0.21	0.23	0.26	0.28	0.30	0.32	0.34	0.35	0.35	0.36	0.36
19	1.00	0.07	0.06	0.06	0.08	0.12	0.15	0.19	0.22	0.25	0.29	0.32	0.36	0.39	0.42	0.45	0.48	0.51	0.53	0.55	0.57	0.59	0.60	0.61	0.62	0.62
		0.05	0.04	0.04	0.03	0.02	0.03	0.05	0.07	0.09	0.11	0.13	0.15	0.18	0.20	0.23	0.25	0.27	0.30	0.32	0.34	0.35	0.36	0.37	0.38	0.38
20	1.06	0.10	0.10	0.11	0.13	0.16	0.20	0.23	0.26	0.30	0.33	0.36	0.40	0.43	0.46	0.49	0.52	0.55	0.57	0.59	0.61	0.62	0.64	0.65	0.66	0.66
		0.04	0.04	0.04	0.02	0.03	0.04	0.06	0.08	0.10	0.12	0.15	0.17	0.19	0.22	0.24	0.27	0.29	0.31	0.33	0.35	0.37	0.38	0.39	0.39	0.39
21	1.11	0.15	0.15	0.16	0.18	0.21	0.24	0.27	0.31	0.34	0.37	0.41	0.44	0.47	0.50	0.53	0.56	0.58	0.61	0.63	0.65	0.66	0.68	0.69	0.69	0.70
		0.04	0.03	0.03	0.02	0.03	0.05	0.07	0.09	0.11	0.14	0.16	0.18	0.21	0.23	0.26	0.28	0.31	0.33	0.35	0.37	0.38	0.39	0.40	0.41	0.41
22	1.17	0.20	0.19	0.20	0.23	0.25	0.29	0.32	0.35	0.38	0.41	0.45	0.48	0.51	0.54	0.57	0.60	0.62	0.64	0.67	0.69	0.70	0.71	0.73	0.73	0.74
		0.03	0.02	0.02	0.03	0.04	0.06	0.08	0.10	0.13	0.15	0.17	0.20	0.22	0.25	0.27	0.30	0.32	0.34	0.37	0.38	0.40	0.41	0.42	0.42	0.43
23	1.22	0.24	0.24	0.25	0.27	0.30	0.33	0.36	0.39	0.42	0.46	0.49	0.52	0.55	0.58	0.61	0.63	0.66	0.68	0.70	0.72	0.74	0.75	0.76	0.77	0.77
		0.02	0.01	0.01	0.03	0.05	0.07	0.09	0.12	0.14	0.16	0.19	0.21	0.24	0.26	0.29	0.31	0.34	0.36	0.38	0.40	0.41	0.43	0.43	0.44	0.44
24	1.28	0.28	0.29	0.30	0.32	0.35	0.37	0.40	0.43	0.46	0.50	0.53	0.56	0.59	0.62	0.65	0.67	0.70	0.72	0.74	0.76	0.77	0.79	0.80	0.81	0.81
		0.01	0.01	0.01	0.04	0.06	0.08	0.11	0.13	0.15	0.18	0.20	0.23	0.25	0.28	0.30	0.33	0.35	0.38	0.40	0.41	0.43	0.44	0.45	0.45	0.46
25	1.33	0.33	0.34	0.35	0.37	0.39	0.42	0.45	0.48	0.51	0.54	0.57	0.60	0.63	0.66	0.69	0.71	0.74	0.76	0.78	0.80	0.82	0.83	0.84	0.84	0.85
		0.01	0.01	0.02	0.04	0.07	0.09	0.12	0.14	0.17	0.19	0.22	0.24	0.27	0.29	0.32	0.34	0.37	0.39	0.41	0.43	0.44	0.46	0.46	0.47	0.47
26	1.39	0.38	0.39	0.40	0.41	0.44	0.46	0.49	0.52	0.55	0.58	0.61	0.64	0.67	0.70	0.72	0.75	0.77	0.79	0.81	0.83	0.85	0.86	0.87	0.89	0.89
		0.02	0.02	0.03	0.05	0.08	0.10	0.13	0.15	0.18	0.20	0.23	0.26	0.28	0.31	0.33	0.36	0.38	0.40	0.43	0.44	0.46	0.47	0.48	0.49	0.49
27	1.44	0.43	0.44	0.45	0.46	0.49	0.51	0.54	0.57	0.60	0.63	0.66	0.69	0.72	0.74	0.77	0.79	0.82	0.84	0.86	0.87	0.89	0.90	0.92	0.93	0.93
		0.03	0.03	0.04	0.06	0.09	0.11	0.14	0.17	0.19	0.22	0.24	0.27	0.30	0.32	0.35	0.38	0.40	0.42	0.44	0.46	0.47	0.49	0.50	0.50	0.51
28	1.50	0.47	0.48	0.49	0.51	0.53	0.56	0.58	0.61	0.64	0.67	0.70	0.72	0.75	0.78	0.81	0.83	0.85	0.88	0.90	0.91	0.93	0.94	0.96	0.96	0.97
		0.04	0.04	0.05	0.08	0.10	0.13	0.15	0.18	0.20	0.23	0.26	0.28	0.31	0.34	0.36	0.39	0.41	0.44	0.46	0.47	0.49	0.50	0.51	0.52	0.52
29	1.56	0.52	0.53	0.54	0.56	0.58	0.60	0.63	0.65	0.68	0.71	0.74	0.77	0.79	0.82	0.85	0.87	0.89	0.92	0.93	0.96	0.97	0.98	0.99	1.00	1.01
		0.05	0.05	0.07	0.09	0.11	0.14	0.16	0.19	0.22	0.24	0.27	0.30	0.33	0.35	0.38	0.40	0.43	0.45	0.47	0.49	0.50	0.52	0.53	0.53	0.54
30	1.61	0.57	0.57	0.59	0.61	0.62	0.65	0.67	0.70	0.73	0.75	0.78	0.81	0.84	0.86	0.89	0.91	0.94	0.95	0.97	0.99	1.01	1.02	1.03	1.04	1.05
		0.06	0.07	0.08	0.10	0.12	0.15	0.18	0.20	0.23	0.26	0.28	0.31	0.34	0.37	0.39	0.42	0.44	0.47	0.49	0.50	0.52	0.53	0.54	0.55	0.55

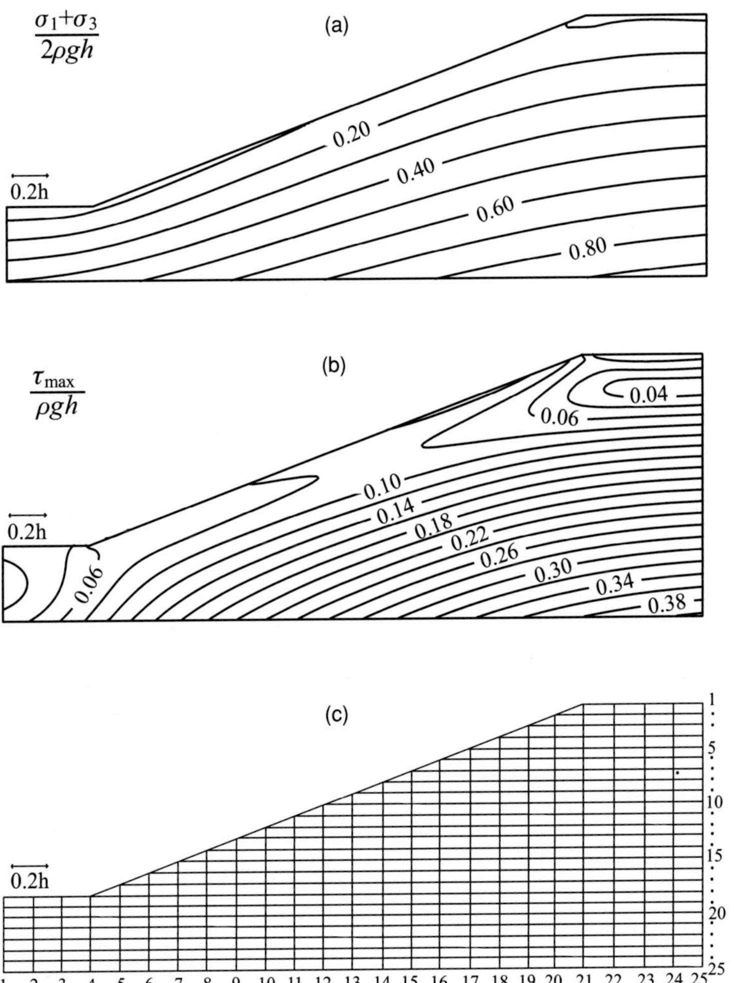

Figure 5.12 Stress contours for a 20° finite slope of height h for a Poisson's ratio of 1/3: (a) normalized mean stress, (b) normalized shear stress, and (c) grid for stress distribution Table 5.3.

can be quantified. Furthermore, if stress profiles at the toe, middle of the slope, and crest locations are desired, the locations can be identified from Figure 5.13c, i.e., the toe is at column 5, the middle of the slope is at column 14, and the crest at column 21. The stress profiles then can be obtained from Table 5.4, where each profile is composed of pairs of mean normal and shear stresses at 25 evenly spaced locations, as shown in Figure 5.21. The stress profiles at different locations are quite different and the differences among them could be significant. For example, the mean stress difference between the toe and crest at the same depth from the slope surface can be as high as 120 kPa or more than 50% difference. The maximum shear stress difference between the toe and middle of the slope can be as much as 30 kPa or more than 100% difference. In summary, Section 5.5 presents the total stress

Table 5.3 Stress distributions for a 20° finite slope of height h for a Poisson's ratio of 1/3

$z(h)$	$x(h)$	1	2	3	4	5	6	7	8	9	10	11	12	13	14	15	16	17	18	19	20	21	22	23	24	25
						Toe																Crest				
		0.00	0.16	0.32	0.48	0.65	0.81	0.97	1.13	1.29	1.45	1.62	1.78	1.94	2.10	2.26	2.42	2.59	2.75	2.91	3.07	3.23	3.39	3.56	3.72	3.88
1	0.00																					0.05	0.09	0.09	0.09	0.09
																						0.05	0.08	0.08	0.08	0.07
2	0.06																				0.10	0.09	0.11	0.11	0.11	0.11
																					0.09	0.06	0.05	0.06	0.06	0.05
3	0.12																			0.11	0.12	0.13	0.14	0.15	0.15	0.14
																				0.10	0.07	0.04	0.03	0.03	0.03	0.03
4	0.18																		0.12	0.13	0.15	0.16	0.17	0.18	0.18	0.18
																			0.11	0.08	0.06	0.03	0.01	0.01	0.01	0.01
5	0.24																	0.12	0.14	0.16	0.18	0.19	0.20	0.21	0.21	0.21
																		0.11	0.09	0.07	0.05	0.03	0.02	0.02	0.02	0.02
6	0.29																0.12	0.14	0.17	0.19	0.21	0.22	0.23	0.24	0.25	0.25
																	0.11	0.09	0.08	0.06	0.05	0.04	0.04	0.04	0.04	0.04
7	0.35															0.11	0.14	0.17	0.20	0.23	0.24	0.26	0.27	0.28	0.28	0.28
																0.10	0.09	0.08	0.07	0.06	0.06	0.06	0.06	0.06	0.06	0.07
8	0.41														0.11	0.14	0.18	0.21	0.23	0.26	0.28	0.29	0.30	0.31	0.32	0.32
															0.10	0.09	0.08	0.08	0.07	0.07	0.08	0.08	0.08	0.08	0.09	0.09
9	0.47													0.11	0.14	0.18	0.21	0.24	0.27	0.29	0.31	0.32	0.34	0.35	0.35	0.35
														0.10	0.09	0.08	0.08	0.08	0.08	0.09	0.09	0.10	0.10	0.11	0.11	0.11
10	0.53												0.10	0.14	0.17	0.21	0.24	0.27	0.30	0.32	0.34	0.36	0.37	0.38	0.38	0.39
													0.09	0.09	0.08	0.08	0.08	0.09	0.10	0.10	0.11	0.12	0.12	0.12	0.13	0.13
11	0.59											0.10	0.13	0.17	0.21	0.25	0.28	0.31	0.33	0.36	0.38	0.39	0.40	0.41	0.42	0.42
												0.09	0.08	0.08	0.08	0.09	0.09	0.10	0.11	0.12	0.13	0.14	0.14	0.14	0.15	0.15
12	0.65										0.09	0.13	0.17	0.21	0.25	0.28	0.32	0.34	0.37	0.39	0.41	0.43	0.44	0.45	0.46	0.46
											0.08	0.08	0.08	0.08	0.09	0.10	0.11	0.12	0.13	0.14	0.15	0.16	0.16	0.17	0.17	0.17

(cont.)

Table 5.3 (cont.)

$z(h)$ \ $x(h)$		1	2	3	4	5	6	7	8	9	10	11	12	13	14	15	16	17	18	19	20	21	22	23	24	25
						Toe																Crest				
		0.00	0.16	0.32	0.48	0.65	0.81	0.97	1.13	1.29	1.45	1.62	1.78	1.94	2.10	2.26	2.42	2.59	2.75	2.91	3.07	3.23	3.39	3.56	3.72	3.88
13	0.71									0.09	0.13	0.17	0.21	0.25	0.29	0.32	0.35	0.38	0.41	0.43	0.45	0.46	0.47	0.48	0.49	0.50
										0.08	0.08	0.08	0.08	0.09	0.10	0.11	0.12	0.14	0.15	0.16	0.17	0.17	0.18	0.18	0.19	0.19
14	0.76								0.08	0.12	0.17	0.21	0.25	0.29	0.32	0.36	0.39	0.41	0.44	0.46	0.48	0.49	0.51	0.52	0.53	0.53
									0.07	0.07	0.08	0.08	0.09	0.10	0.11	0.13	0.14	0.15	0.16	0.18	0.19	0.19	0.20	0.20	0.21	0.21
15	0.82							0.07	0.12	0.17	0.21	0.25	0.29	0.33	0.36	0.39	0.42	0.45	0.47	0.50	0.52	0.53	0.54	0.55	0.56	0.57
								0.07	0.07	0.08	0.08	0.09	0.10	0.12	0.13	0.14	0.16	0.17	0.18	0.19	0.20	0.21	0.22	0.22	0.23	0.23
16	0.88						0.07	0.12	0.16	0.21	0.25	0.29	0.33	0.37	0.40	0.43	0.46	0.49	0.51	0.53	0.55	0.57	0.58	0.59	0.60	0.61
							0.06	0.07	0.07	0.08	0.09	0.10	0.12	0.13	0.15	0.16	0.18	0.19	0.20	0.21	0.22	0.23	0.24	0.24	0.24	0.25
17	0.94					0.07	0.12	0.16	0.21	0.25	0.30	0.33	0.37	0.41	0.44	0.47	0.50	0.53	0.55	0.57	0.59	0.60	0.62	0.63	0.64	0.65
						0.06	0.07	0.07	0.08	0.09	0.11	0.12	0.13	0.15	0.16	0.18	0.19	0.21	0.22	0.23	0.24	0.25	0.25	0.26	0.26	0.27
18	1.00	0.03	0.03	0.03	0.09	0.11	0.16	0.21	0.25	0.30	0.34	0.37	0.41	0.45	0.48	0.51	0.54	0.56	0.58	0.60	0.62	0.64	0.65	0.66	0.67	0.68
		0.02	0.02	0.03	0.07	0.06	0.07	0.08	0.09	0.10	0.12	0.13	0.15	0.16	0.18	0.19	0.21	0.22	0.24	0.25	0.26	0.26	0.27	0.28	0.28	0.28
19	1.06	0.08	0.08	0.09	0.13	0.17	0.21	0.26	0.30	0.34	0.38	0.42	0.45	0.49	0.52	0.55	0.57	0.60	0.62	0.64	0.66	0.68	0.69	0.70	0.71	0.72
		0.02	0.03	0.04	0.06	0.07	0.08	0.09	0.10	0.12	0.13	0.15	0.17	0.18	0.20	0.21	0.23	0.24	0.25	0.26	0.27	0.28	0.29	0.29	0.30	0.30
20	1.12	0.14	0.14	0.16	0.19	0.22	0.26	0.31	0.35	0.39	0.42	0.46	0.50	0.53	0.56	0.59	0.61	0.64	0.66	0.68	0.70	0.71	0.73	0.74	0.75	0.76
		0.02	0.02	0.04	0.06	0.07	0.09	0.10	0.12	0.13	0.15	0.16	0.18	0.20	0.21	0.23	0.24	0.26	0.27	0.28	0.29	0.30	0.31	0.31	0.32	0.32
21	1.18	0.19	0.20	0.22	0.24	0.28	0.32	0.36	0.39	0.43	0.47	0.50	0.54	0.57	0.60	0.63	0.65	0.68	0.70	0.72	0.74	0.75	0.77	0.78	0.79	0.80
		0.01	0.02	0.04	0.06	0.08	0.10	0.11	0.13	0.15	0.16	0.18	0.20	0.21	0.23	0.25	0.26	0.27	0.29	0.30	0.31	0.32	0.32	0.33	0.33	0.34
22	1.24	0.25	0.26	0.27	0.30	0.33	0.37	0.40	0.44	0.48	0.51	0.55	0.58	0.61	0.64	0.67	0.69	0.72	0.74	0.76	0.78	0.79	0.80	0.82	0.83	0.84
		0.01	0.02	0.04	0.06	0.09	0.11	0.12	0.14	0.16	0.18	0.20	0.21	0.23	0.25	0.26	0.28	0.29	0.30	0.32	0.33	0.33	0.34	0.35	0.35	0.35
23	1.29	0.30	0.31	0.33	0.35	0.38	0.41	0.45	0.48	0.52	0.55	0.59	0.62	0.65	0.68	0.70	0.73	0.75	0.77	0.79	0.81	0.83	0.84	0.85	0.86	0.87
		0.01	0.03	0.05	0.07	0.09	0.11	0.13	0.15	0.17	0.19	0.21	0.23	0.24	0.26	0.28	0.29	0.31	0.32	0.33	0.34	0.35	0.36	0.36	0.37	0.37
24	1.35	0.35	0.37	0.38	0.41	0.43	0.47	0.50	0.53	0.57	0.60	0.63	0.66	0.69	0.72	0.75	0.77	0.79	0.81	0.83	0.85	0.87	0.88	0.89	0.90	0.91
		0.02	0.03	0.05	0.08	0.10	0.12	0.15	0.17	0.19	0.21	0.22	0.24	0.26	0.28	0.29	0.31	0.32	0.33	0.35	0.36	0.37	0.37	0.38	0.38	0.39
25	1.41	0.41	0.42	0.44	0.46	0.49	0.52	0.55	0.58	0.61	0.64	0.68	0.71	0.73	0.76	0.79	0.81	0.83	0.85	0.87	0.89	0.90	0.92	0.93	0.94	0.95
		0.03	0.04	0.06	0.08	0.11	0.13	0.16	0.18	0.20	0.22	0.24	0.26	0.27	0.29	0.31	0.32	0.34	0.35	0.36	0.37	0.38	0.39	0.40	0.40	0.40

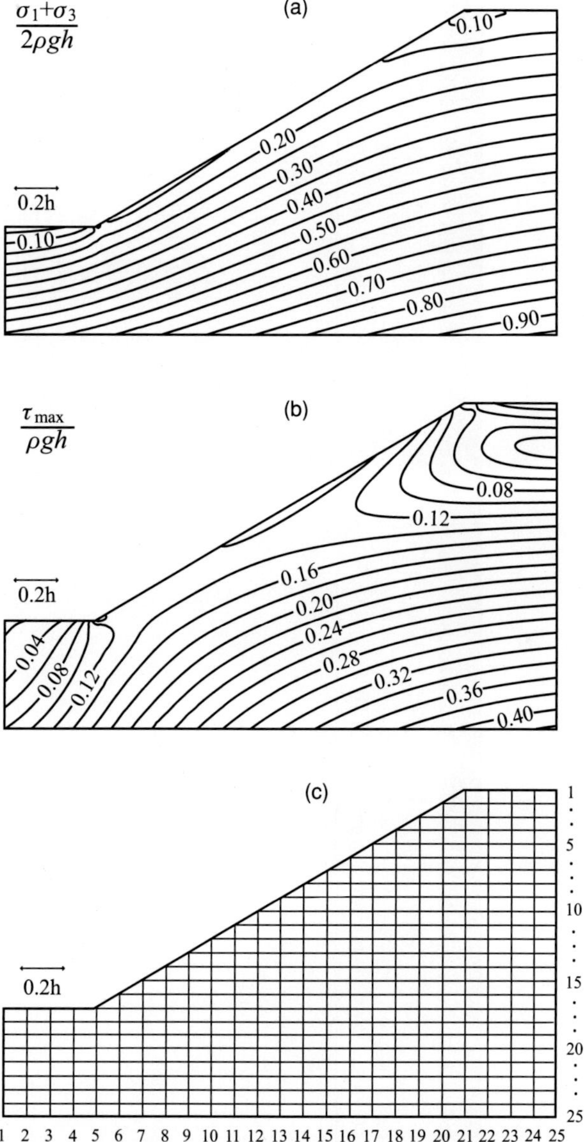

Figure 5.13 Stress contours for a 30° finite slope of height h for a Poisson's ratio of 1/3: (a) normalized mean stress, (b) normalized shear stress, and (c) grid for stress distribution Table 5.4.

distributions resulting from gravity in two-dimensional finite slopes inclined at various angles. Graphical representations (Figures 5.7–5.9) of analytical solutions and graphical representations (Figures 5.10–5.20) and tables (Tables 5.1–5.11) of finite-element solutions are presented in dimensionless form allowing for the assessment of total stress distributions in two-dimensional slopes. These stress fields provide the necessary background information for stress-based stability analyses introduced in Chapter 10.

Table 5.4 Stress distributions for a 30° finite slope of height h for a Poisson's ratio of 1/3

$z(h)$ \ $x(h)$		1	2	3	4	5	6	7	8	9	10	11	12	13	14	15	16	17	18	19	20	21	22	23	24	25
						Toe																Crest				
		0.00	0.11	0.22	0.32	0.43	0.54	0.65	0.76	0.87	0.97	1.08	1.19	1.30	1.41	1.52	1.62	1.73	1.84	1.95	2.06	2.17	2.27	2.38	2.49	2.60
1	0.00																					0.04	0.09	0.10	0.11	0.11
																						0.04	0.08	0.09	0.09	0.10
2	0.06																				0.11	0.09	0.11	0.12	0.13	0.13
																					0.10	0.06	0.06	0.06	0.07	0.07
3	0.13																			0.13	0.13	0.13	0.14	0.15	0.16	0.16
																				0.12	0.09	0.06	0.04	0.04	0.04	0.04
4	0.19																		0.14	0.15	0.16	0.16	0.17	0.18	0.19	0.19
																			0.13	0.10	0.07	0.05	0.04	0.02	0.01	0.01
5	0.25																	0.15	0.16	0.18	0.19	0.19	0.20	0.21	0.22	0.22
																		0.14	0.12	0.09	0.07	0.05	0.04	0.03	0.02	0.02
6	0.31																0.16	0.17	0.19	0.21	0.22	0.23	0.24	0.25	0.25	0.26
																	0.15	0.13	0.11	0.09	0.07	0.06	0.05	0.05	0.04	0.04
7	0.38															0.16	0.18	0.20	0.22	0.23	0.25	0.26	0.27	0.28	0.28	0.29
																0.15	0.13	0.12	0.10	0.09	0.08	0.07	0.07	0.07	0.07	0.07
8	0.44														0.16	0.18	0.21	0.23	0.25	0.27	0.28	0.29	0.30	0.31	0.32	0.33
															0.15	0.14	0.12	0.11	0.10	0.10	0.09	0.09	0.09	0.09	0.09	0.09
9	0.50													0.16	0.18	0.21	0.24	0.26	0.28	0.30	0.31	0.32	0.33	0.34	0.35	0.36
														0.15	0.14	0.13	0.12	0.12	0.11	0.11	0.11	0.11	0.11	0.11	0.11	0.11
10	0.56												0.15	0.18	0.21	0.24	0.27	0.29	0.31	0.33	0.34	0.36	0.37	0.38	0.39	0.39
													0.15	0.14	0.13	0.13	0.13	0.13	0.13	0.13	0.13	0.13	0.13	0.13	0.14	0.14
11	0.63											0.15	0.18	0.22	0.25	0.27	0.30	0.32	0.34	0.36	0.38	0.39	0.40	0.41	0.42	0.43
												0.14	0.14	0.13	0.13	0.13	0.14	0.14	0.14	0.14	0.15	0.15	0.15	0.15	0.16	0.16
12	0.69										0.15	0.18	0.22	0.25	0.28	0.31	0.34	0.36	0.38	0.40	0.41	0.43	0.44	0.45	0.46	0.47
											0.14	0.14	0.14	0.14	0.14	0.15	0.15	0.15	0.16	0.16	0.17	0.17	0.17	0.18	0.18	0.18

13	0.75									0.14	0.18	0.22	0.25	0.29	0.32	0.34	0.37	0.39	0.41	0.43	0.45	0.46	0.47	0.48	0.49	0.50
										0.13	0.14	0.14	0.14	0.15	0.15	0.16	0.17	0.17	0.18	0.18	0.19	0.19	0.19	0.20	0.20	0.20
14	0.81								0.14	0.18	0.22	0.26	0.29	0.32	0.35	0.38	0.41	0.43	0.45	0.47	0.48	0.50	0.51	0.52	0.53	0.54
									0.13	0.13	0.14	0.14	0.15	0.16	0.17	0.17	0.18	0.19	0.20	0.20	0.21	0.21	0.21	0.22	0.22	0.22
15	0.88							0.13	0.17	0.22	0.26	0.29	0.33	0.36	0.39	0.42	0.44	0.46	0.48	0.50	0.52	0.53	0.54	0.56	0.57	0.57
								0.13	0.13	0.14	0.15	0.15	0.16	0.17	0.18	0.19	0.20	0.21	0.21	0.22	0.22	0.23	0.23	0.24	0.24	0.24
16	0.94						0.13	0.18	0.22	0.26	0.30	0.34	0.37	0.40	0.43	0.46	0.48	0.50	0.52	0.54	0.55	0.57	0.58	0.59	0.60	0.61
							0.13	0.13	0.14	0.15	0.16	0.17	0.18	0.19	0.20	0.21	0.22	0.22	0.23	0.24	0.24	0.25	0.25	0.26	0.26	0.26
17	1.00	0.02	0.02	0.03	0.06	0.22	0.18	0.22	0.27	0.31	0.34	0.38	0.41	0.44	0.47	0.49	0.52	0.54	0.56	0.57	0.59	0.60	0.62	0.63	0.64	0.65
		0.01	0.02	0.03	0.05	0.15	0.13	0.14	0.15	0.16	0.17	0.18	0.19	0.20	0.21	0.22	0.23	0.24	0.25	0.26	0.26	0.27	0.27	0.28	0.28	0.28
18	1.06	0.08	0.09	0.11	0.14	0.20	0.24	0.28	0.32	0.35	0.39	0.42	0.45	0.48	0.51	0.53	0.56	0.58	0.60	0.61	0.63	0.64	0.66	0.67	0.68	0.69
		0.02	0.03	0.04	0.07	0.11	0.13	0.14	0.16	0.17	0.18	0.19	0.21	0.22	0.23	0.24	0.25	0.26	0.27	0.27	0.28	0.29	0.29	0.29	0.30	0.30
19	1.13	0.14	0.16	0.17	0.20	0.25	0.29	0.33	0.36	0.40	0.43	0.46	0.49	0.52	0.55	0.57	0.59	0.61	0.63	0.65	0.67	0.68	0.69	0.71	0.72	0.73
		0.02	0.04	0.05	0.08	0.11	0.13	0.15	0.17	0.18	0.20	0.21	0.22	0.23	0.25	0.26	0.27	0.28	0.28	0.29	0.30	0.30	0.31	0.31	0.32	0.32
20	1.19	0.21	0.22	0.24	0.27	0.31	0.34	0.38	0.41	0.44	0.48	0.51	0.54	0.56	0.59	0.61	0.63	0.65	0.67	0.69	0.71	0.72	0.73	0.75	0.76	0.77
		0.03	0.04	0.06	0.09	0.12	0.14	0.16	0.18	0.19	0.21	0.22	0.24	0.25	0.26	0.27	0.28	0.29	0.30	0.31	0.32	0.32	0.33	0.33	0.33	0.34
21	1.25	0.27	0.28	0.30	0.33	0.36	0.39	0.43	0.46	0.49	0.52	0.55	0.58	0.60	0.63	0.65	0.67	0.69	0.71	0.73	0.74	0.76	0.77	0.78	0.79	0.80
		0.03	0.05	0.07	0.10	0.13	0.15	0.17	0.19	0.20	0.22	0.24	0.25	0.26	0.28	0.29	0.30	0.31	0.32	0.32	0.33	0.34	0.34	0.35	0.35	0.36
22	1.31	0.33	0.35	0.37	0.39	0.42	0.45	0.48	0.51	0.54	0.57	0.60	0.62	0.65	0.67	0.69	0.71	0.73	0.75	0.77	0.78	0.80	0.81	0.82	0.83	0.84
		0.04	0.06	0.08	0.11	0.13	0.16	0.18	0.20	0.22	0.23	0.25	0.27	0.28	0.29	0.30	0.31	0.32	0.33	0.34	0.35	0.36	0.36	0.37	0.37	0.37
23	1.38	0.39	0.41	0.42	0.45	0.47	0.50	0.53	0.56	0.58	0.61	0.64	0.66	0.69	0.71	0.73	0.75	0.77	0.79	0.81	0.82	0.84	0.85	0.86	0.87	0.88
		0.05	0.07	0.09	0.12	0.14	0.17	0.19	0.21	0.23	0.25	0.26	0.28	0.29	0.31	0.32	0.33	0.34	0.35	0.36	0.36	0.37	0.38	0.38	0.39	0.39
24	1.44	0.45	0.47	0.48	0.51	0.53	0.56	0.58	0.61	0.63	0.66	0.69	0.71	0.73	0.76	0.78	0.80	0.82	0.83	0.85	0.86	0.88	0.89	0.90	0.91	0.92
		0.07	0.08	0.10	0.13	0.15	0.18	0.20	0.22	0.24	0.26	0.28	0.29	0.31	0.32	0.33	0.34	0.35	0.36	0.37	0.38	0.39	0.39	0.40	0.40	0.41
25	1.50	0.51	0.52	0.54	0.56	0.58	0.61	0.63	0.66	0.68	0.71	0.73	0.75	0.78	0.80	0.82	0.84	0.86	0.87	0.89	0.90	0.92	0.93	0.94	0.95	0.96
		0.08	0.10	0.12	0.14	0.17	0.19	0.21	0.23	0.25	0.27	0.29	0.30	0.32	0.33	0.35	0.36	0.37	0.38	0.39	0.40	0.40	0.41	0.41	0.42	0.42

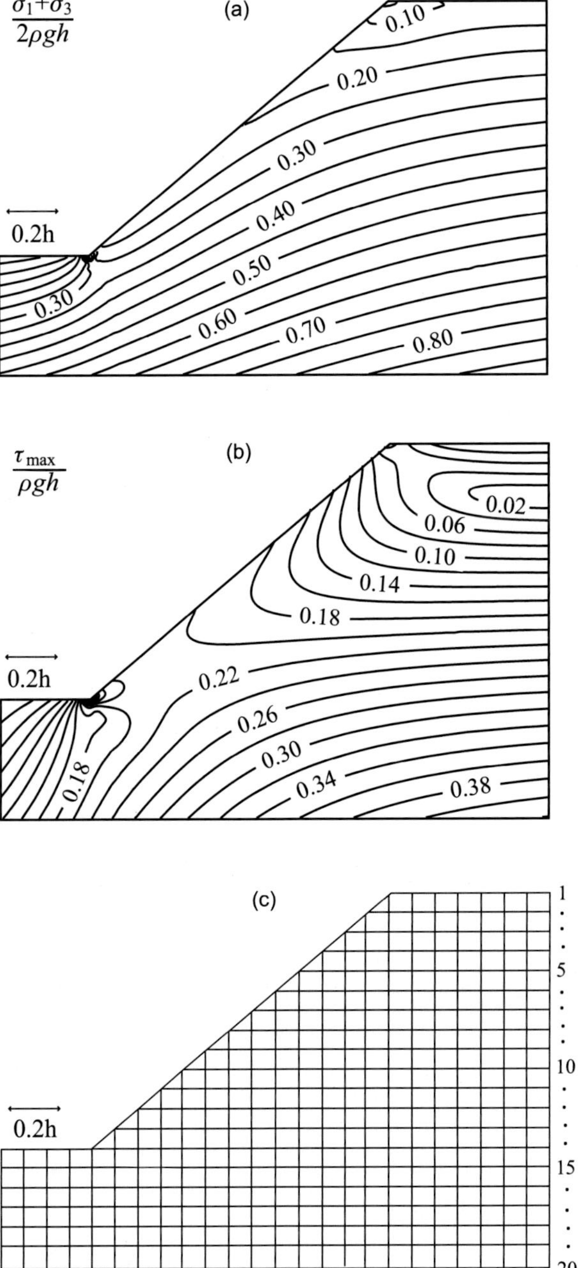

Figure 5.14 Stress contours for a 40° finite slope of height *h* for a Poisson's ratio of 1/3: (a) normalized mean stress, (b) normalized shear stress, and (c) grid for stress distribution Table 5.5.

Table 5.5 Stress distributions for a 40° finite slope of height h for a Poisson's ratio of 1/3

$z(h)$ \ $x(h)$		1	2	3	4	5	6	7	8	9	10	11	12	13	14	15	16	17	18	19	20	21	22	23	24	25
							Toe												Crest							
		0.00	0.09	0.18	0.28	0.37	0.46	0.55	0.64	0.73	0.83	0.92	1.01	1.10	1.19	1.28	1.38	1.47	1.56	1.65	1.74	1.83	1.93	2.02	2.11	2.20
1	0.00																		0.03	0.08	0.10	0.11	0.11	0.11	0.11	0.11
																			0.03	0.06	0.08	0.09	0.10	0.10	0.10	0.10
2	0.08																	0.10	0.09	0.10	0.12	0.13	0.13	0.14	0.14	0.14
																		0.09	0.06	0.05	0.05	0.06	0.06	0.06	0.06	0.06
3	0.15																0.14	0.13	0.14	0.14	0.16	0.16	0.17	0.17	0.17	0.18
																	0.13	0.09	0.07	0.05	0.04	0.03	0.03	0.02	0.03	0.03
4	0.23															0.16	0.16	0.17	0.17	0.18	0.19	0.20	0.21	0.21	0.22	0.22
																0.15	0.12	0.09	0.07	0.05	0.04	0.03	0.02	0.01	0.01	0.01
5	0.31														0.18	0.18	0.20	0.20	0.21	0.22	0.23	0.24	0.24	0.25	0.26	0.26
															0.17	0.14	0.12	0.10	0.08	0.07	0.06	0.05	0.05	0.04	0.04	0.04
6	0.38													0.19	0.20	0.22	0.23	0.24	0.25	0.26	0.27	0.28	0.29	0.29	0.30	0.30
														0.18	0.16	0.14	0.12	0.11	0.09	0.09	0.08	0.08	0.08	0.07	0.07	0.07
7	0.46												0.20	0.21	0.23	0.25	0.26	0.28	0.29	0.30	0.31	0.32	0.32	0.33	0.34	0.34
													0.19	0.17	0.16	0.14	0.13	0.12	0.11	0.11	0.11	0.10	0.10	0.10	0.10	0.10
8	0.54											0.21	0.22	0.25	0.27	0.29	0.30	0.32	0.33	0.34	0.35	0.36	0.27	0.37	0.38	0.39
												0.20	0.18	0.17	0.16	0.15	0.14	0.14	0.14	0.13	0.13	0.13	0.13	0.13	0.13	0.13
9	0.62										0.21	0.23	0.26	0.28	0.30	0.32	0.34	0.35	0.37	0.38	0.39	0.40	0.41	0.41	0.42	0.43
											0.20	0.19	0.18	0.17	0.17	0.16	0.16	0.16	0.16	0.16	0.16	0.16	0.16	0.16	0.16	0.16
10	0.69									0.21	0.24	0.27	0.30	0.32	0.34	0.36	0.38	0.39	0.41	0.42	0.43	0.44	0.45	0.46	0.46	0.47
										0.20	0.20	0.19	0.19	0.18	0.18	0.18	0.18	0.18	0.18	0.18	0.18	0.18	0.18	0.18	0.19	0.19
11	0.77								0.21	0.24	0.28	0.31	0.34	0.36	0.38	0.40	0.42	0.44	0.45	0.46	0.48	0.48	0.49	0.50	0.51	0.52
									0.20	0.20	0.20	0.20	0.20	0.20	0.20	0.20	0.20	0.20	0.21	0.21	0.21	0.21	0.21	0.21	0.21	0.21

(cont.)

Table 5.5 (cont.)

$z(h)$ \ $x(h)$		1	2	3	4	5	6	7	8	9	10	11	12	13	14	15	16	17	18	19	20	21	22	23	24	25
							Toe												Crest							
		0.00	0.09	0.18	0.28	0.37	0.46	0.55	0.64	0.73	0.83	0.92	1.01	1.10	1.19	1.28	1.38	1.47	1.56	1.65	1.74	1.83	1.93	2.02	2.11	2.20
12	0.85							0.21	0.25	0.29	0.32	0.35	0.38	0.40	0.42	0.44	0.46	0.47	0.49	0.50	0.52	0.53	0.54	0.54	0.56	0.56
								0.21	0.21	0.21	0.21	0.21	0.21	0.21	0.22	0.22	0.22	0.23	0.23	0.23	0.23	0.23	0.23	0.24	0.24	0.24
13	0.92						0.23	0.26	0.30	0.34	0.37	0.39	0.42	0.44	0.47	0.49	0.51	0.52	0.53	0.55	0.56	0.57	0.58	0.59	0.60	0.61
							0.22	0.22	0.21	0.21	0.22	0.22	0.23	0.23	0.24	0.24	0.24	0.25	0.25	0.25	0.26	0.26	0.26	0.26	0.26	0.26
14	1.00	0.04	0.05	0.06	0.13	0.43	0.29	0.32	0.35	0.38	0.41	0.44	0.47	0.49	0.51	0.53	0.55	0.56	0.58	0.59	0.61	0.62	0.63	0.64	0.64	0.65
		0.02	0.03	0.05	0.10	0.30	0.22	0.21	0.21	0.22	0.23	0.24	0.24	0.25	0.25	0.26	0.26	0.27	0.27	0.28	0.28	0.28	0.28	0.29	0.29	0.29
15	1.08	0.11	0.13	0.16	0.21	0.30	0.34	0.37	0.40	0.43	0.46	0.49	0.51	0.53	0.55	0.58	0.59	0.61	0.62	0.64	0.65	0.66	0.67	0.68	0.69	0.70
		0.04	0.05	0.09	0.14	0.18	0.19	0.20	0.22	0.23	0.24	0.25	0.26	0.27	0.27	0.28	0.28	0.29	0.29	0.30	0.30	0.30	0.31	0.31	0.31	0.31
16	1.15	0.20	0.22	0.25	0.29	0.35	0.40	0.42	0.45	0.48	0.51	0.54	0.56	0.58	0.60	0.62	0.64	0.66	0.67	0.68	0.70	0.71	0.72	0.73	0.73	0.75
		0.05	0.08	0.11	0.15	0.18	0.19	0.21	0.23	0.24	0.25	0.26	0.27	0.28	0.29	0.30	0.30	0.31	0.32	0.32	0.32	0.33	0.33	0.33	0.33	0.34
17	1.23	0.28	0.30	0.33	0.37	0.41	0.45	0.48	0.51	0.53	0.56	0.59	0.61	0.63	0.65	0.67	0.68	0.70	0.72	0.73	0.74	0.76	0.76	0.78	0.78	0.79
		0.07	0.09	0.13	0.16	0.19	0.21	0.22	0.24	0.25	0.27	0.28	0.29	0.30	0.31	0.32	0.32	0.33	0.33	0.34	0.34	0.35	0.35	0.35	0.36	0.36
18	1.31	0.36	0.38	0.40	0.44	0.47	0.50	0.53	0.56	0.59	0.61	0.63	0.66	0.68	0.70	0.72	0.73	0.75	0.76	0.78	0.79	0.80	0.81	0.82	0.83	0.84
		0.08	0.11	0.14	0.17	0.20	0.22	0.23	0.25	0.27	0.28	0.29	0.31	0.32	0.33	0.33	0.34	0.35	0.35	0.36	0.36	0.37	0.37	0.37	0.38	0.38
19	1.38	0.44	0.46	0.48	0.51	0.54	0.57	0.59	0.62	0.64	0.67	0.69	0.71	0.73	0.75	0.77	0.78	0.80	0.81	0.83	0.84	0.85	0.86	0.87	0.88	0.89
		0.10	0.12	0.15	0.18	0.21	0.23	0.25	0.26	0.28	0.30	0.31	0.32	0.33	0.34	0.35	0.36	0.37	0.37	0.38	0.38	0.39	0.39	0.39	0.40	0.40
20	1.46	0.50	0.52	0.55	0.57	0.60	0.62	0.65	0.68	0.70	0.72	0.74	0.76	0.78	0.80	0.82	0.83	0.85	0.86	0.88	0.89	0.90	0.91	0.92	0.93	0.94
		0.12	0.14	0.17	0.19	0.22	0.24	0.26	0.28	0.30	0.31	0.32	0.34	0.35	0.36	0.37	0.38	0.38	0.39	0.40	0.40	0.41	0.41	0.41	0.42	0.42

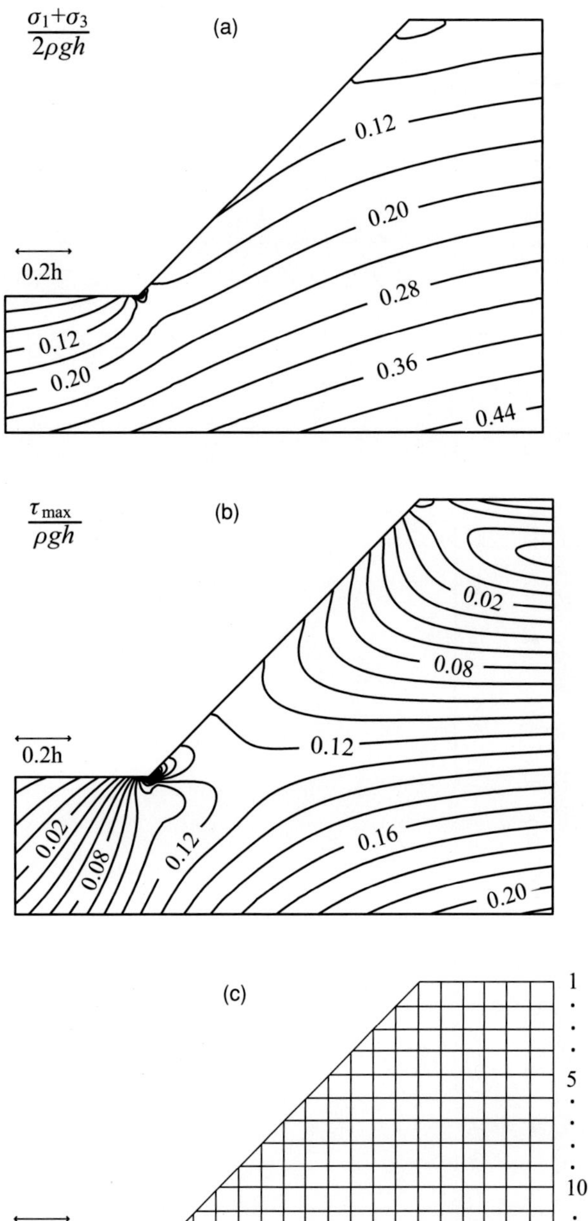

Figure 5.15 Stress contours for a 45° finite slope of height h for a Poisson's ratio of 1/3: (a) normalized mean stress, (b) normalized shear stress, and (c) grid for stress distribution Table 5.6.

Table 5.6 Stress distributions for a 45° finite slope of height h for a Poisson's ratio of 1/3

$z(h)$	$x(h)$	1	2	3	4	5	6	7	8	9	10	11	12	13	14	15	16	17	18	19	20	21	22	23	24	25
									Toe											Crest						
		0.00	0.08	0.17	0.25	0.33	0.42	0.50	0.58	0.67	0.75	0.83	0.92	1.00	1.08	1.17	1.25	1.33	1.42	1.50	1.58	1.67	1.75	1.83	1.92	2.00
1	0.00																			0.01	0.03	0.04	0.05	0.05	0.05	0.06
																				0.01	0.03	0.04	0.04	0.05	0.05	0.05
2	0.08																		0.05	0.04	0.05	0.06	0.06	0.07	0.07	0.07
																			0.05	0.03	0.03	0.02	0.02	0.03	0.03	0.03
3	0.17																	0.07	0.07	0.07	0.07	0.08	0.08	0.09	0.09	0.09
																		0.07	0.05	0.04	0.03	0.02	0.02	0.01	0.01	0.01
4	0.25																0.08	0.08	0.09	0.09	0.09	0.10	0.10	0.10	0.11	0.11
																	0.08	0.06	0.05	0.04	0.03	0.03	0.02	0.02	0.01	0.01
5	0.33															0.10	0.10	0.10	0.11	0.11	0.11	0.12	0.12	0.13	0.13	0.13
																0.09	0.08	0.07	0.06	0.05	0.04	0.04	0.03	0.03	0.03	0.03
6	0.42														0.10	0.11	0.12	0.12	0.13	0.13	0.14	0.14	0.14	0.15	0.15	0.15
															0.10	0.09	0.08	0.07	0.06	0.06	0.05	0.05	0.05	0.05	0.05	0.04
7	0.50													0.11	0.12	0.12	0.13	0.14	0.15	0.15	0.16	0.16	0.16	0.17	0.17	0.18
														0.11	0.10	0.09	0.08	0.08	0.07	0.07	0.06	0.06	0.06	0.06	0.06	0.06
8	0.58												0.11	0.12	0.13	0.14	0.15	0.16	0.17	0.17	0.18	0.18	0.19	0.19	0.19	0.20
													0.11	0.10	0.10	0.09	0.09	0.08	0.08	0.08	0.08	0.08	0.08	0.08	0.07	0.07
9	0.67											0.12	0.13	0.14	0.17	0.16	0.17	0.18	0.19	0.19	0.20	0.20	0.21	0.21	0.22	0.22
												0.11	0.11	0.11	0.10	0.10	0.10	0.09	0.09	0.09	0.09	0.09	0.09	0.09	0.09	0.09
10	0.75										0.12	0.14	0.15	0.16	0.17	0.18	0.19	0.20	0.21	0.21	0.22	0.23	0.23	0.24	0.24	0.24
											0.12	0.12	0.11	0.11	0.11	0.11	0.10	0.10	0.10	0.10	0.10	0.10	0.10	0.10	0.10	0.10
11	0.83									0.13	0.14	0.16	0.17	0.18	0.20	0.21	0.22	0.22	0.23	0.24	0.24	0.25	0.25	0.26	0.26	0.27
										0.12	0.12	0.12	0.12	0.12	0.11	0.11	0.11	0.11	0.11	0.11	0.12	0.12	0.12	0.12	0.12	0.12

12	0.92								0.14	0.15	0.17	0.18	0.20	0.21	0.22	0.23	0.24	0.25	0.25	0.26	0.27	0.27	0.28	0.28	0.29	0.29
									0.14	0.13	0.12	0.12	0.12	0.12	0.12	0.12	0.12	0.12	0.13	0.13	0.13	0.13	0.13	0.13	0.13	0.13
13	1.00	0.01	0.01	0.01	0.02	0.03	0.06	0.32	0.18	0.18	0.19	0.21	0.22	0.23	0.24	0.25	0.26	0.27	0.28	0.28	0.29	0.30	0.30	0.31	0.31	0.32
		0.00	0.01	0.01	0.02	0.03	0.06	0.20	0.13	0.12	0.12	0.12	0.12	0.13	0.13	0.13	0.13	0.13	0.14	0.14	0.14	0.14	0.14	0.14	0.14	0.14
14	1.08	0.05	0.06	0.07	0.08	0.10	0.13	0.18	0.20	0.21	0.22	0.23	0.25	0.26	0.27	0.28	0.29	0.29	0.30	0.31	0.32	0.32	0.33	0.33	0.34	0.34
		0.01	0.02	0.03	0.04	0.06	0.09	0.11	0.11	0.11	0.12	0.13	0.13	0.13	0.14	0.14	0.14	0.14	0.15	0.15	0.15	0.15	0.15	0.15	0.16	0.16
15	1.17	0.10	0.11	0.12	0.13	0.15	0.17	0.20	0.22	0.23	0.25	0.26	0.27	0.28	0.29	0.30	0.31	0.32	0.33	0.33	0.34	0.35	0.35	0.36	0.36	0.37
		0.02	0.03	0.04	0.05	0.07	0.09	0.11	0.11	0.12	0.12	0.13	0.14	0.14	0.15	0.15	0.15	0.15	0.16	0.16	0.16	0.16	0.16	0.17	0.17	0.17
16	1.25	0.14	0.15	0.16	0.17	0.19	0.21	0.23	0.25	0.26	0.27	0.29	0.30	0.31	0.32	0.33	0.34	0.34	0.35	0.36	0.37	0.37	0.38	0.38	0.39	0.39
		0.03	0.04	0.05	0.06	0.08	0.10	0.11	0.12	0.12	0.13	0.14	0.14	0.15	0.15	0.16	0.16	0.16	0.17	0.17	0.17	0.17	0.18	0.18	0.18	0.18
17	1.33	0.18	0.19	0.20	0.21	0.23	0.24	0.26	0.28	0.29	0.30	0.31	0.32	0.34	0.35	0.35	0.36	0.37	0.38	0.38	0.39	0.40	0.40	0.41	0.41	0.42
		0.04	0.05	0.06	0.07	0.09	0.10	0.11	0.12	0.13	0.14	0.14	0.15	0.16	0.16	0.17	0.17	0.17	0.18	0.18	0.18	0.18	0.19	0.19	0.19	0.19
18	1.42	0.23	0.23	0.24	0.25	0.27	0.28	0.29	0.31	0.32	0.33	0.34	0.35	0.36	0.37	0.38	0.39	0.40	0.40	0.41	0.42	0.42	0.43	0.44	0.44	0.44
		0.05	0.05	0.07	0.08	0.09	0.11	0.12	0.13	0.14	0.15	0.15	0.16	0.16	0.17	0.17	0.18	0.18	0.19	0.19	0.19	0.19	0.20	0.20	0.20	0.20
19	1.50	0.26	0.27	0.28	0.29	0.30	0.31	0.33	0.34	0.35	0.36	0.37	0.38	0.39	0.40	0.41	0.42	0.42	0.43	0.44	0.44	0.45	0.46	0.46	0.47	0.47
		0.05	0.06	0.07	0.09	0.10	0.11	0.12	0.14	0.14	0.15	0.16	0.17	0.17	0.18	0.18	0.19	0.19	0.20	0.20	0.20	0.20	0.21	0.21	0.21	0.21

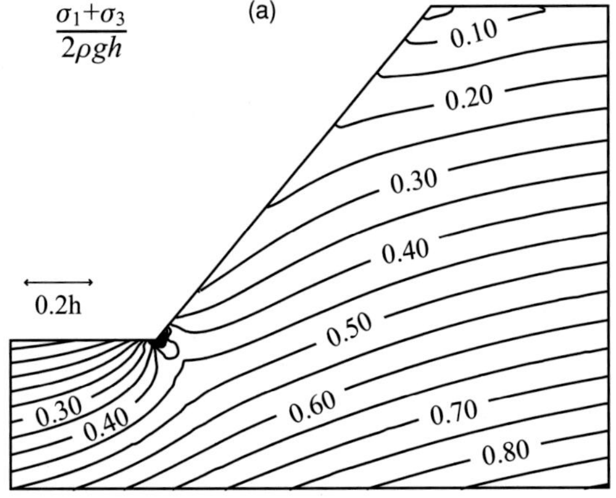

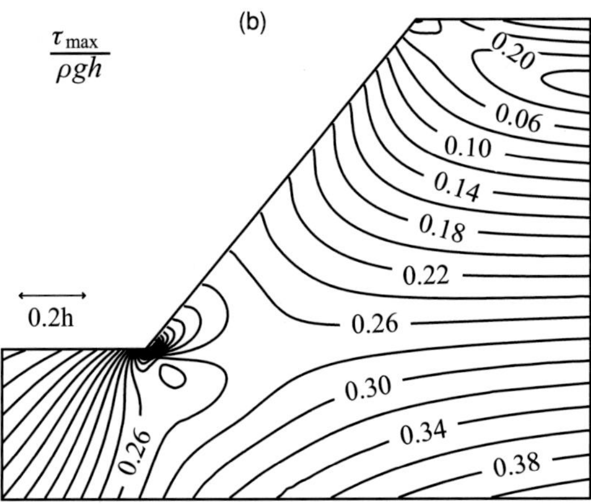

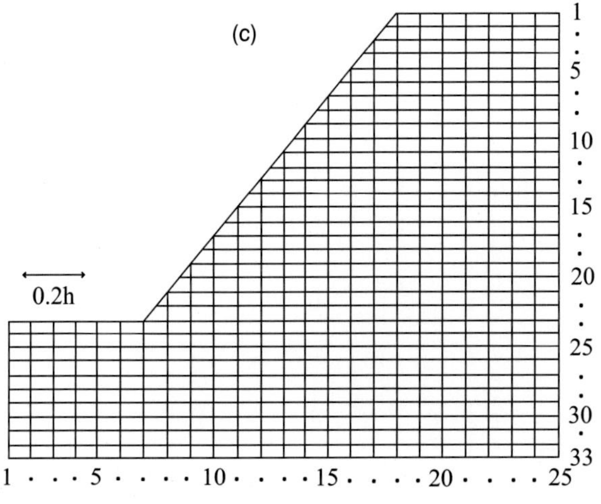

Figure 5.16 Stress contours for a 50° finite slope of height h for a Poisson's ratio of 1/3: (a) normalized mean stress, (b) normalized shear stress, and (c) grid for stress distribution Table 5.7.

Table 5.7 Stress distributions for a 50° finite slope of height h for a Poisson's ratio of 1/3

$z(h)$ \ $x(h)$		1	2	3	4	5	6	7	8	9	10	11	12	13	14	15	16	17	18	19	20	21	22	23	24	25
									Toe										Crest							
		0.00	0.08	0.15	0.23	0.31	0.38	0.46	0.53	0.61	0.69	0.76	0.84	0.92	0.99	1.07	1.14	1.22	1.30	1.37	1.45	1.53	1.60	1.68	1.75	1.83
1	0.00																		0.02	0.06	0.08	0.09	0.10	0.10	0.11	0.11
																			0.02	0.04	0.07	0.08	0.09	0.09	0.09	0.09
2	0.05																		0.06	0.07	0.09	0.10	0.11	0.11	0.12	0.12
																			0.04	0.04	0.05	0.06	0.06	0.07	0.07	0.07
3	0.09																	0.09	0.09	0.10	0.11	0.12	0.13	0.13	0.14	0.14
																		0.09	0.06	0.05	0.04	0.04	0.04	0.05	0.05	0.05
4	0.14																	0.11	0.11	0.12	0.13	0.14	0.15	0.16	0.16	0.16
																		0.10	0.07	0.06	0.04	0.03	0.03	0.03	0.03	0.03
5	0.18																0.14	0.14	0.14	0.15	0.15	0.16	0.17	0.18	0.18	0.18
																	0.13	0.10	0.08	0.06	0.05	0.04	0.03	0.02	0.01	0.01
6	0.23																0.15	0.16	0.16	0.17	0.18	0.18	0.19	0.20	0.20	0.21
																	0.14	0.11	0.09	0.07	0.06	0.05	0.04	0.03	0.02	0.02
7	0.27															0.17	0.17	0.18	0.19	0.19	0.20	0.21	0.21	0.22	0.23	0.23
																0.17	0.14	0.11	0.10	0.08	0.07	0.06	0.05	0.04	0.04	0.03
8	0.32															0.19	0.19	0.20	0.21	0.21	0.22	0.23	0.24	0.24	0.25	0.25
																0.17	0.14	0.12	0.11	0.09	0.08	0.07	0.07	0.06	0.06	0.05
9	0.36														0.20	0.21	0.21	0.22	0.23	0.24	0.25	0.25	0.26	0.27	0.27	0.28
															0.19	0.17	0.15	0.13	0.12	0.10	0.09	0.09	0.08	0.08	0.07	0.07
10	0.41														0.21	0.22	0.23	0.24	0.25	0.26	0.27	0.27	0.28	0.29	0.29	0.30
															0.20	0.17	0.15	0.14	0.13	0.12	0.11	0.10	0.10	0.09	0.09	0.09
11	0.45													0.22	0.23	0.24	0.25	0.26	0.27	0.28	0.29	0.30	0.31	0.31	0.32	0.32
														0.21	0.20	0.18	0.16	0.15	0.14	0.13	0.12	0.12	0.11	0.11	0.11	0.10
12	0.50													0.23	0.25	0.26	0.27	0.28	0.29	0.30	0.31	0.32	0.33	0.34	0.34	0.35
														0.22	0.20	0.18	0.17	0.16	0.15	0.14	0.13	0.13	0.13	0.12	0.12	0.12
13	0.55												0.24	0.25	0.27	0.29	0.30	0.31	0.32	0.33	0.34	0.35	0.35	0.36	0.37	0.37
													0.23	0.22	0.20	0.19	0.18	0.17	0.16	0.15	0.15	0.15	0.14	0.14	0.14	0.14
14	0.59												0.25	0.27	0.29	0.31	0.32	0.33	0.34	0.35	0.36	0.37	0.38	0.39	0.39	0.40
													0.24	0.22	0.21	0.20	0.19	0.18	0.17	0.17	0.16	0.16	0.16	0.16	0.16	0.16
15	0.64											0.26	0.27	0.29	0.31	0.33	0.34	0.35	0.36	0.38	0.39	0.39	0.40	0.41	0.42	0.42
												0.25	0.24	0.23	0.21	0.20	0.20	0.19	0.18	0.18	0.18	0.18	0.17	0.17	0.17	0.17
16	0.68											0.27	0.30	0.32	0.33	0.35	0.36	0.38	0.39	0.40	0.41	0.42	0.43	0.43	0.44	0.45
												0.25	0.24	0.23	0.22	0.21	0.21	0.20	0.20	0.19	0.19	0.19	0.19	0.19	0.19	0.19

(cont.)

Table 5.7 (cont.)

z(h)	x(h)	1	2	3	4	5	6	7	8	9	10	11	12	13	14	15	16	17	18	19	20	21	22	23	24	25
									Toe										Crest							
		0.00	0.08	0.15	0.23	0.31	0.38	0.46	0.53	0.61	0.69	0.76	0.84	0.92	0.99	1.07	1.14	1.22	1.30	1.37	1.45	1.53	1.60	1.68	1.75	1.83
17	0.73										0.27	0.29	0.32	0.34	0.36	0.37	0.39	0.40	0.41	0.42	0.43	0.44	0.45	0.46	0.47	0.47
											0.26	0.26	0.25	0.24	0.23	0.22	0.22	0.21	0.21	0.21	0.20	0.20	0.20	0.20	0.20	0.20
18	0.77										0.29	0.32	0.34	0.36	0.38	0.40	0.41	0.42	0.44	0.45	0.46	0.47	0.48	0.48	0.49	0.50
											0.27	0.26	0.25	0.24	0.24	0.23	0.23	0.22	0.22	0.22	0.22	0.22	0.22	0.22	0.22	0.22
19	0.82									0.29	0.31	0.34	0.36	0.38	0.40	0.42	0.43	0.45	0.46	0.47	0.48	0.49	0.50	0.51	0.52	0.52
										0.28	0.28	0.27	0.26	0.25	0.24	0.24	0.24	0.23	0.23	0.23	0.23	0.23	0.23	0.23	0.23	0.23
20	0.86									0.31	0.34	0.37	0.39	0.41	0.43	0.44	0.46	0.47	0.48	0.50	0.51	0.52	0.53	0.54	0.54	0.55
										0.29	0.28	0.27	0.26	0.25	0.25	0.25	0.25	0.24	0.24	0.24	0.24	0.24	0.24	0.24	0.25	0.25
21	0.91								0.32	0.35	0.37	0.39	0.41	0.43	0.45	0.47	0.48	0.50	0.51	0.52	0.53	0.54	0.55	0.56	0.57	0.58
									0.32	0.30	0.28	0.27	0.26	0.26	0.26	0.26	0.26	0.26	0.26	0.26	0.26	0.26	0.26	0.26	0.26	0.26
22	0.95								0.38	0.38	0.40	0.42	0.44	0.46	0.48	0.49	0.51	0.52	0.53	0.55	0.56	0.57	0.58	0.59	0.60	0.60
									0.34	0.30	0.28	0.27	0.27	0.26	0.26	0.26	0.26	0.27	0.27	0.27	0.27	0.27	0.27	0.27	0.27	0.27
23	1.00	0.01	0.02	0.03	0.05	0.08	0.15	0.78	0.42	0.41	0.43	0.45	0.47	0.48	0.50	0.52	0.53	0.55	0.56	0.57	0.58	0.59	0.60	0.61	0.62	0.63
		0.01	0.02	0.03	0.05	0.08	0.15	0.49	0.31	0.28	0.27	0.27	0.27	0.27	0.27	0.27	0.27	0.28	0.28	0.28	0.28	0.28	0.28	0.29	0.29	0.29
24	1.05	0.07	0.08	0.10	0.12	0.17	0.25	0.43	0.45	0.45	0.46	0.48	0.49	0.51	0.53	0.55	0.56	0.58	0.59	0.60	0.61	0.62	0.63	0.64	0.65	0.66
		0.02	0.03	0.05	0.08	0.12	0.20	0.26	0.24	0.26	0.26	0.27	0.27	0.27	0.28	0.28	0.28	0.29	0.29	0.29	0.29	0.30	0.30	0.30	0.30	0.30
25	1.09	0.12	0.14	0.16	0.19	0.23	0.30	0.41	0.45	0.47	0.48	0.50	0.52	0.54	0.56	0.57	0.59	0.60	0.61	0.63	0.64	0.65	0.66	0.67	0.68	0.69
		0.03	0.05	0.07	0.10	0.15	0.21	0.25	0.24	0.25	0.26	0.27	0.27	0.28	0.29	0.29	0.29	0.30	0.30	0.30	0.31	0.31	0.31	0.31	0.31	0.32
26	1.14	0.18	0.19	0.21	0.24	0.29	0.35	0.42	0.47	0.49	0.51	0.53	0.55	0.57	0.58	0.60	0.61	0.63	0.64	0.65	0.67	0.68	0.69	0.70	0.70	0.71
		0.04	0.06	0.09	0.12	0.17	0.22	0.25	0.24	0.25	0.26	0.27	0.28	0.29	0.29	0.30	0.30	0.31	0.31	0.32	0.32	0.32	0.32	0.33	0.33	0.33
27	1.18	0.23	0.24	0.26	0.30	0.34	0.39	0.45	0.49	0.51	0.54	0.56	0.58	0.59	0.61	0.63	0.64	0.66	0.67	0.68	0.69	0.70	0.71	0.72	0.73	0.74
		0.06	0.07	0.10	0.13	0.18	0.22	0.25	0.25	0.26	0.27	0.28	0.29	0.29	0.30	0.31	0.31	0.32	0.32	0.33	0.33	0.33	0.34	0.34	0.34	0.34
28	1.23	0.27	0.29	0.31	0.35	0.38	0.43	0.47	0.51	0.54	0.56	0.58	0.60	0.62	0.64	0.65	0.67	0.68	0.70	0.71	0.72	0.73	0.74	0.75	0.76	0.77
		0.07	0.09	0.11	0.15	0.18	0.22	0.25	0.26	0.27	0.27	0.29	0.29	0.30	0.31	0.32	0.32	0.33	0.33	0.34	0.34	0.34	0.35	0.35	0.35	0.35
29	1.27	0.32	0.34	0.36	0.39	0.43	0.46	0.50	0.54	0.56	0.59	0.61	0.63	0.65	0.66	0.68	0.70	0.71	0.72	0.74	0.75	0.76	0.77	0.78	0.79	0.79
		0.08	0.10	0.12	0.16	0.19	0.22	0.25	0.26	0.27	0.28	0.29	0.30	0.31	0.32	0.33	0.33	0.34	0.34	0.35	0.35	0.36	0.36	0.36	0.36	0.37
30	1.32	0.37	0.39	0.41	0.44	0.47	0.50	0.54	0.57	0.59	0.62	0.64	0.66	0.68	0.69	0.71	0.72	0.74	0.75	0.76	0.77	0.79	0.80	0.81	0.81	0.82
		0.09	0.11	0.14	0.17	0.20	0.23	0.25	0.27	0.28	0.29	0.30	0.31	0.32	0.33	0.34	0.34	0.35	0.35	0.36	0.36	0.37	0.37	0.37	0.38	0.38
31	1.36	0.41	0.43	0.45	0.48	0.51	0.54	0.57	0.60	0.62	0.65	0.67	0.69	0.70	0.72	0.74	0.75	0.77	0.78	0.79	0.80	0.81	0.82	0.83	0.84	0.85
		0.10	0.12	0.15	0.18	0.21	0.23	0.26	0.28	0.29	0.30	0.31	0.32	0.33	0.34	0.35	0.35	0.36	0.36	0.37	0.37	0.38	0.38	0.38	0.39	0.39
32	1.41	0.46	0.48	0.50	0.52	0.55	0.57	0.60	0.63	0.65	0.68	0.70	0.72	0.73	0.75	0.77	0.78	0.79	0.81	0.82	0.83	0.84	0.85	0.86	0.87	0.88
		0.11	0.13	0.16	0.18	0.21	0.24	0.26	0.28	0.30	0.31	0.32	0.33	0.34	0.35	0.35	0.36	0.37	0.37	0.38	0.38	0.39	0.39	0.40	0.40	0.40
33	1.45	0.50	0.52	0.54	0.56	0.58	0.61	0.64	0.66	0.68	0.71	0.73	0.75	0.76	0.78	0.79	0.81	0.82	0.84	0.85	0.86	0.87	0.88	0.89	0.90	0.91
		0.12	0.14	0.17	0.19	0.22	0.24	0.27	0.29	0.30	0.32	0.33	0.34	0.35	0.36	0.36	0.37	0.38	0.38	0.39	0.40	0.40	0.40	0.41	0.41	0.41

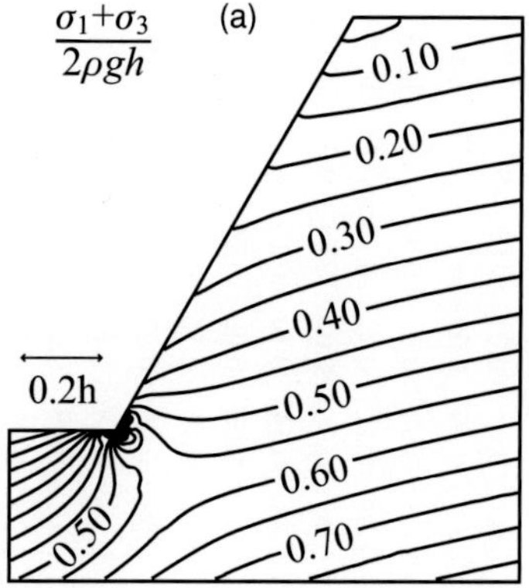

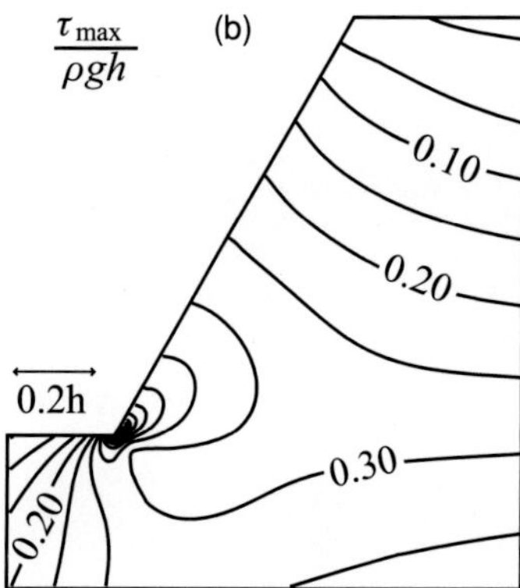

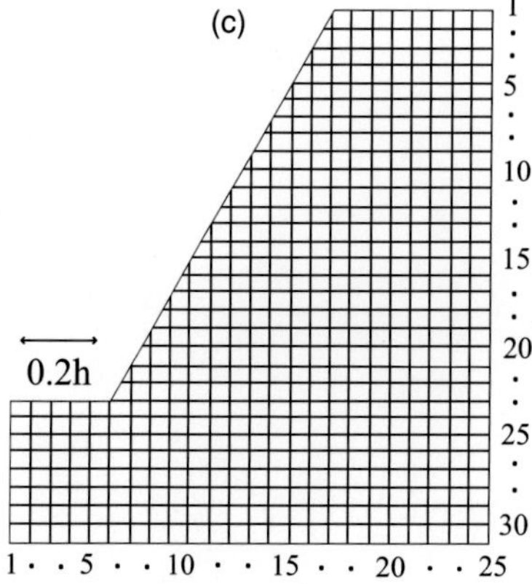

Figure 5.17 Stress contours for a 60° finite slope of height h for a Poisson's ratio of 1/3: (a) normalized mean stress, (b) normalized shear stress, and (c) grid for stress distribution Table 5.8.

Table 5.8 Stress distributions for a 60° finite slope of height h for a Poisson's ratio of 1/3

$z(h)$ \ $x(h)$		1	2	3	4	5	6	7	8	9	10	11	12	13	14	15	16	17	18	19	20	21	22	23	24	25
								Toe										Crest								
		0.00	0.05	0.10	0.16	0.21	0.26	0.31	0.37	0.42	0.47	0.52	0.58	0.63	0.68	0.73	0.79	0.84	0.89	0.94	1.00	1.05	1.10	1.15	1.20	1.26
1	0.00																	0.01	0.03	0.05	0.06	0.07	0.08	0.08	0.09	0.09
																		0.01	0.02	0.04	0.05	0.06	0.06	0.07	0.07	0.08
2	0.05																	0.04	0.05	0.06	0.07	0.08	0.09	0.09	0.10	0.10
																		0.03	0.03	0.03	0.03	0.04	0.04	0.05	0.05	0.06
3	0.09																0.07	0.07	0.08	0.09	0.09	0.10	0.11	0.11	0.12	0.12
																	0.07	0.05	0.04	0.04	0.03	0.03	0.03	0.03	0.03	0.03
4	0.14																0.10	0.10	0.10	0.11	0.12	0.12	0.13	0.14	0.14	0.14
																	0.09	0.07	0.06	0.05	0.04	0.04	0.03	0.03	0.02	0.02
5	0.18															0.12	0.12	0.12	0.13	0.14	0.14	0.15	0.15	0.16	0.16	0.17
																0.12	0.10	0.09	0.07	0.06	0.06	0.05	0.04	0.04	0.03	0.02
6	0.23															0.14	0.14	0.15	0.15	0.16	0.16	0.17	0.18	0.18	0.18	0.19
																0.13	0.11	0.10	0.09	0.08	0.07	0.06	0.06	0.05	0.04	0.04
7	0.27														0.16	0.16	0.17	0.17	0.18	0.18	0.19	0.19	0.20	0.20	0.21	0.21
															0.16	0.14	0.13	0.11	0.10	0.09	0.09	0.08	0.07	0.07	0.06	0.06
8	0.32														0.18	0.19	0.19	0.19	0.20	0.21	0.21	0.22	0.22	0.23	0.23	0.23
															0.17	0.15	0.14	0.13	0.12	0.11	0.10	0.09	0.09	0.08	0.08	0.07
9	0.36													0.20	0.20	0.21	0.21	0.22	0.22	0.23	0.23	0.24	0.24	0.25	0.25	0.26
														0.19	0.18	0.16	0.15	0.14	0.13	0.12	0.12	0.11	0.10	0.10	0.09	0.09
10	0.41													0.21	0.22	0.23	0.23	0.24	0.25	0.25	0.26	0.26	0.27	0.27	0.28	0.28
														0.20	0.19	0.18	0.16	0.15	0.15	0.14	0.13	0.12	0.12	0.11	0.11	0.11
11	0.45												0.23	0.23	0.24	0.25	0.26	0.26	0.27	0.27	0.28	0.29	0.29	0.30	0.30	0.31
													0.23	0.21	0.20	0.19	0.18	0.17	0.16	0.15	0.14	0.14	0.13	0.13	0.13	0.12
12	0.50												0.25	0.26	0.26	0.27	0.28	0.29	0.29	0.30	0.30	0.31	0.31	0.32	0.32	0.33
													0.24	0.22	0.21	0.20	0.19	0.18	0.17	0.16	0.16	0.15	0.15	0.14	0.14	0.14
13	0.55											0.26	0.27	0.28	0.29	0.29	0.30	0.31	0.31	0.32	0.33	0.33	0.34	0.34	0.35	0.35
												0.25	0.24	0.23	0.22	0.21	0.20	0.19	0.19	0.18	0.17	0.17	0.16	0.16	0.16	0.15
14	0.59											0.28	0.29	0.30	0.31	0.32	0.32	0.33	0.34	0.35	0.35	0.36	0.36	0.37	0.37	0.38
												0.26	0.25	0.24	0.23	0.22	0.21	0.21	0.20	0.19	0.19	0.18	0.18	0.17	0.17	0.17
15	0.64										0.29	0.30	0.31	0.32	0.33	0.34	0.35	0.36	0.36	0.37	0.38	0.38	0.39	0.39	0.40	0.40
											0.28	0.28	0.27	0.25	0.24	0.23	0.23	0.22	0.21	0.20	0.20	0.19	0.19	0.19	0.18	0.18

16	0.68									0.31	0.32	0.33	0.34	0.35	0.36	0.37	0.38	0.39	0.39	0.40	0.41	0.41	0.42	0.42	0.43
										0.30	0.29	0.28	0.27	0.26	0.24	0.24	0.23	0.22	0.22	0.21	0.21	0.20	0.20	0.20	0.20
17	0.73								0.32	0.33	0.35	0.36	0.37	0.38	0.39	0.40	0.41	0.41	0.42	0.43	0.43	0.44	0.44	0.45	0.46
									0.31	0.31	0.30	0.29	0.28	0.27	0.26	0.25	0.24	0.23	0.23	0.22	0.22	0.22	0.21	0.21	0.21
18	0.77								0.34	0.36	0.37	0.39	0.40	0.41	0.41	0.42	0.43	0.44	0.45	0.45	0.46	0.46	0.47	0.48	0.48
									0.33	0.32	0.31	0.30	0.28	0.27	0.26	0.26	0.25	0.24	0.24	0.24	0.23	0.23	0.23	0.23	0.22
19	0.82							0.35	0.38	0.39	0.40	0.41	0.42	0.43	0.44	0.45	0.46	0.46	0.47	0.48	0.48	0.49	0.49	0.50	0.51
								0.35	0.35	0.33	0.32	0.30	0.29	0.28	0.27	0.26	0.26	0.25	0.25	0.25	0.24	0.24	0.24	0.24	0.24
20	0.86							0.39	0.41	0.42	0.43	0.44	0.45	0.46	0.47	0.47	0.48	0.49	0.50	0.50	0.51	0.51	0.52	0.53	0.53
								0.38	0.36	0.34	0.32	0.31	0.29	0.28	0.28	0.27	0.27	0.26	0.26	0.26	0.25	0.25	0.25	0.25	0.25
21	0.91						0.42	0.44	0.45	0.45	0.46	0.47	0.48	0.48	0.49	0.50	0.51	0.51	0.52	0.53	0.53	0.54	0.55	0.55	0.56
							0.42	0.40	0.37	0.34	0.32	0.31	0.29	0.29	0.28	0.28	0.27	0.27	0.27	0.27	0.27	0.26	0.26	0.26	0.26
22	0.95						0.52	0.49	0.48	0.48	0.49	0.49	0.50	0.51	0.52	0.52	0.53	0.54	0.55	0.55	0.56	0.57	0.57	0.58	0.58
							0.48	0.40	0.36	0.33	0.31	0.30	0.29	0.29	0.28	0.28	0.28	0.28	0.28	0.28	0.28	0.28	0.28	0.28	0.28
23	1.00	0.04	0.06	0.09	0.13	0.31	1.27	0.59	0.53	0.52	0.51	0.52	0.52	0.53	0.53	0.54	0.55	0.56	0.56	0.57	0.58	0.59	0.59	0.60	0.60	0.61
		0.04	0.06	0.09	0.14	0.28	0.71	0.41	0.35	0.32	0.31	0.30	0.30	0.29	0.29	0.29	0.29	0.29	0.29	0.29	0.29	0.29	0.29	0.29	0.29	0.29
24	1.05	0.12	0.14	0.19	0.25	0.37	0.58	0.60	0.56	0.54	0.54	0.54	0.55	0.55	0.56	0.57	0.58	0.58	0.59	0.60	0.60	0.61	0.62	0.62	0.63	0.64
		0.07	0.10	0.14	0.21	0.33	0.36	0.29	0.29	0.29	0.29	0.29	0.29	0.29	0.29	0.29	0.29	0.29	0.29	0.30	0.30	0.30	0.30	0.30	0.30	0.30
25	1.09	0.18	0.21	0.26	0.32	0.41	0.52	0.56	0.56	0.56	0.56	0.57	0.57	0.58	0.59	0.59	0.60	0.61	0.62	0.62	0.63	0.64	0.64	0.65	0.66	0.66
		0.10	0.13	0.18	0.24	0.31	0.33	0.28	0.28	0.28	0.28	0.29	0.29	0.29	0.30	0.30	0.30	0.30	0.30	0.31	0.31	0.31	0.31	0.31	0.31	0.31
26	1.14	0.25	0.28	0.32	0.37	0.44	0.51	0.55	0.56	0.57	0.58	0.59	0.59	0.60	0.61	0.62	0.63	0.63	0.64	0.65	0.66	0.66	0.67	0.68	0.68	0.69
		0.13	0.16	0.21	0.25	0.30	0.31	0.30	0.29	0.28	0.29	0.29	0.30	0.30	0.30	0.31	0.31	0.31	0.31	0.31	0.32	0.32	0.32	0.32	0.32	0.32
27	1.18	0.30	0.33	0.37	0.41	0.47	0.51	0.55	0.57	0.59	0.60	0.61	0.62	0.63	0.64	0.64	0.65	0.66	0.67	0.68	0.68	0.69	0.70	0.70	0.71	0.72
		0.15	0.18	0.22	0.26	0.29	0.31	0.31	0.30	0.29	0.30	0.30	0.30	0.31	0.31	0.31	0.32	0.32	0.32	0.32	0.33	0.33	0.33	0.33	0.33	0.34
28	1.23	0.35	0.38	0.42	0.45	0.50	0.53	0.57	0.59	0.61	0.62	0.63	0.64	0.65	0.66	0.67	0.68	0.69	0.70	0.70	0.71	0.72	0.72	0.73	0.74	0.74
		0.16	0.19	0.23	0.26	0.29	0.31	0.31	0.31	0.31	0.31	0.31	0.31	0.32	0.32	0.32	0.33	0.33	0.33	0.34	0.34	0.34	0.34	0.34	0.35	0.35
29	1.27	0.40	0.43	0.46	0.49	0.53	0.56	0.59	0.61	0.63	0.64	0.66	0.67	0.68	0.69	0.70	0.70	0.71	0.72	0.73	0.74	0.74	0.75	0.76	0.76	0.77
		0.18	0.20	0.24	0.26	0.29	0.30	0.31	0.32	0.32	0.32	0.32	0.32	0.33	0.33	0.33	0.34	0.34	0.34	0.35	0.35	0.35	0.35	0.36	0.36	0.36
30	1.32	0.45	0.47	0.50	0.53	0.56	0.58	0.61	0.63	0.65	0.66	0.68	0.69	0.70	0.71	0.72	0.73	0.74	0.75	0.76	0.76	0.77	0.78	0.79	0.79	0.80
		0.19	0.21	0.24	0.27	0.29	0.30	0.32	0.32	0.32	0.33	0.33	0.33	0.34	0.34	0.34	0.35	0.35	0.35	0.36	0.36	0.36	0.36	0.37	0.37	0.37
31	1.36	0.50	0.52	0.54	0.57	0.59	0.62	0.64	0.66	0.68	0.69	0.71	0.72	0.73	0.74	0.75	0.76	0.77	0.78	0.79	0.79	0.80	0.81	0.82	0.82	0.83
		0.20	0.22	0.25	0.27	0.29	0.31	0.32	0.33	0.33	0.34	0.34	0.34	0.35	0.35	0.35	0.36	0.36	0.36	0.37	0.37	0.37	0.38	0.38	0.38	0.38

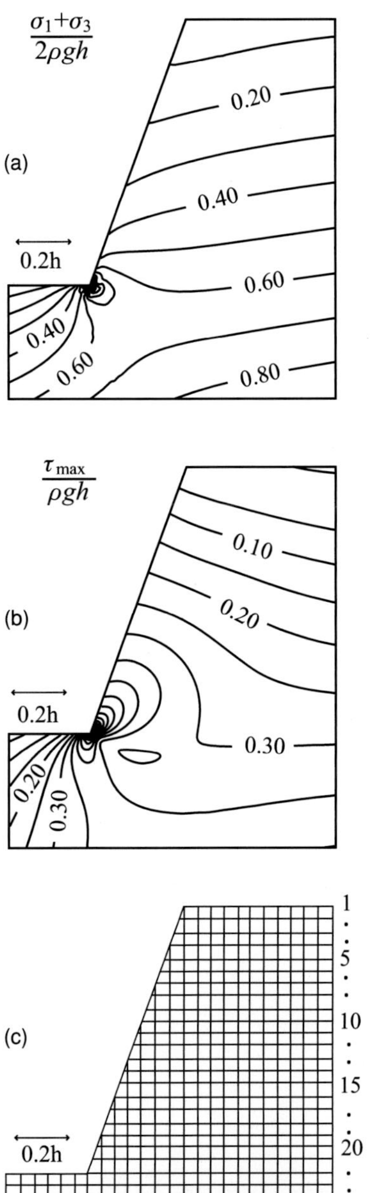

Figure 5.18 Stress contours for a 70° finite slope of height *h* for a Poisson's ratio of 1/3: (a) normalized mean stress, (b) normalized shear stress, and (c) grid for stress distribution Table 5.9.

Table 5.9 Stress distributions for a 70° finite slope of height h for a Poisson's ratio of 1/3

z(h)	x(h)	1	2	3	4	5	6	7	8	9	10	11	12	13	14	15	16	17	18	19	20	21	22	23	24	25
									Toe						Crest											
		0.00	0.05	0.10	0.16	0.21	0.26	0.31	0.36	0.42	0.47	0.52	0.57	0.62	0.68	0.73	0.78	0.83	0.88	0.94	0.99	1.04	1.09	1.14	1.20	1.25
1	0.00														0.01	0.02	0.03	0.04	0.05	0.05	0.06	0.06	0.07	0.07	0.07	0.07
															0.00	0.01	0.02	0.03	0.03	0.04	0.04	0.05	0.05	0.06	0.06	0.06
2	0.05														0.03	0.04	0.05	0.05	0.06	0.06	0.07	0.07	0.08	0.08	0.08	0.09
															0.02	0.02	0.01	0.02	0.02	0.02	0.03	0.03	0.03	0.04	0.04	0.04
3	0.10														0.06	0.07	0.07	0.08	0.08	0.09	0.10	0.10	0.10	0.11	0.11	0.11
															0.05	0.04	0.03	0.03	0.02	0.02	0.02	0.02	0.02	0.02	0.02	0.02
4	0.14													0.08	0.08	0.09	0.10	0.10	0.11	0.11	0.12	0.12	0.13	0.13	0.13	0.14
														0.08	0.07	0.06	0.05	0.05	0.04	0.04	0.03	0.03	0.02	0.02	0.02	0.01
5	0.19													0.11	0.11	0.12	0.12	0.13	0.13	0.14	0.14	0.15	0.15	0.16	0.16	0.16
														0.10	0.09	0.08	0.08	0.07	0.06	0.06	0.05	0.05	0.04	0.04	0.03	0.03
6	0.24													0.13	0.14	0.14	0.15	0.15	0.16	0.16	0.17	0.17	0.18	0.18	0.18	0.19
														0.12	0.11	0.10	0.09	0.09	0.08	0.08	0.07	0.06	0.06	0.06	0.05	0.05
7	0.29												0.15	0.16	0.16	0.17	0.17	0.18	0.18	0.19	0.19	0.20	0.20	0.21	0.21	0.21
													0.15	0.14	0.13	0.12	0.12	0.11	0.10	0.10	0.09	0.09	0.08	0.08	0.07	0.07
8	0.33												0.18	0.18	0.19	0.19	0.20	0.21	0.21	0.22	0.22	0.23	0.23	0.23	0.24	0.24
													0.17	0.16	0.15	0.14	0.14	0.13	0.12	0.12	0.11	0.10	0.10	0.09	0.09	0.09
9	0.38												0.20	0.20	0.21	0.22	0.23	0.23	0.24	0.24	0.25	0.25	0.26	0.26	0.26	0.27
													0.19	0.18	0.17	0.16	0.15	0.15	0.14	0.13	0.13	0.12	0.12	0.11	0.11	0.10
10	0.43											0.22	0.22	0.23	0.24	0.25	0.25	0.26	0.26	0.27	0.27	0.28	0.28	0.29	0.29	0.29
												0.21	0.21	0.20	0.19	0.18	0.17	0.17	0.16	0.15	0.15	0.14	0.13	0.13	0.13	0.12
11	0.48											0.24	0.25	0.25	0.26	0.27	0.28	0.28	0.29	0.29	0.30	0.30	0.31	0.31	0.32	0.32
												0.23	0.22	0.22	0.21	0.20	0.19	0.18	0.18	0.17	0.16	0.16	0.15	0.15	0.14	0.14
12	0.52											0.26	0.27	0.28	0.29	0.30	0.31	0.31	0.32	0.32	0.33	0.33	0.34	0.34	0.34	0.35
												0.25	0.24	0.24	0.23	0.22	0.21	0.20	0.19	0.19	0.18	0.17	0.17	0.16	0.16	0.15
13	0.57										0.27	0.29	0.30	0.31	0.32	0.32	0.33	0.34	0.34	0.35	0.35	0.36	0.36	0.37	0.37	0.37
											0.27	0.27	0.26	0.25	0.24	0.24	0.22	0.22	0.21	0.20	0.19	0.19	0.18	0.18	0.17	0.17
14	0.62										0.30	0.32	0.33	0.34	0.35	0.35	0.36	0.37	0.37	0.38	0.38	0.39	0.39	0.39	0.40	0.40
											0.30	0.29	0.28	0.27	0.26	0.25	0.24	0.23	0.22	0.21	0.21	0.20	0.20	0.19	0.19	0.18
15	0.67										0.33	0.34	0.36	0.36	0.37	0.38	0.39	0.39	0.40	0.40	0.41	0.41	0.42	0.42	0.43	0.43
											0.32	0.31	0.30	0.29	0.28	0.27	0.25	0.24	0.24	0.23	0.22	0.21	0.21	0.20	0.20	0.20

(cont.)

Table 5.9 (cont.)

$z(h)$ \ $x(h)$		1	2	3	4	5	6	7	8	9	10	11	12	13	14	15	16	17	18	19	20	21	22	23	24	25
									Toe						Crest											
		0.00	0.05	0.10	0.16	0.21	0.26	0.31	0.36	0.42	0.47	0.52	0.57	0.62	0.68	0.73	0.78	0.83	0.88	0.94	0.99	1.04	1.09	1.14	1.20	1.25
16	0.71									0.34	0.36	0.38	0.39	0.40	0.40	0.41	0.42	0.42	0.43	0.43	0.44	0.44	0.45	0.45	0.45	0.46
										0.34	0.34	0.33	0.32	0.31	0.29	0.28	0.27	0.26	0.25	0.24	0.23	0.23	0.22	0.22	0.22	0.21
17	0.76									0.38	0.40	0.41	0.42	0.43	0.43	0.44	0.44	0.45	0.45	0.46	0.46	0.47	0.47	0.48	0.48	0.48
										0.37	0.37	0.35	0.34	0.32	0.30	0.29	0.27	0.26	0.26	0.25	0.24	0.24	0.23	0.23	0.23	0.23
18	0.81									0.42	0.44	0.45	0.46	0.46	0.47	0.47	0.47	0.48	0.48	0.49	0.49	0.50	0.50	0.51	0.51	0.51
										0.40	0.39	0.37	0.35	0.33	0.31	0.30	0.28	0.27	0.27	0.26	0.25	0.25	0.25	0.24	0.24	0.24
19	0.86								0.44	0.47	0.48	0.49	0.49	0.49	0.50	0.50	0.50	0.51	0.51	0.52	0.52	0.52	0.53	0.53	0.54	0.54
									0.44	0.44	0.41	0.38	0.35	0.33	0.31	0.30	0.29	0.28	0.27	0.27	0.26	0.26	0.26	0.25	0.25	0.25
20	0.90								0.53	0.54	0.53	0.53	0.53	0.53	0.53	0.53	0.53	0.54	0.54	0.54	0.55	0.55	0.56	0.56	0.57	0.57
									0.51	0.47	0.42	0.38	0.35	0.33	0.31	0.30	0.29	0.28	0.28	0.28	0.27	0.27	0.27	0.27	0.27	0.26
21	0.95								0.66	0.60	0.58	0.56	0.56	0.56	0.56	0.56	0.56	0.56	0.57	0.57	0.58	0.58	0.58	0.59	0.59	0.60
									0.59	0.46	0.40	0.36	0.34	0.32	0.31	0.30	0.29	0.29	0.29	0.28	0.28	0.28	0.28	0.28	0.28	0.28
22	1.00	0.03	0.04	0.06	0.09	0.15	0.38	1.88	0.80	0.66	0.62	0.60	0.59	0.59	0.59	0.59	0.59	0.59	0.60	0.60	0.60	0.61	0.61	0.62	0.62	0.63
		0.03	0.05	0.07	0.10	0.16	0.34	0.92	0.46	0.38	0.35	0.33	0.32	0.31	0.30	0.30	0.30	0.29	0.29	0.29	0.29	0.29	0.29	0.29	0.29	0.29
23	1.05	0.11	0.13	0.16	0.22	0.30	0.45	0.80	0.75	0.68	0.65	0.63	0.62	0.61	0.61	0.61	0.62	0.62	0.62	0.63	0.63	0.64	0.64	0.64	0.65	0.65
		0.06	0.09	0.12	0.17	0.25	0.41	0.47	0.33	0.31	0.31	0.31	0.30	0.30	0.30	0.30	0.30	0.30	0.30	0.30	0.30	0.30	0.30	0.30	0.30	0.30
24	1.10	0.18	0.21	0.25	0.31	0.38	0.50	0.63	0.67	0.66	0.65	0.64	0.64	0.64	0.64	0.64	0.64	0.65	0.65	0.66	0.66	0.67	0.67	0.67	0.68	0.68
		0.10	0.12	0.16	0.22	0.30	0.39	0.41	0.35	0.31	0.30	0.30	0.30	0.30	0.30	0.30	0.31	0.31	0.31	0.31	0.31	0.31	0.31	0.31	0.31	0.31
25	1.14	0.25	0.28	0.31	0.37	0.43	0.52	0.59	0.64	0.65	0.66	0.66	0.66	0.66	0.66	0.67	0.67	0.67	0.68	0.68	0.69	0.69	0.70	0.70	0.71	0.71
		0.12	0.15	0.19	0.25	0.31	0.37	0.38	0.36	0.33	0.31	0.31	0.31	0.31	0.31	0.31	0.31	0.32	0.32	0.32	0.32	0.32	0.32	0.32	0.33	0.33
26	1.19	0.31	0.34	0.38	0.42	0.48	0.54	0.59	0.63	0.65	0.67	0.67	0.68	0.68	0.69	0.69	0.70	0.70	0.71	0.71	0.72	0.72	0.73	0.73	0.74	0.74
		0.15	0.18	0.22	0.27	0.31	0.35	0.37	0.36	0.34	0.33	0.32	0.32	0.32	0.32	0.32	0.32	0.33	0.33	0.33	0.33	0.33	0.34	0.34	0.34	0.34
27	1.24	0.36	0.39	0.43	0.47	0.51	0.56	0.60	0.63	0.66	0.68	0.69	0.70	0.70	0.71	0.72	0.72	0.73	0.73	0.74	0.74	0.75	0.75	0.76	0.76	0.77
		0.17	0.20	0.23	0.28	0.31	0.34	0.36	0.36	0.35	0.34	0.34	0.33	0.33	0.33	0.33	0.34	0.34	0.34	0.34	0.34	0.35	0.35	0.35	0.35	0.35
28	1.29	0.42	0.45	0.48	0.51	0.55	0.59	0.62	0.65	0.68	0.69	0.71	0.72	0.73	0.73	0.74	0.75	0.75	0.76	0.77	0.77	0.78	0.78	0.79	0.79	0.80
		0.18	0.21	0.24	0.28	0.31	0.34	0.36	0.36	0.36	0.36	0.35	0.35	0.35	0.35	0.35	0.35	0.35	0.35	0.35	0.36	0.36	0.36	0.36	0.36	0.36
29	1.33	0.47	0.50	0.52	0.55	0.59	0.62	0.65	0.67	0.70	0.71	0.73	0.74	0.75	0.76	0.77	0.78	0.78	0.79	0.80	0.80	0.81	0.81	0.82	0.82	0.83
		0.20	0.22	0.25	0.29	0.31	0.34	0.35	0.36	0.37	0.37	0.36	0.36	0.36	0.36	0.36	0.36	0.36	0.36	0.37	0.37	0.37	0.37	0.37	0.38	0.38
30	1.38	0.52	0.54	0.56	0.59	0.62	0.65	0.67	0.70	0.72	0.74	0.75	0.76	0.77	0.78	0.79	0.80	0.81	0.81	0.82	0.83	0.83	0.84	0.85	0.85	0.86
		0.21	0.23	0.26	0.29	0.32	0.34	0.35	0.36	0.37	0.37	0.37	0.37	0.37	0.37	0.37	0.37	0.38	0.38	0.38	0.38	0.38	0.38	0.39	0.39	0.39
31	1.43	0.56	0.58	0.61	0.63	0.66	0.68	0.70	0.72	0.74	0.76	0.78	0.79	0.80	0.81	0.82	0.83	0.84	0.84	0.85	0.86	0.86	0.87	0.88	0.88	0.89
		0.22	0.24	0.27	0.30	0.32	0.34	0.35	0.37	0.37	0.38	0.38	0.38	0.38	0.38	0.39	0.39	0.39	0.39	0.39	0.39	0.40	0.40	0.40	0.40	0.40

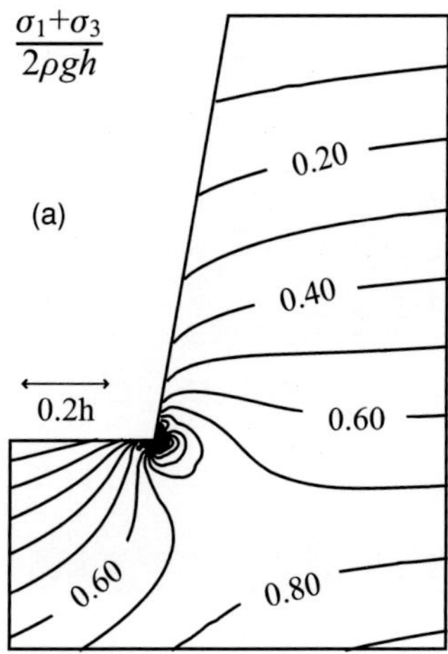

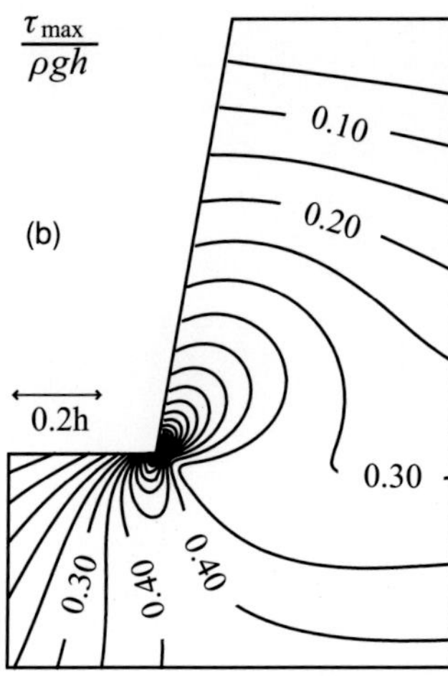

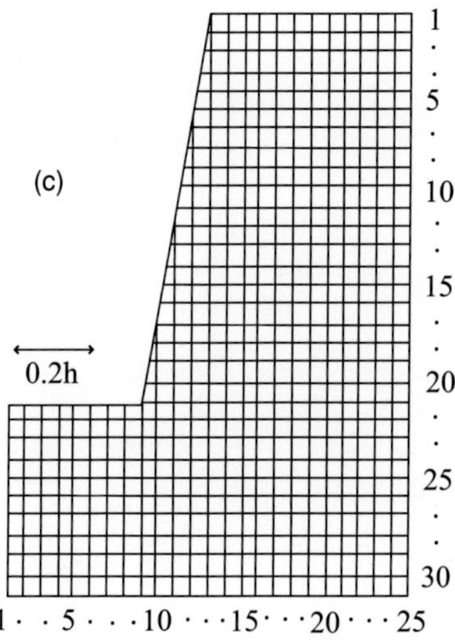

Figure 5.19 Stress contours for a 80° finite slope of height h for a Poisson's ratio of 1/3: (a) normalized mean stress, (b) normalized shear stress, and (c) grid for stress distribution Table 5.10.

Table 5.10 Stress distributions for an 80° finite slope of height h for a Poisson's ratio of 1/3

$z(h)$	$x(h)$	1	2	3	4	5	6	7	8	9	10	11	12	13	14	15	16	17	18	19	20	21	22	23	24	25
											Toe			Crest												
		0.00	0.04	0.09	0.13	0.18	0.22	0.26	0.31	0.35	0.40	0.44	0.48	0.53	0.57	0.62	0.66	0.70	0.75	0.79	0.84	0.88	0.92	0.97	1.01	1.06
1	0.00													0.01	0.01	0.02	0.02	0.02	0.02	0.02	0.03	0.03	0.03	0.03	0.04	0.04
														0.00	0.00	0.00	0.00	0.01	0.01	0.01	0.01	0.02	0.02	0.02	0.02	0.03
2	0.05													0.03	0.03	0.03	0.03	0.04	0.04	0.04	0.05	0.05	0.05	0.05	0.06	0.06
														0.03	0.02	0.02	0.02	0.01	0.01	0.01	0.01	0.01	0.01	0.01	0.01	0.01
3	0.10													0.05	0.05	0.06	0.06	0.06	0.07	0.07	0.07	0.08	0.08	0.08	0.08	0.09
														0.05	0.05	0.04	0.04	0.04	0.04	0.03	0.03	0.03	0.02	0.02	0.02	0.02
4	0.15													0.08	0.08	0.08	0.09	0.09	0.10	0.10	0.10	0.10	0.11	0.11	0.11	0.12
														0.07	0.07	0.07	0.06	0.06	0.06	0.06	0.05	0.05	0.05	0.04	0.04	0.04
5	0.20													0.10	0.11	0.11	0.12	0.12	0.12	0.13	0.13	0.13	0.14	0.14	0.14	0.15
														0.10	0.09	0.09	0.09	0.09	0.08	0.08	0.08	0.07	0.07	0.07	0.06	0.06
6	0.25												0.12	0.13	0.13	0.14	0.14	0.15	0.15	0.15	0.16	0.16	0.17	0.17	0.17	0.17
													0.12	0.12	0.12	0.11	0.11	0.11	0.11	0.10	0.10	0.09	0.09	0.09	0.08	0.08
7	0.30												0.14	0.15	0.16	0.16	0.17	0.17	0.18	0.18	0.19	0.19	0.19	0.20	0.20	0.20
													0.14	0.14	0.14	0.14	0.14	0.13	0.13	0.13	0.12	0.12	0.11	0.11	0.10	0.10
8	0.35												0.17	0.18	0.18	0.19	0.20	0.20	0.21	0.21	0.22	0.22	0.22	0.23	0.23	0.23
													0.17	0.17	0.17	0.16	0.16	0.16	0.15	0.15	0.14	0.14	0.13	0.13	0.12	0.12
9	0.40												0.19	0.20	0.21	0.22	0.23	0.23	0.24	0.24	0.25	0.25	0.25	0.26	0.26	0.26
													0.19	0.19	0.19	0.19	0.19	0.18	0.18	0.17	0.17	0.16	0.15	0.15	0.14	0.14
10	0.45												0.22	0.23	0.24	0.25	0.26	0.26	0.27	0.27	0.28	0.28	0.28	0.29	0.29	0.29
													0.22	0.22	0.22	0.21	0.21	0.20	0.20	0.19	0.19	0.18	0.17	0.17	0.16	0.16
11	0.50											0.24	0.25	0.26	0.27	0.28	0.29	0.29	0.30	0.30	0.31	0.31	0.32	0.32	0.32	0.32
												0.24	0.24	0.24	0.24	0.24	0.23	0.23	0.22	0.21	0.21	0.20	0.19	0.19	0.18	0.17
12	0.55											0.26	0.28	0.29	0.30	0.31	0.32	0.32	0.33	0.33	0.34	0.34	0.35	0.35	0.35	0.35
												0.26	0.27	0.27	0.27	0.26	0.26	0.25	0.24	0.23	0.22	0.22	0.21	0.20	0.19	0.19
13	0.60											0.29	0.31	0.32	0.33	0.34	0.35	0.36	0.36	0.37	0.37	0.37	0.38	0.38	0.38	0.39
												0.29	0.30	0.30	0.29	0.29	0.28	0.27	0.26	0.25	0.24	0.23	0.22	0.22	0.21	0.20
14	0.65											0.33	0.35	0.36	0.37	0.38	0.39	0.39	0.40	0.40	0.40	0.41	0.41	0.41	0.41	0.42
												0.32	0.33	0.33	0.32	0.31	0.30	0.29	0.28	0.27	0.26	0.25	0.24	0.23	0.22	0.22
15	0.70											0.37	0.38	0.40	0.41	0.42	0.42	0.43	0.43	0.43	0.44	0.44	0.44	0.44	0.45	0.45
												0.36	0.36	0.36	0.35	0.34	0.32	0.31	0.29	0.28	0.27	0.26	0.25	0.24	0.24	0.23
16	0.75										0.38	0.41	0.43	0.44	0.45	0.46	0.46	0.46	0.47	0.47	0.47	0.47	0.47	0.48	0.48	0.48
											0.39	0.40	0.40	0.39	0.37	0.36	0.34	0.32	0.30	0.29	0.28	0.27	0.26	0.25	0.25	0.24

17	0.80										0.43	0.46	0.48	0.49	0.49	0.50	0.50	0.50	0.50	0.50	0.50	0.50	0.51	0.51	0.51	0.51
											0.43	0.44	0.43	0.41	0.39	0.37	0.35	0.33	0.31	0.30	0.29	0.28	0.27	0.26	0.25	0.25
18	0.85										0.50	0.53	0.54	0.54	0.54	0.54	0.54	0.54	0.54	0.54	0.54	0.54	0.54	0.54	0.54	0.54
											0.49	0.49	0.47	0.43	0.40	0.37	0.35	0.33	0.31	0.30	0.29	0.28	0.27	0.27	0.26	0.26
19	0.90										0.61	0.62	0.61	0.60	0.59	0.58	0.58	0.57	0.57	0.57	0.57	0.57	0.57	0.57	0.57	0.57
											0.58	0.55	0.49	0.43	0.39	0.37	0.34	0.32	0.31	0.30	0.29	0.29	0.28	0.28	0.27	0.27
20	0.95										0.79	0.72	0.69	0.66	0.64	0.63	0.62	0.61	0.61	0.60	0.60	0.60	0.60	0.60	0.60	0.60
											0.70	0.56	0.46	0.41	0.37	0.35	0.33	0.32	0.31	0.30	0.29	0.29	0.29	0.28	0.28	0.28
21	1.00	0.02	0.02	0.04	0.05	0.07	0.11	0.18	0.48	2.58	1.06	0.84	0.76	0.71	0.68	0.66	0.65	0.64	0.64	0.64	0.63	0.63	0.63	0.63	0.63	0.63
		0.02	0.03	0.04	0.06	0.09	0.13	0.21	0.42	1.09	0.51	0.41	0.37	0.35	0.33	0.32	0.31	0.31	0.30	0.30	0.30	0.29	0.29	0.29	0.29	0.29
22	1.05	0.10	0.11	0.14	0.17	0.21	0.28	0.40	0.56	0.95	0.93	0.83	0.77	0.73	0.71	0.69	0.68	0.67	0.67	0.66	0.66	0.66	0.66	0.66	0.66	0.66
		0.05	0.07	0.09	0.12	0.16	0.23	0.34	0.52	0.56	0.38	0.32	0.31	0.31	0.31	0.31	0.30	0.30	0.30	0.30	0.30	0.30	0.30	0.30	0.30	0.30
23	1.10	0.17	0.20	0.22	0.26	0.31	0.38	0.47	0.60	0.73	0.79	0.78	0.76	0.74	0.72	0.71	0.70	0.70	0.69	0.69	0.69	0.69	0.69	0.69	0.69	0.69
		0.08	0.11	0.14	0.18	0.22	0.30	0.39	0.47	0.49	0.42	0.35	0.32	0.31	0.30	0.30	0.30	0.30	0.31	0.31	0.31	0.31	0.31	0.31	0.31	0.31
24	1.15	0.25	0.27	0.30	0.34	0.39	0.45	0.52	0.60	0.68	0.73	0.74	0.74	0.74	0.73	0.73	0.72	0.72	0.72	0.72	0.72	0.72	0.72	0.72	0.72	0.72
		0.12	0.14	0.18	0.22	0.26	0.32	0.39	0.44	0.45	0.43	0.38	0.35	0.33	0.32	0.32	0.32	0.31	0.31	0.32	0.32	0.32	0.32	0.32	0.32	0.32
25	1.20	0.31	0.34	0.37	0.40	0.45	0.50	0.55	0.61	0.66	0.70	0.72	0.74	0.74	0.74	0.74	0.74	0.74	0.74	0.74	0.74	0.75	0.75	0.75	0.75	0.75
		0.14	0.17	0.20	0.25	0.29	0.33	0.38	0.42	0.43	0.42	0.40	0.38	0.36	0.34	0.34	0.33	0.33	0.33	0.33	0.33	0.33	0.33	0.33	0.33	0.33
26	1.25	0.37	0.40	0.43	0.46	0.50	0.54	0.58	0.63	0.67	0.70	0.72	0.74	0.75	0.76	0.76	0.76	0.76	0.77	0.77	0.77	0.77	0.77	0.78	0.78	0.78
		0.17	0.19	0.23	0.26	0.30	0.34	0.37	0.40	0.41	0.41	0.40	0.39	0.38	0.36	0.36	0.35	0.35	0.35	0.34	0.34	0.34	0.34	0.35	0.35	0.35
27	1.30	0.43	0.45	0.48	0.51	0.54	0.58	0.61	0.65	0.68	0.71	0.73	0.75	0.76	0.77	0.78	0.78	0.79	0.79	0.79	0.80	0.80	0.80	0.81	0.81	0.81
		0.18	0.21	0.24	0.28	0.31	0.34	0.37	0.39	0.41	0.41	0.41	0.40	0.39	0.38	0.38	0.37	0.37	0.36	0.36	0.36	0.36	0.36	0.36	0.36	0.36
28	1.35	0.48	0.51	0.53	0.56	0.58	0.61	0.64	0.67	0.70	0.73	0.75	0.76	0.78	0.79	0.80	0.80	0.81	0.81	0.82	0.82	0.83	0.83	0.83	0.84	0.84
		0.20	0.23	0.26	0.29	0.31	0.34	0.37	0.39	0.40	0.41	0.41	0.41	0.40	0.40	0.39	0.39	0.38	0.38	0.38	0.38	0.38	0.37	0.38	0.38	0.38
29	1.40	0.53	0.55	0.58	0.60	0.63	0.65	0.68	0.70	0.73	0.75	0.77	0.78	0.80	0.81	0.82	0.83	0.83	0.84	0.84	0.85	0.85	0.86	0.86	0.87	0.87
		0.22	0.24	0.27	0.29	0.32	0.34	0.37	0.38	0.40	0.41	0.41	0.41	0.41	0.41	0.40	0.40	0.40	0.39	0.39	0.39	0.39	0.39	0.39	0.39	0.39
30	1.45	0.58	0.60	0.62	0.64	0.66	0.69	0.71	0.73	0.75	0.77	0.79	0.81	0.82	0.83	0.84	0.85	0.86	0.86	0.87	0.88	0.88	0.89	0.89	0.89	0.90
		0.23	0.25	0.28	0.30	0.32	0.35	0.37	0.38	0.40	0.41	0.41	0.42	0.42	0.41	0.41	0.41	0.41	0.41	0.41	0.41	0.40	0.40	0.40	0.40	0.40
		0.63	0.65	0.66	0.68	0.70	0.72	0.74	0.76	0.78	0.80	0.82	0.83	0.84	0.86	0.87	0.88	0.88	0.89	0.90	0.90	0.91	0.91	0.92	0.92	0.93
		0.24	0.26	0.29	0.31	0.33	0.35	0.37	0.38	0.40	0.41	0.41	0.42	0.42	0.42	0.42	0.42	0.42	0.42	0.42	0.42	0.42	0.42	0.42	0.42	0.42

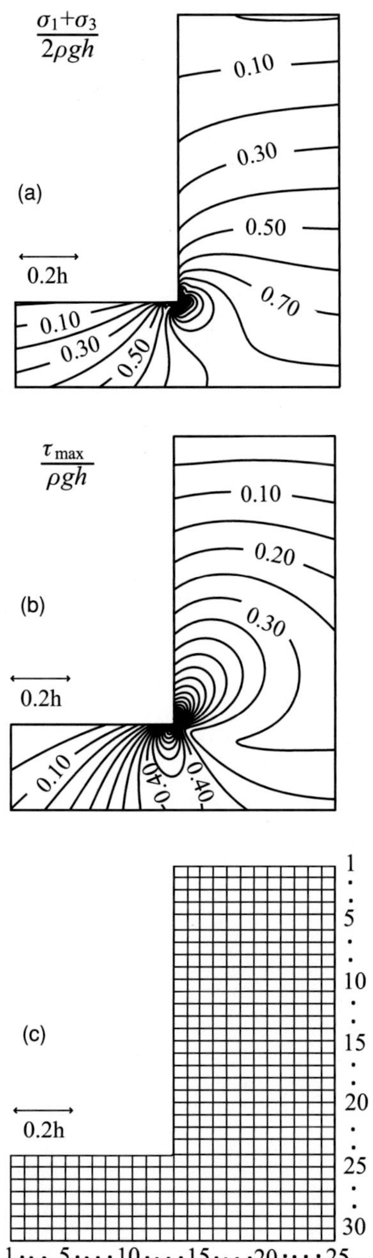

Figure 5.20 Stress contours for a 90° finite slope of height h for a Poisson's ratio of 1/3: (a) normalized mean stress, (b) normalized shear stress, and (c) grid for stress distribution Table 5.11.

Table 5.11 Stress distributions for a 90° finite slope of height h for a Poisson's ratio of 1/3

$z(h)$ \ $x(h)$		1	2	3	4	5	6	7	8	9	10	11	12	13	14	15	16	17	18	19	20	21	22	23	24	25
		0.00	0.05	0.10	0.14	0.19	0.24	0.29	0.34	0.38	0.43	0.48	0.53	Toe 0.58	Crest 0.62	0.67	0.72	0.77	0.81	0.86	0.91	0.96	1.01	1.05	1.10	1.15
1	0.00													0.01	0.01	0.01	0.00	0.00	0.00	−0.01	−0.01	−0.01	−0.01	−0.01	−0.01	−0.01
														0.00	0.00	0.01	0.01	0.01	0.01	0.02	0.02	0.02	0.02	0.02	0.02	0.02
2	0.04													0.02	0.02	0.02	0.02	0.02	0.01	0.01	0.01	0.01	0.01	0.01	0.01	0.02
														0.02	0.02	0.02	0.03	0.03	0.03	0.03	0.03	0.03	0.03	0.03	0.03	0.03
3	0.09													0.04	0.04	0.04	0.04	0.04	0.04	0.04	0.04	0.04	0.04	0.04	0.05	0.05
														0.04	0.04	0.04	0.05	0.05	0.05	0.05	0.05	0.05	0.05	0.05	0.04	0.04
4	0.13													0.06	0.06	0.06	0.06	0.06	0.06	0.06	0.07	0.07	0.07	0.07	0.07	0.07
														0.06	0.06	0.06	0.07	0.07	0.07	0.07	0.07	0.07	0.06	0.06	0.06	0.06
5	0.17													0.08	0.08	0.08	0.09	0.09	0.09	0.09	0.09	0.10	0.10	0.10	0.10	0.10
														0.08	0.08	0.08	0.09	0.09	0.09	0.09	0.09	0.08	0.08	0.08	0.08	0.08
6	0.22													0.10	0.10	0.11	0.11	0.11	0.12	0.12	0.12	0.12	0.13	0.13	0.13	0.13
														0.10	0.10	0.11	0.11	0.11	0.11	0.11	0.11	0.10	0.10	0.10	0.10	0.09
7	0.26													0.11	0.12	0.13	0.13	0.14	0.14	0.14	0.15	0.15	0.15	0.16	0.16	0.16
														0.12	0.12	0.13	0.13	0.13	0.13	0.13	0.13	0.12	0.12	0.12	0.11	0.11
8	0.30													0.13	0.14	0.15	0.16	0.16	0.17	0.17	0.18	0.18	0.18	0.19	0.19	0.19
														0.14	0.14	0.15	0.15	0.15	0.15	0.15	0.15	0.14	0.14	0.14	0.13	0.13
9	0.35													0.15	0.16	0.17	0.18	0.19	0.19	0.20	0.21	0.21	0.21	0.22	0.22	0.22
														0.16	0.16	0.17	0.17	0.18	0.18	0.17	0.17	0.17	0.16	0.16	0.15	0.14
10	0.39													0.17	0.18	0.19	0.20	0.21	0.22	0.23	0.23	0.24	0.24	0.24	0.25	0.25
														0.17	0.18	0.19	0.20	0.20	0.20	0.19	0.19	0.18	0.18	0.17	0.17	0.16
11	0.43													0.19	0.21	0.22	0.23	0.24	0.25	0.26	0.26	0.27	0.27	0.27	0.28	0.28
														0.19	0.21	0.21	0.22	0.22	0.22	0.22	0.21	0.20	0.20	0.19	0.18	0.18
12	0.48													0.21	0.23	0.25	0.26	0.27	0.28	0.29	0.29	0.30	0.30	0.30	0.31	0.31
														0.22	0.23	0.24	0.25	0.25	0.24	0.24	0.23	0.23	0.22	0.21	0.20	0.19
13	0.52													0.23	0.26	0.27	0.29	0.30	0.31	0.32	0.32	0.33	0.33	0.33	0.34	0.34
														0.24	0.25	0.26	0.27	0.27	0.27	0.26	0.25	0.24	0.23	0.22	0.21	0.21
14	0.56													0.26	0.28	0.30	0.32	0.33	0.34	0.35	0.35	0.36	0.36	0.36	0.37	0.37
														0.26	0.28	0.29	0.30	0.30	0.29	0.28	0.27	0.26	0.25	0.24	0.23	0.22
15	0.61													0.28	0.31	0.33	0.35	0.36	0.38	0.38	0.39	0.39	0.40	0.40	0.40	0.40
														0.29	0.31	0.32	0.33	0.33	0.32	0.31	0.29	0.28	0.27	0.25	0.24	0.23

(cont.)

Table 5.11 (cont.)

z(h) \ x(h)		1	2	3	4	5	6	7	8	9	10	11	12	13	14	15	16	17	18	19	20	21	22	23	24	25
														Toe	Crest											
		0.00	0.05	0.10	0.14	0.19	0.24	0.29	0.34	0.38	0.43	0.48	0.53	0.58	0.62	0.67	0.72	0.77	0.81	0.86	0.91	0.96	1.01	1.05	1.10	1.15
16	0.65													0.31	0.34	0.37	0.39	0.40	0.41	0.42	0.42	0.42	0.43	0.43	0.43	0.43
														0.32	0.34	0.35	0.36	0.35	0.34	0.32	0.31	0.29	0.28	0.27	0.25	0.24
17	0.69													0.34	0.38	0.41	0.43	0.44	0.45	0.45	0.46	0.46	0.46	0.46	0.46	0.46
														0.35	0.38	0.39	0.39	0.38	0.36	0.34	0.33	0.31	0.29	0.28	0.26	0.25
18	0.74													0.38	0.43	0.45	0.47	0.48	0.49	0.49	0.50	0.50	0.49	0.49	0.49	0.49
														0.39	0.42	0.43	0.42	0.41	0.38	0.36	0.34	0.32	0.30	0.28	0.27	0.26
19	0.78													0.41	0.47	0.50	0.52	0.53	0.53	0.53	0.53	0.53	0.53	0.52	0.52	0.52
														0.42	0.46	0.47	0.45	0.43	0.40	0.37	0.34	0.32	0.30	0.29	0.28	0.27
20	0.82													0.46	0.54	0.57	0.58	0.58	0.58	0.57	0.57	0.56	0.56	0.56	0.55	0.55
														0.48	0.52	0.52	0.48	0.45	0.41	0.37	0.35	0.32	0.31	0.29	0.28	0.27
21	0.87													0.53	0.63	0.65	0.65	0.64	0.63	0.62	0.61	0.60	0.59	0.59	0.59	0.58
														0.55	0.60	0.57	0.50	0.45	0.40	0.37	0.34	0.32	0.30	0.29	0.28	0.27
22	0.91													0.64	0.75	0.74	0.72	0.70	0.67	0.66	0.64	0.63	0.62	0.62	0.61	0.61
														0.65	0.68	0.60	0.50	0.44	0.39	0.35	0.33	0.31	0.30	0.29	0.28	0.28
23	0.95													1.03	0.91	0.87	0.80	0.75	0.72	0.69	0.68	0.67	0.66	0.65	0.64	0.64
														0.91	0.76	0.56	0.45	0.40	0.36	0.33	0.32	0.31	0.30	0.29	0.28	0.28
24	1.00	−0.02	−0.02	−0.01	−0.01	0.00	0.00	0.02	0.03	0.05	0.08	0.14	0.49	3.20	1.27	0.98	0.86	0.80	0.76	0.73	0.71	0.70	0.69	0.68	0.67	0.67
		0.02	0.02	0.01	0.01	0.00	0.01	0.02	0.04	0.07	0.11	0.18	0.45	1.39	0.52	0.42	0.37	0.34	0.32	0.31	0.30	0.30	0.29	0.29	0.29	0.28
25	1.04	0.03	0.04	0.04	0.05	0.07	0.08	0.10	0.13	0.18	0.24	0.38	0.62	1.27	1.10	0.97	0.87	0.82	0.78	0.75	0.73	0.72	0.71	0.70	0.70	0.69
		0.01	0.01	0.01	0.02	0.03	0.05	0.07	0.09	0.14	0.20	0.33	0.60	0.73	0.39	0.33	0.31	0.31	0.30	0.30	0.30	0.30	0.29	0.29	0.29	0.29
26	1.08	0.09	0.10	0.11	0.12	0.14	0.16	0.19	0.23	0.28	0.36	0.48	0.67	0.91	0.95	0.91	0.86	0.82	0.79	0.77	0.76	0.74	0.74	0.73	0.72	0.72
		0.02	0.03	0.04	0.05	0.06	0.09	0.12	0.15	0.21	0.29	0.42	0.57	0.60	0.47	0.36	0.32	0.31	0.30	0.30	0.30	0.30	0.30	0.30	0.30	0.30
27	1.13	0.15	0.16	0.17	0.19	0.21	0.23	0.27	0.31	0.37	0.44	0.54	0.68	0.79	0.86	0.86	0.84	0.82	0.80	0.79	0.78	0.77	0.76	0.75	0.75	0.75
		0.04	0.05	0.06	0.08	0.10	0.12	0.16	0.20	0.26	0.34	0.44	0.52	0.54	0.49	0.41	0.36	0.33	0.32	0.31	0.31	0.31	0.31	0.31	0.31	0.31
28	1.17	0.20	0.21	0.23	0.25	0.27	0.29	0.33	0.37	0.42	0.49	0.57	0.67	0.75	0.81	0.82	0.82	0.82	0.81	0.80	0.79	0.78	0.78	0.77	0.77	0.77
		0.05	0.06	0.08	0.10	0.12	0.15	0.19	0.23	0.29	0.36	0.43	0.49	0.51	0.48	0.44	0.39	0.36	0.34	0.33	0.33	0.32	0.32	0.32	0.32	0.32
29	1.21	0.26	0.27	0.29	0.31	0.33	0.35	0.39	0.43	0.48	0.53	0.60	0.68	0.73	0.78	0.80	0.81	0.82	0.81	0.81	0.81	0.80	0.80	0.80	0.79	0.79
		0.07	0.08	0.10	0.12	0.15	0.18	0.22	0.26	0.31	0.37	0.43	0.47	0.48	0.47	0.45	0.41	0.39	0.37	0.35	0.35	0.34	0.34	0.34	0.33	0.33
30	1.26	0.31	0.32	0.34	0.36	0.38	0.41	0.45	0.48	0.53	0.57	0.63	0.69	0.73	0.77	0.80	0.81	0.82	0.82	0.82	0.82	0.82	0.82	0.82	0.82	0.82
		0.09	0.10	0.12	0.14	0.17	0.20	0.24	0.28	0.33	0.37	0.42	0.45	0.47	0.47	0.45	0.43	0.41	0.39	0.38	0.37	0.36	0.35	0.35	0.35	0.35
31	1.30	0.36	0.37	0.39	0.41	0.43	0.46	0.49	0.52	0.56	0.60	0.65	0.70	0.74	0.77	0.80	0.81	0.82	0.83	0.83	0.84	0.84	0.84	0.84	0.84	0.84
		0.10	0.12	0.14	0.16	0.19	0.22	0.25	0.29	0.33	0.37	0.41	0.44	0.46	0.46	0.45	0.44	0.42	0.41	0.39	0.38	0.38	0.37	0.37	0.36	0.36

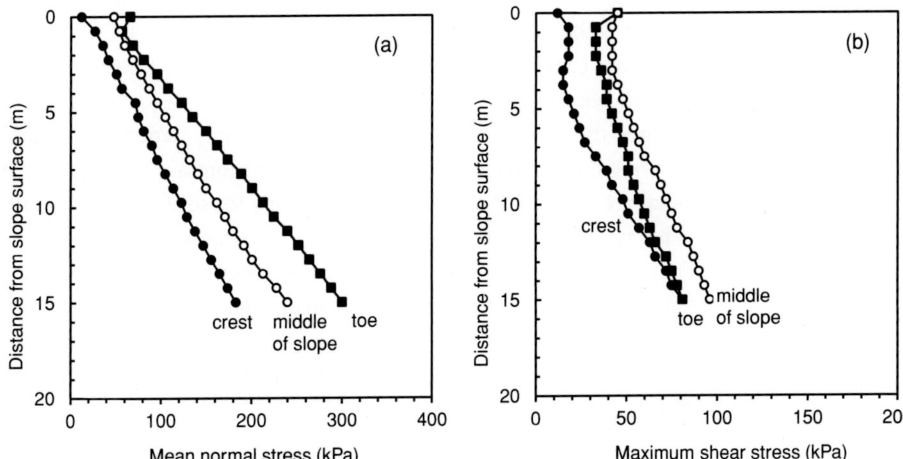

Figure 5.21 Stress profiles at the toe, middle and crest locations of the slope for a 30° finite slope of height $h = 15$ m when Poisson's ratio is 1/3 and unit weight $\rho g = 20$ kN/m^2: (a) mean normal stress, and (b) maximum shear stress.

5.6 Problems

1. What are the units of body force? Is it a force or a stress? What is its direction on earth?
2. What are the differences and links between stress and traction?
3. In a three-dimensional world, how many independent stress components act on a body at each point?
4. What is the graphical interpretation of shear strain?
5. In linear elasticity theory, how many independent parameters are needed in order to define the stress–strain relationship?
6. What are the physical meanings of the parameters E, G, λ, and ν?
7. What is the parameter that describes the link between volumetric strain and volumetric stress?
8. In two-dimensional space, how many independent stress components act at a point? What are the principal stresses?
9. If you know the stress components at a point with orthogonal planes A and B, as shown in Figure 5.5, how do you find the principal stresses and the angle between the maximum principal stress and the plane A?
10. If you know the principal stress components at a point, how do you find the shear and normal stresses on a plane rotated α degrees from the maximum principal stress?
11. If you know the normal stress components σ_{xx} and σ_{yy}, and τ_{xy} at a point, how do you calculate the maximum shear stress τ_{max} at that point?
12. How many dependent variables, how many independent variables, and how many equations are there in the equations of motion shown in Equation (5.33)?

13 If a body of matter with a density ρ is moving very rapidly with a constant velocity c, what is the magnitude of the inertial force acting on this body?

14 Equations (5.37)–(5.42) describe the stress–strain relationships for elastic materials. Why do we call them "linear" elastic materials?

15 In three-dimensional linear elasticity, how many dependent variables are there and what are they? How many independent equations are there and what are they?

16 What are the common boundary conditions for solving stress–strain problems?

17 What is a plane stress problem? What is a plane strain problem?

18 For a plane strain problem for which only stresses are considered, how many equations are needed and what are they? How many elastic parameters are needed?

19 For a plane stress problem for which only stresses are considered, how many equations are needed and what are they? How many elastic parameters are needed?

20 If you know the shear modulus G and Poisson's ratio ν, how do you estimate the elastic modulus E?

21 Consider two hillslopes with the same geometry but very different elastic moduli. The elastic modulus of one is 10 times that of the other. Would you expect the stress field distributions in these two hillslopes to be different? Why?

22 A point in a hillslope has the following state of stress:

$$\sigma = \begin{bmatrix} \sigma_{xx} & \sigma_{xy} & \sigma_{xz} \\ \sigma_{yx} & \sigma_{yy} & \sigma_{yz} \\ \sigma_{zx} & \sigma_{zy} & \sigma_{zz} \end{bmatrix} = \begin{bmatrix} 100 & 50 & 90 \\ 50 & 200 & 60 \\ 90 & 60 & 300 \end{bmatrix} \text{(kPa)}$$

where x and y are the horizontal coordinates and z is the vertical coordinate. Find the principal stress components and directions. Plot the Mohr circles to represent the state of stress at this point. Find the maximum shear stress and direction of its action at this point.

23 For a finite slope with a 15° slope angle, a slope height of 50 m, and Poisson's ratio of 1/3, calculate the mean stress and maximum shear stress profiles beneath the toe, middle, and the crest of the slope using: (a) Savage's analytical solutions, and (b) FE solutions. Compare the results by plotting the profiles. What are your conclusions from this exercise?

24 For a finite slope with a 45° slope angle, a slope height of 50 m, and Poisson's ratio of 1/3, calculate the mean stress and maximum shear stress profiles beneath the toe, middle, and the crest of the slope by: (a) Savage's analytical solutions, and (b) FE solutions (presented in Section 5.5.2). Compare the results by plotting the profiles. Draw your conclusions from this exercise. Draw conclusions from comparisons with the results for the 15° finite slope (previous problem).

6 Effective stress in soil

6.1 Terzaghi's and Bishop's effective stress theories

The definition of effective stress in compressible porous media is complicated due to the coupling effect between the pore fluid and the media skeleton (e.g., Skempton, 1960a; 1960b; Iverson and Reid, 1992). This coupling effect requires that the compressibility of the media constituents be considered. However, in shallow hillslope environments materials are generally loose and partially saturated and the coupling can be ignored. For most soil mechanics problems Terzaghi's effective stress, which explicitly ignores this coupling, has been shown to be practical and sufficient.

In the early 1950s, some twenty years or so after the introduction of Terzaghi's effective stress relation (Terzaghi, 1936, 1943), the engineering community began to recognize that the understanding of foundation problems required the consideration of partially saturated soil conditions. Terzaghi's effective stress principle states that the stress going through soil's skeleton σ' is the state variable directly controlling the strength and deformation behavior of saturated soil and can be deduced from the total stress σ and pore-water pressure u_w (shown in Figure 6.1a):

$$\sigma' = \sigma - u_w \tag{6.1}$$

The direct use of Terzaghi's effective stress relation in assessing the state of stress in partially saturated foundation soils was considered inadequate and several new effective stress relations were proposed. Among them, Bishop's (1954, 1959) relation has drawn the most attention over time. Bishop recognized the importance of pore-water pressure to effective stress. Under partially saturated conditions, pore-water pressure does not contribute to effective stress without modification and a scaling factor varying between zero and unity was introduced to express effective stress, i.e.,

$$\sigma' = \sigma - u_a + \chi(u_a - u_w) \tag{6.2}$$

with the effective stress parameter χ given a value of 0 when the soil is dry and unity when the soil is saturated. Bishop further suggested that the effective stress parameter χ should be a function of the degree of saturation. Bishop's effective stress relation, in tensor notation, is

$$\sigma'_{ij} = \begin{Bmatrix} \sigma_{xx} - u_a & \tau_{xy} & \tau_{xz} \\ \tau_{xy} & \sigma_{yy} - u_a & \tau_{yz} \\ \tau_{xz} & \tau_{yz} & \sigma_{zz} - u_a \end{Bmatrix} + \begin{Bmatrix} \chi(u_a - u_w) & 0 & 0 \\ 0 & \chi(u_a - u_w) & 0 \\ 0 & 0 & \chi(u_a - u_w) \end{Bmatrix} \tag{6.2}$$

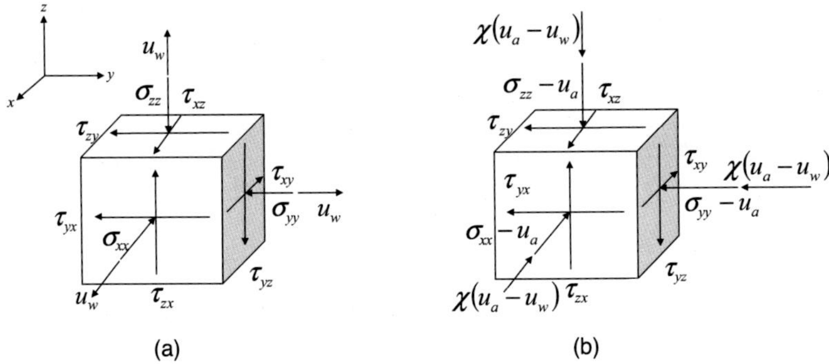

Figure 6.1 Illustrations of (a) Terzaghi's effective stress as a stress tensor at the REV level in saturated porous media, and (b) Bishop's effective stress as a stress tensor at the REV level in unsaturated porous media.

Figure 6.1b shows Bishop's effective stress for a representative elementary volume or REV. Bishop's effective stress relation is attractive in that it retains the simplicity of Terzaghi's effective stress relation and reduces to Terzaghi's relation in fully saturated soil. Bishop's relation is also intuitively correct in that it predicts the absence of any contribution of pore-water pressure to effective stress in dry soil. However, as described in detail in this chapter, Lu and his co-workers show that such a limitation is fundamentally flawed.

The validation of Bishop's effective stress relation has been a long journey. Immediately after its conception, several researchers questioned its suitability for describing the mechanical behavior of soil under unsaturated conditions. Prior to the 1990s, because of the limitations of experimental techniques to quantify mechanical behavior in general, and the measurement and control of soil suction in particular, sufficient experimental confirmation of Bishop's effective stress relation was not possible. The most damaging argument, among others, challenging the validity of Bishop's relation was the work of Jennings and Burland (1962). Jennings and Burland performed imbibition tests on several remolded soils using an oedometer. An oedometer is a loading device in which a cylindrical soil sample is radially confined with no displacement allowed while axis loading is applied for time–displacement measurements. The soils were wetted from the base of the specimens. Contrary to predictions based on Bishop's relation (6.2), the volume of the soils reduced dramatically or the soils "collapsed." According to Bishop's effective stress relation (Equation (6.2)), as a soil wets, soil suction diminishes, leading to a reduction in effective stress. This reduction in effective stress should result in an increase in the volume of the soil, counter to the experimental observation of collapse. Thus, Jennings and Burland concluded that Bishop's effective stress relation is not valid for describing the mechanical behavior of soil under unsaturated conditions. The logic of this argument appeared sound at the time of its publication and was largely accepted by the soil mechanics community until the late 1990s. In the intervening period, Coleman's (1962) two independent stress variables theory drew considerable attention, and until recently became the leading approach used to describe the state of stress in unsaturated soils (Fredlund and Morgenstern, 1977). The essentials of Coleman's two independent stress variables theory will be discussed in Section 6.2.

Figure 1.4 Debris flow on the flank of Mt. St. Helens triggered by an eruption in March 1982 (photo by Tom Casadevall, USGS).

Figure 1.5 Rockslides triggered by the 2008 M7.9 Wenchuan, China, earthquake in Beichuan town (photo by Yueping Yin, China Geological Survey).

Figure 1.6 Landslides triggered by heavy rainfall in Brazil in January 2011 (AP photo).

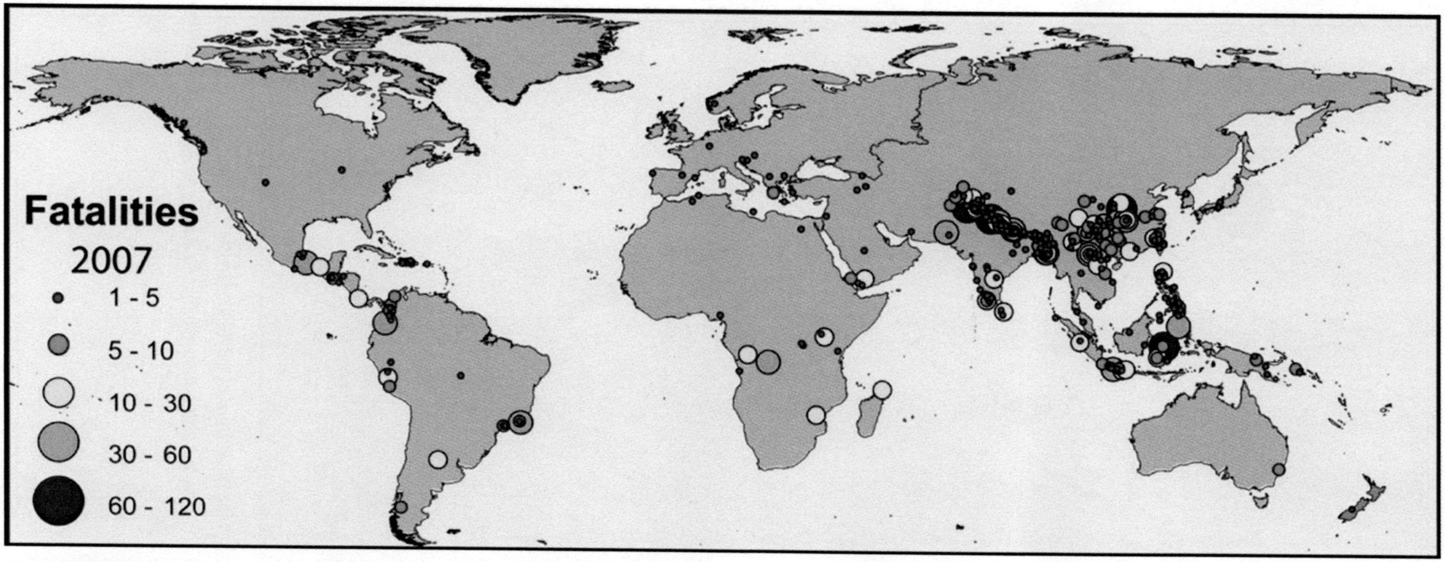

Figure 1.8 Distribution of landslide fatalities in 2007 (from Petley, 2008).

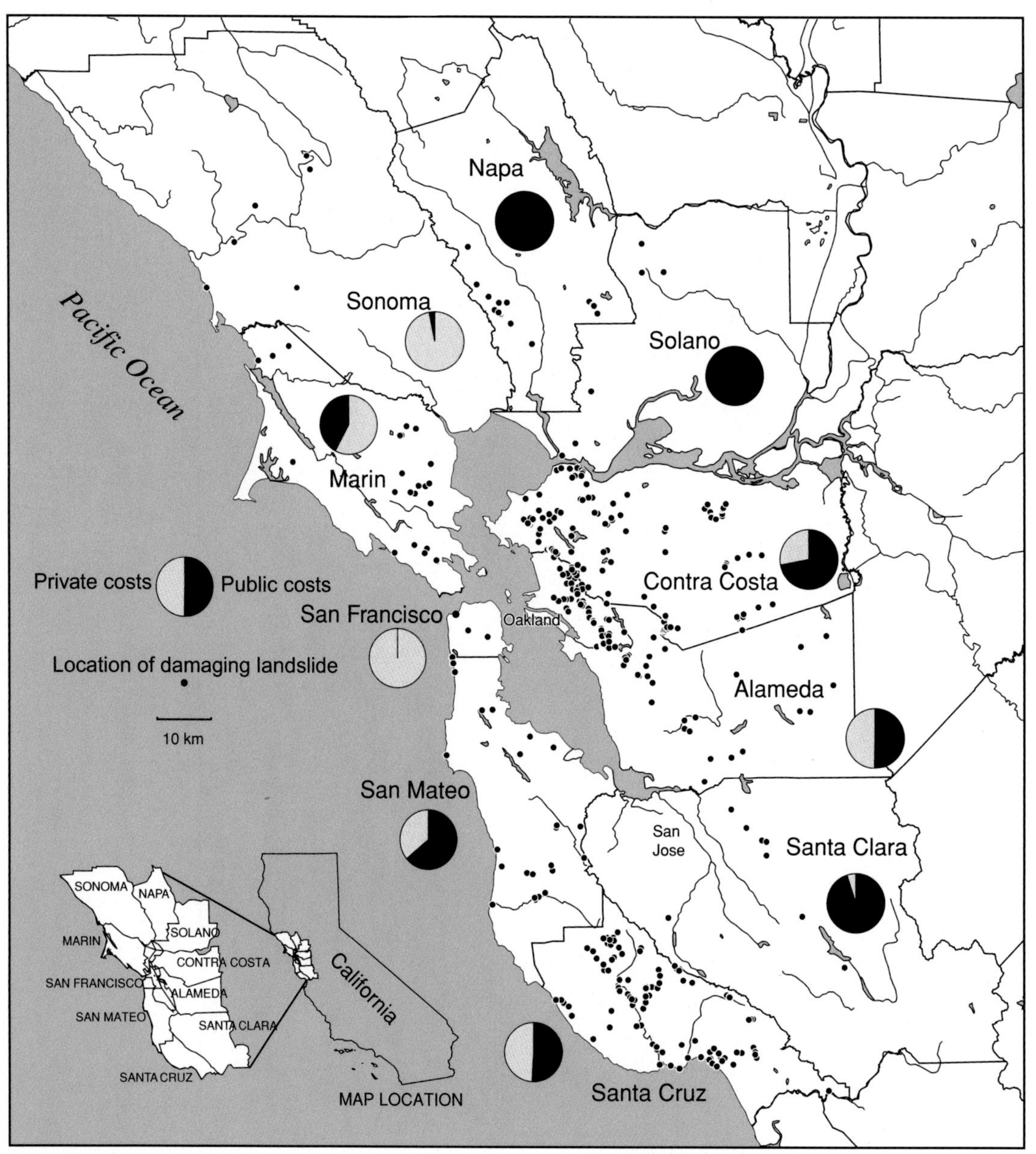

Figure 1.9 Map showing the location of damaging landslides in the San Francisco Bay region during the winter season of 1997–8. Pie charts show the distribution of public (black) vs. private (grey) costs (from Godt and Savage, 1999).

Figure 1.10 Photograph showing abundant shallow landslides and debris flows near Valencia, Los Angeles County, California. The light, un-vegetated scars are shallow failures of hillslope material caused by heavy rainfall during the winter of 2005 (photo by the authors).

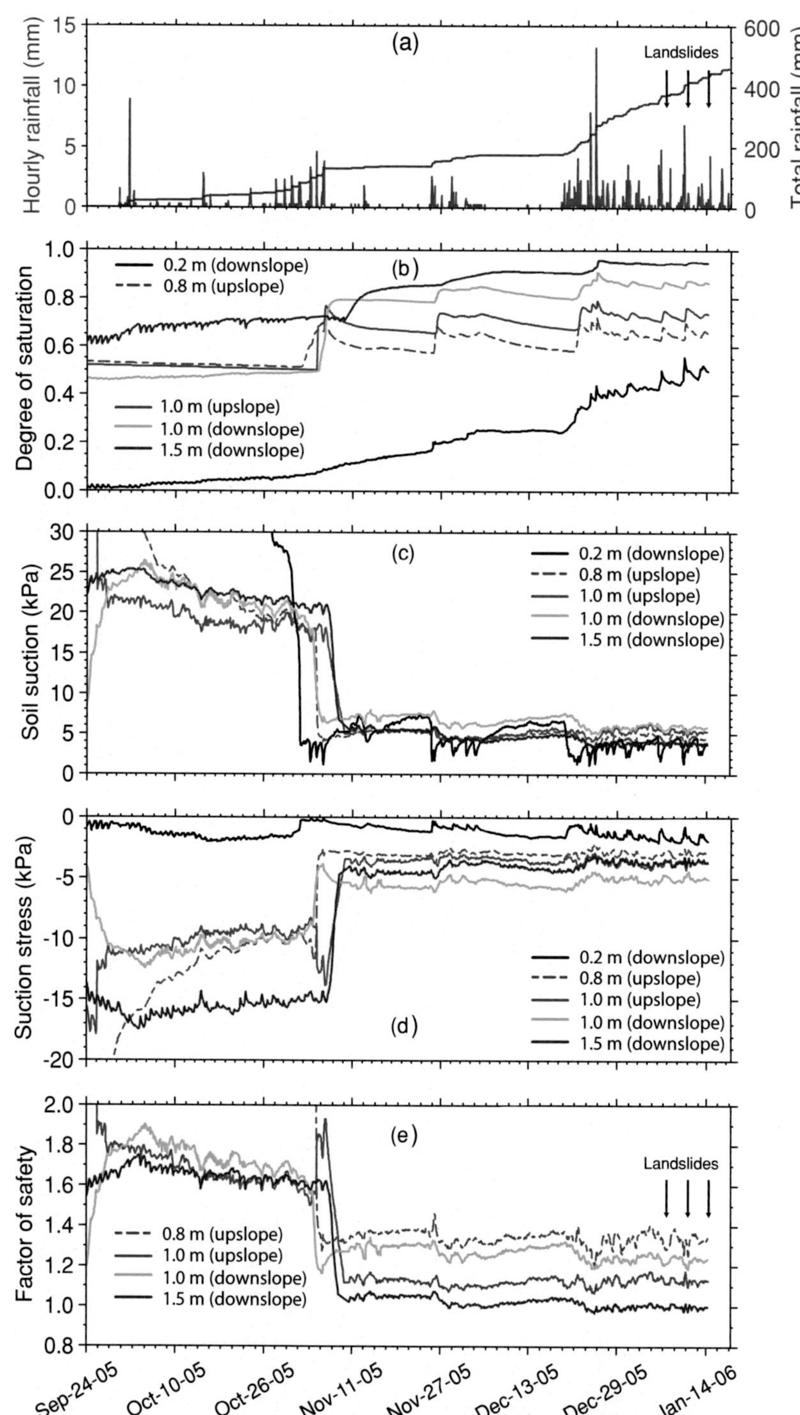

Figure 1.11 Shallow landslides under partially saturated conditions at the Edmonds, Washington, site. (a) Hourly and cumulative rainfall, (b) hourly soil saturation, (c) hourly soil suction, (d) suction stress, and (e) factor of safety for the period 24 September 2005 to 16 January 2006 at various depths from an upslope and downslope instrument array. Arrays are separated by about 3 m (from Godt *et al.*, 2009).

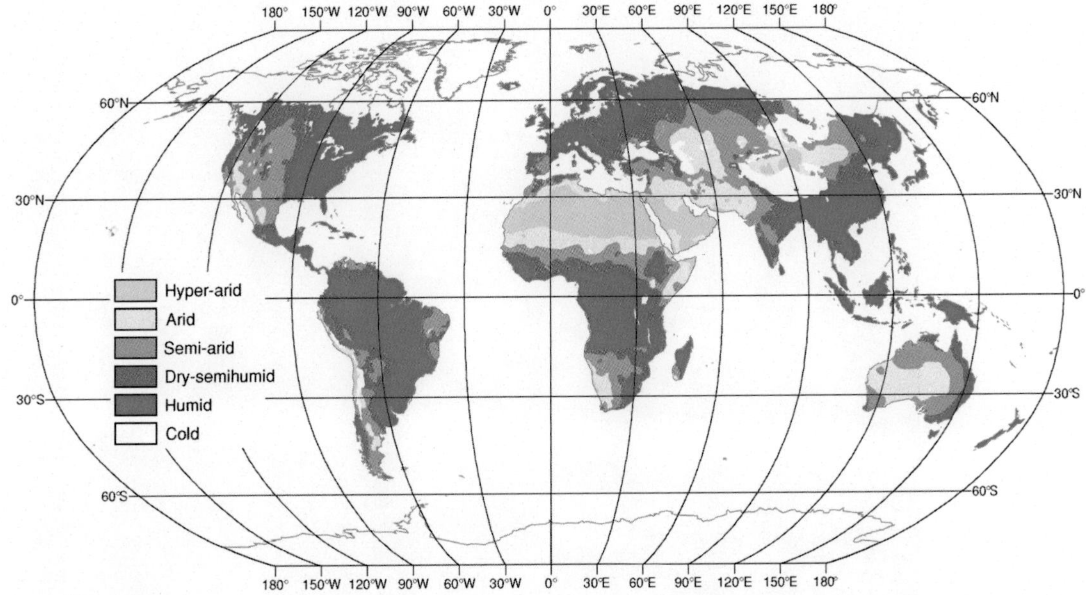

Figure 2.4 Global humidity index map (adapted from GRID/UNEP, Office of Arid Land Studies, University of Arizona).

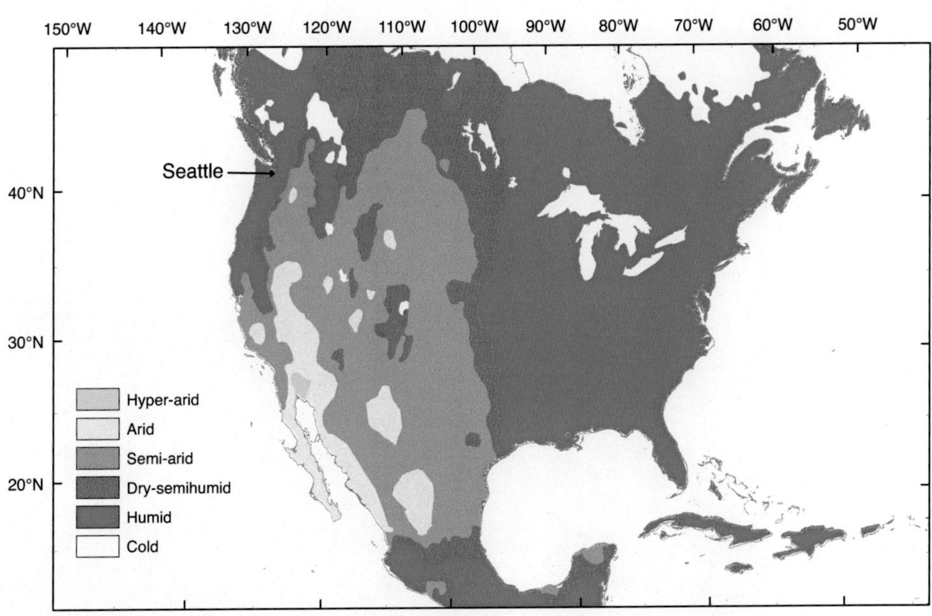

Figure 2.5 Humidity index map of North America (adapted from GRID/UNEP, Office of Arid Land Studies, University of Arizona).

Figure 2.33 Tensile cracking developed near the top of a hillslope in Jefferson County, Colorado.

Figure 2.34 Local-scale cracks or cavitations due to swelling and shrinking along the Qinghai-Tibet Highway.

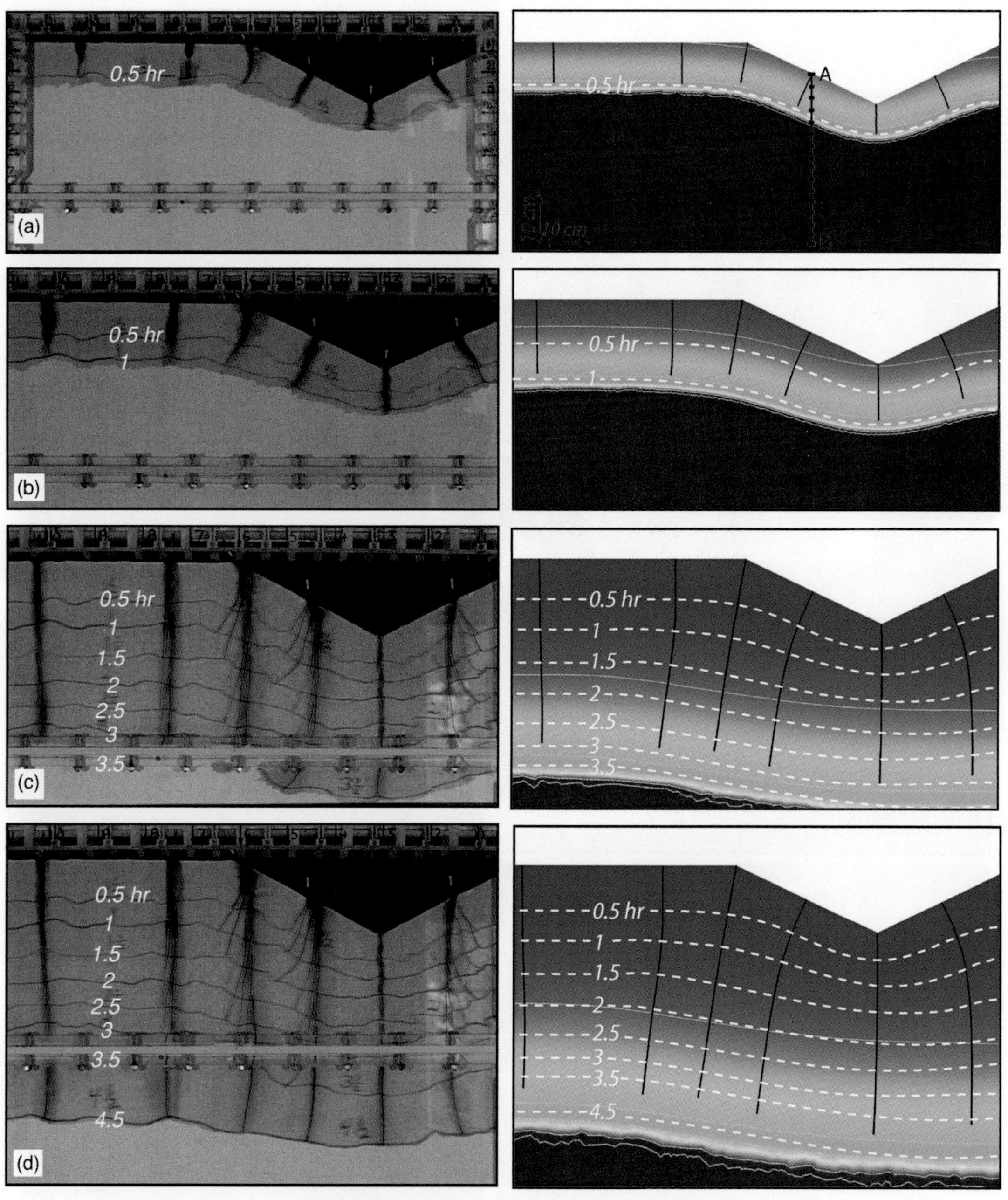

Figure 4.15 Comparison of flow field subject to a constant rainfall intensity of 50 mm/hr between experimental results (left column) and numerical simulations (right column) at (a) 30 minutes after rainfall, (b) 1 hour after rainfall, (c) 3.5 hours after rainfall, and (d) 4.5 hours after rainfall. Black lines in experimental results (left column) are streak lines. Black lines in numerical simulations (right column) are pathlines. Contour lines are wetting front locations at labeled times. See text for explanation of the differences between streak lines and pathlines. The saturated moisture content, θ_s, for the simulations was 0.37 (from Lu *et al.*, 2011).

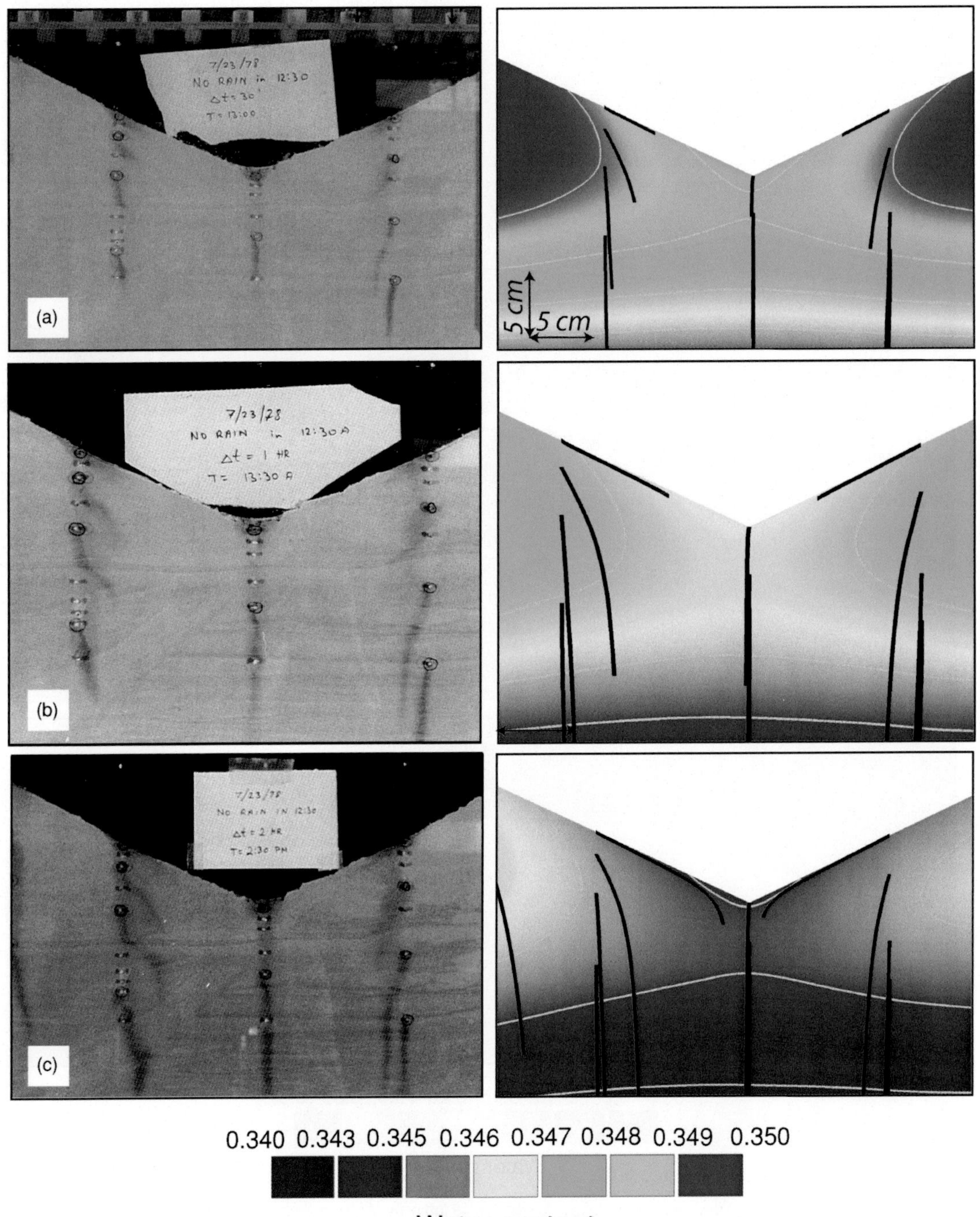

Figure 4.17 Comparison of flow fields after cessation of steady rainfall intensity of 100 mm/hr between experimental results (left column) and numerical simulations (right column) at (a) 30 minutes after rainfall ended, (b) 1 hour after rainfall ended, and (c) 2 hours after rainfall ended (from Lu *et al.*, 2011).

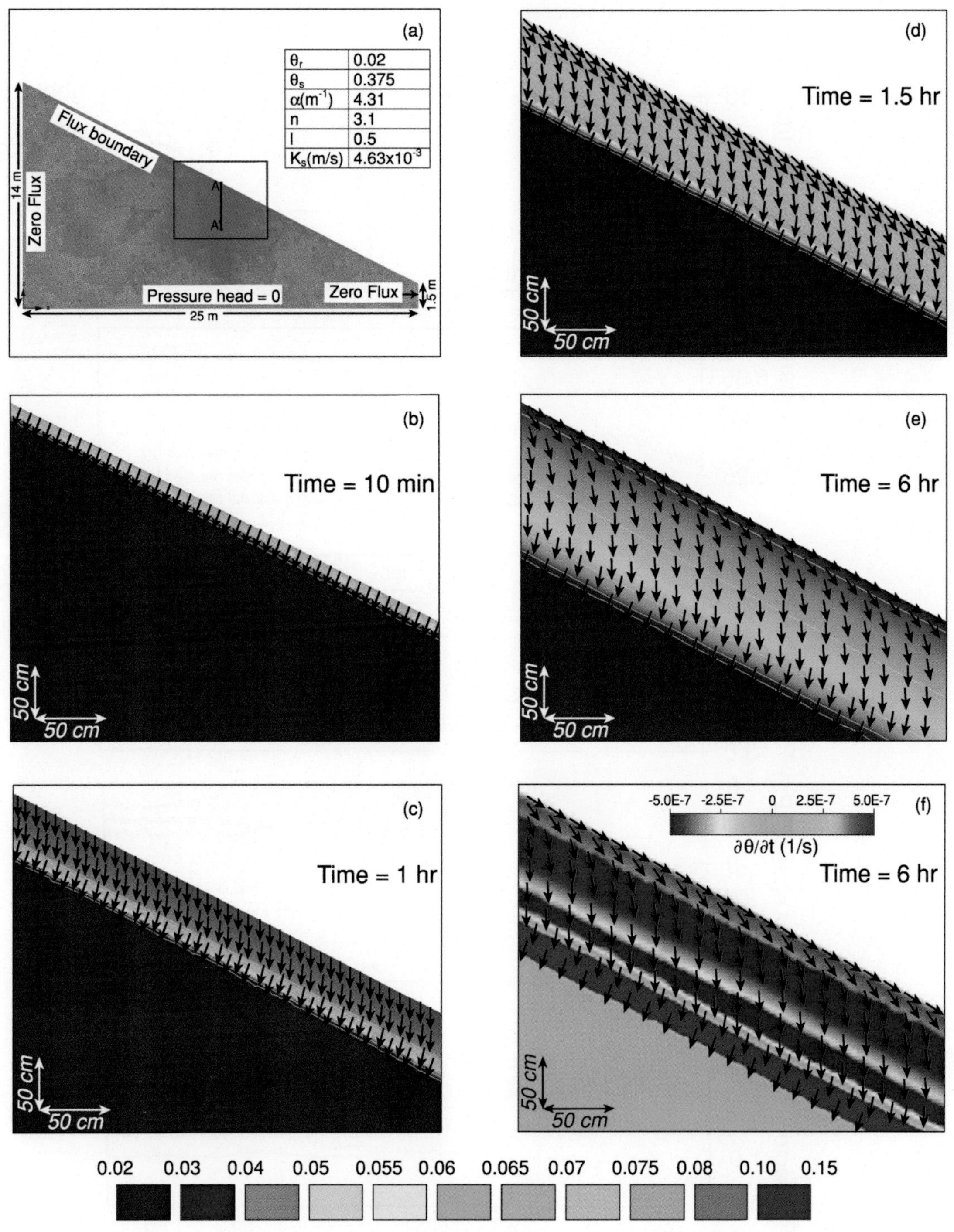

Figure 4.19 Simulation results from a 40 mm/hr step-increase-then-cease rainfall intensity boundary condition: (a) simulated domain, (b) velocity direction and moisture content at 10 minutes, (c) at 1 hour, (d) at 1.5 hours, (e) at 6 hours and (f) velocity direction and $\partial\theta/\partial t$ at 6 hours (from Lu *et al.*, 2011).

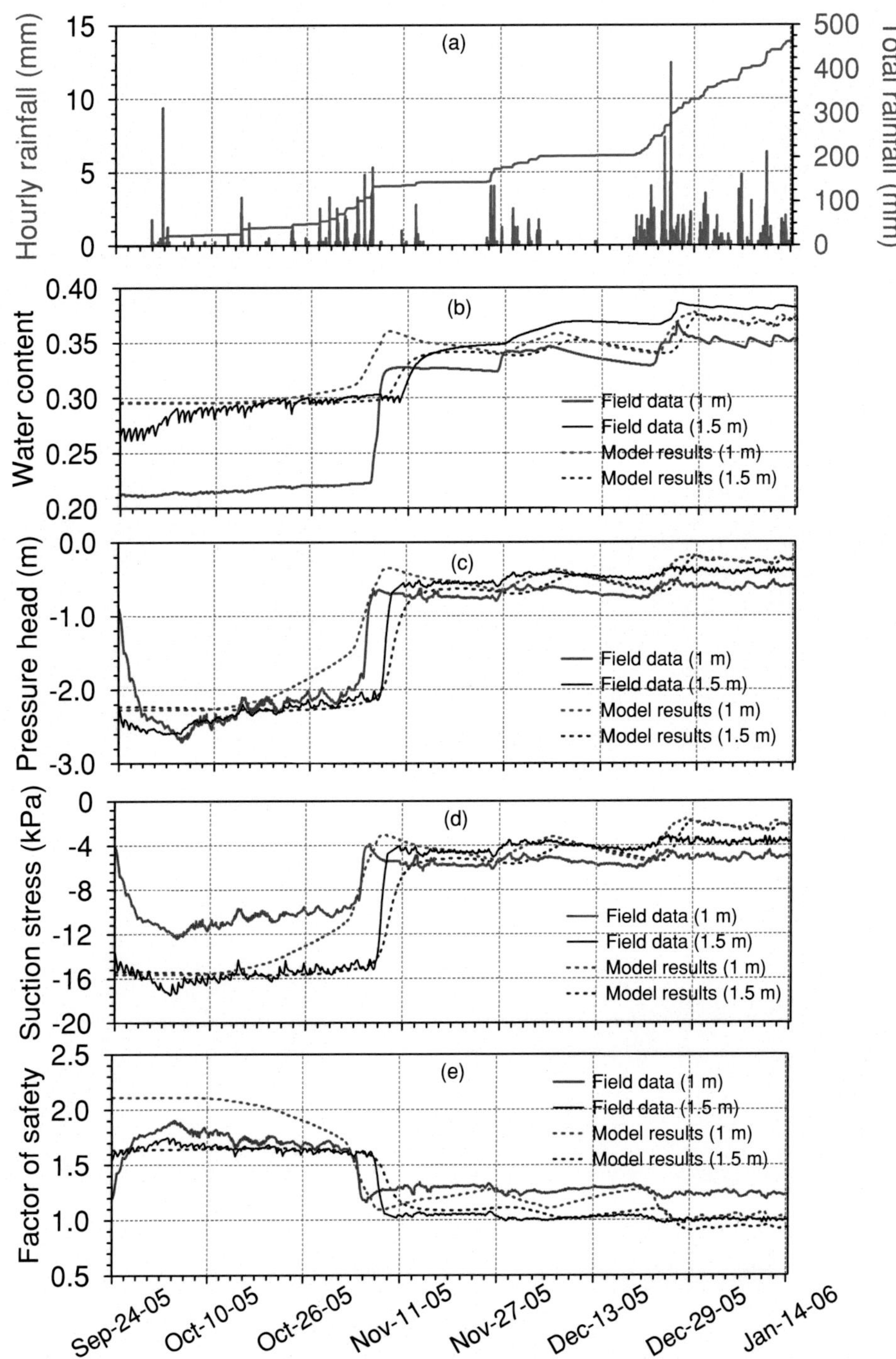

Figure 9.22 (a) Hourly and cumulative rainfall, (b) modeled and observed soil water content, (c) observed and modeled pressure head, (d) modeled and calculated suction stress, and (e) modeled and calculated factor of safety for the period 24 September 2005 to 14 January 2006 at 1.0 and 1.5 m depths. The landslide at the site occurred on 14 January 2006 (from Godt *et al.*, 2012).

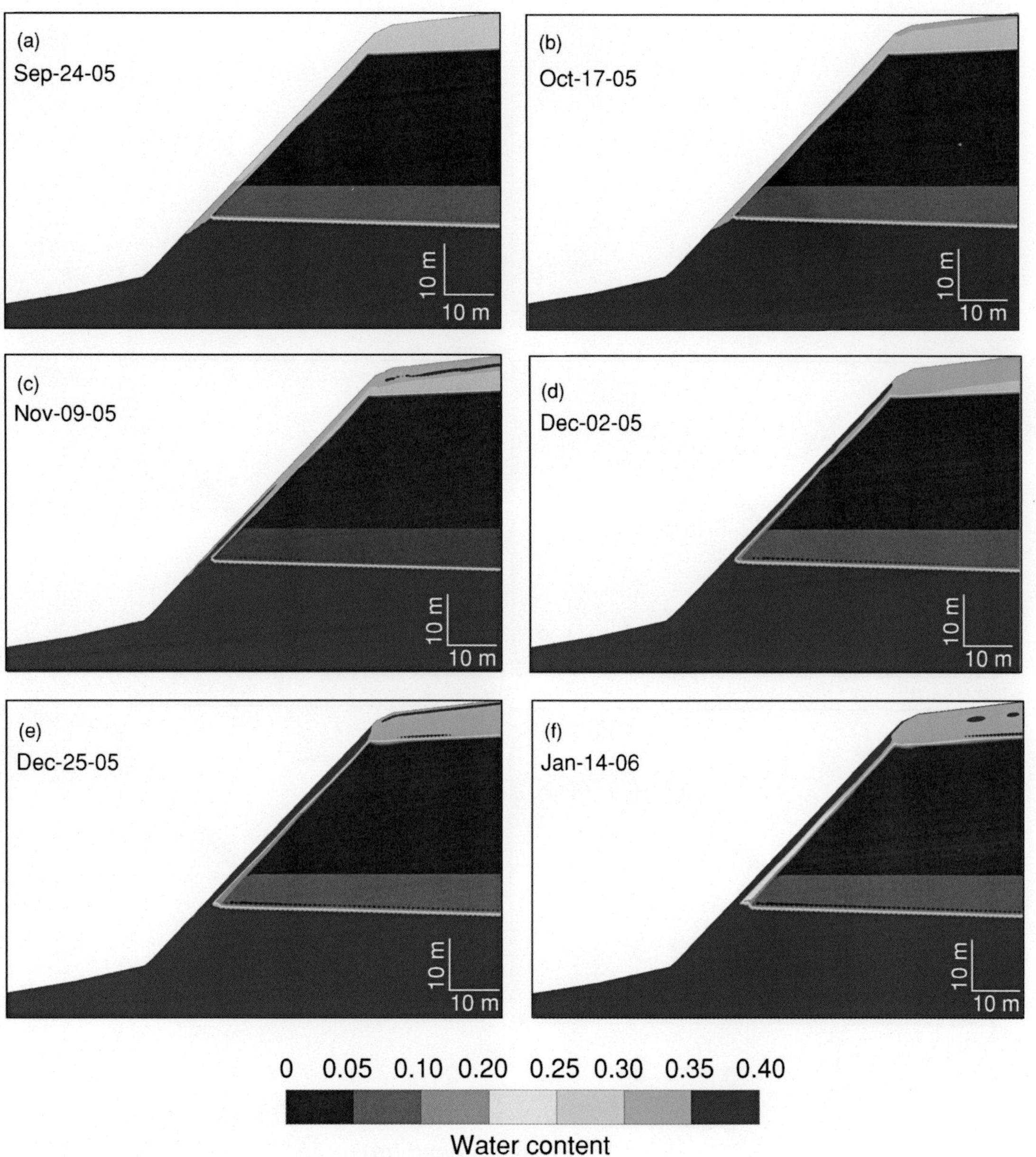

Figure 10.19 Evolution of simulated soil water content resulting from the rainfall shown in Figure 10.17.

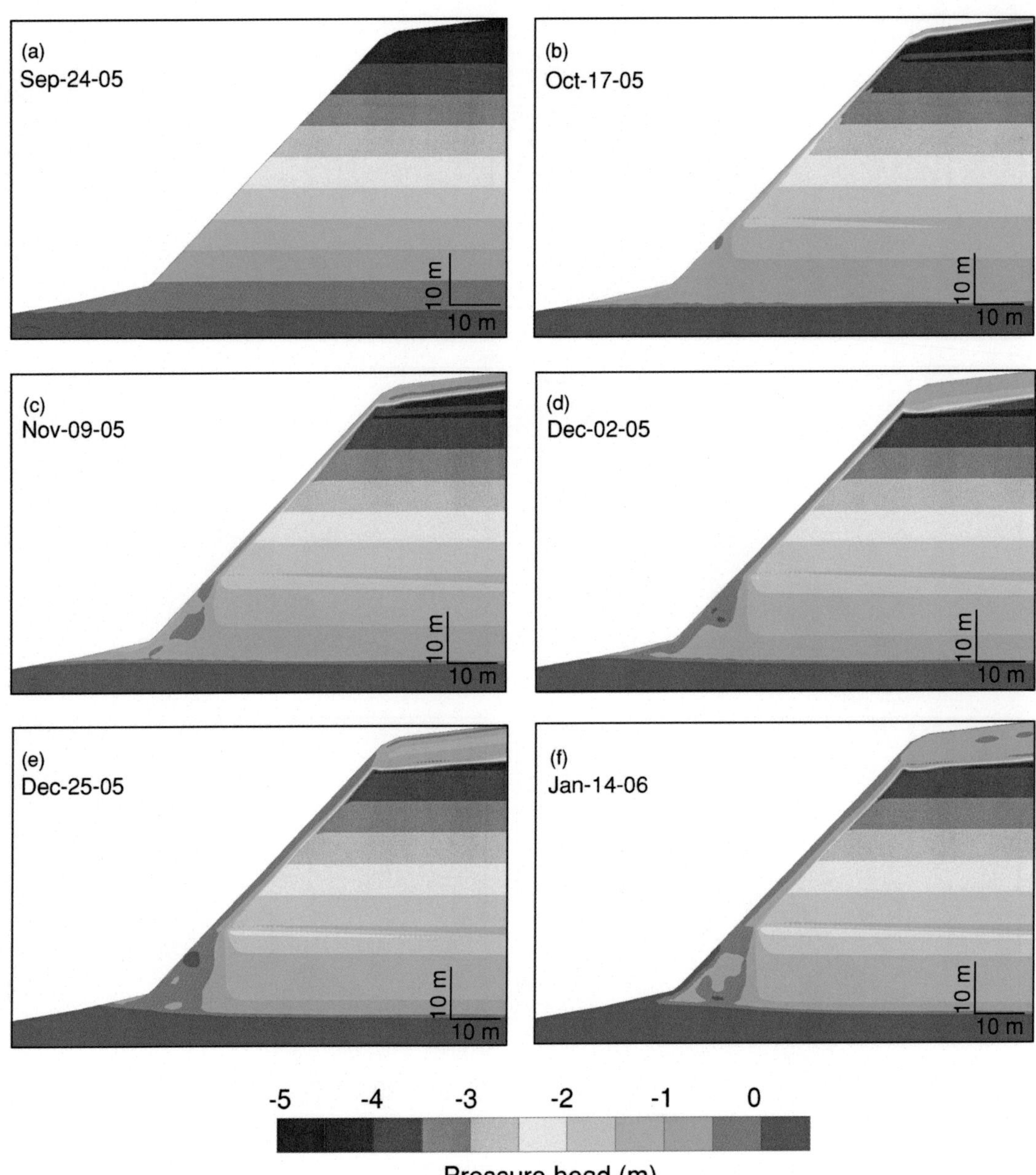

Figure 10.20 Evolution of simulated pressure head resulting from the rainfall shown in Figure 10.17.

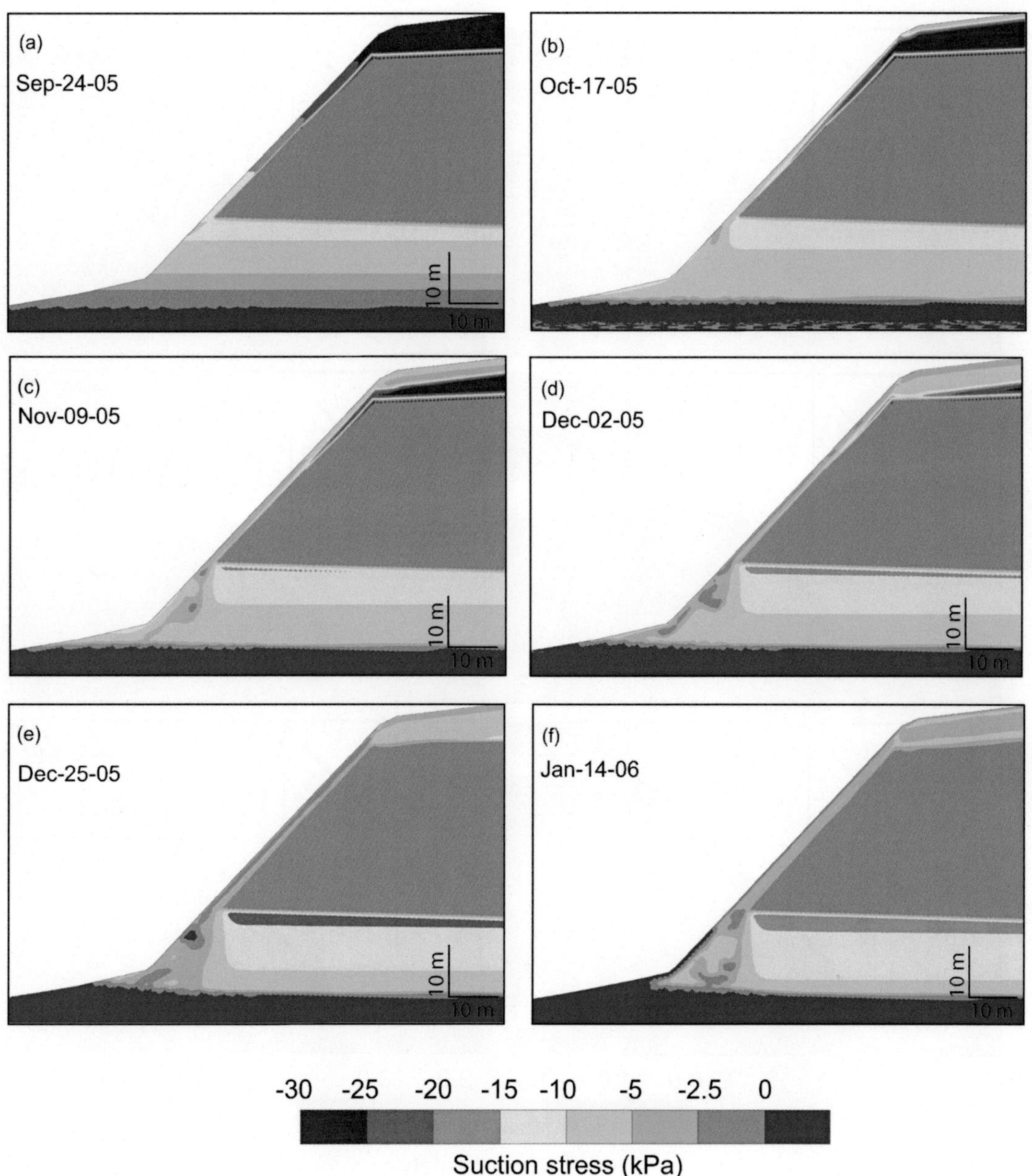

Figure 10.22 Evolution of simulated suction stress resulting from the rainfall shown in Figure 10.17.

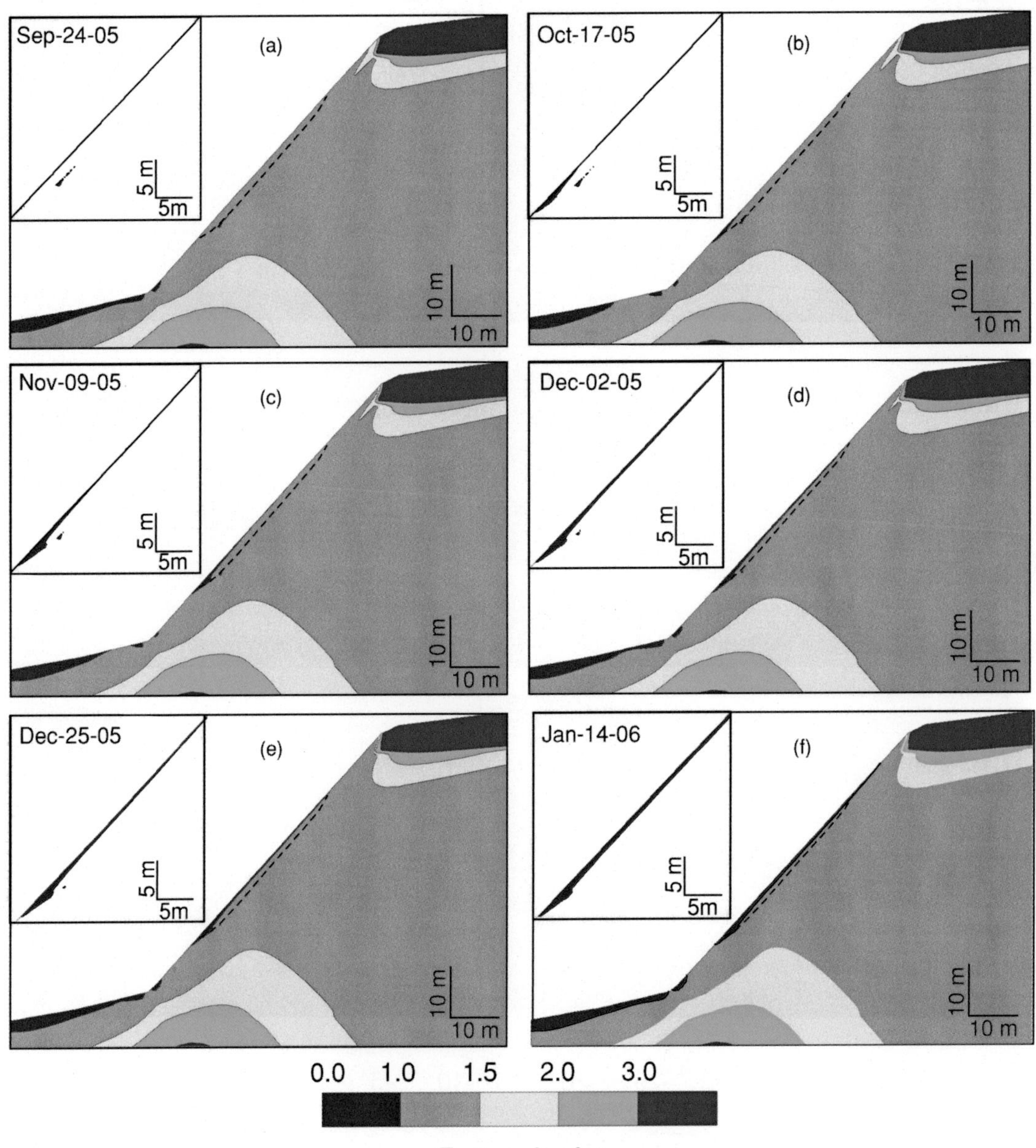

Figure 10.24 Evolution of FS for the rainfall shown in Figure 10.17. The inset in the upper left corner of each figure shows the area where the landslide occurred. The black dashed line shows the appoximate location of the failure surface from Godt *et al.*, (2009).

Owing to advances in experimental testing, considerable new understanding of the mechanical behavior of unsaturated soil has accumulated over the past two decades. Khalili *et al.* (2004) re-examined the validity of Bishop's effective stress relationship using a suite of recent experimental results and showed that Bishop's effective stress relationship is, in fact, generally valid. The observations of collapse behavior used by Jennings and Burland to invalidate Bishop's effective stress relation have also been reinterpreted by several researchers (Khalili *et al.*, 2004; Lu, 2011). Recent work by Lu and Likos (2004b, 2006) and Lu (2008) demonstrates the conceptual validity of Bishop's effective stress but resolves several key deficiencies in Bishop's effective stress relation, such as the need for zero effective stress in nearly dry soil, where matric suction is high. This work is systematically introduced in this section.

6.2 Coleman's independent stress variables theory

A barrier to the greater acceptance of Bishop's effective stress relation is the elusive nature of the effective stress parameter χ. For example, in a letter addressed to the journal *Géotechnique*, Coleman (1962) stated, "It would appear, therefore, that the factor χ used by Jennings and Burland (1962) must to a certain extent depend upon current stress and stress history." Morgenstern (1979) further stated, "The parameter χ when determined for volume change behavior was found to differ when determined for shear strength." Today, this stress behavior dependence on wetting history can be resolved by the concept of hysteresis and should not be used to dismiss the effective stress principle. Rather, one should recognize that the effective stress parameter χ, like many other material parameters such as permeability and water content, is non-unique and state dependent. In the three decades following the conception of Bishop's effective stress relation, an independent stress-state variable concept emerged, but it provided little improvement in the representation of effective stress in unsaturated soils.

The earliest use of the independent stress-state variables concept in unsaturated soil behavior can be traced back to the work of Coleman (1962), who proposed circumventing Bishop's effective stress parameter χ by using the variables $(u_a - u_w)$, $(\sigma - u_a)$, or $(\sigma - u_w)$ to describe the volumetric change in unsaturated soils. Fredlund and Morgenstern (1977) endorsed this concept by providing a theoretical stress analysis based on multiphase continuum mechanics, as illustrated in Figure 6.2. In this work, and many others that followed, it is clear, as shown in Figure 6.2, that the two variables $((u_a - u_w)$ and $(\sigma - u_a))$ are used as not only stress-state variables, but also as stress variables. In tensor notation under Cartesian coordinates aligned to the principal directions, they are the net normal stress $(\sigma_{ij} - u_a)$:

$$\left(\sigma_{ij} - u_a\delta_{ij}\right) = \begin{Bmatrix} \sigma_{xx} - u_a & 0 & 0 \\ 0 & \sigma_{yy} - u_a & 0 \\ 0 & 0 & \sigma_{zz} - u_a \end{Bmatrix} \tag{6.3a}$$

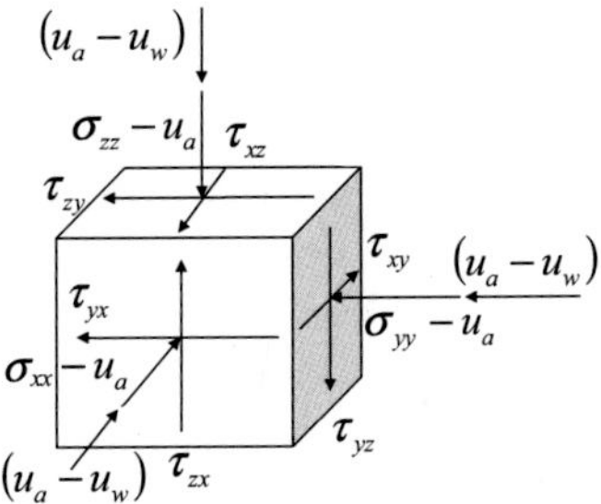

Figure 6.2 Coleman's two independent stress variables interpreted by Fredlund and Morgenstern (1977) at the REV level in unsaturated porous media (after Fredlund and Morgenstern, 1977).

and matric suction $(u_a - u_w)$:

$$(u_a - u_w)\delta_{ij} = \begin{Bmatrix} u_a - u_w & 0 & 0 \\ 0 & u_a - u_w & 0 \\ 0 & 0 & u_a - u_w \end{Bmatrix} \tag{6.3b}$$

Recent work by Lu (2008) reasoned that the consideration of both variables as stress-state variables is fundamentally sound; however, consideration of them as stress variables is flawed in that they are not stress quantities at the REV and macroscopic levels. The two independent stress-state variables theory has led to wide use of the variables as stress variables in the development of several theoretical frameworks for unsaturated soil mechanics since the 1990s (e.g., Alonso *et al.*, 1990; Fredlund *et al.*, 1995; Vanapalli *et al.*, 1996). Unlike the effective stress approach, the two independent stress-state variables approach demands re-creation or modification of shear strength criteria for saturated or dry conditions. The re-creation or modification of shear strength criteria under unsaturated conditions has recently been shown to be physically unsound and conceptually flawed in the definition of stress at the REV and macroscopic scales (Lu, 2008).

6.3 Lu *et al.*'s suction stress theory

In addition to the challenge presented by the hysteresis of Bishop's effective stress parameter χ, many works, like that of Bishop, have sought relations among the parameter χ, the degree of saturation, and the surface tension. Such approaches tend to overemphasize the role of surface tension and overlook the other physical and chemical mechanisms that operate at the inter-particle scale that are independent of external stresses. Specifically, Bishop's effective stress relation predicts zero change in effective stress when soil is dry or the parameter χ is zero. This situation may be true for sand, but not for silt or clay, for which the magnitude of inter-particle stress or effective stress that results from van der Waals attraction could

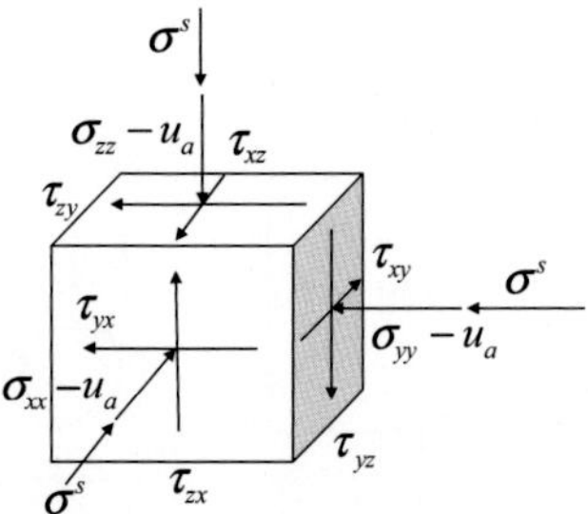

Figure 6.3 Illustrations of effective stress relationship of Lu and Likos (2004b, 2006) as a stress tensor at the REV level in unsaturated porous media. The suction stress σ^s, shown above, is the mechanical equivalent of the internal inter-particle stresses (see caption of Figure 6.6 for a proof).

be as large as several hundred kPa. The convention of setting the parameter χ equal to zero when soil is dry introduces mathematical complications as well. Assuming that the χ parameter is a continuous function, as the moisture condition of a soil approaches the dry state, presumably the parameter χ will approach zero too. On the other hand, soil suction or matric suction could approach very high values under the same drying conditions. This leads to a fundamental question: what is the product of the χ parameter and matric suction, or effective stress?

Lu and Likos (2004b, 2006) investigated all the possible physical-chemical mechanisms that contribute to inter-particle stress or effective stress in soils. Instead of using the χ parameter, they defined all the inter-particle stresses under a concept called suction stress σ^s. The suction stress concept expands Bishop's effective stress relation and unifies Terzaghi's effective stress relationship into the following form:

$$\sigma' = \sigma - u_a - \sigma^s \tag{6.4}$$

where suction stress σ^s is a characteristic function of soils that possesses hysteretic or state-dependent behavior based on the wetting or drying history. In tensor notation, effective stress in terms of suction stress can be expressed in the following form and is illustrated in Figure 6.3:

$$\sigma'_{ij} = \sigma_{ij} - u_a\delta_{ij} - \sigma^s\delta_{ij} = \begin{Bmatrix} \sigma_{xx}-u_a & \tau_{xy} & \tau_{xz} \\ \tau_{xy} & \sigma_{yy}-u_a & \tau_{yz} \\ \tau_{xz} & \tau_{yz} & \sigma_{zz}-u_a \end{Bmatrix} - \begin{Bmatrix} \sigma^s & 0 & 0 \\ 0 & \sigma^s & 0 \\ 0 & 0 & \sigma^s \end{Bmatrix} \tag{6.4}$$

Lu *et al.* (2010) further proposed a closed-form equation to describe suction stress in all soils under variably saturated conditions. The validity of the effective stress principle for unsaturated soil and the unified equation by Lu and Likos (2004b) and Lu *et al.* (2010) is discussed in the next section.

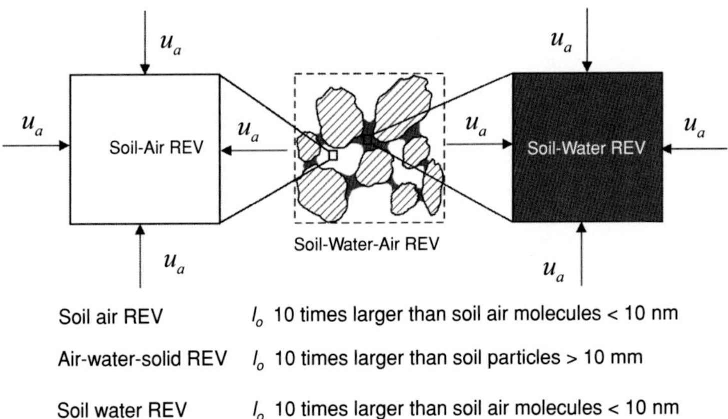

Figure 6.4 Illustrations of differences in the scale of REVs for the different stress concepts. Because the REV for stresses such as air pressure and pore-water pressure are different than that of the soil–water mixture, the use of air pressure and pore-water pressure in defining stresses at the REV of soil–water mixtures requires up-scaling these processes.

6.4 Unified effective stress representation

6.4.1 Unified effective stress principle

The REV concept provides a reference to examine the relations among several key concepts involved in variably saturated porous media, specifically those among matric suction, inter-particle stress, and total stress. Figure 6.4 illustrates the differences in the REV definitions for the basic stresses considered in variably saturated porous media; pore-air pressure, pore-water pressure, and the stress concept in a soil–water mixture REV. As shown, in a thermodynamically equilibrated system, both the air pressure and pore-water pressure concepts can be established with an REV at the pore scale, typically on the order of tens of nanometers. Therefore, defining the pore-air pressure and pore-water pressure at a scale greater than tens of nanometers but less than that of the water menisci and air pockets is statistically meaningful. On the other hand, inter-particle stresses based on pore-water pressure and air pressure need to be established at a much larger scale in order for the REV to be statistically meaningful. For clay soils, the size of the REV l_o is on the order of tens of micrometers; for sandy soils, the size of the REV is on the order of tens of millimeters. Pore air and water pressures, together with other inter-particle physical-chemical forces, along with the size and geometry of the water meniscus, must first be converted to inter-particle forces at the particle scale to calculate the magnitude of the forces at each individual particle's contacts within the REV. These forces must then be normalized by the size of the soil–water REV. In continuum mechanics, the normalization is often rigorously done by the so-called "homogenization" technique (e.g., Andrade and Borja, 2006; Berdichevsky, 2009). Therefore, using pore-water and air pressures as macroscopic stress quantities, as

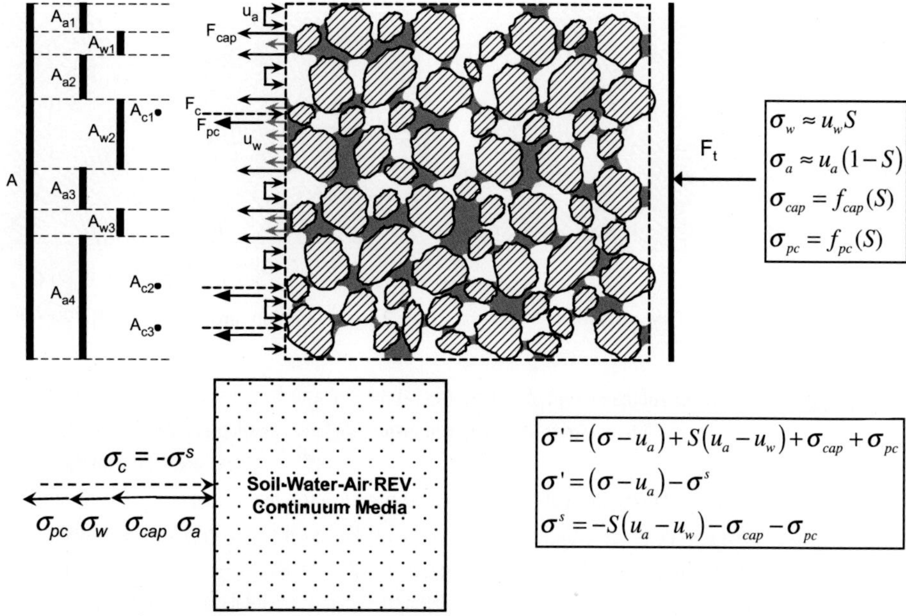

Figure 6.5 Illustrations of the concept of internal forces and stress components at the REV scale: (top) each of the inter-particle forces, including the contributions from pore-air pressure and pore-water pressure, can be upscaled to the stress quantities at the REV level, and (bottom) these inter-particle stresses are completely counterbalanced by the inter-atomic Born stress without involvement of any external force (Lu, 2008).

is done in the theoretical development of the two independent stress variables concept, is physically incorrect.

One way to unify different stresses, such as pore-water pressure, pore-air pressure, and various physical-chemical forces that act at particle contacts, is to use a mechanically equivalent stress at the REV level of a soil–water–air mixture called suction stress (Lu and Likos, 2004b, 2006). Starting from the pore or particle-contact scale, there are five well-understood forces: (1) a tensile inter-particle attractive force that results from the generally negative pore-water pressure in the soil matrix, (2) a surface tension attractive or capillary force that acts at the air–liquid interface, (3) an electric double-layer force, which is mostly repulsive in nature, (4) a van der Waals attractive force, and (5) the Born inter-atomic repulsive forces, as illustrated in Figure 6.5. Forces (1) and (2) are only present in unsaturated soils and result in inter-particle stresses $S(u_a - u_w)$ and σ_{cap}, respectively. Forces (3) and (4) are commonly called physical-chemical forces, which exist at all degrees of saturation and result in inter-particle stress σ_{pc}. All these capillary and physical-chemical forces are completely balanced by the Born repulsive forces that result in stress σ_C at the particle contact level. The Born repulsive force is a short-distance (atomic scale) force (e.g., Verwey and Overbeek, 1948; Rosen, 1989; Israelachvili, 1992) that prevents two contacting particles from penetrating into each other. It is important to recognize that

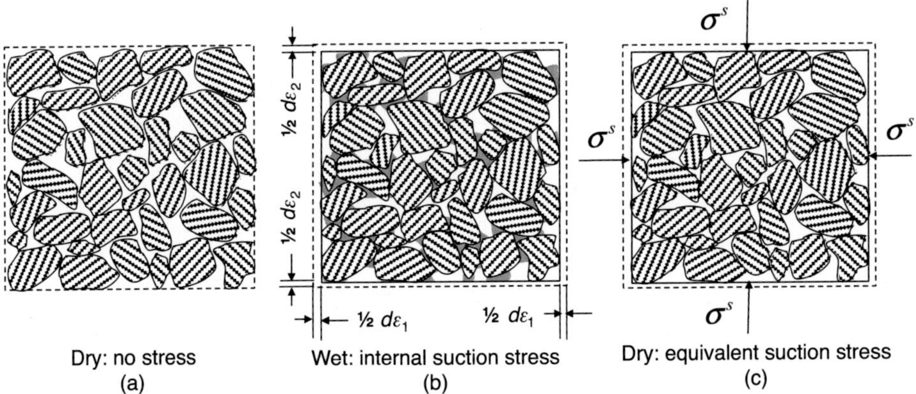

Figure 6.6 Illustrations of the concept of suction stress at the REV scale by the principle of virtual work: (a) initially there is no external and internal stress, (b) internal inter-particle forces develop as the soil is wet, causing volumetric strain $d\varepsilon_i$ (the work done on the soil skeleton for an REV is then $dW' = \sigma' d\varepsilon_i$), and (c) the mechanically equivalent external stress equal to the summation of all the inter-particle forces normalized (upscaled) by the REV size is the suction stress (the work done due to the suction stress σ^s on an REV is $dW^s_\sigma = \sigma^s d\varepsilon_i$). The energy balance under the work-conjugate principle without energy loss is $dW' = \sigma' d\varepsilon_i = -dW^s_\sigma = -\sigma^s d\varepsilon_i$, or, $\sigma' = -\sigma^s$. Therefore, suction stress is the effective stress (skeleton stress) with zero total (external) stress.

the Born repulsive force is passive in nature, meaning that its existence and magnitude are completely in response to the other inter-particle forces. Therefore, the summation of forces (1)–(4) is equal to the the Born force but in the opposite direction. Until recently, the existence of the Born repulsive force and its role in balancing the internal stresses has been largely ignored by the soil mechanics community.

By upscaling forces (1)–(4) to the REV scale, an internal stress can be defined (Figure 6.5):

$$\sigma^s = -\sigma_C = -\sigma_{cap} - \sigma_{pc} - S(u_a - u_w) \tag{6.5}$$

where σ^s is suction stress, σ_{cap} is capillary stress, S is the degree of saturation defined as the ratio of volume of pore water to pore volume, $(u_a - u_w)$ is the difference between pore-air pressure and pore-water pressure and is called matric suction, and σ_C is the Born repulsive stress.

An important conclusion drawn from Equation (6.5) is that Bishop's effective stress relation (6.2) is not sufficient to describe effective stress in unsaturated soils. The inter-particle physical-chemical and capillarity stresses conceptualized in suction stress should be considered because the magnitude of these inter-particle stresses, as elaborated below, is such that they cannot be ignored.

If the principle of virtual work is applied at the REV level, suction stress is mechanically equivalent to the inter-particle stresses that cause soil deformation at the REV scale, as illustrated in Figure 6.6.

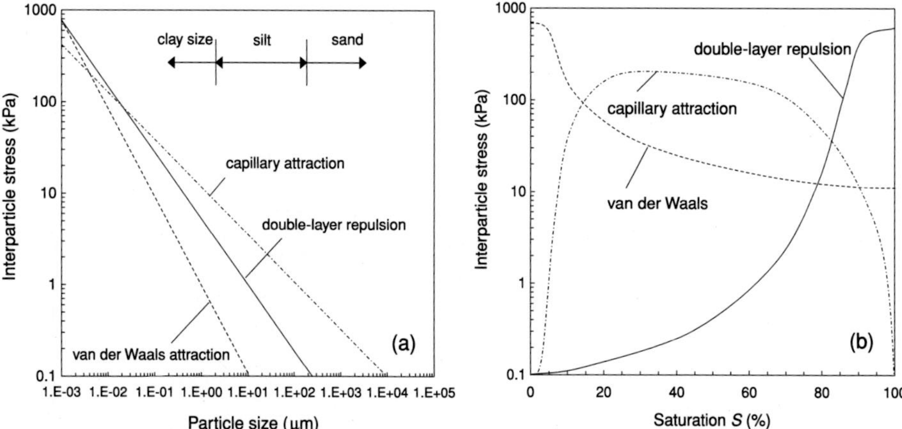

Figure 6.7 Illustrations of the dependence of internal stresses (suction stress) on (a) particle size, and (b) degree of saturation (after Lu and Likos, 2006).

The relative importance of the inter-particle stresses as a function of particle size was first described by Ingles (1962), and was first described as a function of the degree of saturation by Lu and Likos (2006). As shown in Figure 6.7a, the magnitude of all three inter-particle stresses could be as high as several hundred kPa for sub-clay-sized materials, but diminish in different fashions with increasing particle size. Stress due to the van der Waals attractive force is negligible among silt and sand-size particles, double-layer repulsive stress is negligible among sand-size particles, and capillary attractive stress becomes insignificant among particles greater than sand size. The dependence of inter-particle stresses on the degree of saturation follows quite different patterns, as shown in Figure 6.7b. The stress due to the van der Waals attractive force is strongest in dry soil; it can reach several hundred kPa in dry clays, but decreases as the degree of saturation increases. On the other hand, double-layer repulsive stress is virtually non-existent in dry soil, but increases as the degree of saturation increases. It can reach several hundred kPa in fully saturated clays, providing the source for drained cohesion. The capillary attractive force varies with the degree of saturation in a non-monotonic fashion; it is not present in dry soil, increases to a maximum of several hundred kPa at some saturation, maintains a high value for some range of saturation, and then drastically decreases. It completely vanishes when the degree of saturation reaches the capillary water retention regime, which typically is greater than the 95% degree of saturation for most soils.

Both the magnitude and patterns shown in Figure 6.7 provide insight into the possible form of suction stress as a function of degree of saturation in different soils. Suction stress is not zero in clay soils that are either dry or saturated, but the opposite case is true for sands. Figure 6.8 illustrates the possible patterns with respect to degree of saturation and the typical order of magnitude of suction stress for all soils. A realistic suction stress, therefore, should confirm the following patterns for all soils under all saturations. Starting from full saturation, all soils have a suction stress equal to the pore-water pressure, as the

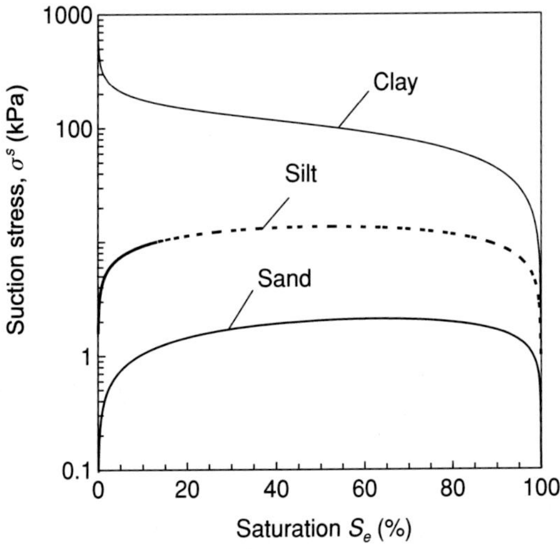

Figure 6.8 Illustrations of variations of suction stress as a function of the degree of saturation for different soils (after Lu *et al.*, 2010).

upscaling of pore-water pressure to the REV level of a soil–water mixture leads to a value equal to the pore-water pressure. This regime for which pore-water pressure and suction stress correspond one-to-one, as illustrated in Figure 6.9b, can extend from tensile (negative in value) to compressive (positive value), just as in Terzaghi's effective stress relation. At some point, as pore-water pressure becomes increasingly negative, the soil begins to desaturate leading to the breakdown of the one-to-one relation between pore-water pressure and suction stress. The breakthrough pore-pressure value is called the air-entry pressure as illustrated in Figure 6.7a and is dependent on soil type or particle-size distribution and fluid properties. Following the inter-particle stress dependence on saturation as shown in Figure 6.8, different soils have different suction stress-saturation relations, as shown in Figure 6.8. In sandy soils, the magnitude of suction stress is on the order of a few kPa, in silt soils it is on the order of tens of kPa, and in clay soils it is on the order of hundreds of kPa, as illustrated in Figure 6.8. A fundamental difference in the suction stress-saturation relation occurs when the degree of saturation approaches zero or the residual moisture content of the soil. For sandy and some silt soils, suction stress diminishes to zero as the soil dries, whereas for clays and some silt soils, suction stress continues to increase, approaching a limiting value at the residual moisture content of the soil. Lu *et al.* (2010) identified a single mathematical equation that fully describes these patterns for all soils under all degrees of saturation, which will be introduced in Sub-section 6.4.3.

6.4.2 Experimental validation and determination of suction stress

Incorporating the concept of suction stress, a generalized effective stress for variably saturated soil can be described by expanding Equation (6.4):

$$\sigma'_{ij} = \sigma_{ij} - u_a \delta_{ij} - \sigma^s \delta_{ij} \tag{6.4}$$

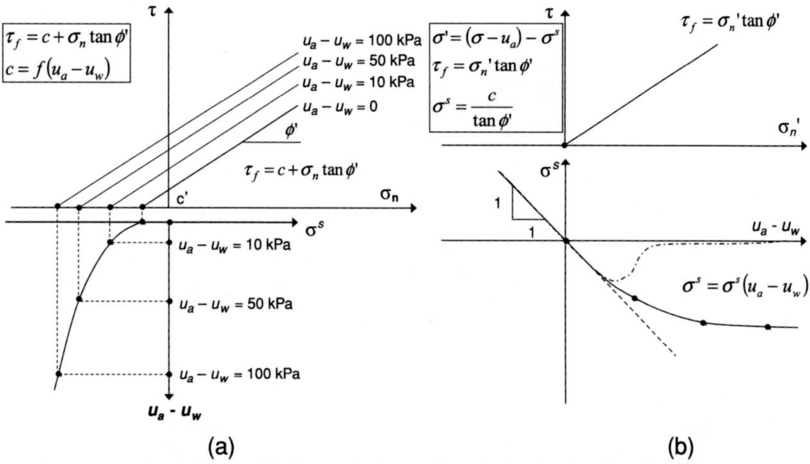

Figure 6.9 Illustrations of the determination of suction stress from shear strength tests: (a) calculations of suction stress from apparent cohesion, and (b) unified representation of the Mohr–Coulomb failure criterion with the SSCC.

by considering suction stress as a characteristic curve depicted in Equation (6.5) and Figure 6.8. Suction stress extends the role of pore-water pressure, u_w, to unsaturated conditions:

$$\sigma^s = u_w \qquad \text{for} \quad (u_a - u_w) \leq 0 \tag{6.6}$$

$$\sigma^s = f_1(u_a - u_w) = f_2(S) \qquad \text{for} \quad (u_a - u_w) \geq 0 \tag{6.7}$$

For a given soil, the functional dependence of suction stress on either matric suction or saturation, as conceptually depicted in Equation (6.7), can be evaluated from either direct shear or triaxial strength tests under variably saturated conditions. The link between suction stress and shear strength can be established by examining the physical origin of apparent cohesion, c, as illustrated through a hypothetical case of a series of shear strength tests in Figure 6.9a. When soil is sheared under saturated and zero-matric suction conditions, the drained cohesion c' and the internal friction angle ϕ' can be identified (Figure 6.9a). Experimental evidence indicates that for many soils, as matric suction increases, the apparent cohesion of soil increases, as indicated in the upward shift of the Mohr–Coulomb failure envelope in the shear stress–total stress representation. Experimental evidence also indicates that the friction angle ϕ' does not change with varying suction. A better description of apparent cohesion is the shear strength of a soil for zero normal stress (the intercept of the shear stress axis). This description does not allude to any bonding force in the soil that the term cohesion implies. "True" cohesion would be the cohesive (normal) stress among soil particles, and in this case the intercept on the normal stress axis, or suction stress. Therefore, the apparent cohesion is the mobilization of suction stress to shear resistance under the shear failure of soils. In mathematical terms, it is (see Figure 6.9a)

$$\sigma^s = f(u_a - u_w) = -\frac{c}{\tan\phi'} \tag{6.8}$$

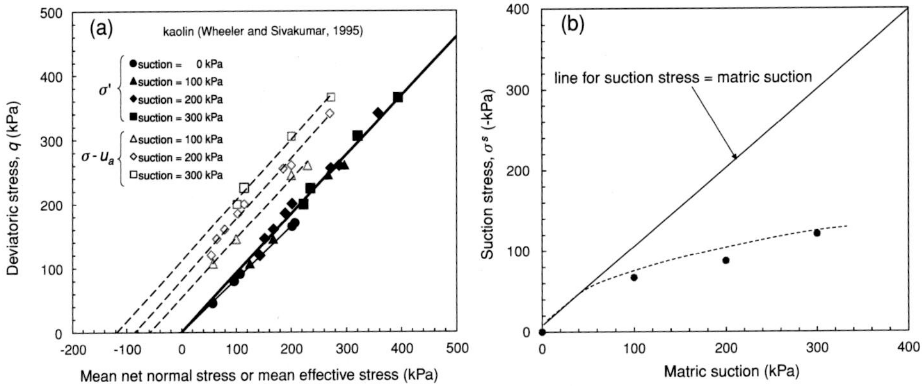

Figure 6.10 Illustrations of the determination of suction stress from shear strength tests: (a) shear strength data in total stress p-q space and effective stress p'-q space for a kaolin (from triaxial shear strength tests by Wheeler and Sivakumar, 1995), and (b) the deduced SSCC.

With the knowledge of the shear strength parameters (drained cohesion c', friction angle ϕ') identified under saturated conditions and the c intercept for a given matric suction value, the suction stress characteristic curve (SSCC) can be identified directly using Equation (6.8), as illustrated in Figure 6.9. It is now possible to portray the shear strength behavior under variably saturated conditions with one unique Mohr–Coulomb envelope under a unified effective stress equation (6.4) (Figure 6.9b).

The validity of the effective stress equation (6.4) using the concept of SSCC (Equation (6.5)) can be examined using published shear strength data accumulated over the past two decades. Lu and Likos (2006) and Lu *et al.* (2010) have examined shear strength data for a broad range of soils and found that the effective stress concept under SSCC is generally valid. To illustrate how the shear strength data can be interpreted using the SSCC as well as the validity of the SSCC, three data sets obtained from triaxial shear strength tests are presented here. The first set of data (from Wheeler and Sivakumar, 1995) is for remolded kaolin, a commercially available clay soil that belongs to the 1:1 clay-mineral family. Test results are shown in the p-q space (defined below) in Figure 6.10a.

Rather than portraying shear strength data in the shear stress–normal stress space that is commonly done for direct shear test results, triaxial test results are commonly portrayed in the deviatoric stress $q = (\sigma_1 - \sigma_3)$ and mean normal stress $p = (\sigma_1 + 2\sigma_3)/3$ (or the mean effective normal stress $p' = (p - u_a) - \sigma^s$) space as the applied stresses are the principal stresses σ_1, and $\sigma_2 = \sigma_3$. Correspondingly, suction stress values (the intercepts on the mean normal stress axis) for a given matric suction can be obtained by the following equation:

$$\sigma^s = -\frac{d + M(p - u_a)_f - q_f}{M} \tag{6.9}$$

where d is the intercept on the deviatoric stress q axis when p is zero, M is defined by the internal friction angle ϕ', and the subscript f indicates the failure state. In a conventional triaxial setting, the mathematical definitions of d and M in terms of the friction angle ϕ'

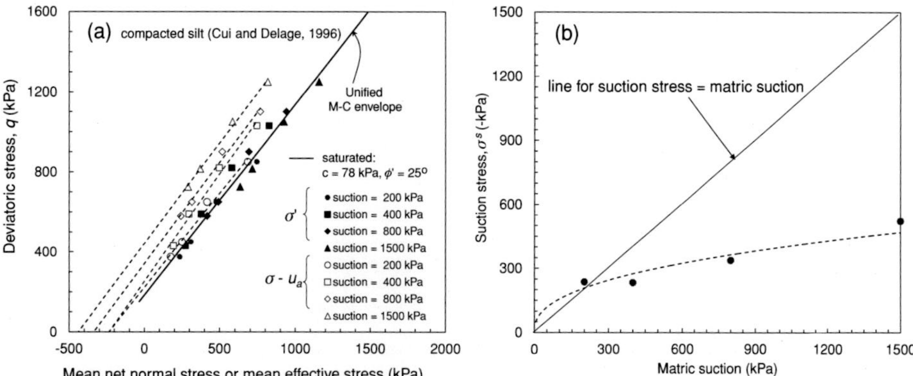

Figure 6.11 Illustrations of the determination of suction stress from shear strength tests: (a) shear strength data in total stress p-q space and effective stress p'-q space for a compacted silt (from triaxial shear strength tests by Cui and Delage, 1996), and (b) the deduced SSCC.

and apparent cohesion c are

$$M = \frac{6 \sin \phi'}{3 - \sin \phi'} \tag{6.10}$$

$$d = \frac{cM}{\tan \phi'} \tag{6.11}$$

The calculated suction stress values as a function of matric suction are shown in Figure 6.10b and the corresponding effective stress values are shown in Figure 6.10a in solid symbols. Shear strengths closely follow the saturated Mohr–Coulomb envelope for all test data under various saturation conditions, implying that the effective stress representation of shear strength under the SSCC for this soil is unique and valid.

Test results for the second soil described here are those from a compacted silt from Cui and Delage (1996), shown in Figure 6.11a under various matric suction values. Suction stress values (the intercept of the failure envelopes on the mean normal stress axis) are first calculated and shown as a function of matric suction in Figure 6.11b. Shear strengths follow the saturated Mohr–Coulomb envelope for all test data in the p'-q space, implying that the effective stress representation of shear strength under the SSCC for this soil is unique and valid.

Test results for the third soil described here are those for a glacial till from Vanapalli *et al.* (1996), shown in Figure 6.12a for both saturated and unsaturated failure envelopes in the p-q space. Suction stress values (intercept of the failure envelopes on the mean normal stress axis) are first calculated and shown as a function of matric suction in Figure 6.12b. Shear strengths closely follow the saturated Mohr–Coulomb envelope for all test data in the p'-q space, again implying that the effective stress representation of shear strength under the SSCC for this soil is unique and valid.

Note that testing shear strength under unsaturated conditions is a complicated and lengthy process that could require several months to complete. Obtaining each data point may require several weeks for the soil to reach equilibrium under a given applied matric suction.

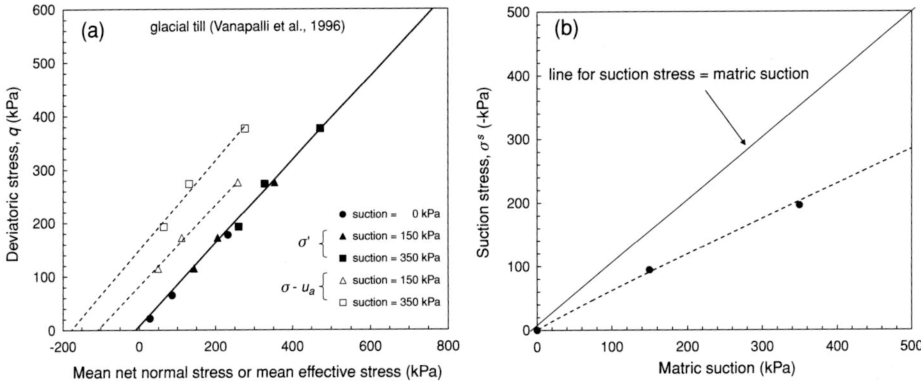

Figure 6.12 Illustrations of the determination of suction stress from shear strength test results: (a) shear strength data in total stress *p*-*q* space and effective stress *p*′-*q* space for a glacial till (data from Vanapalli *et al.*, 1996), and (b) the deduced SSCC.

It is generally not possible to maintain matric suctions above 1,500 kPa because of the difficulty of imposing those conditions in a testing environment, and the time for a soil to come to equilibrium could be much longer than several months. Multiple samples are often required in order to obtain one failure envelope for a given matric suction, which often leads to uncertainties in the data reduction. Nevertheless, until recently, shear strength tests, both direct shear and triaxial, remained the primary way to assess the mechanical behavior of soil under unsaturated conditions. The next section describes a recent theoretical development by which effective stress and soil water retention behavior can be intrinsically linked. The implication of such a linkage is that effective stress can be quantified without relying on unsaturated shear strength tests.

6.4.3 Unified equation for effective stress

The functional form of the dependence of suction stress on matric suction or saturation proposed by Lu and Likos (2004b) is

$$\sigma^s = -S_e(u_a - u_w) = -\frac{S - S_r}{1 - S_r}(u_a - u_w) = -\frac{\theta - \theta_r}{\theta_s - \theta_r}(u_a - u_w) \qquad (6.12)$$

where S is the degree of saturation, defined as the volume of liquid relative to volume of void in an REV, θ is the volumetric water content, defined as the volume of liquid to the total volume of soil in an REV, S_r is the residual saturation, θ_r is the residual water content, θ_s is the saturated water content, and S_e is the equivalent degree of saturation. A thermodynamic justification of the suction stress equation (6.12) provided by Lu *et al.* (2010) postulates that suction stress represents the energy stored by the soil water responsible for inter-particle stresses. Graphically, as illustrated in Figure 6.13, it is the shaded area under the soil water retention curve (SWRC). It is important to recognize that the *equivalent* degree of saturation, instead of the degree of saturation that has been widely used in the literature, is used in Equation (6.12). The physical interpretation is that only part of the energy stored

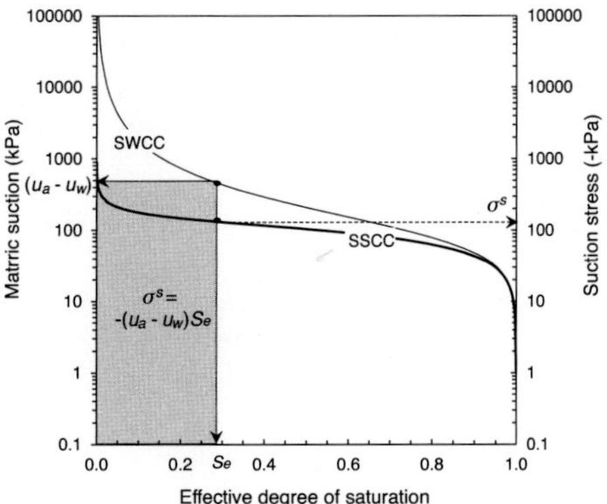

Figure 6.13 Illustration of suction stress from the viewpoint of energy stored in soil water. Matric suction is the mechanical energy stored in soil water per unit volume of soil water, whereas suction stress is the mechanical energy stored in soil water per unit REV.

in the surface hydration water, or the water that makes up the so-called residual moisture content, is available for inter-particle stresses.

If a soil water retention function is introduced, a closed-form expression for suction stress can be described as a sole function of either matric suction or saturation. van Genuchten (1980) presented a closed-form equation for the SWRC:

$$S_e = \frac{S - S_r}{1 - S_r} = \left\{ \frac{1}{1 + [\alpha(u_a - u_w)]^n} \right\}^{1-1/n} \tag{6.13}$$

where α and n are fitting parameters. The parameter α is correlated with the air-entry pressure and the n parameter is correlated with the pore size distribution. Substituting Equation (6.13) into Equation (6.12) leads to a closed-form equation for suction stress either in terms of matric suction $(u_a - u_w)$ (Lu *et al.*, 2010):

$$\sigma^s = -\frac{(u_a - u_w)}{(1 + [\alpha(u_a - u_w)]^n)^{(n-1)/n}} \tag{6.14}$$

or in terms of the equivalent degree of saturation S_e (Lu *et al.*, 2010):

$$\sigma^s = -\frac{S_e}{\alpha}\left(S_e^{\frac{n}{1-n}} - 1\right)^{\frac{1}{n}} \tag{6.15}$$

For typical sand, silt, and clay soils, matric suction (Equation (6.13)) and suction stress (Equation (6.14)) as intrinsically related functions of saturation are shown in Figure 6.14. The corresponding SSCC in terms of matric suction (Equation (6.14)) is shown in Figure 6.15.

For soil water retention, Figure 6.14a shows that saturation varies greatly for matric suction less than tens of kPa for sandy soil, for matric suction less than tens of hundreds of kPa for silt, and for matric suction less than tens of thousands of kPa for clay. For suction stress, Figure 6.14b shows that suction stress varies non-monotonically over several kPa for sand, over several tens of kPa for silt, and over several hundred kPa for clay. The variation in suction stress can also be illustrated by portraying it as a function of matric

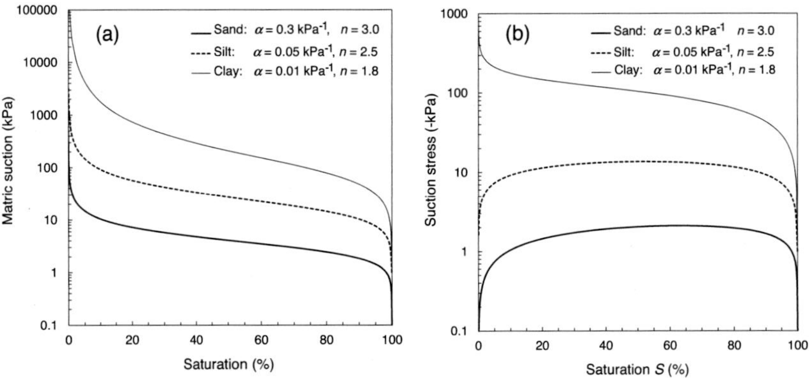

Figure 6.14 Illustrations of SWRC (a) and SSCC (b) for hypothetical sand, silt, and clay soils.

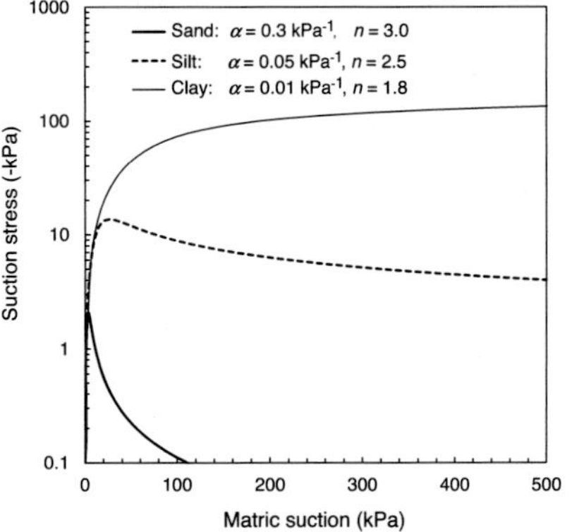

Figure 6.15 Illustrations of SSCC as a function of matric suction for hypothetical sand, silt, and clay soils.

suction, as shown in Figure 6.15. The magnitude of effective stress variation as a function of soil saturation provides a good indication of the effect of rainfall on hillslope stability for different soils. For example, infiltration will cause relatively large changes in stress in sandy hillslopes in the upper few meters near the ground surface (1 kPa of suction stress is equivalent to the gravity-induced normal stress of ~5 cm of soil). Infiltration will likewise affect the stability of the upper several meters in silt hillslopes, and the upper several tens of meters in clay hillslopes.

6.4.4 Validity of unified equation for effective stress

There are several far-reaching theoretical and practical implications of Equation (6.14) or Equation (6.15). The complete correlation between the matric suction function (Equation (6.13)) and the suction stress function (Equation (6.14)) indicates that hydrologic and mechanical behavior of variably saturated soil share common physical controls, namely, the type of fluid, gas, and solid involved, and the pore structure. Because effective stress

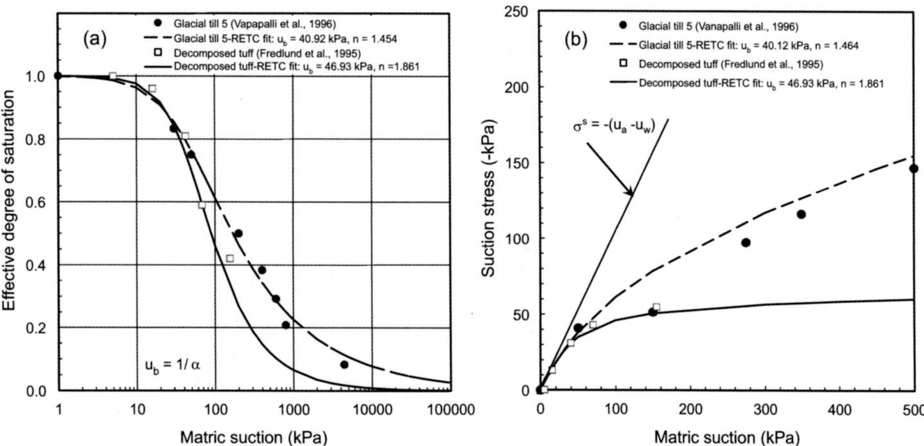

Figure 6.16 Experimental validation of the intrinsic relation between the SWRC and the SSCC for a glacial till (data from Vanapalli *et al.*, 1996) and a decomposed tuff (data from Fredlund *et al.*, 1995).

due to variation in soil saturation, as expressed by Equation (6.15), can be completely characterized by the same set of parameters that define the soil water retention curve, results from one experiment, either hydraulic or mechanical, are sufficient to quantify either curve. Lu *et al.* (2010) conducted an extensive analysis of published shear strength or/and soil water retention data for 20 soils varying from sand, to silt, to clay to examine the validity of Equation (6.14). Their analysis showed that, to the accuracy of experimental methods, Equation (6.14) is valid. Both mechanical and hydraulic test results from four soils are used here to illustrate how to identify the controlling parameters for experimental confirmation of Equation (6.14). The first two soils are silt and clay soils with SSCC that are monotonic functions of matric suction, whereas the last two soils are granular and sandy soils with SSCC that are non-monotonic functions of matric suction.

Figure 6.16a shows the SWRC for two soils; a glacial till (Vanapalli *et al.*, 1996) and a decomposed tuff (Fredlund *et al.*, 1995). Using these data and van Genuchten's (1980) model (Equation (6.13)), parameters α and n are identified: $u_b = 1/\alpha = 40.92$ kPa and $n = 1.454$, and the SWRC are plotted in Figure 6.16a. Figure 6.16b shows the suction stress data deduced from the triaxial shear strength data following the procedure described earlier in this section (Equations (6.9)–(6.11)). Using these data and Lu *et al.*'s (2010) model (Equation (6.14)), parameters α and n are identified: $u_b = 1/\alpha = 40.12$ kPa and $n = 1.464$. As shown, for the glacial till soil, the differences between parameters α and n from hydraulic and mechanical tests are small; within 2% for α and within 1% for n. For the decomposed tuff, the parameters from the hydraulic and mechanical tests are identical. These test results show the intrinsic relationship between the SWRC and the SSCC.

The non-monotonic suction stress behavior of granular and sandy soils is often displayed in either tensile strength or shear strength test results. Figure 6.17a shows the water retention data from hydraulic tests for Ottawa sand (Lu *et al.*, 2009) and limestone agglomerates (Schubert, 1984). Using these data and Equation (6.13), parameters α and n are

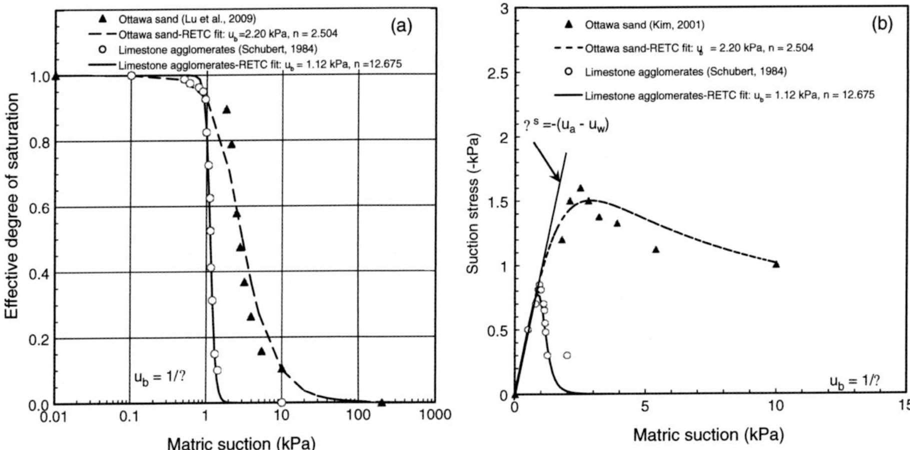

Figure 6.17 Experimental validation of the intrinsic relation between the SWRC and the SSCC for Ottawa sand (data from Lu *et al.*, 2009) and a limestone agglomerate (data from Schubert, 1984).

identified: $u_b = 1/\alpha = 2.20$ kPa and $n = 2.504$ for Ottawa sand and $u_b = 1/\alpha = 1.12$ kPa and $n = 12.675$ for the limestone agglomerate. Using these parameters and the suction stress equation (6.14), the SSCC are plotted in Figure 6.17b as functions of matric suction along with experimental tensile strength test data. As shown, the SSCC deduced from the hydraulic tests agree with the tensile strength data, establishing the validity of Equation (6.14) for soils with non-monotonic suction stress behavior.

6.5 Suction stress profile in hillslopes

6.5.1 Steady-state profiles in one dimension: single layer

A general equation for the vertical profile of suction stress under steady infiltration/evaporation conditions can be obtained by applying the suction stress expression (Equation (6.12)) with the analytical solutions for suction and water content profiles developed in previous Equations (3.43)–(3.46). Substituting Equations (3.43)–(3.46) into Equation (6.12) leads to

$$\sigma^s(z, q) = -\frac{\gamma_w}{\beta}\left[\left(\frac{q}{k_s}+1\right)\exp(-\beta z)-\frac{q}{k_s}\right]^{1/n}\ln\left[\left(\frac{q}{k_s}+1\right)\exp(-\beta z)-\frac{q}{k_s}\right] \tag{6.16}$$

The one-dimensional unsaturated zone and all the notation are the same as those defined in Section 3.5.

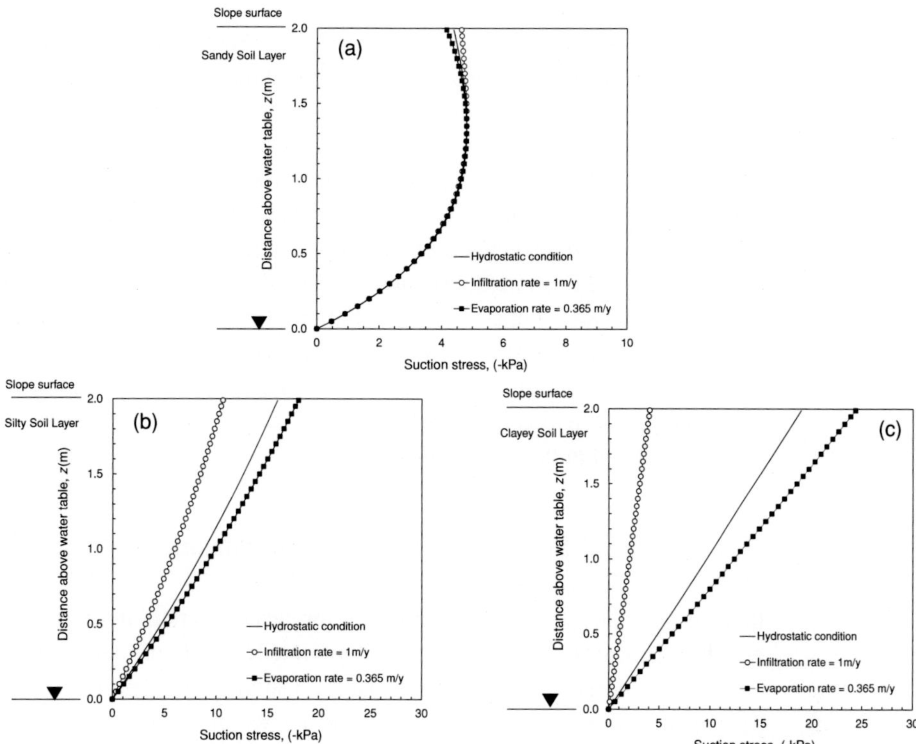

Figure 6.18 Profiles of suction stress under varying infiltration/evaporation conditions in a 2 m unsaturated (a) sand, (b) silt, and (c) clay soil layers. The corresponding profiles of water content and suction are shown in Figure 3.18 for the sand, Figure 3.19 for the silt, and Figure 3.20 for the clay.

Cases i, ii, and iii: single layer sand, silt, or clay system

To examine the variability of the suction stress profiles for different soils and infiltration/evaporation conditions, unsaturated hydro-mechanical properties defined in Table 3.4 will be used. The steady-state suction stress profiles in a 2 m unsaturated layer for three hypothetical cases of sand, silt, and clay soils under different infiltration conditions are shown in Figure 6.18.

Several interesting aspects of the suction stress profiles can be observed from Figure 6.18. In the 2 m unsaturated sand layer (Case i in sub-section 3.5.1) under hydrostatic conditions (i.e., $q = 0$), suction stress varies from zero at the water table, reaches its minimum value of -4.8 kPa at $\sim$1.25 m above the water table, and then increases to ~ -4.4 kPa at the ground surface. Suction stress is insensitive to the infiltration/evaporation rate; it only deviates from the hydrostatic profile near the ground surface by about 0.3 kPa. In contrast, in the silt layer (Case ii in sub-section 3.5.1) under hydrostatic conditions, suction stress varies monotonically from zero at the water table to -16.1 kPa at the ground surface. Under 1 m/year of steady infiltration, suction stress is -10.7 kPa at the ground surface; under a 0.365 m/year steady evaporation rate, suction stress decreases to -18.1 kPa at the ground surface. In the clay layer (Case iii in Section 3.5.1), the patterns are similar to those in the silty layer, but the magnitude in variation is higher. Suction stress varies nearly linearly from zero at the water table to −19.0 kPa at the ground surface under hydrostatic conditions. The 2 m clay layer is nearly saturated under the hydrostatic condition (Figure 3.20a). Suction stress increases to −4.1 kPa at the ground surface under 1 m/year of steady infiltration, and decreases to −24.5 kPa at the ground surface under 0.365 m/year of steady evaporation. For thicker soil layers, the general pattern in the suction stress profile remains similar, but its magnitude could be quite different. The analytical solutions for suction stress (Equation (6.16)), suction (Equation (3.45)), and water content (Equation (3.46)) provide convenient, order of magnitude estimates of the variation of pore-water conditions and suction stress useful for assessing hydro-mechanical conditions in a single layer in hillslopes under steady-state conditions. Two-material systems are often encountered in natural hillslopes and represent typical field conditions for heterogeneity. Analytical solutions for these settings will be described in the next section.

6.5.2 Steady-state profiles in one dimension: multiple layers

In sub-section 3.5.2, analytical solutions for steady suction and water content profiles in one-dimensional two-layer system were presented (Equations (3.47)–(3.54)). Following the unified effective stress principle, an analytical solution for profiles of suction stress can be developed as well. The two-layer system is defined in Figure 3.21. Substituting Equations (3.47)–(3.50) into Equation (6.12) yields an analytical expression for the suction stress in the lower layer bounded by a water table ($0 \le z \le L_1$):

$$\sigma^s(z,q) = -\frac{\gamma_w}{\beta_1}\left[\left(\frac{q}{K_{s1}}+1\right)\exp\left(-\beta_1 z\right)-\frac{q}{K_{s1}}\right]^{1/n_1}\ln\left[\left(\frac{q}{K_{s1}}+1\right)\exp\left(-\beta_1 z\right)-\frac{q}{K_{s1}}\right] \tag{6.17}$$

Similarly, substituting Equations (3.52) and (3.53) into Equation (6.12) yields an analytical expression for the suction stress profile in the upper layer bounded by the atmosphere ($L_1 \le z \le L_1 + L_2$):

$$\sigma^s(z,q) = -\frac{\gamma_w}{\theta_{s2}}\theta\left(z\right)h_m\left(z\right) \tag{6.18}$$

where $\theta(z)$ and $h_m(z)$ can be evaluated from Equations (3.48)–(3.54).

Equations (6.17) and (6.18) provide useful limiting conditions (steady state) to assess suction stress variations under various infiltration/evaporation rates. Depending on the hydro-mechanical properties, infiltration/evaporation rates, and thickness of each of the two layers in the system, profiles of suction stress differ in terms of magnitude and pattern. In the following, six hypothetical combinations of sand, silt, and clay layers, each 1 m thick in two-layer, unsaturated systems are described.

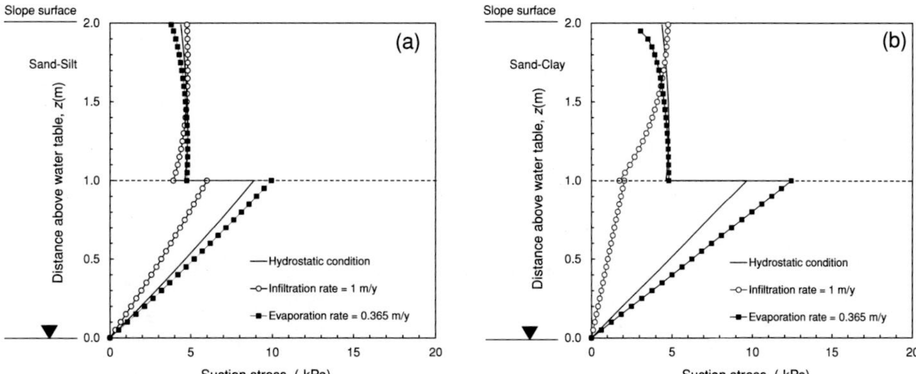

Figure 6.19 Profiles of suction stress under varying infiltration/evaporation conditions in a 2 m thick unsaturated (a) sand–silt system, and (b) sand–clay system.

Case iv: sand–silt system

The steady-state profiles of water content and suction are shown in Figure 3.22 (Case iv in sub-section 3.5.2). Under the steady-state conditions, suction stress remains relatively constant with depth in the upper sand layer, at around −4.7 kPa (Figure 6.19a). Evaporation causes some decrease (∼1.0 kPa) in suction stress near the ground surface, but no change at the sand–silt interface ($z = 1.0$ m). Infiltration increases suction stress near the ground (∼0.4 kPa) and some increase (∼0.85 kPa) at the layer interface. Because of the contrast in material properties at the interface, suction stress decreases significantly (∼4.0 kPa under hydrostatic, 0.7 kPa under 1 m/year infiltration, and 5.2 kPa under 0.365 m/year evaporation conditions) from the sand layer to the silt layer. An infiltration rate of 1 m/year will cause an increase of 3.1 kPa in suction stress, whereas an evaporation rate of 0.365 m/year will lead to a decrease of 1.47 kPa in suction stress. Because an increase in effective stress or a reduction in suction stress will tend to strengthen the interface, the above analysis indicates that evaporation will tend to increase stability along the interface whereas infiltration will tend to decrease stabilty along the interface.

Case v: sand–clay system

The steady-state profiles of water content and suction are shown in Figure 3.23 (Case v in sub-section 3.5.2). The variation in the suction stress profiles in this system is fundamentally the same as those described for the sand–silt system, but the magnitude of the suction stress change is greater at the interface (Figure 6.19b). An infiltration rate of 1 m/year will cause an increase of 7.6 kPa in suction stress, whereas an evaporation rate of 0.365 m/year will lead to a reduction of 2.78 kPa in suction stress. Therefore, a finer-grained underlying layer, such as clay, will tend to further increase stability along the interface under evaporation conditions compared to the sand–silt case, whereas infiltration will tend to further decrease stability.

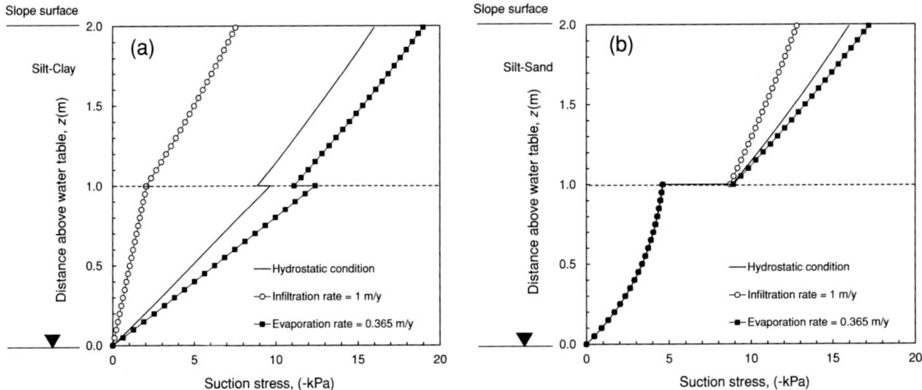

Figure 6.20 Profiles of suction stress under varying infiltration/evaporation conditions in a 2 m thick unsaturated (a) silt–clay system, and (b) silt–sand system.

Case vi: silt–clay system

For this case the steady-state profiles of water content and suction are shown in Figure 3.24. Here, suction stress generally increases with increasing depth, except at the interface where suction stress abruptly changes (Figure 6.20a). For this 2 m thick unsaturated system, the suction stress profile varies nearly linearly within each layer. Suction stress is sensitive to the rate of infiltration/evaporation throughout the entire 2 m layer. A steady infiltration rate of 1 m/year will result in an increase of 8.65 kPa in suction stress at the ground surface, leading to an overall decrease in stability of the entire 2 m layer. A steady evaporation rate of 0.365 m/year will lead to a reduction of 3.0 kPa in suction stress at the ground surface, leading to an overall stabilizing effect for the entire 2 m layer.

Case vii: silt–sand system

For this case the steady-state profiles of water content and suction are shown in Figure 3.25. Here, suction stress generally increases nearly linearly with depth in the upper silt layer, but non-linearly in the lower sand layer (Figure 6.20b). There is an abrupt change in suction stress (4.3 kPa) at the interface. The suction stress profiles are sensitive to the steady infiltration/evaporation rate in the silt layer, but insensitive to the steady infiltration/evaporation rate in the sand layer. A steady infiltration rate of 1 m/year will result in an increase of 3.22 kPa in suction stress at the ground surface, leading to a decrease in stabilty of the upper 1 m silt layer. A steady evaporation rate of 0.365 m/year will lead to a reduction of 1.2 kPa in suction stress at the ground surface, leading to a stabilizing effect on the upper 1 m silt layer.

Case viii: clay–sand system

For this case the steady-state profiles of water content and suction are shown in Figure 3.26. Here, suction stress varies similarly to that of the silt–sand system, but with a greater

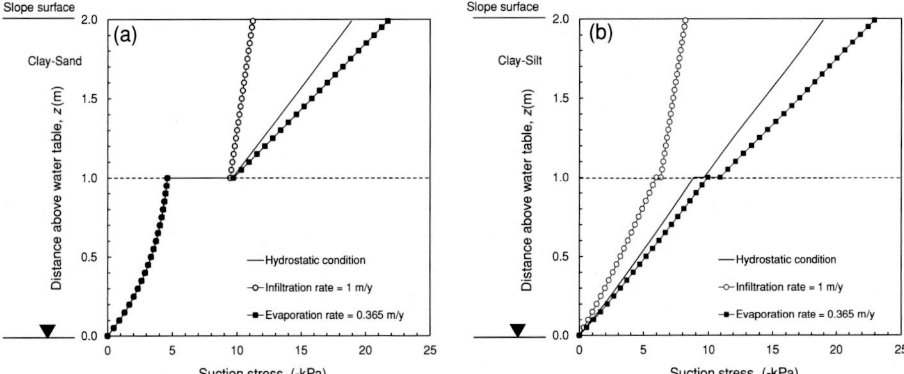

Figure 6.21 Profiles of suction stress under varying infiltration/evaporation conditions in a 2 m thick unsaturated (a) clay–sand system, and (b) clay–silt system.

magnitude of variation in the 1 m clay layer (Figure 6.21a). There is an abrupt change in suction stress (5.0 kPa) at the layer interface. The suction stress profiles are sensitive to the steady infiltration/evaporation rate in the clay layer, but insensitive to the steady infiltration/evaporation rate in the sand layer. A steady infiltration rate of 1 m/year will result in an increase of 7.68 kPa in suction stress at the ground surface, leading to a decrease in the stabilty of the upper 1 m clay layer. A steady evaporation rate of 0.365 m/year will lead to a reduction of 3.1 kPa in suction stress at the ground surface, leading to an overall stabilizing effect on the upper 1 m clay layer.

Case ix: clay–silt system

For this case the steady-state profiles of water content and suction are shown in Figure 3.27. Here, suction stress generally increases with increasing depth, except at the interface where suction stress abruptly decreases (Figure 6.21b). For this 2 m thick unsaturated system, the suction stress profiles vary nearly linearly within each layer. Suction stress is sensitive to the rate of infiltration/evaporation throughout the entire 2 m system. A steady infiltration rate of 1 m/year results in an increase of 10.8 kPa in suction stress at the ground surface, leading to a destabilizing effect on the entire 2 m thick system. A steady evaporation rate of 0.365 m/year will lead to a reduction of 4.0 kPa in suction stress at the ground surface, leading to an overall stabilizing effect on the entire 2 m system.

The one-dimensional suction stress profiles described above for steady-state infiltration/evaporation conditions provide an estimate of the impact of suction stress in an "average time" sense. In principle, Equations (6.17) and (6.18) can be used to assess suction stress profiles in two-layer systems with any dimensions. However, transient analyses are needed to assess the full range of changes in suction stress that result from infiltration. To this end, the following section describes the transient profiles of water content and suction stress using the Green–Ampt vertical infiltration model.

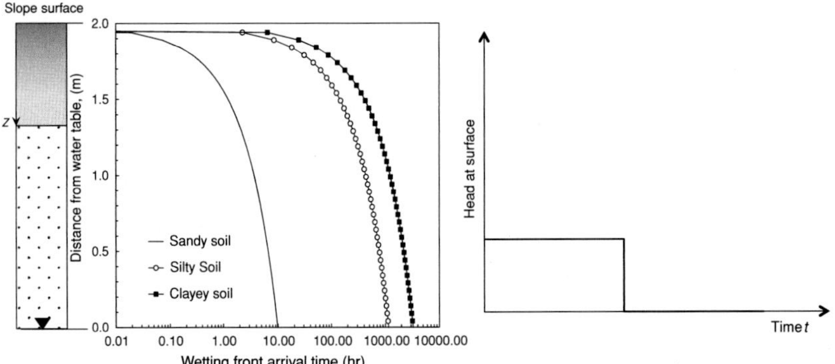

Figure 6.22 Illustration of the sharp wetting front progression into an initially dry soil simulated using the Green–Ampt model. The system parameters for the hypothetical soil are: $h_o - h_i = 1.0$ m, $\theta_o - \theta_i = 0.30$, and $k_o = 1.0^{-5}$ m/s for the sand layer, $h_o - h_i = 1.0$ m, $\theta_o - \theta_i = 0.35$, and, $k_o = 1.0^{-7}$ m/s for the silt layer, and $h_o - h_i = 1.0$ m, $\theta_o - \theta_i = 0.40$, and, $k_o = 1.0^{-8}$ m/s for the clay layer.

6.5.3 Transient state suction stress profiles in one dimension: single layer

In a soil layer of initially low water content conditions, which might result from a long period of no rain, vertically downward infiltration can be described using the so-called Green–Ampt model as described in detail in sub-section 4.2.2. This model simulates the progression of a sharp wetting front under both gravity and a gradient in matric suction as shown in Figure 6.22. In such a model, the moisture content behind the wetting front is a constant near saturation and an implicit analytical expression between the wetting front position z and time t since the onset of infiltration is (Equation (4.34))

$$\frac{K_o}{\theta_o - \theta_i} t = z - (h_o - h_i) \ln \left(1 + \frac{z}{h_o - h_i} \right) \tag{6.19}$$

where K_o is the hydraulic conductivity that corresponds to the volumetric water content θ_o, and h_o is the hydraulic head behind the wetting front, θ_i is the initial volumetric water content, and h_i is the initial hydraulic head. The dynamics of the wetting front progression for different soils is illustrated in Figure 6.22 for a 2 m thick unsaturated zone composed of hypothetical sand, silt, and clay soils. The three soils have the same hydrologic properties as those listed in Table 3.4. Initially, the soil layer is assumed to be at the hydrostatic condition such that the suction stress profile can be calculated from Equation (6.16). The progression of the wetting front from the ground surface to the water table will take 10 hours in the 2 m thick sand layer, ~1000 hours in the 2 m thick silt layer, and several thousand hours or several months in the 2 m thick clay layer (Figure 6.22).

The mechanical consequences of the simulated infiltration process can be assessed using the concept of suction stress or the unified effective stress described in Section 6.4. As illustrated in Figures 6.14b and 6.15, suction stress is zero for all soils when they are saturated if the suction stress at the saturated state is used as the zero reference point. Therefore, the maximum increase in suction stress behind the wetting front can be assessed

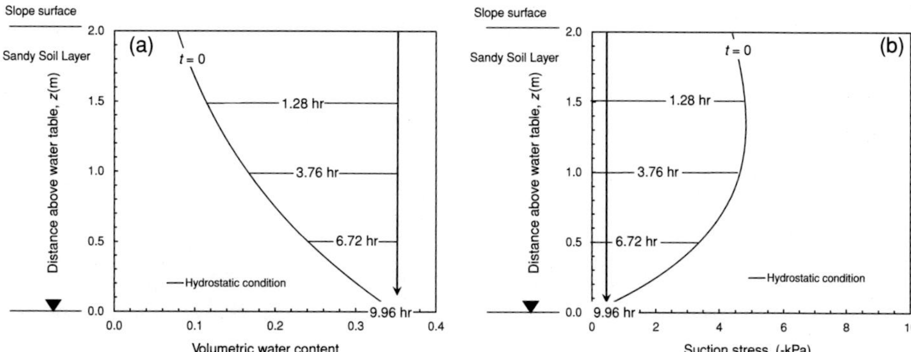

Figure 6.23 Illustrations of transient moisture content variation (a) and suction stress variation (b) in a 2 m thick unsaturated sand layer initially under hydrostatic conditions.

if the suction stress profile prior to infiltration is known. Assuming the profile of initial suction stress is consistent with hydrostatic conditions, the dynamics of the variation of soil water content and diminishing suction stress in the hypothetical 2 m thick sand layer is shown in Figure 6.23. Initially under hydrostatic conditions ($t = 0$), the water content increases with the depth from 0.08 to its saturated value of 0.30 at the water table. The corresponding suction stress profile is characterized by zero suction stress at the water table, reaches its minimum of −4.81 kPa at ~1.3 m above the water table, and increases to −4.38 kPa at the ground surface. Imposing a sustaining head 1.0 m above the initial head at the ground surface, the wetting front propagates downward and the profile of soil water content is characterized by a sharp wetting front behind, which the soil is nearly saturated (Figure 6.23a). The wetting front reaches 0.5 m below the ground surface at time $t = 1.28$ hours, 1.0 m at $t = 3.76$ hours, 1.5 m at $t = 6.72$ hours, and finally reaches the water table at $t = 9.96$ hours. By then, the entire sand layer is nearly saturated and the corresponding suction stress in the entire layer disappears (Figure 6.23b). This example illustrates that effective stress in this sand layer can be reduced by 4.81 kPa (equivalent to a total stress due to 0.25 m of soil weight) when the wetting front reaches ~0.7 m below the slope surface, this is sufficient to induce or trigger a landslide if the slope is close to failure. Chapter 10 describes a case study under such a scenario.

For a 2 m thick layer of silt, the hydrostatic water content profile prior to infiltration is nearly linear (according to the theory described in sub-section 3.5.1), varying from 0.37 at the ground surface to 0.45 at the water table (Figure 6.24a). The corresponding initial suction stress profile increases monotonically from −15.98 kPa at the ground surface to zero at the water table (Figure 6.24b). Imposing a constant head at the ground surface, the wetting front progresses at the rate shown in Figure 6.23a; it takes ~149 hours to reach 0.5 m below the ground surface, ~438 hours to reach 1.0 m below the ground surface, 784 hours to reach 1.5 m below the ground surface, and finally 1,162 hours to reach the water table (Figure 6.24a). The dynamics of the disappearance of suction stress behind the wetting front are shown in Figure 6.24b. The greatest change in suction stress (15.98 kPa) occurs at the ground surface. As illustrated here, it takes over 1.5 months for the wetting front to

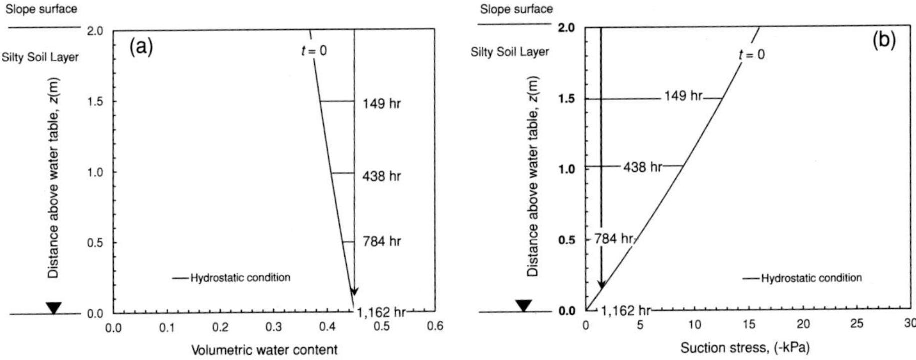

Figure 6.24 Illustrations of transient moisture content variation (a) and suction stress variation (b) in a 2 m thick unsaturated silt layer initially under hydrostatic conditions.

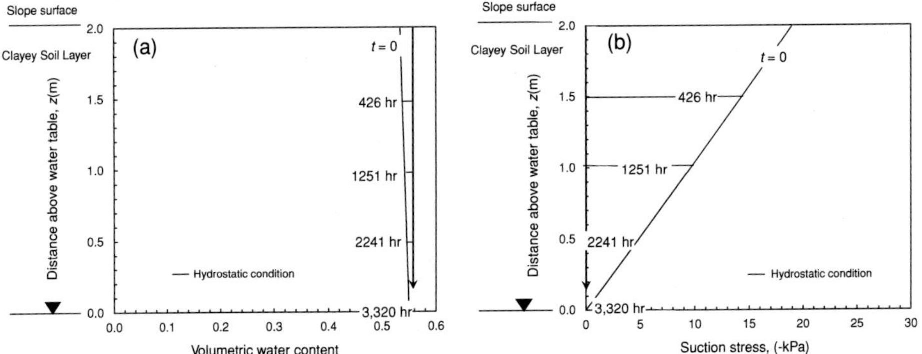

Figure 6.25 Illustrations of transient moisture content variation (a) and suction stress variation (b) in a 2 m thick unsaturated clay layer under initially hydrostatic conditions.

reach the water table. Nevertheless, the possibility of tens of kPa reduction in effective stress near the slope surface (<1.0 m) upon wetting in silty soils is significant in comparison with the total stress due to the self-weight of the soil. Such a significant reduction in effective stress could be responsible for the "collapsing" phenomenon observed in many loess soils.

For a 2 m thick layer of clay, the hydrostatic water content prior to infiltration is nearly saturated and varies linearly from 0.53 at the ground surface to 0.55 at the water table (Figure 6.25a). The corresponding initial suction stress increases linearly from −18.93 kPa at the ground surface to zero at the water table (Figure 6.25b). After imposing a constant head at the ground surface, the wetting front progresses slowly at the rate shown in Figure 6.25; it takes ~426 hours to reach 0.5 m below the ground surface, ~1,251 hours to reach 1.0 m below the ground surface, 2,241 hours to reach 1.5 m below the ground surface, and finally 3,320 hours to reach the water table (Figure 6.25a). The dynamics of the disappearance of suction stress behind the wetting front are shown in Figure 6.25b. The greatest change in suction stress (18.93 kPa) occurs at the ground surface. As illustrated here, it takes over 4.5 months for the wetting front to reach the water table. Nevertheless, the possibility of tens of kPa reduction of effective stress near the slope surface (<1.0 m) upon wetting in clay soils is significant in comparison with the total stress due to the self-weight of the soil.

Suction stress distributions in multi-dimensional hillslopes under both steady-state and transient conditions will be described in Chapters 9 and 10.

6.6 Problems

1 What is the mathematical form of Bishop's effective stress?
2 Why in Terzaghi's effective stress is there a negative sign in front of the pore-water pressure whereas in Bishop's effective stress there is a positive sign in front of pore-water pressure?
3 What are the units of the product of χ times matric suction in Bishop's effective stress?
4 What is the range of Bishop's effective stress parameter χ?
5 What is the value of χ in saturated soil? What is the value of χ if the soil is dry?
6 What is the basic concept of Coleman's independent stress variable theory?
7 How many independent stress variables are there in Coleman's theory? What are they?
8 What is the physical meaning of matric suction? Is matric suction a stress variable?
9 At the REV level shown in Figure 6.1, is it correct to consider matric suction as a stress quantity? Why?
10 What are the main reasons that have been cited to discount Bishop's effective stress?
11 What is the mathematical form of Lu *et al.*'s effective stress?
12 What is the relationship between suction stress and pore-water pressure when soil becomes saturated?
13 What are the major physical mechanisms in Lu *et al.*'s suction stress?
14 In Terzaghi's effective stress, does pore-water pressure depend on total stress? In Lu *et al.*'s effective stress, does suction stress depend on total stress?
15 Is suction stress an external stress or internal stress? What is the stress that counterbalances the suction stress internally?
16 What is the order of magnitude value of suction stress in sand, silt, and clay soils?
17 What are the major differences between suction stress characteristic curves for sand and clay?
18 What is the relationship between suction stress and apparent cohesion?
19 Does the apparent cohesion reflect the internal frictional or shear resistance of soils? Does the suction stress reflect the internal frictional or shear resistance of soils?
20 What in Figures 6.10–6.12 shows the validity of Lu *et al.*'s unified effective stress principle?
21 What is the physical meaning or thermodynamic basis of suction stress?
22 What in Figures 6.16 and 6.17 shows the intrinsic relationship between SWCC and SSCC?

23 If one knows a soil's SWCC, can one know its SSCC, or vice versa?

24 What is the range of suction stress under varying infiltration and evaporation in a 2 m thick sand layer above a water table? What is the range of suction variation? What is the range of water content variation?

25 What is the range of suction stress under varying infiltration and evaporation in a 2 m thick layer of silt above a water table? What is the range of suction variation? What is the range of water content variation?

26 What is the range of suction stress under varying infiltration and evaporation in a 2 m thick layer of clay above a water table? What is the range of suction variation? What is the range of water content variation?

27 Consider a sand layer similar to that described in Problem 24, except that the thickness is 4 m. What is the approximate range of suction stress variation? What is the approximate range of suction variation? What is the approximate range of water content variation?

28 Consider a silt layer similar to that described in Problem 25, except the thickness is 4 m. What is the approximate range of suction stress variation? What is the approximate range of suction variation? What is the approximate range of water content variation?

29 Consider a clay layer similar to that described in Problem 26, except the thickness is 4 m. What is the approximate range of suction stress variation? What is the approximate range of suction variation? What is the approximate range of water content variation?

30 Consider the two-layer soil system shown in Figures 6.19–21. Which system has the greatest change in suction stress at the soil interface?

31 Consider a 2 m thick sand layer above a water table under transient rainfall conditions. What is the corresponding change in suction stress? What is the corresponding change in suction? What is the corresponding change in water content?

32 Consider a 2 m thick silt layer above a water table under transient rainfall conditions. What is the corresponding change in suction stress? What is the corresponding change in suction? What is the corresponding change in water content?

33 Consider a 2 m thick clay layer with a thickness above a water table under transient rainfall conditions. What is the corresponding change in suction stress? What is the corresponding change in suction? What is the corresponding change in water content?

34 Consider a flat surface underlain by a two-layer system of sand overlying silt. The thickness of the sand layer is 1 m, and the silty layer is 1 m thick. The following three infiltration conditions are then applied to the system: $q = -10^{-8}$ m/s, 0.0 m/s, and 10^{-8} m/s. The unsaturated hydraulic properties are as follows: $\beta_{sand} = 3\ \text{m}^{-1}$, $\beta_{silt} = 0.3\ \text{m}^{-1}$, $n_{sand} = 4.0$, $n_{silt} = 3.0$, $\theta_{sand} = \theta_{silt} = 0.4$, $K_{s\text{-}sand} = 10^{-5}$ m/s, and $K_{s\text{-}silt} = 10^{-7}$ m/s.

(a) Calculate and plot steady suction profiles.
(b) Calculate and plot steady water content profiles.
(c) Calculate and plot steady suction stress profiles.
(d) Draw your major conclusions from this exercise.

35 Consider a silt soil layer with a 20 m thick unsaturated zone (from the surface to the water table) that has the following hydraulic properties: $\theta_o = 0.4$, $\theta_i = 0.05$, and $K_o = 10^{-7}$ m/s. Assume that the Green–Ampt model is valid and the suction head across the wetting front is $h_o - h_i = 1.0$ m.

(a) Estimate the arrival time at the water table of a downwardly infiltrating wetting front after heavy rainfall.

(b) Plot arrival distance vs. arrival time.

(c) Estimate and plot the suction and moisture content profiles under hydrostatic conditions.

(d) Estimate and plot the suction stress profile under hydrostatic conditions.

(e) Plot the contour of the wetting front arrival times over plots of (c) and (d).

(f) Draw your major conclusions from this exercise.

PART IV

HILLSLOPE MATERIAL PROPERTIES

7 Strength of hillslope materials

7.1 Failure modes and failure criteria

7.1.1 Definition of strength

The failure of hillslope materials refers to the abrupt and irreversible downward and outward movement of the surficial soil and rock on a hillside. However, a consensus analytic criterion describing the failure of hillslope materials is not available. Stress and strength concepts are used to define the failure state, as are stress and/or strain, rate of stress/strain and associated strength parameters. Furthermore, the definition of the strength of hillslope materials varies widely dependent on the discipline and perspective. The strength of hillslope materials, as summarized by Selby (1993), has been defined in three general ways: (1) the ability of materials to resist deformation by compressive, tensile, or shear stresses, (2) the ability of materials to resist abrasion, and (3) the ability of materials to resist being transported by a fluid. In soil mechanics, strain is often included as an additional variable that is used to define the strength of soils, such as in the application of kinematics-based limit analysis (e.g., Chen, 1975; Michalowski, 1995) or analyses within the framework of critical state soil mechanics (e.g., Schofield and Wroth, 1968). In this book, strength refers to the ability of hillslope materials to resist strain that arises from the three common modes of tensile, compressional, and shear stresses, or by combinations of any two of these modes.

A simple experiment to measure the strength of cohesive porous materials is a uniaxial compression test. As shown in Figure 7.1, a principal maximum stress is applied in the vertical direction to a cylindrical soil specimen while no stress is applied in the horizontal direction. The state of stress can be illustrated in the shear stress-normal stress plane shown as broken circles in Figure 7.1. Failure, commonly marked by the development of cracks or shear bands, occurs when the applied stress reaches a certain level defined as the uniaxial compressive yield strength, shown in Figure 7.1 as a solid circle. The magnitude of the uniaxial strength of hillslope materials can vary from zero for dry and non-cohesive sandy soil to several thousand kPa for dry and overly consolidated clay or cemented sands. The direction of the failure plane is characteristically along $45^\circ + \phi/2$, where ϕ is called the internal friction angle and will be discussed further in this section.

If tensile stress, rather than compressional stress, is applied in the vertical direction to a cohesive or wet soil specimen with a cylindrical shape, tensile failure or strength can be obtained when the applied tensile stress reaches the tensile strength of the material, as illustrated in Figure 7.1. The tensile strength of soil is largely overlooked in the geotechnical literature, but can, in fact, be substantial and play an important role in the stability of

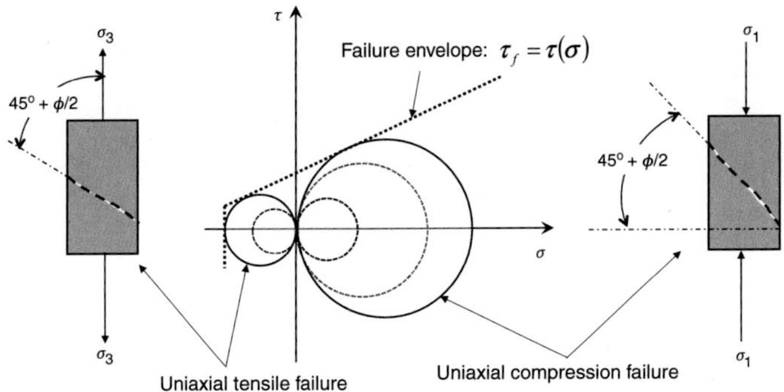

Figure 7.1 Illustration of the state of stress in uniaxial compression and tensile strength tests.

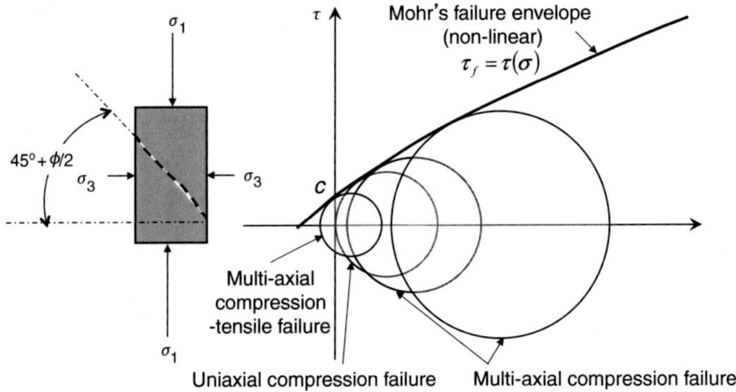

Figure 7.2 Illustration of the state of stress in bi-axial or conventional compression strength tests.

hillslopes. For wet non-cohesive sandy soils, the inter-particle tensile stress that arises from capillarity can provide tensile strength on the order of several kPa. Stress or strength changes of several kPa may determine the stability of the upper few meters of a sandy hillslope. This point is illustrated fully in Chapter 9. In clay hillslopes, tensile strength can be on the order of several hundred kPa and is highly dependent of the degree of saturation. This strength may provide a critical contribution to the stability of hillslopes under variably saturated conditions. This chapter describes the intrinsic relation between tensile strength and uniaxial compression strength.

The state of stress of hillslopes is generally more complicated than can be described by the two simple uniaxial stress states. In hillslopes, the state of stress could be multi-axial compressional or multi-axial compression–tensile conditions, as illustrated in Figure 7.2. In general, the greater the minimum principal stress, the greater the maximum principal stress will be. The failure states under different combinations of principal stresses form an envelope called the Mohr failure envelope, which can be characterized by a non-linear smooth curve shown in Figure 7.2. The non-linear behavior is generally more pronounced at low stress levels. However, for most soils with stress levels less than several thousand kPa, a linear relation, called the Mohr–Coulomb failure envelope, sufficiently describes the state of failure. The linear relation implies that the strength of soils can be characterized

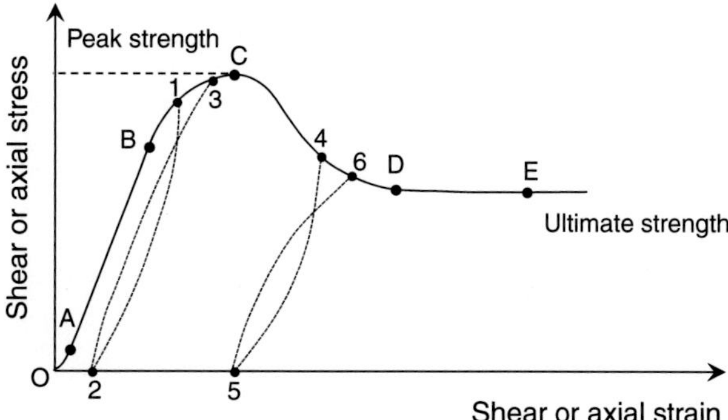

Figure 7.3 Illustration of stress–strain relation for soils.

by two parameters: the internal friction angle ϕ and cohesion c. Graphically, it states that the failure plane, invariant of the ratio of the applied principal stresses, remains the same at $45^\circ + \phi/2$, as shown in Figure 7.2a.

7.1.2 Stress–strain relation

The failure state of soil can be defined in terms of resisting strain or the stress–strain relation, as illustrated in Figure 7.3. In general, the stress–strain relation can be divided into two stages; pre- and post-failure. The pre-failure stage can be subdivided into three segments; OA, AB, and BC, and the post-failure can be subdivided into two segments: CD and DE, each marked by differing characteristics. When a soil specimen is subjected to a small load, the stress–strain relation is non-linear and slightly convex upward. It follows the path OA and the strain is considered to be elastic, meaning that upon unloading the strain is fully recoverable or stress–strain follows the curve AO. As the load increases, the stress–strain relation behaves elastically and linearly (AB). The slope of AB is called the elastic shear modulus (for shear stress vs. shear strain) or Young's modulus (for axial stress vs. axial strain). Beyond a certain stress level (B), the stress–strain relation follows a non-linear path that is concave downward. Plastic or permanent strain is observed upon unloading (e.g., 1→2). Reloading or increasing stress (e.g., 2→3) will follow a different stress–strain path than that of unloading. This phenomenon is called stress–strain hysteresis. The non-recoverable stress–strain is also accompanied by a continuous decrease in Young's modulus, reflected by the concave downward stress–strain path. When the Young's modulus reaches zero (C), the stress reaches a maximum that is defined as the peak strength. This point (C) is often used to define failure state.

The post-failure phase of the stress–strain relation begins at point C and follows the negative slope of the path CD, as illustrated in Figure 7.3. A physical consequence of the zero or negative modulus along this path is that the material will not absorb any additional mechanical energy from the applied external stress. In fact, during this negative modulus stage, further deformation of the material will result in a release of mechanical energy from the stressed material. If the material is brittle, such as overly consolidated clay or dense sand, the negative slope of the modulus curve may be very steep and the residual strength

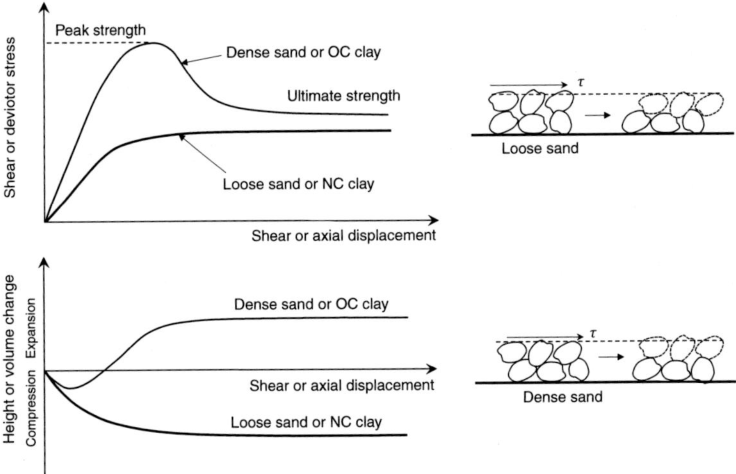

Figure 7.4 Illustration of the stress–strain relation for normally consolidated and over-consolidated soils.

(DE) very small. On the other hand, for soft materials such as normally consolidated clay or loose sand, the negative slope of the stress–strain relation may be less steep and the residual strength (DE) could be substantial. The mechanical behavior of materials in the residual strength regime is marked by very little cohesion along the failure surface, but may be important in many natural hillslopes and existing landslides in which a few kPa of strength may determine stability. Along the path CD, unloading (e.g., 4→5) will give rise to substantial permanent or unrecoverable strain. Upon reloading (e.g., 5→6), hysteresis gives rise to a decrease in shear stress (point 6) compared to the shear stress prior to unloading (point 4).

The stress–strain relation is unique for each hillslope material and is dependent on the history of soil and rock formation, and the history of deformation driven by processes such as uplift and failure. Nonetheless, generalizations developed by the field of soil mechanics are widely accepted in foundation design and slope stability analysis. In general, loose sand and normally consolidated (NC) clay are considered to have similar stress–strain patterns, similarly dense sand and overly consolidated (OC) materials also share similar patterns, as illustrated in Figure 7.4. NC refers to the state of stress when the current stress (mostly as a result of vertical depth of overburden) is the maximum stress a soil has ever experienced, whereas OC refers to the state of stress when the current stress is less than the maximum stress a soil has ever experienced. For example, glacial deposits may have experienced much higher stress levels in the past when they were under ice, and thus are in an OC state. In contrast, the load on contemporary sediment deposits may be increasing and thus this material is in a NC state. Upon the application of an external load, the volume and void space of loose sand/NC clay will generally decrease as the material adjusts to a more stable configuration, eventually reaching a constant ultimate strength and volume. On the other hand, dense sand/OC clay tends to dilate slightly upon initial loading, which is accompanied by an increase in volume. The expansion in volume results from particle-level dilative movement, as illustrated in Figure 7.4. The strength of these materials is

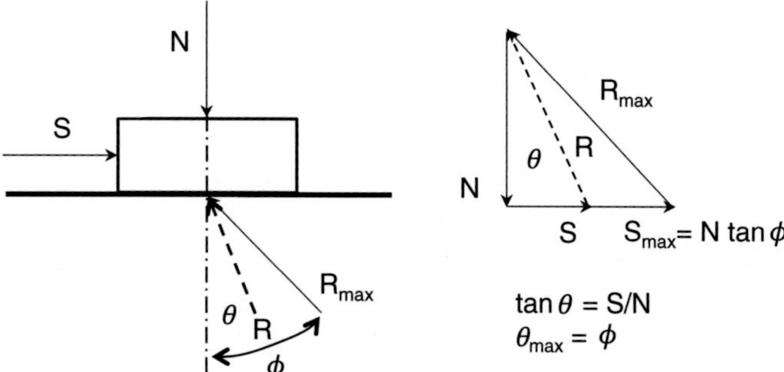

Figure 7.5 Illustration of the friction angle concept.

characterized by a peak strength followed by a sharp decline before reaching the ultimate or residual strength. This behavior is called "brittle" behavior in contrast to the "ductile" behavior often observed in shear strength tests performed on loose sand/NC clay.

In conventional design and analysis of soils in engineered slopes, it is generally assumed that loose sand and NC clay do not possess cohesion, whereas only OC silt and clay have cohesion. This assumption is not generally suitable for hillslope stability analysis, particularly for natural hillslopes, as NC soils often exhibit shear strength due to cohesion as large as several tens of kPa. This contribution to shear strength may determine the stability of the upper several tens of meters of hillslope materials. This chapter presents a general framework that unifies the classic cohesion concept with inter-particle stress or tensile stress under variably saturated conditions. The common thread used to unify all the associated strength concepts is internal frictional behavior. Frictional strength generally falls into two categories: frictional shear strength resulting from external stresses and frictional shear strength resulting from the internal stresses of cohesion.

7.2 Shear strength due to frictional resistance

7.2.1 Friction angle concept

Shear strength that results from external stresses can be understood from the frictional resistance at the base of a sliding block that sits on a rough flat surface, illustrated in Figure 7.5. For a constant normal force **N** and increasing shear force **S**, the contact surface of the block will develop a resistant force **R** with components equal, but opposite to the applied forces **N** and **S**. Based on the force equilibrium principle, the applied shear force is equal to the product of the normal force **N** times the tangent of the angle θ in which the resistant force **R** acts. Shear resistance defined in such a manner is called frictional shear resistance and reaches a threshold when static equilibrium breaks down. The angle at this threshold is called the friction angle of the interface and is denoted by ϕ. In soil or rock, such frictional shear resistance originates at the contacts among particles. The sliding-block

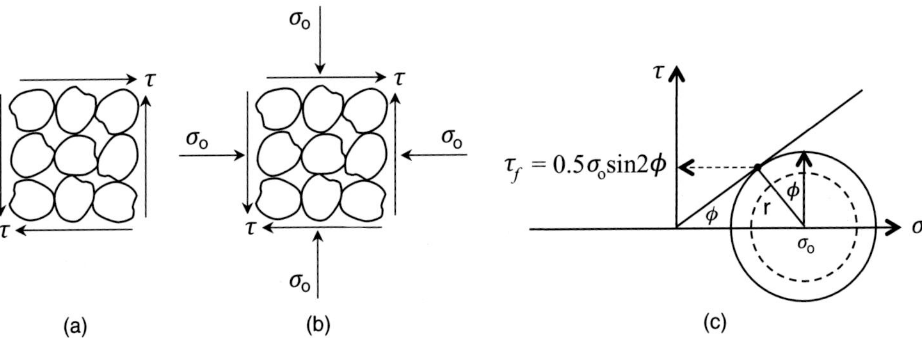

Figure 7.6 Illustration of shear strength that arises from frictional resistance for particles with point-to-point contacts: (a) under pure shear stress with no normal stress, (b) under a constant normal stress with frictional shear stress, and (c) stress path (vertical arrowed line) and shear strength under (b). This illustration provides an explanation of where the residual shear strength of a soil comes from after the failure state has been reached.

model shown in Figure 7.5 can be used as an analog for the overall shear resistance of a soil or rock specimen under external normal and shear stress, such as those illustrated in Figure 7.1. The characteristic angle that describes the maximum ratio of the applied normal stress σ and shear stress τ , and the orientation of the failure surface shown in Figure 7.1 is called the internal friction angle and is typically denoted by ϕ.

Shear resistance that arises from the frictional behavior of the inter-particle contacts in soil or rock can be illustrated by considering an assemblage of non-cohesive particles subject to external stresses shown in Figure 7.6. If the assemblage is subject to pure shear stress and no external normal stress (Figure 7.6a), the assemblage will not resist the shear stress and will fail. However, if normal stress σ_o is applied prior to shear stress τ, as shown in Figure 7.6b, the assemblage will resist the shear stress until the state of stress (Mohr circle) reaches the failure criterion called the Coulomb friction law:

$$\tau = \sigma \tan \phi \tag{7.1}$$

The point where the Mohr circle under the state of stress shown in Figure 7.6b intercepts the failure criterion (7.1) can be determined mathematically as follows. From the geometry shown in Figure 7.6c, the radius of the failure Mohr circle $r = \sigma_o \sin \phi$ and the shear strength $\tau_f = r \cos \phi$, which lead to (Figure 7.6c):

$$\tau_f = \frac{\sigma_o}{2} \sin 2\phi \tag{7.2}$$

From the geometry shown in Figure 7.6c, the normal stress on the failure plane is $\sigma_f = \tau_f \tan \phi$. Substituting Equation (7.2) for τ_f leads to

$$\sigma_f = \sigma_o \cos^2 \phi \tag{7.3}$$

Equation (7.1) can be deduced from Equations (7.2) and (7.3). Thus, for ideal non-cohesive granular materials, shear resistance calculated by Equation (7.1) or (7.2) can only exist with a prior application of normal stress.

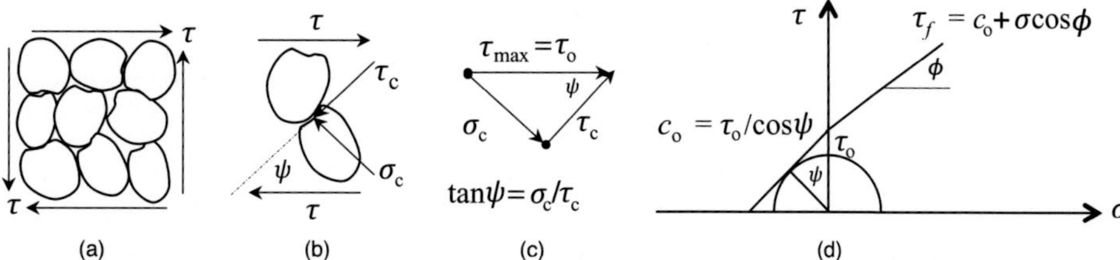

Figure 7.7 Illustration of shear strength due to inter-particle locking: (a) area-to-area contacts, (b) interlocking at particle contact, (c) force equilibrium at particle contact, and (d) resulting apparent cohesion and shear strength due to interlocking.

7.2.2 Apparent cohesion concept

Experimental evidence indicates that most granular materials, such as dry sand, are not the ideal non-cohesive materials described above, i.e., no normal stress and no shear resistance behavior. Instead, results from shear strength tests on most granular materials show some non-zero intercept of the failure envelop on the shear stress axis (see Figure 7.7). This non-zero intercept is generally referred to as "apparent cohesion" or "interlocking cohesion" and can be explained using the following model.

For the point-to-point inter-particle contacts shown in Figure 7.6, no shear resistance is present if no normal stress is applied, as shear resistance is fully dependent on the normal contact stress shown in Equation (7.2). However, in real soils, the inter-particle contacts are more complicated. The contacts are area-to-area contacts rather than point-to-point, as shown in Figure 7.7a. The external shear stress itself, as shown in Figure 7.7b, will cause stress normal to the contact area. This is the interlocking mechanism under which there are two possibilities for the failure of contacts: frictional failure or crushing failure. Crushing failure occurs when the shear stress mobilized by friction is greater than the induced inter-particle shear resistance. This failure results in the breaking of an inter-granular "tooth" when the inter-particle shear or compressive stress exceeds the strengths of the particles.

Under an applied shear stress τ (Figure 7.7b), both normal and shear stresses will be induced at particle contacts:

$$\sigma_c = \tau \sin\psi \tag{7.4}$$

$$\tau_c = \tau \cos\psi \tag{7.5}$$

with ψ being the local composite contact angle that controls the relative inter-particle movement or local dilation. When the external shear stress reaches some threshold value τ_o, failure will occur. The composite contact angle ψ approaches the local contact friction angle ϕ_c, and local equilibrium illustrated in Figure 7.7(c, d) will lead to the apparent cohesion c_o:

$$c_o = \tau_o/\cos\phi_c \tag{7.6}$$

The local friction angle ϕ_c depends on the degree of interlocking, and for angular sands also depends on particle geometries and surface morphology. The local friction angle is usually larger than the internal friction angle ϕ, and can be related to the dilation angle determined from shear stress–displacement data shown in Figure 7.4b. Considering the apparent cohesion due to interlocking, the macroscopic failure criterion in the compressional stress regime can be written as

$$\tau_f = c_o + \sigma \tan \phi \tag{7.7}$$

Experimental data from both direct shear and triaxial tests indicate that for dry sandy soils the apparent cohesion c_o could be as much as several kPa. However, in dealing with engineered slopes, this apparent cohesion has been traditionally ignored on the grounds that, at most, the accuracy of designs is on the order of a several kPa. But in many natural hillslope environments, several kPa of stress can determine the shallow stability of materials as is illustrated in Chapters 9 and 10.

7.2.3 Internal friction angle of sand

For loose or unconsolidated sandy or granular materials, the macroscopic behavior of the inter-particle scale mechanisms of interlocking (i.e., evolution from point-to-point contacts to area-to-area contacts) can be conceptualized by examining the macroscopic changes in the internal friction angle. Often, such behavior is observed in the strong dependence of the friction angle on the relative density D_r or relative dry density RDD in results from density or void ratio-controlled shear strength tests. The relative density is commonly defined in terms of the void ratio e, the maximum void ratio $e_{\max}$, and the minimum void ratio $e_{\min}$:

$$D_r = \frac{e_{\max} - e}{e_{\max} - e_{\min}} \tag{7.8}$$

The RDD is commonly defined in terms of the dry unit weight γ_d, the maximum dry unit weight $\gamma_{d\max}$, and the minimum dry unit weight $\gamma_{d\min}$:

$$\text{RDD} = \frac{\gamma_d - \gamma_{d\min}}{\gamma_{d\max} - \gamma_{d\min}} \tag{7.9}$$

From the definitions of void ratio and unit weight, it can be shown that RDD can be written in terms of D_r as

$$\text{RDD} = D_r \frac{\gamma_d}{\gamma_{d\max}} \tag{7.10}$$

The dry maximum and minimum unit weights of earthen materials can be determined by conducting the standard ASTM tests (D4253 for the maximum dry unit weight and D4254 for the minimum dry unit weight).

Figure 7.8 shows the peak and residual shear strengths as functions of RDD from two types of shear strength tests for a sand: conventional triaxial and plane strain (Cornforth, 1964, 2005). The peak shear strength follows a distinct decreasing functional relation as RDD decreases under each stress condition. The peak shear strength in plane strain is always greater than the peak shear strength in triaxial tests and the friction angle could increase by as much as 16° from RDD = 0 to RDD = 100%. The residual shear strengths, however, are insensitive to RDD and remain approximately the same as they are in the

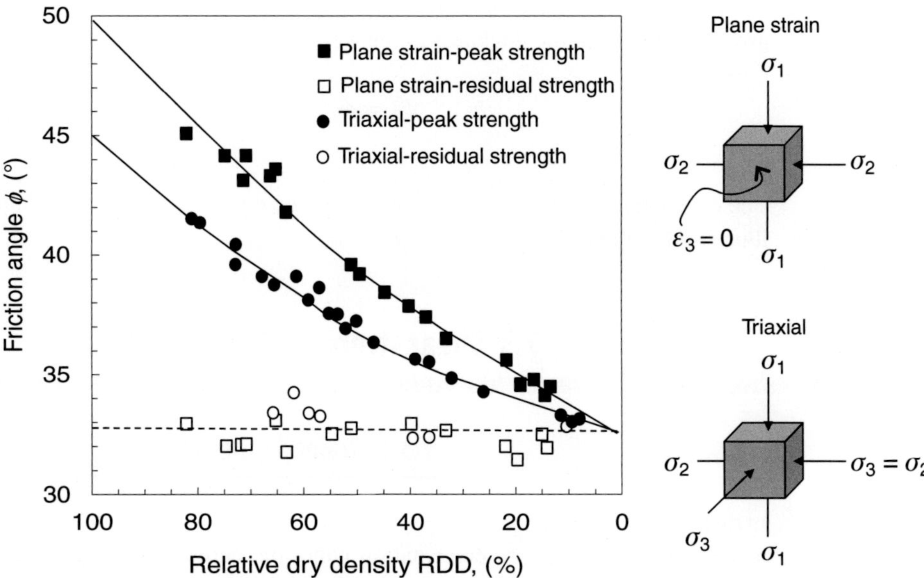

Figure 7.8 Dependence of the friction angle on relative dry density (RDD) for Brasted sand in conventional triaxial ($\sigma_2 = \sigma_3$) and plane strain compression tests (data from Cornforth, 2005).

loosest state (RDD = 0%). Other researchers have observed similar patterns (e.g., Taylor, 1941; Hafiz, 1950; Kirkpatrick, 1957).

The shear strength of dry cohesionless materials such as dry sand can be expressed in the form of the Coulomb friction law (7.1):

$$\tau = \sigma \tan \phi_d \tag{7.11}$$

where the internal friction angle ϕ_d can be further linked to some basic material properties such as the relative density and relative dry density. Based on previous studies, the peak friction angle ϕ_d of dry materials can be expressed in terms of RDD, the friction angle ϕ_0 at RDD = 0%, the friction angle ϕ_{100} at RDD = 100%, and a fitting parameter b:

$$\phi_d = \phi_0 + (\phi_{100} - \phi_0)\left(\frac{\text{RDD}}{100}\right)^b \tag{7.12}$$

For the data shown in Figure 7.8, the triaxial test results yield $\phi_0 = 32.7°$, $\phi_{100} = 45°$, and $b = 1.54$, or

$$\phi_d = 32.7° + 12.5°\left(\frac{\text{RDD}}{100}\right)^{1.54}$$

And the plane strain test data lead to $\phi_0 = 32.7°$, $\phi_{100} = 50°$, and $b = 1.37$, or

$$\phi_d = 32.7° + 17.5°\left(\frac{\text{RDD}}{100}\right)^{1.37}$$

In sand hillslopes, the RDD can be directly related to the depth of interest H_{ss} from the ground surface and the thickness of the weathered or loose zone z_w (Lu and Godt, 2008):

$$\phi_d = \phi_0 + \frac{\phi_{100} - \phi_0}{1 + \dfrac{z_w}{H_{ss}}} \tag{7.13}$$

The dependence of the dry friction angle on the depth H_{ss} for various thicknesses of z_w is illustrated in Figure 9.7 in Chapter 9.

7.3 Shear strength due to cohesion

7.3.1 Drained cohesion

In this section (7.3), as well as the section that follows (7.4), the commonly used cohesions, namely, drained cohesion, cementation cohesion, capillary cohesion, and root cohesion, will be discussed. The word "cohesion" is used here consistent with the definition that is widely accepted in soil mechanics and other disciplines, i.e., the intercept of the failure envelope on the shear axis, typically denoted by the symbol c in the literature.

Drained cohesion refers to the intercept under saturated drained or constant pore-pressure conditions. As illustrated in Figure 7.9, cohesion defined in such a manner is the shear strength C, rather than the more common and accurate meaning of the word "cohesion," which describes the bonding stress or stress normal to the contact points among particles. As described here, cohesion c originates from a non-shear stress called the isotropic tensile strength σ_{tia}, shown as a point on the far left of the normal stress axis. The state of stress at this point is shown in the top left in Figure 7.9. This bonding stress is the source of cohesion in silt and clay soils under saturated condition (e.g., Lu and Likos, 2006). For the state of stress at isotropic tensile strength σ_{tia}, there is no shear stress at any point in any direction, and failure occurs only when the applied external stress reaches the bonding strength (or tensile strength) provided by inter-particle physical–chemical bonding. This strength exists with or without the presence of external stresses. Because no shear stress develops when a soil fails under isotropic tensile stress, the isotropic tensile strength of soil is independent of the internal friction angle.

In contrast to isotropic tensile strength, uniaxial tensile strength σ_{tua} may be defined for the case where an element of soil fails under tensile stress applied normal to one principal plane, with zero stress applied to corresponding orthogonal planes (circle A in Figure 7.9). This is the tensile strength measured via various forms of direct tension tests in the literature (e.g., Bishop and Garga, 1969; Perkins, 1991; Lu *et al.*, 2007).

Assuming the ratio of shear stress to normal stress in the tensile stress regime remains the same as in the compressive stress regime, i.e., tan ϕ, then uniaxial tensile strength σ_{tua} can logically be considered the mobilization of isotropic bonding stress when the maximum principal stress remains zero. This state of stress is depicted as circle A in Figure 7.9 and is illustrated for a corresponding soil element in the top middle part of the figure. In

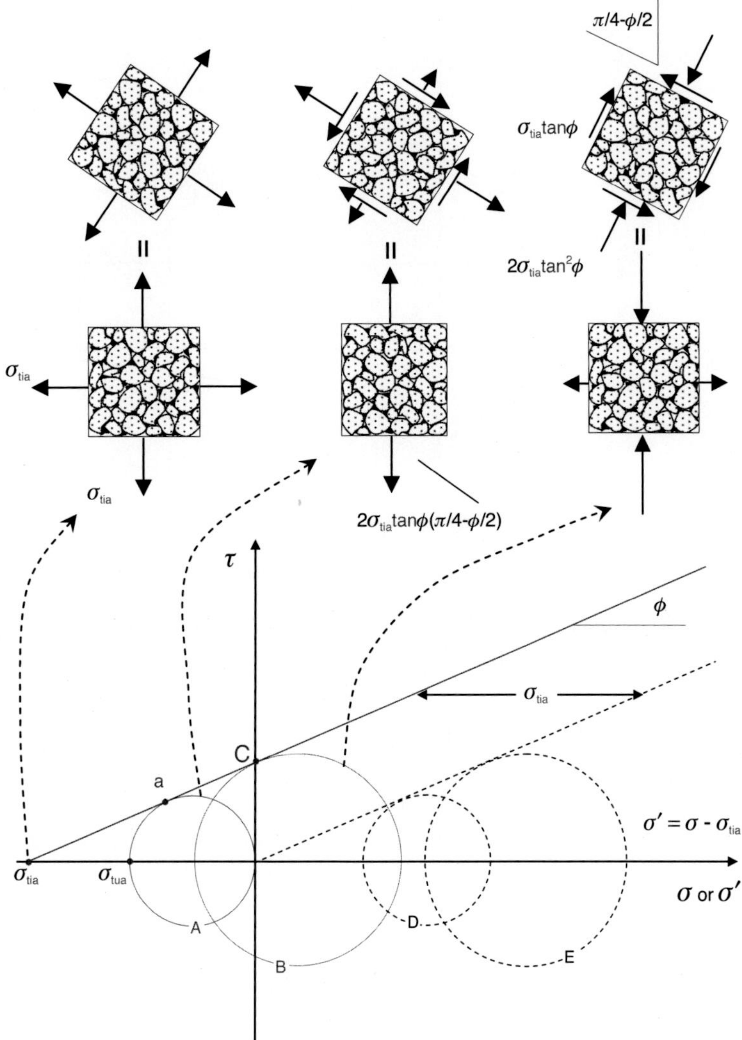

Figure 7.9 Dependence of drained cohesion on isotropic tensile strength (after Lu *et al.*, 2009).

other words, failure occurs not because the applied stress reaches the bonding strength, but because the ratio of shear stress to normal stress, at point *a*, reaches tan ϕ. Thus, the uniaxial tensile strength typically measured in experimental tests is actually a measure of frictional strength resulting from the mobilization of isotropic bonding stress. As depicted in the top middle in Figure 7.9, the magnitude of mobilized shear strength at this state is equal to $\sin\phi\,\tan(\pi/4 - \phi/2)\sigma_{tia}$.

The maximum amount of mobilized shear strength due to the isotropic bonding stress is the intercept of the Mohr–Coulomb criterion with the shear stress axis (point C), or cohesion *c*. Here, the corresponding state of stress is depicted as circle B in Figure 7.9 for the soil element in the top right. As shown, *c* is a shear stress, rather than a cohesive stress

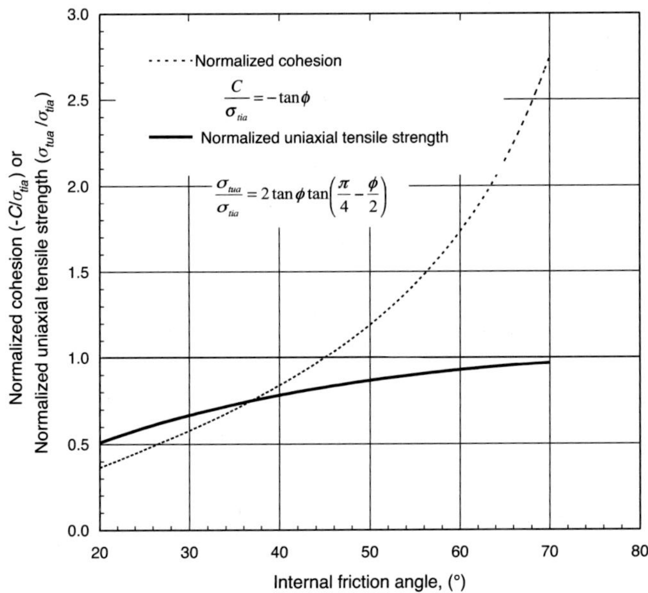

Figure 7.10 Illustration of the ability of uniaxial tensile strength and apparent cohesion to mobilize isotropic tensile stress into shear strength as functions of friction angle (from Lu *et al.*, 2009).

and thus the name "cohesion" is inaccurate and misleading. On the other hand, the isotropic tensile strength is indeed a direct reflection of inter-granular bonding stress, which is closer to the meaning of the word "cohesion."

Mathematical relations among isotropic tensile strength (σ_{tia}), apparent cohesion (c), and uniaxial tensile strength (σ_{tua}) can be established, by considering the geometry shown in Figure 7.9, as

$$\frac{c}{\sigma_{tia}} = -\tan\phi \tag{7.14}$$

$$\frac{c}{\sigma_{tua}} = -\frac{1}{2\tan\left(\frac{\pi}{4}-\frac{\phi}{2}\right)} \tag{7.15}$$

$$\frac{\sigma_{tua}}{\sigma_{tia}} = 2\tan\phi\tan\left(\frac{\pi}{4}-\frac{\phi}{2}\right) \tag{7.16}$$

Figure 7.10 illustrates the above relations. As indicated, the efficiency with which the external stress mobilizes isotropic bonding stress or tensile strength σ_{tia} to shear strength C increases from less than 40% at a friction angle of 20° to over 270% at a friction angle of 70°. At a friction angle of 45°, the mobilized shear strength C (or c) is equal to the isotropic tensile strength. The ability of sand to mobilize isotropic bonding stress to uniaxial tensile strength varies from 51% at a friction angle of 20° to 97% at a friction angle of 70°. This suggests that uniaxial tensile strength, such as that measured in ideal direct tension tests, will be less than or equal to the isotropic tensile strength for Mohr–Coulomb materials.

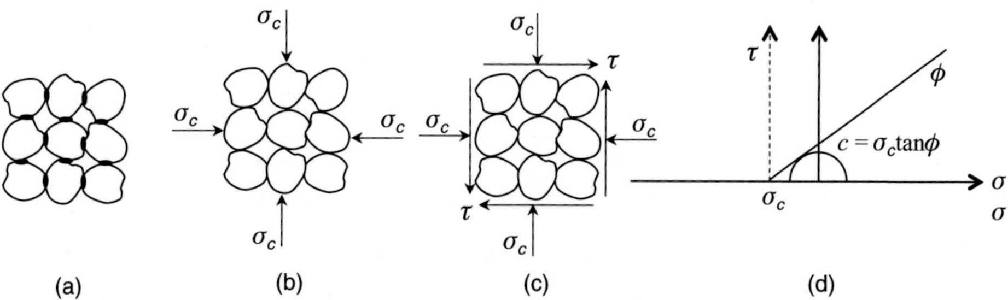

Figure 7.11 Illustration of tensile strength due to cementation.

Circle B (Figure 7.9) touching the Mohr–Coulomb envelope on the shear stress axis at C has some interesting features. First, it can only be attained for the unique pair of principal stresses shown in the top right of Figure 7.9. The maximum principal stress is compressive and the minimum principal stress is tensile. The ratio of the minimum principal stress to the maximum principal stress is equal to the negative coefficient of active earth pressure $\tan^2(\pi/4 - \phi/2)$. Active earth pressure refers to the failure state where a soil's self weight is the sole reason for the failure. Only a single pair of parallel planes exists where zero normal stress occurs at $\pi/4 + \phi/2$ from the plane of the maximum principal stress. However, on the planes orthogonal to this pure shear plane, there is additional compressive normal stress equal to $2\sigma_{tia}\tan^2\phi$.

A generalized effective stress, therefore, can be used to unify cohesion with frictional strength. This is illustrated by the rightward shift of the failure envelope in Figure 7.9. The Mohr–Coulomb criterion can now be rewritten to reflect such generalization:

$$\tau_f = c' + \sigma \tan\phi = \sigma_{tia} \tan\phi + \sigma \tan\phi = (\sigma_{tia} + \sigma)\tan\phi \tag{7.17}$$

As illustrated by the dashed Mohr circles D and E in Figure 7.9, cohesion exhibited by cohesive soils can be treated in an equivalent manner for non-cohesive soils with frictional strength if the above generalized effective stress $(\sigma_{tia} + \sigma)$ is employed.

A detailed analysis and determination of drained cohesion under various drainage conditions will be discussed in Section 7.5.

7.3.2 Cementation cohesion

Cementation in hillslope materials often arises from the precipitation of dissolved minerals and salts. It mostly occurs in the shallow unsaturated environment where evaporation is active. Cementation is often also associated with biological and chemical activities such as colloid transport and formation of iron oxides. Cementation occurs mostly at or around particle contacts (see Figure 7.11), leading to a strengthening of the inter-particle bonds and elevated shear strength of hillslope materials. Tensile strength resulting from cementation varies in magnitude and could be from tens of kPa to hundreds of kPa in soils, substantially increasing the stability of shallow materials on hillslopes (e.g., Sitar and Clough, 1983; Collins and Sitar, 2008). If a mechanically equivalent cohesion, shown in

Figure 7.11b is adopted, cohesion c_c due to a cementation bonding stress of σ_c can be estimated as

$$c_c = \sigma_c \tan\phi \tag{7.18}$$

The Mohr–Coulomb failure criterion considering cementation can be written with an equivalent (effective) stress concept (see Figure 7.11c, d):

$$\tau_f = c_c + \sigma \tan\phi = \sigma_c \tan\phi + \sigma \tan\phi = (\sigma_c + \sigma)\tan\phi \tag{7.19}$$

The maximum cementation cohesion can be experimentally determined by conducting either direct shear or triaxial shear tests on dry materials.

7.3.3 Capillary cohesion

Lu and Likos (2006) defined the isotropic tensile strength under unsaturated conditions as a part of effective stress called "suction stress." Much like positive pore-water pressure in saturated soil, suction stress may be considered as additive with total stress to define an effective stress. This allows unsaturated soils to be considered within the conventional Terzaghi effective stress framework. Because suction stress is generally tensile, it increases the effective stress under unsaturated conditions. According to Lu and Likos (2006), suction stress has four components arising from various physical and physical-chemical mechanisms: van der Waals attractive forces, electric double layer forces, tensile pore-water pressure, and surface tension. Although both van der Waals attractive forces and electric double layer forces are the source of apparent cohesion under saturated conditions, they are highly dependent on saturation. For unsaturated fine-grained materials such as clay, all four components need to be considered over a wide range of saturation. For unsaturated coarse-grained materials such as sand, the latter two – tensile pore-water pressure and surface tension – dominate the generation of suction stress. Under the framework of the suction stress concept (as described in Chapter 6), these stress mechanisms can be unified and the effective stress for both saturated and unsaturated soil can be cast in one unified form:

$$\sigma' = \sigma - u_a - \sigma^s = \sigma - u_a - f(u_a - u_w) = \sigma - u_a - f(S) \tag{7.20}$$

where u_a is the pore air pressure, u_w is the pore-water pressure, S is the degree of saturation, and σ^s is the suction stress. Dependence of suction stress on the equivalent degree of saturation is described in Chapter 6 (Lu and Likos, 2004b; Lu *et al.*, 2010) and for various soils is illustrated in Figure 7.12, and is given by

$$\sigma^s = -\frac{S_e}{\alpha}\left(S_e^{\frac{n}{1-n}} - 1\right)^{\frac{1}{n}} \tag{6.15}$$

Tensile strength is equal to the negative of suction stress. For sandy soils, the peak tensile strength is as large as several kPa and zero for both dry and fully saturated conditions. For silty soils, the peak tensile strength could be as large as several tens of kPa. For clayey soils, tensile strength increases from its minimum at full saturation to its maximum of several

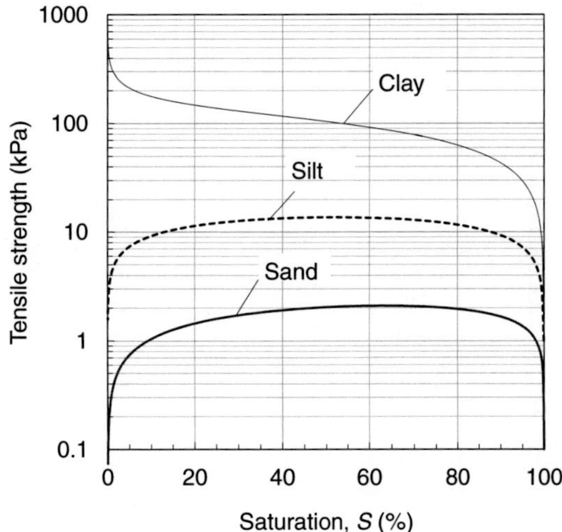

Figure 7.12 Illustration of dependence of tensile strength on saturation for various soils (from Lu *et al.*, 2009).

hundred kPa for dry conditions. Extensive theoretical and experimental treatments of the suction stress characteristic curve for various soils are presented in Chapter 6.

If soil is under compressive or shear stress, suction stress can be mobilized into shear strength or part of apparent cohesion. Suction stress is the isotropic tensile stress illustrated in Figure 7.9. If a linear friction law such as Coulomb's law is applied, the contribution of suction stress to apparent cohesion can be accounted for by Equation (7.14), i.e.,

$$c = -\sigma_{tia} \tan \phi \tag{7.14}$$

7.4 Shear strength due to plant roots

7.4.1 Role of root reinforcement in hillslope stability

The roots of trees, shrubs, grass, and other plants play a significant role in hillslope hydrology and stability. Plant roots contribute to slope hydrology by storing and releasing water in evapo-transpiration processes, and to slope stability by providing tensile strength to hillslope materials. Living plants can contribute as much as tens of kPa to the tensile strength of near surface soils. In hillslope environments, this tensile strength acts to resist tensile cracking or can be converted to shear strength that resists shear stresses. Therefore, it is important to recognize that the tensile strength of plant roots is the common origin of both the tensile and shear strength contribution of roots to the stability of soils. The conversion of the tensile strength of roots to shear strength depends on a number of factors including the type of soil (friction angle) and geomorphologic and climatic setting (slope angle, stratigraphy,

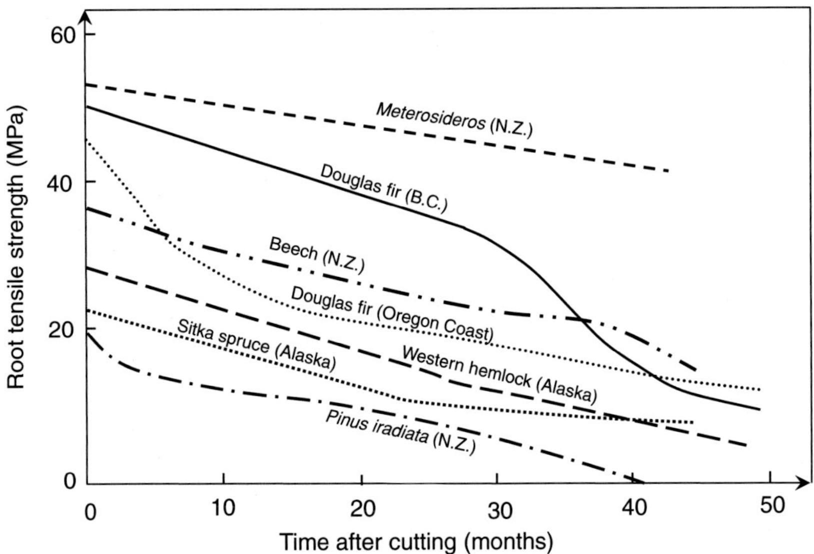

Figure 7.13 Illustration of tensile strength of plant roots as a function of time since cutting (after O'Loughlin and Ziemer, 1982).

water availability, and morphology), and plant species. Quantitative models used to assess stability of vegetated hillslopes need to take account of the tensile strength of plant roots. When trees are harvested, their roots will gradually lose tensile strength over time due to biological decay, as illustrated in Figure 7.13. The rate of decay is primarily dependent on climatic setting and species. This reduction in strength, in conjunction with rainfall, is often sufficient to trigger landslides, and can alter both the susceptibility to landsliding and sediment transport of areas where vegetation is removed (e.g., Montgomery *et al.*, 2000).

7.4.2 Shear strength of rooted soils

Soil strength that results from plant roots is dynamic and can play a significant role in the stability of surficial materials on hillslopes. In addition to the reduction of root strength due to root decay, root strength increases over time as plant roots reestablish and grow (Figure 7.14). Thus, the net root strength decreases following pant removal, then increases after trees or other vegetation are planted, resulting in a period that is susceptible to landslides (Figure 7.14). The principal mechanism by which plant roots reinforce soil is the tensile strength provided by the root network. In hillslope environments, tensile stress in the tree root network can develop under tension stress conditions, but is more likely to develop under shear stress conditions. Understanding the mobilization of tensile stress and its conversion to shear strength of hillslope materials is the key to the development of a quantitative framework that includes the contribution to shear strength from plant roots. As illustrated in Figure 7.15a, different plants have different root structures, leading to a complex spatial and temporal distribution of the root network. The spatial distribution of the root network

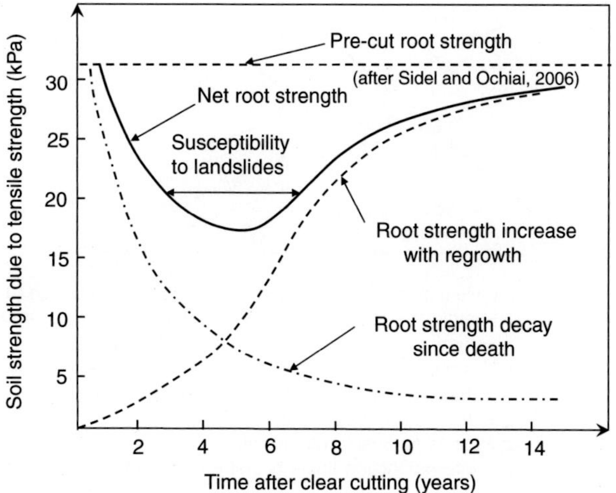

Figure 7.14 Illustration of shear strength (tensile strength) of vegetated soil as a function of time since cutting.

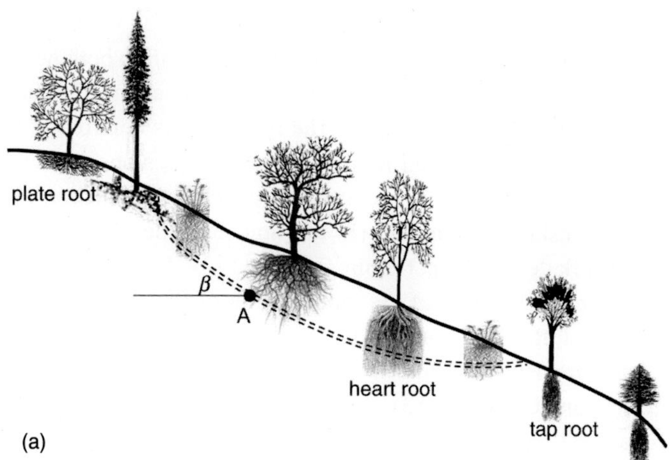

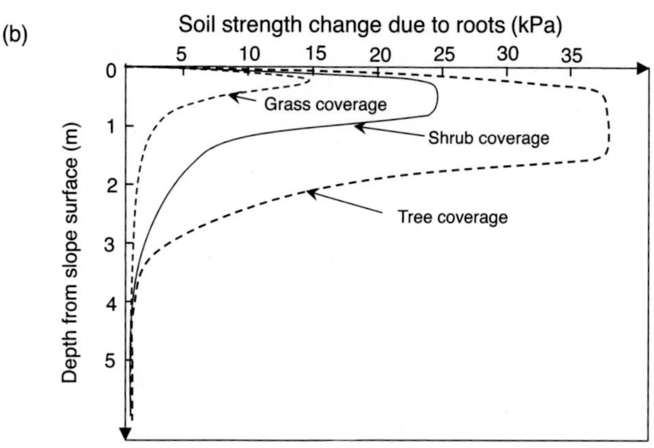

Figure 7.15 (a) Illustration of various root architectures in the near-surface hillslope environment. (b) Variation of soil shear strength in the vertical direction for different vegetation coverage.

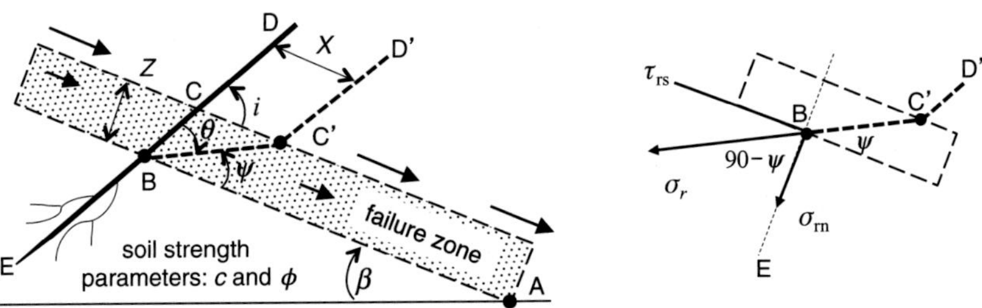

i = initial root orientation with respect to failure surface (0 ~ 180°)
θ = angle of shear distortion with respect to i at root rupture (0 ~ 90°)
ψ = rupture root orientation with respect to failure surface = $i - \theta$ (0 ~ 180°)
β = angle of failure surface at point A with respect to horizontal at failure (0 ~ 90°)
X = maximum displacement of failure zone along failure surface
Z = thickness of failure zone
c = cohesion of soil without roots
ϕ = internal friction angle of soil without roots

Figure 7.16 Illustration of the role of an individual root (DCBE for non-deformed and D′C′BE for deformed) in increasing the shear strength of soil along a failure surface near point A in Figure 7.15a.

leads to shear strength variation in hillslope soil in both horizontal and vertical directions. The variation of soil shear strength in the vertical direction for different vegetation coverage is illustrated in Figure 7.15b. The main factors controlling the shear strength variation in rooted soils are (1) soil friction properties, (2) slope and subsurface geometries, (3) plant root sizes and network architecture, which are a function of plant species and health, (4) root tensile strength, and (5) the time-dependent variations of root tensile strength and network development (Figure 7.14).

The contribution of the mobilization of the tensile strength of roots to shear strength is illustrated in Figure 7.16. The theoretical basis of most of the current theories and measurement methods stems from the work of Waldron (1977), Wu *et al.* (1979), and Gray and Ohashi (1983). Rather than seeking constitutive relations for composite soil and root materials, the theory was established for a specific case of root and slope conditions or shear failure plane, generalized in Figure 7.16. For a single root-soil system at the limit-equilibrium state, both root and soil fail simultaneously. The projection of the fully mobilized root tensile strength σ_r on the failure plane is the shear resistance τ_{rs}:

$$\tau_{rs} = \sigma_r \sin(90^\circ - \psi) \tag{7.21}$$

An additional shear resistance is provided by the normal compressive stress σ_{rn} ($\sigma_{rs} = \sigma_r \cos(90^\circ - \psi)$) through the mobilization of the internal friction angle ϕ:

$$\tau_{rs} = \sigma_r \cos(90^\circ - \psi) \tan\phi \tag{7.22}$$

The total shear resistance $\Delta\tau_r$ arising from the tensile strength of a single root σ_r in terms of shear stress along the local shear failure plane is (e.g., Gray and Ohashi, 1983):

$$\Delta\tau_r = [\sin(90^\circ - \psi) + \cos(90^\circ - \psi)\tan\phi]\,\sigma_r = R_r\sigma_r \tag{7.23}$$

where R_r can be called the "root shear strength conversion factor." In principle, the above analysis should be applied only to a prescribed failure surface penetrated by a single root. However, in all the work along this line, namely, Waldron (1977), Wu *et al.* (1979), and Gray and Ohashi (1983), Equation (7.23) was used to modify the Mohr–Coulomb shear strength criterion without complete experimental validation or independent theoretical verification. Specifically, the contribution to shear strength captured by Equation (7.23) was directly considered as an increase in shear strength due to root tensile strength in a modified Mohr–Coulomb failure criterion, i.e.,

$$\tau_f = c + \Delta\tau_r + \sigma_n \tan\phi \tag{7.24}$$

where c (kPa) is soil cohesion, σ_n is the normal stress (kPa) on the failure plane without roots, and ϕ is the internal friction angle (°) of soil. There are two theoretical and practical challenges when using Equation (7.23) as the basis for root shear strength assessments: (1) quantifying R_r for roots embedded in soils with certain spatial distribution and orientation, and (2) quantifying the mobilized tensile strength with certain root architectures. Several assumptions are made in order to simplify Equation (7.23): (1) roots are perpendicular to the failure plane ($\theta = 90°-\psi$), and (2) the angle of shear distortion θ is in the range of ~48–72°. As shown in Figure 7.16, by the law of sines, inter-dependence among X, Z, i, θ, and ψ can be established:

$$\frac{x}{z} = \frac{\sin\theta}{\sin\psi \sin i} \tag{7.25}$$

The first assumption leads to $i = 90°$ or $\theta = 90° - \psi$, and Equation (7.25) becomes

$$\frac{x}{z} = \frac{\sin\theta}{\sin(90° - \theta)\sin 90°} = \tan\theta \tag{7.26}$$

Substituting $\psi = 90° - \theta$ into Equation (7.23) yields Wu *et al.*'s (1979) equation for shear strength increase:

$$\Delta\tau_r = [\sin\theta + \cos\theta\tan\phi]\,\sigma_r = R_r\sigma_r \tag{7.27}$$

The laboratory direct shear experiments conducted by Jewell (1980) and Gray and Ohashi (1983) on reinforced sand show that soil-fiber composites with randomly orientated fibers have the same shear strength compared with similar composites with vertically orientated fibers. For other orientations, shear strength oscillates with the initial root orientation and reaches maximum when fibers are embedded at $i = 60°$. Shear strengths for this orientation could be more than 100% greater than those for roots oriented at $i = 90°$. Shear strength decreases to a minimum of 20% when fibers are embedded at $i = 120°$. Whereas these experiments confirm the oscillating nature of shear strength variation with the orientation of the reinforced materials and the soil's internal friction angle as predicted by Equation (7.27), they do not indicate the validity of Equation (7.27) in composite materials such as soils embedded with roots. This is because Equation (7.27) was established (Waldron, 1977; Wu *et al.*, 1979) based on the assumption, and supported by field observations, that root tensile strength is fully mobilized, i.e., roots break when the root–soil composite materials fail. This is contrary to the lab experiments conducted by Jewell (1980) and Gray and Ohashi (1983), where embedded fibers were pulled out or intact when failure occurred.

The second assumption states that the angle of shear distortion θ is in the range of $\sim$48–72°. Confirmation of this range for the shear distortion angle θ has not been established because of the difficulty of field measurements. Sensitivity analysis of R_r defined by Equation (7.27) performed by varying the distortion angle θ between 48° and 72° and the soil internal friction angle between 25° and 40° indicates that an average R_r has an approximate value of 1.2, leading to an approximation of shear strength increase described as:

$$\Delta\tau_r \approx 1.2\sigma_r \tag{7.28}$$

Sensitivity analysis of a slightly more general Equation (7.23) was conducted by Pollen-Bankhead and Simon (2010) by examining possible values of initial root orientation i, angle of shear distortion θ, and friction angle ϕ. Results from both deterministic and probabilistic algorithms indicates that $R_r = 1.2$ is not representative and likely overestimates the shear strength increase that results from root tensile strength. To date, this assumption still remains largely untested both experimentally and theoretically.

Recent studies (e.g., Cohen *et al.*, 2011) showed that the above conceptualization blurs the distinction between the failure criteria (i.e., the Mohr–Coulomb rule) and the constitutive model of a rooted soil. Such a constitutive relation is needed to predict the strength increase of soils due to the presence of roots during deformation. To illustrate this point, consider the *in-situ* shear test by Fannin *et al.* (2005) of a cohesionless colluvium, where the shear stress peaks twice, once due to soil mobilized at its peak shear strength and a second time due to the tensile strength mobilization of roots present in the specimen. Note that these peaks occur at different displacements, and thus simply adding root shear resistance (τ_r) to the shear strength of the matrix τ would grossly overestimate the resistance of the root–soil composite to shear deformation and hence its resistance to mobilization in a landslide.

7.4.3 Tensile strength of roots

A common assumption used to further simplify Equation (7.23) for the contribution of roots to soil strength is the postulate that all roots in the unit volume fail at the same time. This assumption leads to an expression for shear strength increase in terms of an average root tensile strength from all roots in a unit area A:

$$\Delta\tau_r = 1.2\sum_{i=1}^{n}\sigma_{ri}\frac{A_{ri}}{A} \tag{7.29}$$

Recognition of the potential error in the simultaneous failure assumption has led to an alternate model that describes the progressive failure of roots along the failure plane. This model, called a "fiber bundle model" or FBM, was recently developed (Pollen and Simon, 2005; Thomas and Pollen, 2010) and incorporates several controlling factors in root tensile strength: the tensile strength dependence on root diameter, root architecture, and force redistribution during progressive root failure. These studies indicate that root

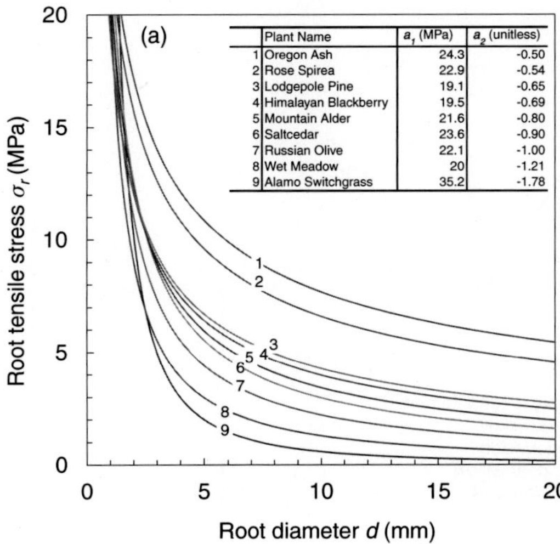

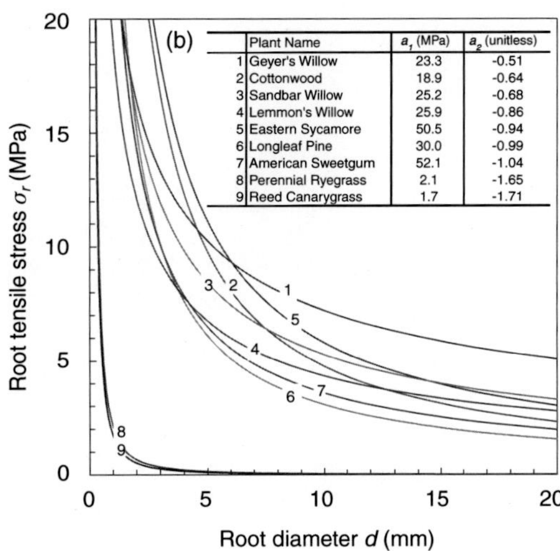

Figure 7.17 Illustration of tensile strength of individual roots as a function of root diameter for different species of plants (parameters from Pollen and Simon, 2005): (a) group 1 and (b) group 2 (grouping here for illustration purpose).

tensile strength defined by Equation (7.29) could overestimate the actual root strength by ~50–60%. The experimental confirmation of this result remains to be established.

Because the tensile strength of individual roots can be measured *in situ* or in the laboratory, many studies have quantified the tensile strength of different plant roots and the variation in tensile strength with root size. In general, the tensile strength of individual roots σ_r decreases non-linearly with increasing root diameter d. This is somewhat counterintuitive and the exact physical reason is yet to be established. A plausible explanation is that smaller roots have larger specific surface areas and root tissue near the surface is stronger than that of the interior of the root. A model for describing such dependency has been proposed and used by several investigators (Waldron and Dakessian, 1981; Pollen and Simon, 2005):

$$\sigma_r = a_1(d)^{a_2} \tag{7.30}$$

where a_1 (MPa m$^{-a_2}$) and a_2 (unitless) are fitting parameters that can be deduced from a species-specific investigation. In Equation (7.30), the root diameter d is in mm. Figure 7.17 illustrates the dependency of individual root tensile strength on diameter for different plant species.

7.4.4 Spatial and temporal variation of root strength

One important character of root strength is its dynamic nature, particularly for forested hillslopes. Root strength is provided by the live biomass of root, leading to a proportional relationship between root mass and root strength in the near-surface soil environment. When living trees die or are harvested, their roots start to decay. Over time, on the scale of several years to decades, the roots will lose their strength. On the other hand, if new

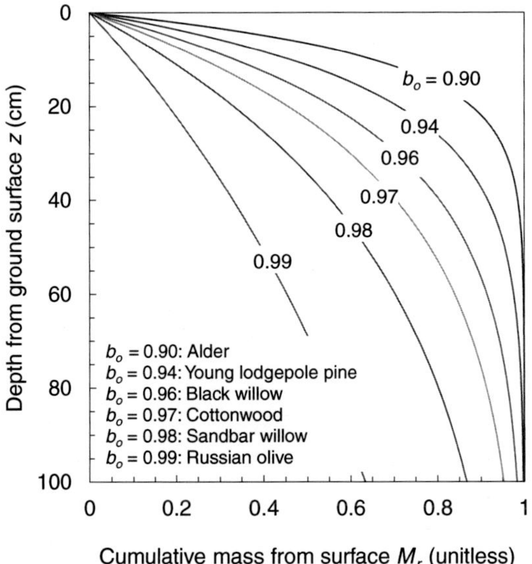

Figure 7.18 Cumulative root mass fraction as a function of depth z from the ground surface (data from Pollen-Bankhead and Simon, 2009).

trees are planted, root strength will gradually increase over time (e.g., Figure 7.14). In several decades, depending on the tree species, maximum root strength will be achieved. An empirical equation capturing the previously mentioned root mass and shear strength relations was proposed by Zeimer (1981), based on shear strength tests of pine roots in sandy soils in coastal northern California:

$$\Delta\tau_r = b_1 + b_2 m_r \tag{7.31}$$

where shear strength is in kPa, b_1, and b_2 are empirical fitting parameters and are 3.13, and 3.31, respectively, for the tested root-reinforced soils, and m_r is the root mass per unit volume of the reinforced soil in kg/m^3.

Root mass distribution in the shallow subsurface is complex and depends on root architecture (Figure 7.15a) and hillslope environment. In general, it is greatest near the ground surface and decays with increasing depth. Jackson *et al.* (1996) suggested that the exponential decrease model proposed by Gale and Grigal (1987) well represents the cumulative mass fraction (M_r)–depth (z) distribution:

$$M_r = 1 - b_o^z \tag{7.32}$$

where b_o is a fitting parameter reflecting the cumulative rate of root mass with depth. As illustrated in Figure 7.18, low b_o values indicate more root mass near the ground surface. As the depth increases, the total cumulative mass fraction M_r approaches unity.

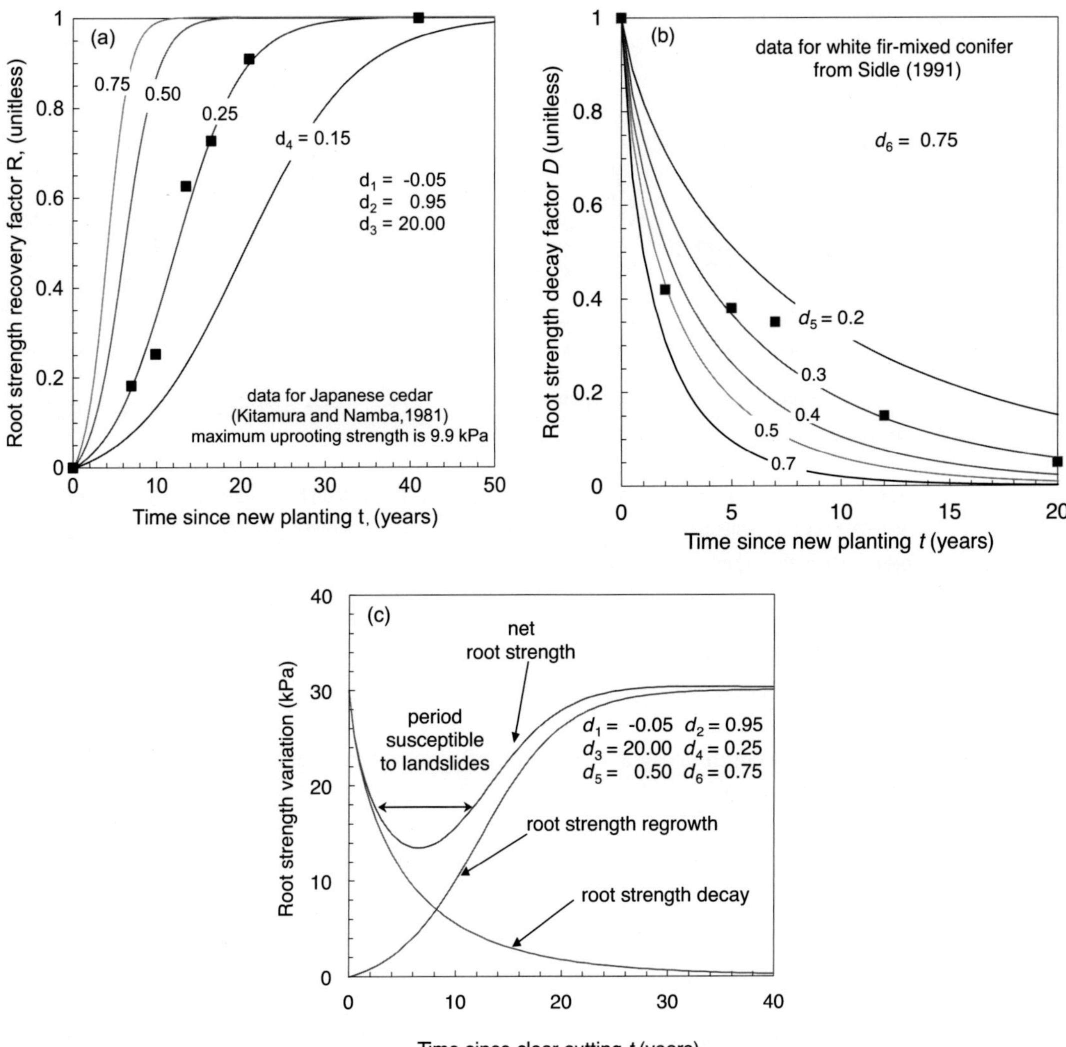

Figure 7.19 Illustration of root strength as function of time: (a) root strength growth, (b) root strength decay, and (c) root strength variation for a hypothetical root.

Root strength growth as a function of time can also be modeled using a sigmoidal function (after Sidle, 1992):

$$\Delta\tau_r = \tau_{r\,\max}\left[d_1 + \left(d_2 + d_3 e^{-d_4 t}\right)^{-1}\right] \tag{7.33}$$

where $\tau_{r\max}$ is the maximum root strength when the root system reaches its full maturity, t is time, and d_1, d_2, d_3, and d_4 are fitting parameters. The time-dependent shear strength increase described by Equation (7.33) is illustrated in Figure 7.19a. Sidle (1992) also offered a useful and uniform procedure to determine the four fitting parameters from field data.

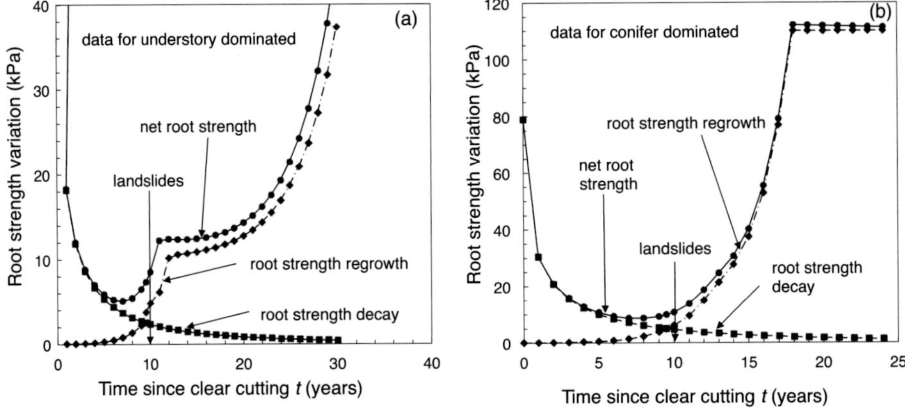

Figure 7.20 Illustration of root strength variations for two sites that were clear-cut logged in 1986 and yielded landslides in 1996: (a) RC#1 site where vegetation was dominated by understory plants, and (b) RC#2 site where vegetation was dominated by coniferous and hardwood trees (Schmidt *et al.*, 2001). Landslides occurred at both sites 2–3 years after net root strength reached the minimum.

It involves identifying the inflection point (t_i, $\Delta\tau_i$) of the sigmoidal curve, the point for twice the inflection time ($2t_i$, $\Delta\tau$), and the initial (minimum) and ultimate (maximum) root strengths.

The root strength decay as a function of time can be modeled by an exponential decay (after Sidle, 1992):

$$\Delta\tau_r = \tau_{r\,\max} e^{-d_5 t^{d_6}} \tag{7.34}$$

where $\tau_{r\max}$ is the maximum root strength before tree harvesting, and d_5 and d_6 are fitting parameters. The time-dependent shear strength decrease described by Equation (7.34) is illustrated in Figure 7.19b. The net change in root strength as a function of time is illustrated in Figure 7.19c where a vulnerable period for landsliding after cutting can be identified.

As a field example, the multi-year variation of root strength and its impact on hillslope stability at two forest harvest sites in the Elliot State Forest in Oregon is shown in Figure 7.20 (from Schmidt *et al.*, 2001). Both sites have similar topography and were clear-cut harvested in 1986 but have distinct vegetation regrowth patterns. One site (RC#1 and root strength data shown in Figure 7.20a) is dominated by the incursion of understory vegetation with little establishment of coniferous vegetation, whereas the adjacent site (RC#2) was vegetated by abundant conifers and hardwoods (root strength data shown in Figure 7.20b).

Root strength at the understory vegetation RC#1 site reaches a maximum of 12 kPa after 12 years. Root strength data were measured in the first 10 years. After that, theoretical projections were made based on the abundance and species composition of vegetation present at the time the root strength measurements were obtained. The two sites show very different root strength recovery patterns. Root strength remains less than 15 kPa for 18 years at RC#1, whereas full strength of 110 kPa at RC#2 was achieved in 18 years. Nonetheless, the growth of root strength is similar in the first 10 years (less than 10 kPa) after timber harvest and landslides occurred in 1996 (10 years after the clear-cut harvest) at both sites.

7.5 Shear strength under various drainage conditions

7.5.1 Shear strength of saturated soils

Both the shear and tensile strengths of soil are mainly controlled by their composition, stress and pore-water pressure history, and the prevailing loading conditions. The presence of water or other fluids in soil marks a fundamental difference in its mechanical behavior compared with other non-porous materials. One unified way to describe the strength behavior of hillslope materials is to use the effective stress principle under variably saturated conditions. The basic postulate is that shear strength of soils can only be provided by mobilized shear stress, and shear stress can only exist in the soil skeleton. Effective stress, in a general sense, is the stress in the soil skeleton or solid framework. For example, for a slope subjected to rainfall-induced seepage under saturated conditions, the shear strength parameters for the same soil, such as cohesion and friction angle, would be different depending on the consideration of pore-water pressure. If a hillslope is well drained, such that there is very little change in the water table or pore-water pressure as a result of external loading, the contribution of pore-water pressure to changes in shear strength can be ignored. For this case the total stress distribution, together with shear strength parameters is sufficient for a stability analysis of the slope. The distribution of effective stress does not change significantly under such circumstances. This represents a limiting case, which is called the "drained" condition. Shear strength tested under such conditions accurately represents the ability of the soil skeleton to resist shear motion. On the other hand, if a hillslope is poorly drained, such that pore-water pressure or the water table change rapidly upon loading, pore-water pressure becomes an important factor in counterbalancing the total stress. This may result in significant changes in effective stress. The limiting case where there is no drainage in the system is called the "undrained" condition. Shear strength parameters identified under such circumstances represent the lower limit of the shear strength of hillslope materials. Because the conditions of hillslopes likely lie somewhere between "drained" and "undrained," effective stress, together with effective shear strength parameters, should be used in slope stability analyses.

While the effective stress principle unifies both drained and undrained shear strength behavior for soils, determining the effective shear strength parameters or drained parameters under saturated conditions may not be practically convenient or necessary. Drained tests typically require explicit knowledge of pore-water pressure (either monitored or controlled) in the soil specimen. Such tests are typically costly and time-consuming. On the other hand, undrained tests often require little or no control or measurement of pore-water pressure, and most importantly, take very little time to conduct. Therefore, for field conditions that can be well represented by undrained conditions, undrained shear strength tests, although conservative in estimating shear resistance, could be preferable.

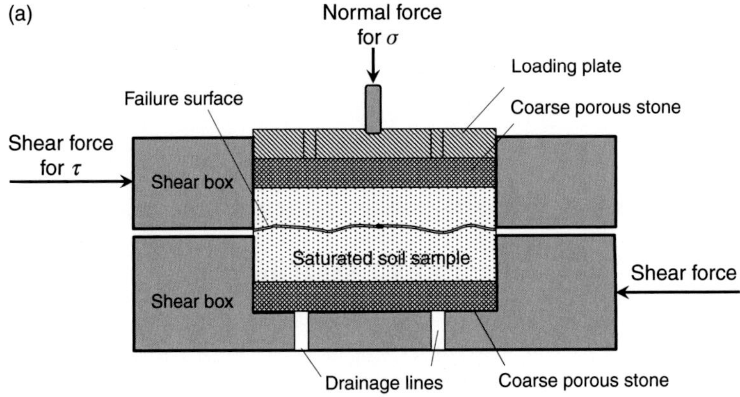

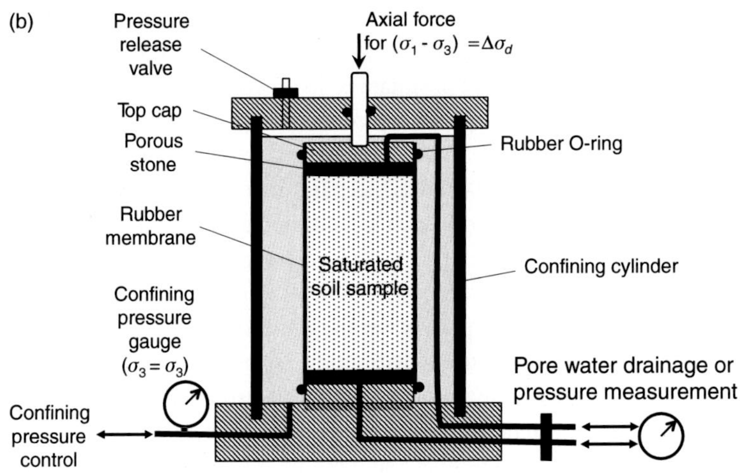

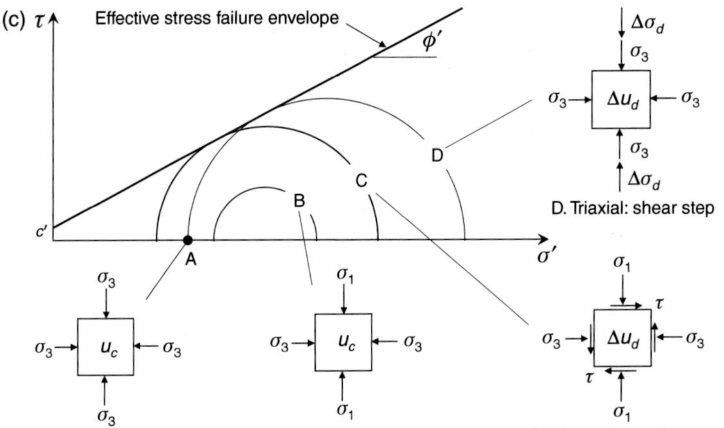

Figure 7.21 Illustration of two conventional shear strength tests: (a) direct shear, (b) triaxial, and (c) two testing steps for stress and drainage controls where the triaxial shear test follows steps A and D and the direct shear test, steps B and C.

Shear strength tests for saturated soils have been standardized and are available commercially around the world. There are principally two types of shear strength tests for obtaining both drained and undrained shear strength parameters: direct shear and triaxial tests, as illustrated in Figure 7.21. In conventional direct shear tests, both normal compressive stress and shear stress are applied independently (Figure 7.21a), while pore-water pressure is measured or maintained at levels close to ambient atmospheric pressure. In the conventional triaxial tests (Figure 7.21b), normal compressive principal stresses along vertical (σ_1) and horizontal (σ_3) directions are applied independently while pore-water pressure (u) is controlled and measured. The shear stress in the triaxial setting is realized through the difference in the applied principal stresses.

Both tests follow two steps in applying stresses in order to mimic stress history and current stresses and pore-water pressure representative of field conditions, as shown in Figure 7.21c. The first step applies a stress level comparable to *in-situ* conditions using a vertical stress in direct shear, or all-around stress in triaxial tests. At this step, the vertical stress step in direct shear tests allows vertical displacement but no horizontal displacement. Such a condition is called the "K_o condition." Triaxial tests allow displacement in all directions in response to the isotropic compressive total stress applied outside of an impermeable cylindrical-shaped membrane. During the compressive loading step, if the pore-water pressure is maintained or controlled, water already in the sample will drain from soil specimens and specimens are considered to be "consolidated." This process can take substantial time to reach steady state, and depends on several factors such as specimen length, permeability, and compressibility of the material. If no drainage is permitted during this step, it is called "undrained." The undrained testing process is usually performed immediately.

The second step applies a shear stress that typically increases at a defined strain rate until shear failure occurs. In direct shear tests, the shear stress is directly applied along a predetermined failure surface (typically in the horizontal direction), whereas in triaxial tests, a vertical stress, over and above the all-around stress, is applied. The geometry of the failure surface that develops in such a manner is not known a-priori and must be determined after failure occurs. The criterion for failure is the peak load that can be determined from the stress–strain data collected from the shear strength tests (e.g., Figure 7.4a). Pore-water pressure during the second step also can be controlled and this procedure is called "drained." If pore-water pressure is measured with no drainage allowed then this procedure is called "undrained." Both cases represent limiting cases of field conditions.

Depending how the previously mentioned two steps are implemented, shear strength tests can be carried out for different conditions: (a) "consolidated-drained" or CD conditions, (b) "consolidated-undrained" or CU conditions, (c) "unconsolidated-undrained" or UU conditions, and (d) "unconfined-compression" or UC conditions. Shear strength criteria and parameters under each of these conditions are defined differently and described below.

7.5.2 Consolidated-drained conditions

Each hillslope material has a unique composition and history of loading and formation that leads to a unique set of shear strength parameters. Nonetheless, some general patterns,

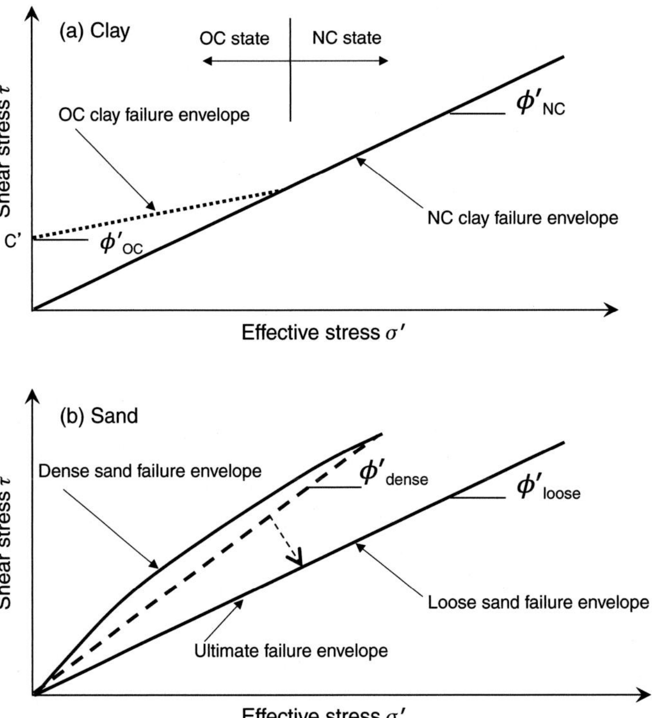

Figure 7.22 Illustration of the failure envelope under consolidated-drained conditions: (a) clay and (b) sand.

drawn from previous investigations, can be instrumental in understanding the shear strength behavior of hillslope materials. CD tests will provide the most general conditions for shear strength as they quantify the shear strength of the soil skeleton or effective stress behavior in shear resistance. They are the most complicated tests, as both steps demand the measurement of pore-water pressure, which must be allowed to fully dissipate between each load increment prior to failure.

Parallel to the stress-displacement behavior described in Section 7.1, the pattern of shear strength behavior in terms of the shear failure envelope shown in Figure 7.22 can be drawn along the lines of the OC state, NC state (defined in Section 7.1), or loose sand and dense sand. Under saturated CD conditions, total stress (i.e., applied external stresses) is equal to effective stress (soil skeleton stress) or modified by a constant offset. For these conditions the shear strength envelope and its parameters can be defined under either total or effective stress, as shown in Figure 7.22. Figure 7.22a highlights the shear strength behavior of cohesive soil such as silt and clay. For silt and clay, if the state of stress at failure is within the soil NC state, the friction angle so identified is called ϕ'_{NC}. The prime symbol in soil mechanics generally denotes that the parameters were determined for drained conditions or for effective stress. In classical soil mechanics theory, cohesion of all soils, including clay and silt under the NC state, is considered negligible or zero for traditional design purposes. This requires that the failure envelope pass through zero on the shear stress axis, as shown

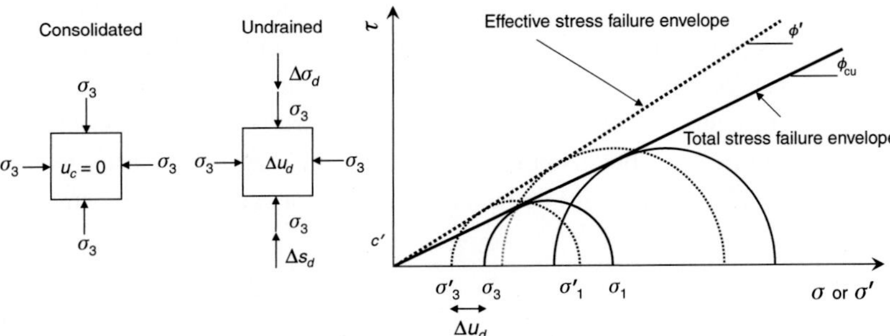

Figure 7.23 Illustration of failure envelopes in terms of total stress and effective stress under consolidated-undrained conditions. Shear strength (parameter ϕ_{CU}) for total stress in lieu of effective stress (parameter ϕ') is often used in practice as it avoids the direct measurement of pore pressure.

in Figure 7.22a. Such neglect of cohesion may not be appropriate for the analysis of natural hillslopes. For failure of silt and clay soils under the OC state of stress, the friction angle thus identified is called ϕ'_{OC}. Interception of the failure envelope for clay and silt under saturated-drained conditions is called effective drained cohesion c', and is non-zero, as shown in Figure 7.22a. For OC soils, the cohesion c' could be as great as several tens of kPa.

Shear strength behavior for sand is illustrated in Figure 7.22b. For loose sand, one envelope usually covers the full stress range. Although apparent cohesion arising from the inter-locking mechanism described in Section 7.2 can be as much as 10 kPa in some sands, this is generally ignored. However, for dense sand, the failure envelope is typically non-linear or the friction angle is not a constant but rather dependent on the normal stress. The friction angle typically is greater at low normal stress, but asymptotically approaches a lower value at high normal stress. For example, dense sand with angular particle shapes has a friction angle ϕ'_{dense} as high as 70° under normal stress on the order of a few kPa, which decreases to a constant value of 40° as the normal stress increases to the range of several hundred kPa. After failure occurs, dense sand still has considerable shear strength that can be characterized by no or little inter-locking apparent cohesion but with a friction angle comparable to loose sand ϕ'_{loose}, as illustrated in Figure 7.22b.

7.5.3 Consolidated-undrained conditions

CU tests are typically conducted in triaxial testing equipment rather than direct shear equipment because pore-water pressure is more easily controlled in a triaxial cell. The first of the two testing steps applies a total stress under drained or controlled pore-water pressure conditions, identical to CD tests. In the second step, drainage is not allowed and shear stress is applied at a defined rate until the soil fails while monitoring the response pore pressure Δu_d, as shown in Figure 7.23. Because the second step is performed without a time-dependent consolidation process, CU tests can generally be performed rapidly. The failure envelope can be obtained from several such tests with different all-around stress (triaxial) in the drained or consolidation step, and results can be plotted in both total and effective stress vs. shear stress space. Therefore, shear strength parameters, such as drained (ϕ' from the effective stress envelope) and undrained (ϕ_{CU} from the total stress envelope) can be deduced. Pore-water pressure change due to the application of the deviator stress (difference in principal stresses) can be used not only for assessing effective stresses at the failure state, but also the state of stress (i.e., OC or NC) prior to shearing.

Skempton (1954) defined a parameter for such deviatoric stress as

$$B = \frac{\Delta u_d}{\Delta \sigma_d} \tag{7.35}$$

Typically, for heavily OC soils, B values that are less than -1.0 indicate a negative pore pressure in response to deviatoric stress, B values between -1.0 and 0 indicate light OC, and B values greater than 0 indicate positive pore pressure development upon shearing in NC soils.

For triaxial tests, the Mohr–Coulomb failure criterion becomes

$$\sigma_{1f} = \sigma_{3f} \tan^2\left(\frac{\pi}{4} + \frac{\phi_{\mathrm{CU}}}{2}\right) + c \tan\left(\frac{\pi}{4} + \frac{\phi_{\mathrm{CU}}}{2}\right) \tag{7.36}$$

where c and ϕ_{CU} are cohesion and friction angle under consolidated-undrained conditions, respectively. Equation (7.36) is the basis for interpreting CU test results to obtain shear strength parameters c and ϕ_{CU}.

In order to obtain effective or drained shear strength parameters c' and ϕ', the Mohr–Coulomb criterion in terms of the principal effective stresses is

$$\sigma'_{1f} = \sigma'_{3f} \tan^2\left(\frac{\pi}{4} + \frac{\phi'}{2}\right) + c' \tan\left(\frac{\pi}{4} + \frac{\phi'}{2}\right) \tag{7.37}$$

With the knowledge of the pore-water pressure parameter B determined from Equation (7.35), Equation (7.37) can be written in terms of total stresses yet still with the effective shear strength parameters:

$$\sigma_{1f} - B\Delta u_d = \left(\sigma_{3f} - B\Delta u_d\right) \tan^2\left(\frac{\pi}{4} + \frac{\phi'}{2}\right) + c' \tan\left(\frac{\pi}{4} + \frac{\phi'}{2}\right) \tag{7.38}$$

In foundation design and geotechnical analysis, the drained shear strength parameters are often used as the design basis for assessing the long-term performance of the systems. For assessment of short-term performance, such as for temporary excavations, the undrained shear strength parameters are often used. In saturated hillslope environments, the stability of shallow poorly drained soils after heavy rainfall could be better represented by undrained conditions, whereas the deep stability of slopes where the water table rises in response to large-scale, long-term precipitation may be better described by drained conditions.

7.5.4 Unconsolidated-undrained conditions

Shear strength behavior under saturated poorly drained conditions can be quickly assessed by UU tests. The two main advantages of such tests are the lack of need to measure or monitor pore-water pressure and the time-consuming consolidation process is not necessary.

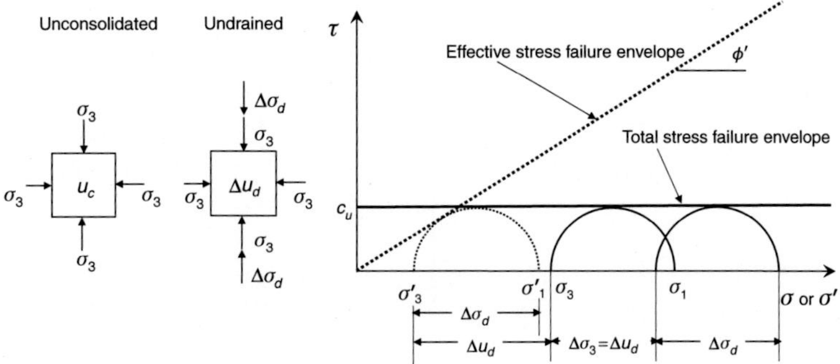

Figure 7.24 Illustration of failure envelope under unconsolidated-undrained conditions. Shear strength (parameter c_u) for total stress in lieu of effective stress (parameter ϕ') is often used in practice as it avoids the direct measurement of pore pressure.

Figure 7.24 depicts the two steps of loading and the stress state at the corresponding failure state for UU tests.

At the failure state, the total stress representation results in a failure envelope that is virtually flat with the shear strength parameters $\phi_u = 0$ and c_u, with c_u being the radius of the Mohr circles (solid circles). The salient feature of the total stress representation at failure is a constant radius of the circles, which are invariant to applied all-around stress σ_3 in the first step. This allows UU shear strength tests to be characterized with one parameter c_u. For consolidated clay, the shear strength c_u could be on the order of several tens of kPa. One way to unify all the Mohr circles under different total stresses is to apply the effective stress principle, shown here as circles with dashed curves in the effective stress vs. shear stress space (Figure 7.23). If the pore-water pressure is measured, experimental evidence indicates that all the circles (solid) would convert to one circle (dash in Figure 7.24). Pore-water pressure increase u_c in the first unconsolidated step is due to isotropic total stress σ_3 and can be defined by the Skempton parameter A:

$$A = \frac{u_c}{\sigma_3} \tag{7.39}$$

Parameter A varies between 0 and 1.00 and is an indication of whether the soil specimen is fully saturated or not. That is, for full saturation ($A = 1$), the confining stress (σ_3) should be equal to the measured pore pressure. The pore-water pressure change at the end of the second step when the soil specimen fails is

$$u = A\sigma_3 + B\Delta\sigma_d \tag{7.40}$$

The Mohr–Coulomb failure criterion in terms of effective stress at failure can be described by

$$\sigma'_{1f} = \sigma'_{3f} \tan^2\left(\frac{\pi}{4} + \frac{\phi'}{2}\right) + c' \tan\left(\frac{\pi}{4} + \frac{\phi'}{2}\right) \tag{7.37}$$

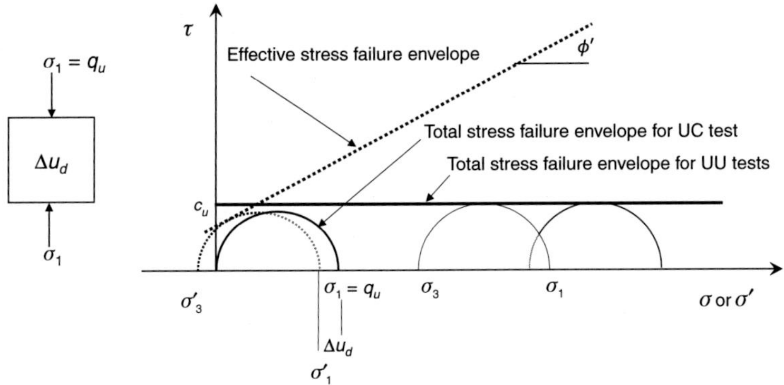

Figure 7.25 Illustration of failure envelope under unconfined-compression conditions. Shear strength (parameter $c_u = q_u/2$) for total stress in lieu of effective stress (parameter ϕ') is often used in practice as it avoids the direct measurement of pore pressure.

With the knowledge of the pore-water pressure parameters A and B determined from Equations (7.35) and (7.39), Equation (7.37) can be written in terms of total stresses:

$$\sigma_{1f} - A\sigma_3 - B\Delta u_d = \left((1 - A)\sigma_{3f} - B\Delta u_d\right)\tan^2\left(\frac{\pi}{4} + \frac{\phi'}{2}\right) + c'\tan\left(\frac{\pi}{4} + \frac{\phi'}{2}\right) \tag{7.41}$$

Given the failure state under different confining stress σ_3, the failure circles (Figure 7.23 with one shown as a dashed curve) can be used with Equation (7.41) to determine the friction angle of the soil under drained conditions (rather than for undrained from the UU test).

Another quick way to obtain undrained shear strength parameters is to conduct unconfined-compression (UC) tests, as shown in Figure 7.25. Note that here the "U" stands for unconfined rather than unconsolidated. For such tests, there is no need to control or measure pore-water pressure or provide confining stress σ_3; rather a vertical stress is applied quickly at increasing levels to the soil specimen until failure occurs. The UC strength is defined as q_u and is typically slightly smaller than the cohesion obtained from UU tests, but provides a quick and simple way to assess undrained shear strength behavior.

7.6 Unified treatment of shear strength of hillslope materials

In the previous sections, it was shown that the shear strength of hillslope materials can be understood using the concepts of cohesive shear resistance and frictional shear resistance. Cohesive shear resistance results from the inter-locking shear resistance c_o (Section 7.2), mobilized cementation c_c (Section 7.3), inter-particle tensile stress (suction stress) c_s (Section 7.3), and tensile strength of tree roots $\Delta\tau_r$ (Section 7.4). The frictional shear resistance results from total stress or external stress σ_n mobilized by the internal or inter-particle

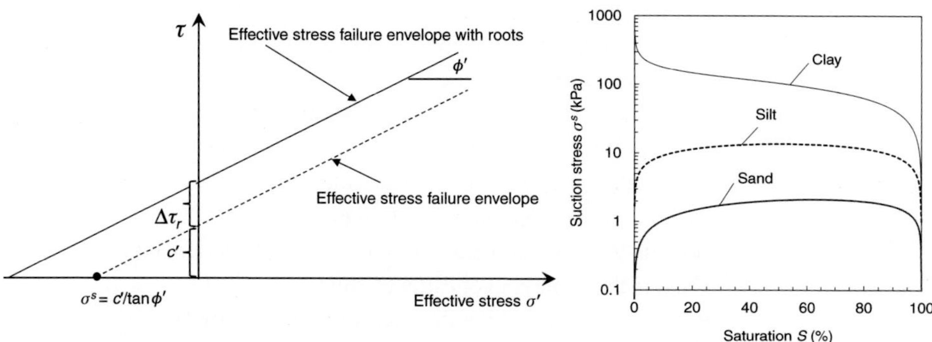

Figure 7.26 Illustration of unified cementation and capillary cohesion resulting from suction stress under variably saturated conditions. Thus the apparent cohesion is the mobilized shear strength by suction stress (inter-particle bonding stress) via internal friction (angle).

friction angle ϕ'. Thus, we can generalize the shear strength of hillslope material under variably saturated condition as

$$\tau_f = c_o + c_c + c_s + \Delta\tau_r + \sigma \tan\phi' \tag{7.42}$$

The first term c_o is formulated in Equation (7.6) for the granular interlocking mechanism in dry sandy soils, the second term c_c is formulated in Equation (7.18) for cementation in all types of soils, the third term c_s results from capillary cohesion or more generally suction stress and is captured in Equation (7.20) (also see Chapter 6 for more details), the fourth term accounts explicitly for the shear strength enhancement due to plant roots and is formulated in Equation (7.28) or (7.29). Furthermore, the first two terms can be lumped into the drained cohesion c', the third term is coupled with the internal friction angle through effective stress or suction stress concept, as illustrated in Figure 7.26. The mobilized cohesion by suction stress is shown in Figure 7.26a and the magnitude of suction stress depends on the type of soil and the effective degree of saturation, as illustrated in Figure 7.26b. Thus, Equation (7.42) covers all the major physical mechanisms for the shear strength of hillslope materials and can be rewritten in a unified shear strength criterion as

$$\tau_f = c' + \Delta\tau_r + (\sigma - \sigma^s)\tan\phi' = c' + \Delta\tau_r + \sigma_n' \tan\phi' \tag{7.43}$$

7.7 Problems

1. How is strength defined in this book?
2. What is the minimum stress for uniaxial compression failure? What is the maximum stress for uniaxial tensile failure?

3. For the multi-axial stress test shown in Figure 7.2, if the Mohr circle at failure goes through the "cohesion" C point, will the maximum and minimum stresses be compressive and/or tensile?
4. In the general description of stress–strain relations for soil shown in Figure 7.3, which segment is most appropriate for estimating the elastic modulus?
5. Which soils, loose sand or dense sand, normally consolidated clay, or overly consolidated clay, will exhibit peak-strength behavior?
6. What is the physical meaning of the internal friction angle?
7. What is the apparent cohesion of soil?
8. What factors control the internal friction angle of sand?
9. What is the relation between apparent cohesion and the particle contact friction angle?
10. For sand with constant relative dry density, which state of stress, plane-strain or triaxial, has a greater internal friction angle?
11. What is the difference between isotropic tensile strength and uniaxial tensile strength?
12. What is the quantitative relation between isotropic tensile strength and uniaxial tensile strength?
13. What is the quantitative relation between isotropic tensile strength and apparent cohesion?
14. What is the maximum amount of uniaxial tensile strength that can be mobilized from isotropic tensile strength?
15. What is the origin of apparent cohesion?
16. What is the maximum amount of apparent cohesion that can be mobilized from isotropic tensile strength?
17. What is the relation between apparent cohesion that arises from cementation and the cementation bonding stress?
18. Does suction stress depend on the internal friction angle of soil?
19. What is the general functional relation between tree root strength and the time since the death of a tree?
20. What is the general functional relation between the strength of vegetated soil and the time since cutting and planting of trees?
21. What type of strength (compression, tensile, or shear) do plant roots contribute to the shear strength of soil?
22. What is the quantitative relation between shear strength of rooted soil and the tensile strength of roots?
23. When the root diameter decreases, does the tensile strength of roots increase?
24. Why does landslide susceptibility increase following forest clearing?
25. What is the consolidated-drained test?
26. What is the major difference between the shear strength behavior of over-consolidated and normally consolidated clay?
27. What is the major difference between the shear strength behavior of loose sand and dense sand?
28. What is the consolidated-undrained test?
29. Will shear stress induce an increase or decrease in pore-water pressure under undrained conditions?

30 What is the Mohr–Coulomb failure criterion in terms of shear stress and normal effective stress?

31 What is the Mohr–Coulomb failure criterion in terms of maximum principal effective stress and minimum principal effective stress?

32 For a saturated soil specimen under isotropic undrained loading σ, what is the corresponding pore-water pressure increase?

33 What is the general form of the Mohr–Coulomb failure criterion under variably saturated conditions for soil with vegetation roots?

34 What is the general form of effective stress under variably saturated conditions?

35 From laboratory shear strength tests, it is found that a dry sand has an apparent cohesion of 5 kPa, and dilation angle of 43°. What is the interlocking shear stress among the sand particles?

36 The dependence of the friction angle on the relative dry density for a sand is shown in Figure 7.8. If the state of stress at a field site is neither plane strain nor conventional triaxial, but somewhere between them, estimate the peak friction angle as a function of relative dry density by averaging the values between the plane strain and triaxial test results.

37 If uniaxial tensile strength is used to represent isotropic tensile strength, estimate the relative error as a function of internal friction angle.

38 For three representative soils, sand, silt, and clay, with tensile strengths shown in Figure 7.12, estimate the hydro-mechanical parameters α and n for each of these three soils using Equation (6.15). Determine the exact values of the peak tensile strength and the saturation for these three soils.

39 Quantitatively assess the relative error of root shear strength given by Equation (7.28) in representing Equation (7.27) for the range of possible variations in friction angle and shear distortion angle.

40 If at a field site the root strength decay and recovery can be described by Equations (7.33) and (7.34) with the measured set of parameters: $d_1 = -0.05$, $d_2 = 0.95$, $d_3 = 25$, $d_4 = 0.20$, $d_5 = 0.50$, and $d_6 = 0.75$, assess the time dependence of root shear of this site. From such an assessment, identify the period susceptible to landslides.

41 A consolidated-undrained test is performed on a normally consolidated clay and the following results are obtained at a failure state: confining stress $\sigma_{3f} = 80$ kPa, deviator stress $\Delta\sigma_d = 60$ kPa, and pore pressure $\Delta u_d = 45$ kPa. Calculate the shear strength parameter friction angle under both consolidated-undrained and drained conditions.

8 Hydro-mechanical properties

8.1 Overall review

8.1.1 Methods for measurement of suction

Determination of the hydraulic and mechanical properties of unsaturated soil, such as the soil water retention curve (SWRC), hydraulic conductivity function (HCF), and suction stress characteristic curve (SSCC) is critical and necessary for analyzing fluid flow and mechanical behavior of unsaturated soils in hillslope environments. Many methods have been developed for measuring unsaturated hydraulic properties using both experimental and theoretical approaches. The common experimental methods for SWRC measurement are (1) axis translation techniques (Tempe cells, pressure plate apparatus, modified triaxial cells (ASTM D6836)), (2) varieties of tensiometers, (3) psychrometers, (4) filter paper, (5) hygrometer, and (6) humidity chamber (e.g., Hilf, 1956; Spanner, 1951; Gee *et al.*, 1992; Houston *et al.*, 1994; Likos and Lu, 2003). Concurrent control and measurement of water content and suction are typically performed using the aforementioned suction techniques and water content measurement techniques such as (1) measurement of inflow and outflow of fluid volume or weight, (2) time-domain reflectometry (TDR probes), and (3) dielectric moisture probes (ECHO-probes). The approximate measurement ranges for various suction or SWRC measurements are shown in Figure 8.1.

In general, existing laboratory techniques for suction measurement and control are complex and time-consuming to use, and limited in the range of suction or water content that can be measured, as shown in Figure 8.1. In addition, most experimental methods, such as Tempe cells and pressure plates, are best suited for determining the SWRC under drying conditions (see the review by Bocking and Fredlund (1980)). However, wetting conditions better describe infiltration and flow in hillslopes shortly following rainfall. Because of the hysteretic behavior typical of many soils, suction values at the same water content but with different wetting or drying histories could be very different, as described in this section. By way of introduction, the aforementioned five methods for measuring or controlling suction are briefly described below.

Tempe cells and pressure plates

These methods (e.g., Hilf, 1956 and see Figure 8.1) rely on a technique called "axis translation," in which suction is applied to a soil specimen by elevating the air pressure in a sealed

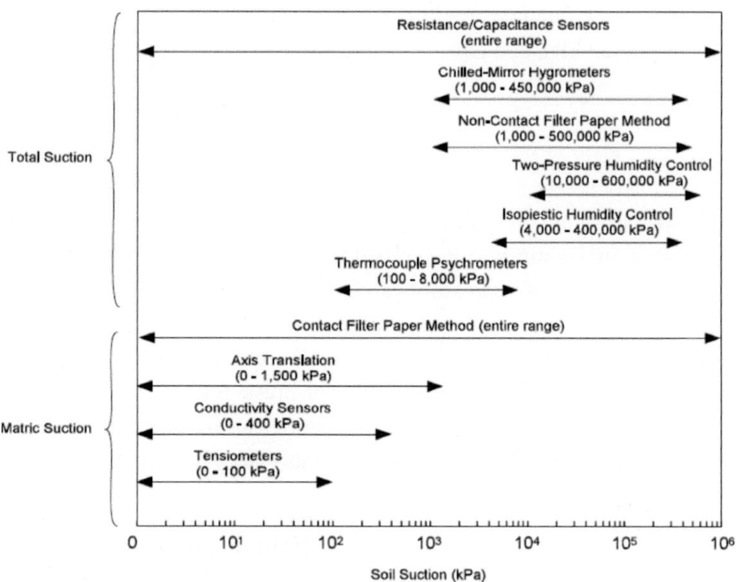

Figure 8.1 Illustration of the approximate measurement ranges for various suction measurement techniques (from Lu and Likos, 2004a).

chamber while pore-water pressure is maintained at the ambient atmospheric value. The practical difference between tempe cell and pressure plate methods is the type and size of their respective sealed chambers. Both techniques require the use of a high-air-entry (HAE) ceramic stone or membrane to maintain pore-water pressure. The HAE material is saturated and placed in contact with the specimen. Pore-water pressure within the material and specimen is held nearly constant at the ambient atmospheric pressure value. Because the flow rate through saturated HAE materials is small, the practical limit of pressures that can be maintained with HAE materials is about 1500 kPa. Commonly available HAE materials typically only maintain suctions less than 500 kPa. Thus, the technique is generally limited to suctions less than 1500 kPa. Another limitation is the time that is needed to bring the pore water in a soil specimen into equilibrium with an applied suction. Even for matric suctions less than 500 kPa, several months or longer may be required to obtain a SWRC for a soil specimen. Because the technique requires the soil specimen to come to equilibrium with a given applied suction, soil water must be continuous within the soil specimen for water to flow in or out of the specimen from the HAE material. Thus these methods are best suited for determining the SWRC for a drying process when soil water is in the capillary or funicular regime. The water retention state in unsaturated soil can be divided into three distinct regimes in order of decreasing saturation: capillary, funicular, and pendular (Figure 3.3). In the capillary regime, air can only exist in bubble form, thus pore water is interconnected. As the air-filled porosity increases, both water phase and air phase are interconnected and this is called the funicular regime. At some point, pore air is inter-connected, but pore water becomes disconnected, and this is called the pendular regime. For sandy, silty, and clayey soils, the pendular regime is reached when soil suctions are greater than 100 kPa, several hundred kPa, and tens of thousands of kPa, respectively.

Tensiometers

Tensiometers are the only device capable of directly measuring the pore-water pressure (suction) in soil without the manipulation of air pressure. Soil suction is sensed through a saturated ceramic stone embedded in direct contact with soil. The saturated ceramic stone is connected to a pressure gauge or electronic sensor with a water-filled tube. Thus, suction present in the soil is transmitted through water that is continuous across the porous stone to a device that measures pressure. Because bubbles will spontaneously form, or the water will "cavitate" or become vapor at pressures less than about -100 kPa (see Figure 8.1) tensiometers are used primarily for measuring soil matric suction less than $\sim$80–100 kPa. Tensiometers are used in both the laboratory and the field.

Humidity methods

The common methods for measuring soil suction based on measuring equilibrium relative humidity are psychrometer (Spanner, 1951), filter paper (e.g., Likos and Lu, 2002), hygrometer (e.g., Gee *et al.*, 1992), and humidity chamber (Hardy, 1992; Likos and Lu, 2003). These techniques (see Figure 8.1) are similarly used to measure the relative humidity of pore air within an unsaturated soil specimen. Assuming local thermodynamic equilibrium, Kelvin's equation (e.g., Lu and Likos, 2004a) can be used along with the measured relative humidity to infer soil suction. Psychrometers and filter paper can be used either under lab or field conditions, whereas the hygrometer and humidity chamber are typically used under laboratory conditions. Other than the contact filter paper method, these techniques are generally suitable for measuring or controlling relative humidity at less than 95% or for suctions greater than 1000 kPa. The time for local thermodynamic equilibrium is very sensitive to temperature fluctuations for relative humidity values greater than 95%.

8.1.2 Methods for measurement of hydraulic conductivity

Methods for measuring the HCF have typically been developed independently from those for measuring the SWRC. Common methods for measuring the HCF are (1) constant head, (2) constant flow, (3) outflow, and (4) instantaneous profiles (e.g., Richards and Weeks, 1953; Corey, 1957; Gribb, 1996; Lu *et al.*, 2006). Reviews of these methods and discussion of their advantages and disadvantages can be found in the literature (e.g., Stannard, 1992; Benson and Gribb, 1997; Lu and Likos, 2004a). Experiments for wetting conditions and HCF determination under laboratory conditions using techniques such as the instantaneous profile method typically require long periods of time and present additional experimental challenges associated with sensors and instrumentation (e.g., Hamilton *et al.*, 1981; Daniel, 1983; Meerdink *et al.*, 1996; Chiu and Shackelford, 1998). Because of the complex nature of these methods, the time-consuming process in conducting permeability measurements, the limitations in controlling wetting and drying histories, and the requirement for suction and/or water content probes, measurements of HCFs are not as commonly performed as measuring SWRCs in current practice. Each of the previously mentioned methods is described in detail in Lu and Likos (2004a). For introduction, they are briefly described below.

Constant head

This laboratory technique (Corey, 1957) is a direct application of Darcy's law (3.17) under steady flow conditions. Different suction heads are maintained at the two ends of the specimen while the flow rate is measured. When the flow rate reaches a steady constant value, the ratio of the measured flow rate to the imposed hydraulic gradient gives the hydraulic conductivity under the imposed suction value. A series of such tests can be performed on a specimen under different suction values in order to obtain the hydraulic conductivity function. Because the range of the feasible applied suctions and suction differences is large, and because testing times run to several months or longer, this technique is mostly used for coarse-grained soils.

Constant flow

Instead of imposing a constant head on one end of a specimen, a constant flow is either injected into or withdrawn from the soil specimen. When steady-state conditions are reached, the head difference is measured and the corresponding hydraulic conductivity is inferred from the ratio of the imposed flow rate to the measured head difference across the specimen according to Darcy's law. Because the imposed flow rate can be extremely small, on the order of 0.01 cm^3/day, this method is suitable for all types of soils from sand to silt and clay soils (Olsen *et al.*, 1991; Lu *et al.*, 2006). Because the technique requires obtaining steady-state conditions under multiple applied flow rates, the testing times for obtaining HCF could be as long as several months. The testing procedure is complicated by the need for imposed flow rates to be estimated prior to the test in order to minimize the head difference within the specimen.

Outflow

This lab method (Gardner, 1956) records the transient response of flow out of a specimen under a constant head or hydraulic gradient condition. Thus, it relies on interpretation or solution of the Richards equation (4.21) to find the controlling parameters of the diffusivity function. Often, in order to simplify the Richards equation, small head differences or hydraulic gradients are applied so that the assumption of constant diffusivity for a given moisture content is valid. Under such an assumption, the Richards equation can be linearized and analytical solutions are available to infer the hydraulic diffusivity from measured outflow volume time series. Most of the early works relied heavily on the analytical solution of outflow in the soil–ceramic stone system, because the accuracy and range in measuring the hydraulic diffusivity are limited. This technique is typically used to determine the HCF for coarse-grained soil under drying conditions. In recent years, with the advent of sensors and numerical modeling, this method has been explored for measuring HCF under different experimental settings, such as implementing suction or/and moisture content probes in soil specimens and applying inverse modeling techniques for hydraulic property identifications. With the new advances in testing procedures and robust inverse numerical modeling, this method has emerged as a leading technique in measuring SWRC, HCF, and SSCC. More detailed and systematic information will be provided in the rest of this chapter.

Instantaneous profiles

This technique (Richards and Weeks, 1953; Watson, 1966; Hamilton *et al.*, 1981) has been applied in both laboratory and field settings. The transient response of both water content and soil suction are monitored at different locations along a profile in which fluid is flowing. Time series of soil suction and water content at each location are collected. The time series of soil suction at two adjacent locations provides an average hydraulic gradient between these two locations. The moisture content time series provides an average moisture content gradient, and thus flow rate, between the two locations. By applying Darcy's law at any given time, the ratio of these two time series (hydraulic gradient and flow rate) provides the HCF. Because (1) a large volume of soil is typically required for this technique, (2) multiple soil suction and water content probes are needed, and (3) there is a difference in wetting or/and drying history at each location, this technique provides only a rough estimate of the HCF of unsaturated soils. This method is also time-consuming (on the order of several months) and the testing procedure is complicated as intrusive emplacement of suction and moisture content sensors is required.

The concurrent measurement of SWRC and HCF has been a research focus over the years. Axis translation is a commonly used laboratory technique for measuring SWRC over the suction range of 0–1,500 kPa. However, depending on the number of data points collected along the SWRC, the testing procedure may require several weeks or even several months. Laboratory measurement of the HCF is generally complicated and time consuming (e.g., Meerdink *et al.*, 1996; Lu *et al.*, 2006). An attractive approach developed to reduce testing time is the one-step outflow method. This method consists of the application of one large increment of matric suction and the monitoring of outflow volume to obtain the hydraulic diffusivity function by analytical solution of the Richards equation. This information is combined with the SWRC so that the HCF may be inferred (Gardner, 1956). The main deficiencies of this method are the assumption that the impedance of the ceramic stone used in the testing apparatus is negligible, and the assumption in the analytical solution that the hydraulic conductivity is constant over each increment in applied suction. These assumptions often lead to inaccurate estimation of the HCF. Over the years, improvements have been made; however, other assumptions and challenges have been introduced. The mathematical procedures involved in reducing the outflow data are complex, and insufficient experimental evidence has been collected to validate the improved methods (e.g., Miller and Elrick, 1958; Kunze and Kirham, 1961; Gupta *et al.*, 1974). Recent advances (Šimůnek *et al.*, 1999) have provided the unsaturated hydraulic properties for soil under drying conditions using inverse modeling techniques based on earlier experimental data by Kool *et al.* (1985). Other inverse modeling studies have used data from suction and moisture content sensors (e.g., Ridley and Way, 1996; Walker *et al.*, 2001; Singh and Kuriyan, 2001). However, these approaches typically involve tests on disturbed soil specimens and the identification of unique hydrologic parameters using inverse methods presents challenges. Until recently, no consistent testing procedure and common inverse modeling algorithm have been established, and few experimental results establishing the validity and generality of these methods have been presented.

In the following section, a technique that combines a simple (no suction or water content probes), fast (several days), accurate measurement of water content change measurement (from outflow or inflow fluid weight), and a robust inverse modeling algorithm for measuring the SWRC and HCF of soils under both drying and wetting conditions is described.

8.2 Transient release and imbibition method (TRIM)

8.2.1 Working principle of TRIM

The principles of the transient release and imbibition (TRIM hereafter) method (Wayllace and Lu, 2012) are illustrated in Figure 8.2. Consider a soil sample to which a sudden large change in suction is applied through the axis translation technique. The transient outflow response from the specimen is a unique function for each soil, which is completely controlled by the diffusivity of the soil, the HAE ceramic stone, and the configuration of the system. Because a HAE ceramic disk is used in the system, the hydraulic properties and impedance of the disk must be explicitly taken into account. The high-resolution transient outflow response obtained experimentally can then be used as an objective function for a numerical model that solves the Richards equation in which the soil parameters that define the SWRC and the HCF can be identified through inverse modeling. This technique can be used for a sudden increase in suction for a drying process and a sudden decrease in suction for a wetting process. If a sudden increase in suction is applied, water outflow is observed; thus, the water content of the soil decreases with time until the hydraulic head gradient in the system approaches zero and steady-state conditions are achieved. Conversely, when the specimen is subjected to a sudden decrease in suction, water is imbibed into the soil sample and the water content of the soil increases. Because an electronic balance is used to record changes in water content, accurate time series of water content and its derivative with respect to time (i.e., flow rate) required by the inverse model can be reliably obtained.

8.2.2 TRIM device

The apparatus used for TRIM tests has five main components, as illustrated in Figure 8.3a: (1) a flow cell in which an undisturbed or remolded soil specimen is placed and matric suction is controlled, (2) a pressure panel to manually control matric suction, (3) a reservoir and bubble trap for sample saturation, air bubble trapping and removal, (4) setup container and electronic balance to record water inflow or outflow, and (5) a computer with

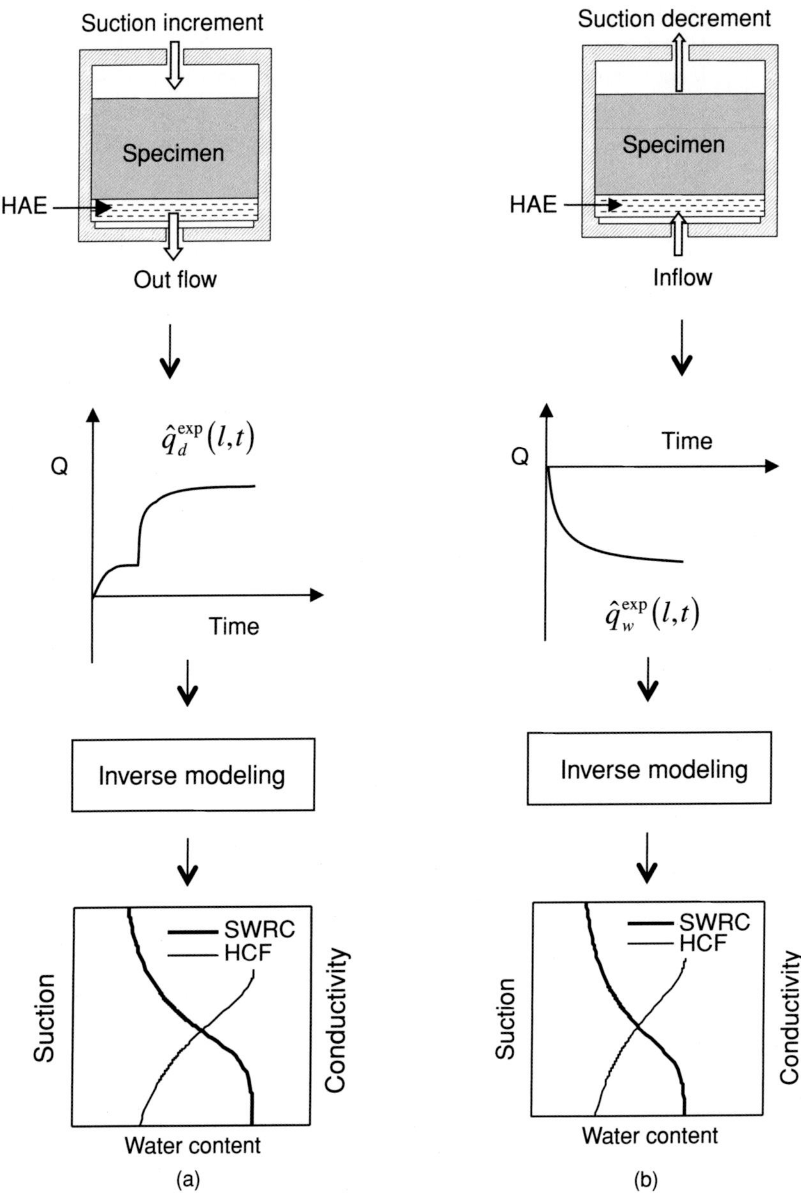

Figure 8.2 Illustration of TRIM method for obtaining the SWRC and HCF (after Wayllace and Lu, 2012).

graphical interface software for real-time graphic display and data logging. Matric suction is controlled by the axis translation technique in which the pore-air pressure is elevated to a controlled value while maintaining the pore-water pressure behind the ceramic stone at the ambient atmospheric pressure.

Details of the main system components are provided in Figures 8.3b–e. The flow cell is made of aluminum; its base is designed to accommodate a HAE ceramic disk (0.32 cm

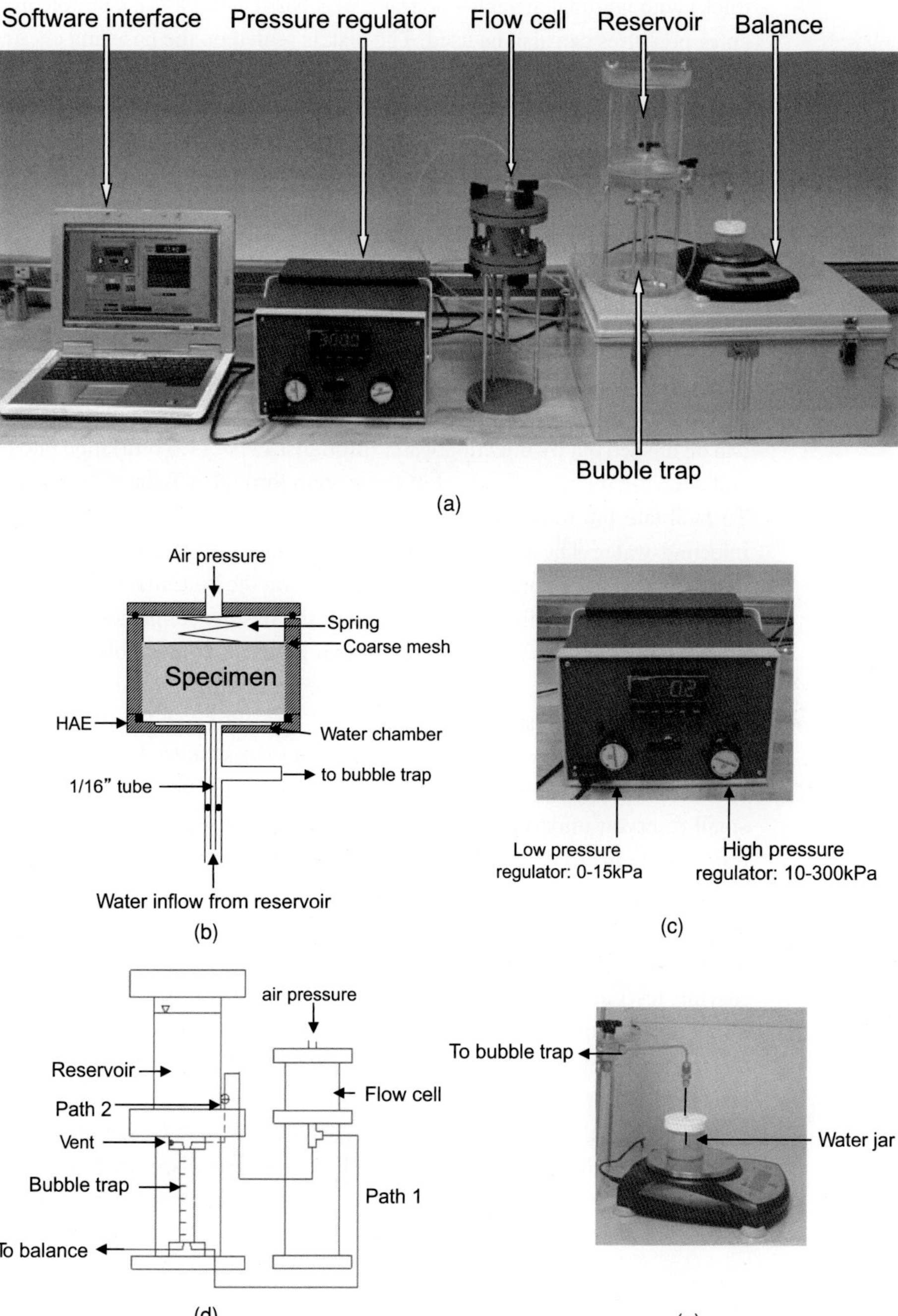

Figure 8.3 Apparatus used in the TRIM method: (a) photograph, (b) flowcell, (c) pressure regulator, (d) bubble trap, and (e) balance and container (after Wayllace and Lu, 2012).

thick) with nominal air-entry pressure of 3 bars (~300 kPa). Ceramic disks with other air-entry pressures can also be used. The disk is seated on the base and encircled with a square O-ring such that the system is sealed between the O-ring and disk and between the O-ring and the base plate. A thin water chamber of 0.07 cm in thickness and 5 cm in diameter is located beneath the disk to maintain the ceramic disk in a saturated condition. The soil specimen may be compacted directly into the cell, or it may be placed in an additional mold to obtain a given diameter for either undisturbed or remolded samples. A coarse mesh and spring are placed on top of the soil specimen to minimize any volume change (Figure 8.3b). The pressure panel to control matric suction has two switchable regulators, one for controlling pressures of 0–15 kPa, while the second one is used for pressures between 10 and 300 kPa. The 15 kPa pressure regulator typically has a resolution of 0.1 kPa, which is needed to identify the air-entry value of coarse-grained soils.

Any air bubbles trapped in the water chamber behind the HAE disk or plumbing system can be flushed out by injecting water through a 1/16″ (1.6 mm) tube into the water chamber and allowing the air to flow out of the system through a T-shaped connection (Figure 8.3b). To facilitate the removal of trapped air, the flow cell can be placed upside down while injecting water. The volume of air is quantified with a bubble trap consisting of a glass reservoir with four ports (two on top and two on the bottom) (Figure 8.3d). The bubble trap is originally filled with water. One port in the base is connected to the main cell while the other base port drains to the balance. As air flows into the bubble trap, it displaces the water in the glass reservoir, thus, providing a means to quantify the volume of air bubbles. The two ports in the top cap of the trap are used for venting and refilling water when necessary. A large water reservoir is used for saturating the system through a three-way valve such that the water can be injected in two ways: (1) water flows through the 1/16″ tube to the small reservoir underneath the ceramic disk, out of the T-shaped connection to the bubble trap, and drains to the balance for system saturation; or (2) water is injected directly to the bubble trap to refill it.

Water flowing in or out of the soil sample is measured using an electronic balance with a 200 g capacity and 0.01 g accuracy. A water container sits on the balance for collecting (drying test) and supplying (wetting test) water flowing in or out of the soil specimen. Evaporation is minimized by drilling a hole in the middle of the water container cover with a diameter slightly larger than 1/16″ so that only a tube less than or equal to 1/16″ can pass through without touching the cover (Figure 8.3e).

The graphical interface and data acquisition software is written in LabVIEW (National Instruments). The software automatically records and displays, at a specified rate, the time-series data of the pressure applied to the sample, the water mass change from the balance, and any remarks from the experimenter(s) during a TRIM test.

8.2.3 Parameter identifications by TRIM

The water content time-series is considered to be a signature function for a given soil specimen with a fixed flow cell configuration. The transient unsaturated flow process leading to this signature function can be described by the Richards equation. In one-dimensional space aligned with the gravity direction z, the Richards equation can be written in the suction head h form as

$$\frac{\partial}{\partial z}\left[K(h)\left(\frac{\partial h}{\partial z}+1\right)\right]=\frac{\partial\theta\,(h)}{\partial h}\frac{\partial h}{\partial t} \tag{4.15}$$

where $\theta(h)$ and $k(h)$ are the SWRC and HCF of the soil, respectively, and t is time. In a typical TRIM test, the initial and boundary conditions for the drying process are

$$h(z, t=0)=0 \tag{8.1a}$$

$$h(z=0, t>0)=h_d \tag{8.1b}$$

$$\frac{\partial h(z=-l, t>0)}{\partial z}=0 \tag{8.1c}$$

where h_d is the applied increase in matric suction head, and l is the sample height l_1 plus the thickness of the HAE ceramic stone l_2. Conversely, the initial and boundary conditions for the wetting process are

$$h(z, t=t_d)=h\,(z) \tag{8.2a}$$

$$h(z=0, t>t_d)=h_w \tag{8.2b}$$

$$\frac{\partial h(z=-l, t>t_d)}{\partial z}=0 \tag{8.2c}$$

where t_d is the time when the drying loop is terminated, and h_w is the applied one-step decrease in matric suction head. Analytical solution of Equation (4.15) under the initial and boundary conditions (8.1) or/and (8.2) are available under two strict assumptions: the hydraulic conductivity is a constant, and the impedance of the HAE ceramic stone can be ignored (e.g., Gardner, 1956; Miller and Elrick, 1958; Kunze and Kirham, 1961; Gupta *et al.*, 1974). These two assumptions are not valid for high impedance ceramic stones and for soil specimens in which suction and moisture content vary substantially. Therefore, analytical approaches for solving this type of problem have been largely abandoned. With the availability of computer power and development of numerical solutions of partial differential equations, Equation (4.15) can be solved numerically using finite-element and finite-difference techniques. The numerical solution of Equation (4.15) under Equation (8.1) conditions for the drying state leads to time series of outflow at the bottom of the HAE ceramic stone:

$$q(0, t)=K_s^c\frac{\partial h(z=0, t)}{\partial z}=\hat{q}_d(l, t) \tag{8.3}$$

and the numerical solution of Equation (4.15) under Equation (8.2) conditions for the wetting state can be expressed:

$$q(0, t)=K_s^c\frac{\partial h(z=0, t)}{\partial z}=\hat{q}_w(l, t) \tag{8.4}$$

where K_s^c is the saturated hydraulic conductivity of the HAE ceramic stone.

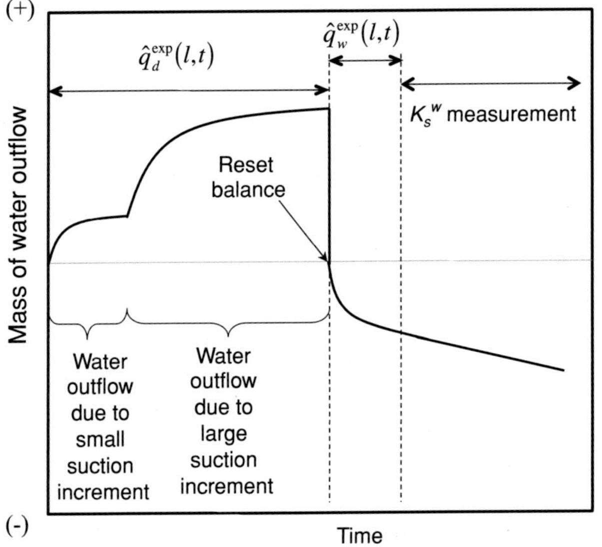

Figure 8.4 Sample transient data obtained from typical TRIM test with complete drying and wetting loops (from Wayllace and Lu, 2012).

Experimental unsaturated flow data measured using the electronic balance for transient water outflow ($\hat{q}_d^{\exp}(l,t)$) or inflow ($\hat{q}_w^{\exp}(l,t)$) as a function of time (illustrated in Figure 8.4) are used as objective functions in an inverse numerical model to obtain the unsaturated hydraulic properties of the soil. This process is called parameter optimization and consists of setting up a forward model with appropriate initial and boundary conditions and an initial estimate of the optimized parameters. The anticipated or predicted system response ($\hat{q}_d(l,t)$ for drying or $\hat{q}_w(l,t)$ for wetting) is calculated using a numerical solution of the governing flow equation (Equation (4.15)) with appropriate initial and boundary conditions (8.1) for the drying loop and (8.2) for the wetting loop. The predicted response is then compared to the real system response, in this case the experimental data of water outflow as a function of time ($\hat{q}_d^{\exp}(l,t)$ for drying or $\hat{q}_w^{\exp}(l,t)$ for wetting). The system parameters to be inversely estimated are then iteratively adjusted and optimized until the differences between observed and expected responses are within the degree of precision desired (van Genuchten, 1981; Kool *et al.*, 1985; van Dam *et al.*, 1994; Toride *et al.*, 1995). To ensure parameter uniqueness, the inverse model is typically run repeatedly using a range of initial parameter estimates to verify that the solution converges to the same or similar results with some preset tolerances.

The results presented below were obtained by using the Hydrus-1D (Šimůnek *et al.*, 2008) code that has an inverse modeling option. The code implements a Levenberg–Marquardt non-linear optimization algorithm. The Levenberg–Marquardt algorithm combines the Gauss–Newton algorithm and the method of gradient descent in order to minimize the deviations between measured ($\hat{q}_d^{\exp}(l,t)$ for drying state, or $\hat{q}_w^{\exp}(l,t)$ for wetting state)

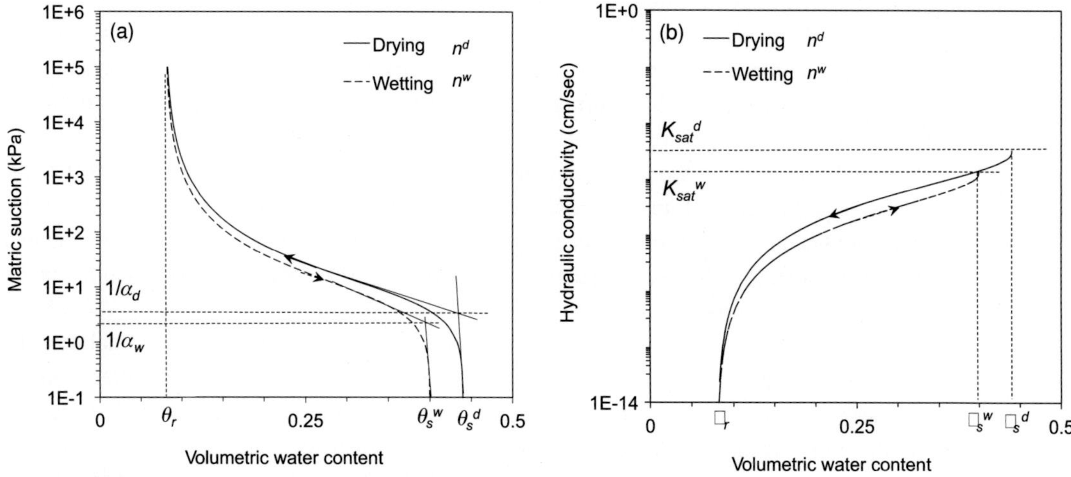

Figure 8.5 Illustration of hysteresis in (a) SWRC and (b) HCF.

and calculated system responses ($\hat{q}_d(l,t)$ or $\hat{q}_w(l,t)$) (Levenberg, 1944; Šimůnek *et al.*, 2008).

Inverse modeling for drying conditions

Water flow into the system is modeled for a one-dimensional variably saturated medium with two materials: the ceramic disk and the soil. The constitutive relations between suction head, water content, and hydraulic conductivity under drying conditions may be represented using the van Genuchten (1980) and Mualem models (1976), described by the following equations:

$$\frac{\theta - \theta_r^d}{\theta_s^d - \theta_r^d} = \left[\frac{1}{1 + \left(\alpha^d |h|\right)^{n^d}}\right]^{1-\frac{1}{n^d}} \tag{8.5}$$

$$K^d = K_s^d \frac{\left\{1 - \left(\alpha^d |h|\right)^{n^d-1}\left[1 + \left(\alpha^d |h|\right)^{n^d}\right]^{\frac{1}{n^d}-1}\right\}^2}{\left\{1 + \left(\alpha^d |h|\right)^{n^d}\right\}^{\frac{1}{2}-\frac{1}{2n^d}}} \tag{8.6}$$

where θ_s is the saturated volumetric water content, θ_r is the residual moisture content, n and α are empirical fitting parameters with α being the inverse of the air-entry pressure head and n the pore size distribution parameter, K_s is the saturated hydraulic conductivity, and the superscript d indicates drying state. The graphic definitions of these parameters are illustrated in Figure 8.5.

As explained in the TRIM testing procedure (Section 8.3), two suction increments are applied after the soil specimen is saturated. The first increment is set to slightly above the air-entry value so that the sample is just in the unsaturated flow regime. Water outflow is then monitored as a function of time (Figure 8.4). Once steady-state conditions are reached, a second large suction increment is applied. The numerical model simulates this drying process, taking into account both suction increments in the objective function. By using two steps the model is better constrained in comparison with a single step (Toorman *et al.*, 1992; van Dam *et al.*, 1994; Durner and Iden, 2011) and the parameters obtained through inverse modeling are more accurate estimates.

Initial conditions in the numerical model are defined through pressure heads, with preset values equal to zero. Boundary conditions are specified as no flow on top and a specified pressure head on the bottom of the simulated flow cell. The value set for the latter changes with time such that the applied suction increments are simulated in an identical manner to the experimental process.

The soil parameters that are estimated using inverse modeling are the residual moisture content θ_r^d, parameters α^d, n^d, and the saturated hydraulic conductivity K_s^d. Initial estimates of these parameters and a range of minimum and maximum possible values are specified by the user. Saturated hydraulic conductivity K_s^d is sensitive to any disturbance of the soil specimen; thus, experimentally obtained values have a certain range of uncertainty. To minimize the uncertainty in the estimated K_s^d only a narrow range around the measured saturated hydraulic conductivity value should be used in the inverse modeling procedure. Hydrologic properties for the ceramic disk are specified by the user. Once all soil properties have been obtained, the SWRC and HCF for drying conditions are fully defined.

Inverse modeling for wetting conditions

The numerical model setup for wetting conditions is similar to that for drying, where both the soil sample and ceramic disk are modeled in one-dimensional flow, and hydrologic constitutive relations may be defined using the Mualem (1976) and van Genuchten (1980) models:

$$\frac{\theta - \theta_r^w}{\theta_s^w - \theta_r^w} = \left[\frac{1}{1 + (\alpha^w |h|)^{n^w}}\right]^{1-\frac{1}{n^w}} \tag{8.7}$$

$$K^w = K_s^w \frac{\left\{1 - (\alpha^w |h|)^{n^w-1}\left[1 + (\alpha^w |h|)^{n^w}\right]^{\frac{1}{n^w}-1}\right\}^2}{\left\{1 + (\alpha^w |h|)^{n^w}\right\}^{\frac{1}{2}-\frac{1}{2n^w}}} \tag{8.8}$$

where the superscript w indicates wetting. The definitions of these parameters are illustrated in Figure 8.5. Initial conditions for the wetting loop are specified using the pressure head distribution at the time when the drying loop was terminated. Boundary conditions are specified as no-flow on the top of the flow cell and a specified head on the bottom.

The soil parameters estimated using an inverse modeling of wetting conditions are θ_s^w, α^w, n^w, and K_s^w. Hysteresis is expected in both the SWRC and HCF, thus, smaller values for θ_s, K_s, and water entry pressure are commonly obtained (Mualem, 1976; Pham *et al.*, 2005). The saturated hydraulic conductivity K_s^w estimated using the inverse model may be

Table 8.1 Steps and testing times for completing a TRIM test

Step	Description	Time (hours)
1	Specimen preparation	2–6
2	System saturation	15–36
3	Data logging	–
4	Application of small suction increment	12–24
5	Application of large suction increment	48
6	Quantification of diffused air	0.5
7	Application of wetting conditions	7–24
8	Obtaining an objective function and performing inverse modeling	3
Total		88–138

verified with the experimental data considering steady flow through two materials in series connection (e.g., Freeze and Cherry, 1979):

$$K_{eq} = \frac{l_s + l_c}{\dfrac{l_s}{K_s^w} + \dfrac{l_c}{K_s^c}} \tag{8.9}$$

where K_{eq} is the equivalent hydraulic conductivity of the system, K_s^c is the saturated hydraulic conductivity of the ceramic disk, l_s is the length of the soil sample, and l_c is the length of the ceramic stone. The residual moisture content for the wetting loop is set to the θ_r value obtained for the drying loop. The estimated saturated volumetric water content for wetting conditions can be verified experimentally by measuring the water content of the sample after the tests are completed.

8.3 TRIM testing procedure

A typical testing program for measuring both drying and wetting SWRC and HCF involves eight steps. The approximate testing time required to complete each step is provided in Table 8.1. Depending on the soil tested, constructing the principal drying and a scanning wetting SWRC and HCF with the TRIM method requires 4 to 9 days.

Step 1. Specimen preparation

The flow cell apparatus can accommodate both undisturbed and remolded samples as neither suction probe nor moisture content sensor is needed. Undisturbed specimens can be placed in their own mold inside the flow cell, while remolded samples can be compacted directly into the flow cell. Special care must be taken so that the ceramic disk is not damaged during the compaction process. The experimenter(s) must then obtain accurate measurements of the porosity n_p and of the saturated hydraulic conductivity. The former can be obtained by

$$n_p = 1 - \frac{m_s}{G_s \rho_w V_t} \tag{8.10}$$

where G_s is the specific gravity, m_s is the mass of solids, V_t is the total volume of the specimen and ρ_w is the density of water. Saturated hydraulic conductivity is measured independently and it may be accomplished by using either the constant head or the falling head method after the specimen is fully saturated (in Step 2).

Step 2. Saturation of the system

Prior to testing, the experimenter(s) must saturate the entire system including the ceramic disk, the thin water chamber underneath the disk, the bubble trap, the plumbing, and the soil specimen. Saturation of the disk is accomplished by partially and then fully submerging it in de-aired water for 12–24 hours while maintaining a vacuum of about 80 kPa in a glass desiccator. A similar procedure is used to saturate the soil specimen; vacuum is applied on the top of the sample while de-aired water is imbibed through the bottom. Change in volume is minimized by placing a coarse mesh and a spring between the top of the soil sample and the top cap of the flow cell. Depending on the soil, saturation of the sample may take 3 to 12 hours. Generally, at the end of this step some excess water is observed at the top of the soil sample. Saturation of the plumbing is accomplished by flowing water from the large reservoir through the flow cell to the bubble trap and draining to the water jar that sits on the balance. The bubble trap is filled with water using the two ports on its top cap.

Step 3. Data logging

Data are logged using a graphical interface program written in LabVIEW. The experimenter specifies the interval to record the mass and applied pressure time series. It is recommended to log data every 10 seconds right after any changes in matric suction are applied to the soil sample and every 10 minutes when conditions are closer to steady state. An example of typical data obtained during a TRIM test showing Steps 4 through 8 is provided in Figure 8.6.

Step 4. Application of small suction increment

After saturation of the specimen, any excess water on top of the sample is allowed to drain by gravity; this process may be accelerated by applying a small pressure increment lower than the air-entry pressure of the soil. Then, a small suction increment slightly above air entry is applied to the sample to ensure suction is beyond air entry of the specimen. Water outflow is monitored for steady-state conditions (Figure 8.4). It takes approximately 12 to 24 hours to complete this step. This small increment in suction (close to the air-entry value) is set to the magnitude beyond which water outflow is observed when matric suction is increased in small increments (0.1 kPa for sands, 0.5 kPa for silts). Typical values for sands

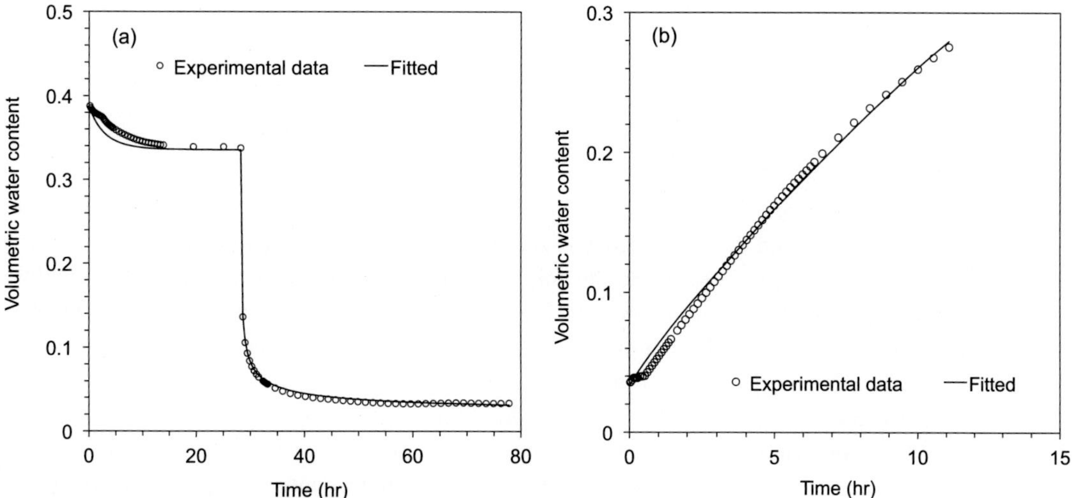

Figure 8.6 Example of experimental and fitted objective functions for (a) drying conditions, (b) wetting conditions (after Wayllace and Lu, 2012).

are 1 to 2 kPa and for silts are 6 to 8 kPa. The flow rate observed is larger at times closer to zero due to the head gradient in the soil. As the total head distribution in the soil becomes constant, the observed flow rate decreases to zero.

Step 5. Application of large suction increment

Next, the experimenter(s) must apply a large suction increment to the specimen while measuring the water outflow as a function of time. The combination of data from Steps 4 and 5 is used as the objective function for the drying state $\hat{q}_d^{\mathrm{exp}}(l, t)$ (Figure 8.4). The limit of matric suction applied during this step is set by the air-entry pressure of the ceramic disk. For example, if a three-bar ceramic stone is employed, it limits the largest suction increment to about 300 kPa. If needed, the disk may be replaced with higher or lower air-entry pressure. Water outflow due to the suction increment is monitored and recorded until steady-state or close to steady-state conditions are met. This step requires approximately 48 hours to be completed.

Step 6. Quantification of diffused air through ceramic disk

Any diffused air needs to be accounted for in order to accurately measure the water mass changes on the electronic balance. As explained in the description of the apparatus, air bubbles diffused through the ceramic disk are quantified by flushing them to the air bubble trap. The flow cell may be turned on its side and tapped to help flush all bubbles out. For 290 kPa of air pressure applied during 2 days, typically, 1 to 2 ml of diffused air are observed. When correcting the function of flow rate versus time for drying conditions, it is assumed that air diffuses at a constant rate. The time required to complete this step is about 30 minutes.

Step 7. Application of wetting conditions

After all diffused air has been quantified, the experimenter must ensure that all plumbing is saturated and that the balance is zeroed. Next, a large suction decrement is applied so that the water is imbibed by the soil specimen. The mass of water inflow is monitored and recorded in a similar way to the drying loop. If desired, a positive pressure head can be applied to the bottom of the sample by decreasing the applied air pressure to zero and adjusting the elevation of the flow cell relative to the elevation of the water jar placed on the balance. Due to the hydraulic gradient created in this manner, water will flow from the water jar on the balance to the flow cell even after the soil has reached the wetting saturated water content θ_s^w. A typical response for application of wetting conditions is depicted in Figure 8.4. The rate of water imbibition is first increasing then, as the total head distribution becomes constant, steady-state conditions are reached. If a gradient is created between the water reservoir and the soil sample, the water mass in the balance decreases linearly with time after saturated wetting conditions are reached. The tests presented in this chapter were performed with 0 kPa of applied air pressure and 8 cm of elevation difference between the water jar and the ceramic disk (approximately -0.8 kPa of matric suction). In the numerical model, the bottom boundary condition for the wetting path is equal to the pressure head at the base of the ceramic disk (i.e., 8 cm). The transient data considered for the objective function $\hat{q}_w^{\exp}(l, t)$ must correspond only to the unsaturated flow regime where moisture content of the sample increases with time. Depending on the type of soil and specimen dimensions, this step takes 7 to 24 hours to complete.

Step 8. Obtaining objective function and performing inverse modeling

The objective function is defined as the water volume flowing in or out of the soil as a function of time. Two numerical models that implement the Levenberg–Marquardt algorithm are set up, one for drying and the other one for wetting conditions. To ensure uniqueness of the results, the experimenter may check that the same parameters are obtained for different initial estimated values of the parameters that are identified through inverse modeling. Since the numerical model simulates one-dimensional unsaturated flow, an equivalent three-dimensional solution can be found by multiplying the water inflow or outflow given by the model with the cross-sectional area of the soil sample. Figure 8.6 displays an example of change in volumetric water content of a soil for both drying and wetting conditions; the hollow circles are experimental data ($\hat{q}_d^{\exp}(l, t)$ for drying or $\hat{q}_w^{\exp}(l, t)$ for wetting) and the smooth lines are the responses predicted by the model. The time required to complete this step is approximately 3 hours. Once all the hydrologic properties are obtained, the SWRC and HCF for both drying and wetting conditions may be plotted.

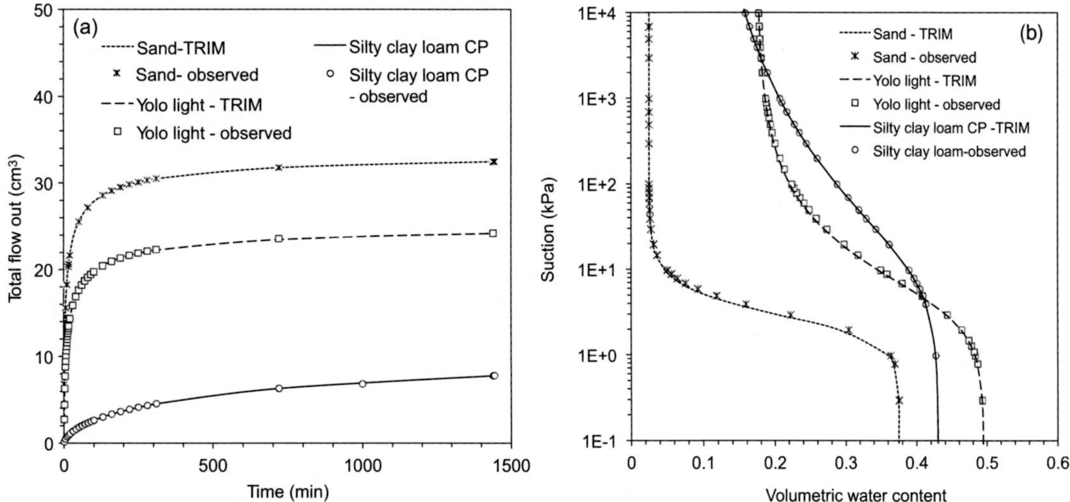

Figure 8.7 Results obtained for verification of uniqueness using transient response data and inverse modeling for sand, silt, and clay: (a) transient response, (b) obtained SWRC (from Wayllace and Lu, 2012).

8.4 Validation of the TRIM method

Three main approaches have been used to demonstrate the validity of the TRIM technique: (1) verification that the parameters obtained through inverse modeling are independent of the initial values assigned to them, (2) repeatability of results during the experimental portion, and (3) comparisons of the SWRC results for both drying and wetting with other traditional methods (Wayllace and Lu, 2012).

8.4.1 Uniqueness of results obtained by inverse modeling

The working principle for this technique is that when a large suction change is applied to a soil subjected to fixed initial and boundary conditions, the transient response is unique to that soil. Thus, if information of water outflow or inflow as a function of time is provided, the hydrologic properties of the soil may be calculated with an inverse model. The results obtained must be independent of the initial estimates for the variables calculated this way. For verification of uniqueness of results, a forward model for a given soil with known properties was executed and the expected transient response obtained. These data were then treated as "observed data" and a numerical inverse model was performed providing random initial estimates for θ_r, α, and n; it was then verified that the results obtained with the inverse model converge to the actual soil properties. The procedure described above was repeated for 16 different soils ranging from sand to clay. Typical transient response and SWRC for sand, silt, and clay specimens are provided in Figure 8.7; the circles, squares, and crosses represent the "observed data" (data obtained with the forward model) while the solid lines

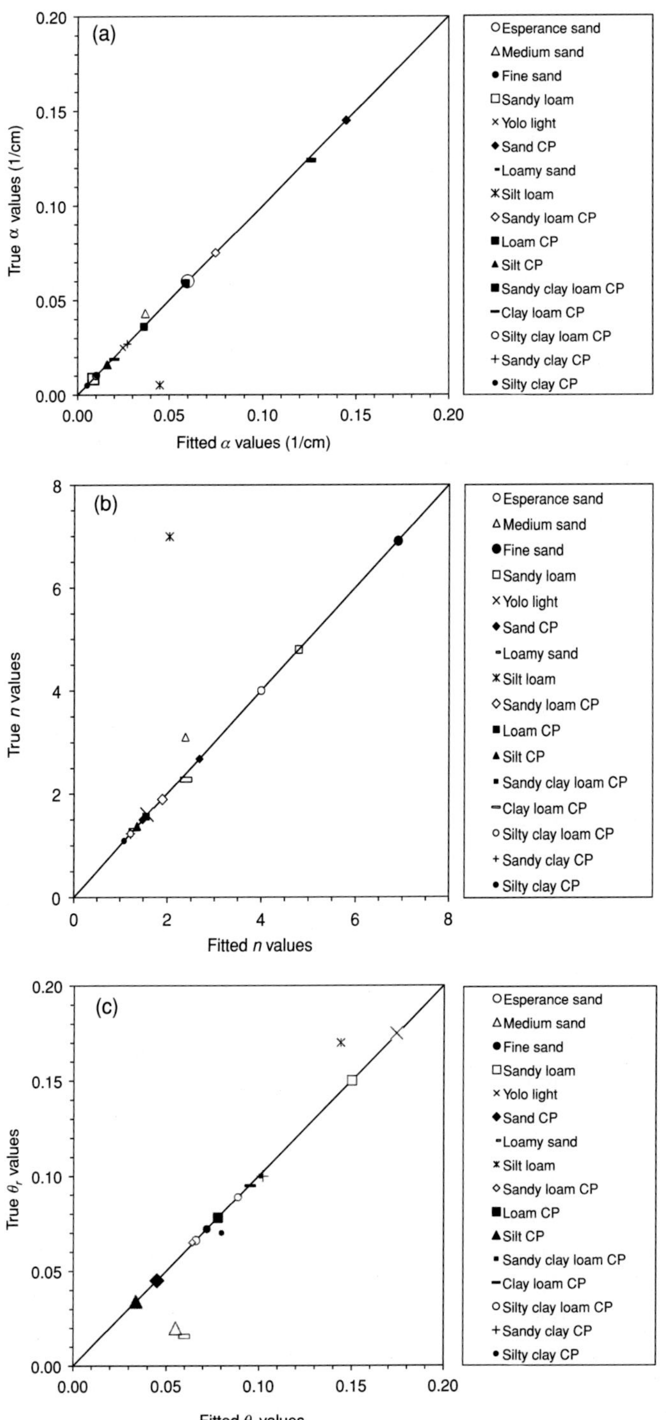

Figure 8.8 Comparison of results obtained with inverse modeling and given model input values for 16 different soils: (a) α parameter, (b) n parameter, and (c) residual moisture content θ_r (from Wayllace and Lu, 2012).

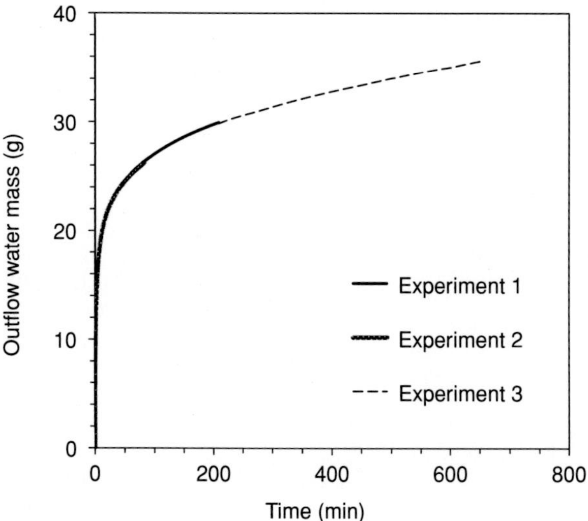

Figure 8.9 Transient response for three different tests performed on the same sample to illustrate the repeatability of TRIM tests (from Wayllace and Lu, 2012).

represent the results obtained with inverse modeling. A comparison of the soil parameters obtained through inverse modeling and the true parameters (used in the forward model) is provided in Figure 8.8. With the exception of silt loam, the α and n calculated parameters compare well with their corresponding "observed" values, having a maximum difference of 8% and 17% respectively. The scatter for the comparison of θ_r values is slightly larger than the other two parameters, with the largest difference of 20% between calculated and "observed" results.

8.4.2 Repeatability of TRIM tests

Repeatability of the outflow data obtained experimentally was verified by performing three independent trials on the same soil sample. A sandy soil (SP) was compacted to a porosity of 0.4 and saturated as specified in the procedure section. Then 2 kPa of suction was applied. Next, 300 kPa of suction was applied to the specimen and the transient outflow response was recorded. The soil was then re-saturated and an identical procedure was repeated two more times; the data from the three trials are reported in Figure 8.9. Results indicate that for an identical soil sample and fixed initial and boundary conditions, the data are repeatable. For the three trials recorded, all points are within 0.01% of each other.

8.4.3 Independent experimental confirmation

Two remolded soils, a poorly graded sand and a silty clay, were tested using both the Tempe cell and the TRIM method. Soil properties for both soils and the ceramic disk are reported in Table 8.2. For this validation, the soil samples were first saturated and tested with the TRIM method, then re-saturated and tested with a Tempe cell. The apparatus used for both

Table 8.2 Hydro-mechanical properties of tested materials

	Classification	Remolded sandy colluvium SP	Remolded silty clay CL–ML	Undisturbed silty clay CL–ML	Ceramic disk
TRIM	G_s	2.65	2.7	2.7	
	θ_s^d	0.39	0.48	0.44	0.34
	K_s^d (cm/s)	3.1E-04	1.0E+05	1.1E-05	2.5E-07
	α^d (kPa^{-1})	0.18	0.025	0.12	0.0015
	n^d	3	1.75	1.44	7
	θ_r	0.018	0.13	0.1	0.07
	θ_s^w	0.29	0.33	0.33	
	α^w (kPa^{-1})	0.41	0.32	0.13	
	n^w	3.5	1.8	1.52	
	K_s^w (cm/s)	2.1E-04	7.3E-06	3.9E-07	
Tempe cell	α^d (kPa^{-1})	0.2	0.03		
	n^d	2.8	1.6		
	θ_r	0.024	0.13		
	θ_s^w	0.29			
	α^w (kPa^{-1})	0.41			
	n^w	2.5			

tests is the same; the method differs in applying one large suction change for the former and several small suction changes and waiting for steady-state conditions in the latter case.

The SWRC and HCF (wetting and drying) obtained for the poorly graded sand are provided in Figure 8.10. The SWRC for wetting and drying obtained with the TRIM method are plotted with solid lines, data points measured with Tempe cells are represented with circles, and the best fit to the Tempe cell data using the van Genuchten model and RETC least-squared regression algorithm (van Genuchten *et al.*, 1991) are plotted with dashed lines (Figure 8.10a). Results obtained with both methods for the drying loop compare well, with the greatest difference for a given value of suction of about 10%. For the wetting process, the results are similar for moisture contents near saturation, while a difference of about 35% exists for data close to the residual moisture content. Hydrologic soil parameters obtained with both methods are reported in Table 8.2. The differences in values for the van Genuchten parameters α and n obtained by the two methods are 10% and 7% for the drying loop and 0% and 28% for the wetting loop, respectively. Comparison of the HCF obtained with the TRIM method and Mualem (1976) model applied to the Tempe cell data is presented in Figure 8.10b. The solid lines represent drying and wetting HCF acquired with TRIM while the dashed lines correspond to functions calculated using the Mualem model based on Tempe cell data and parameters calculated using a least-square fitting algorithm. The HCFs obtained for the drying loop using both methods are similar, however,

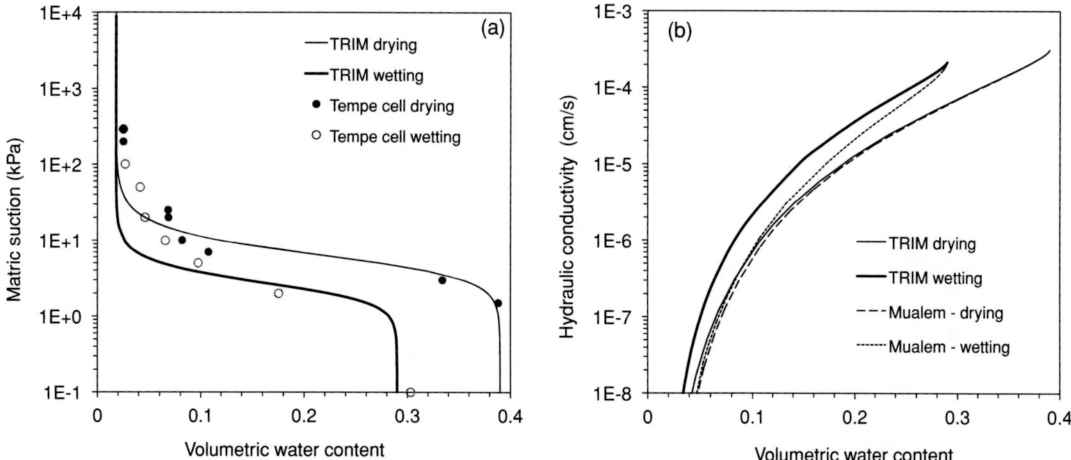

Figure 8.10 Comparison of the SWRC and HCF obtained with the TRIM method and Tempe cell under drying and wetting conditions for a remolded poorly graded sand: (a) SWRC and (b) HCF (from Wayllace and Lu, 2012).

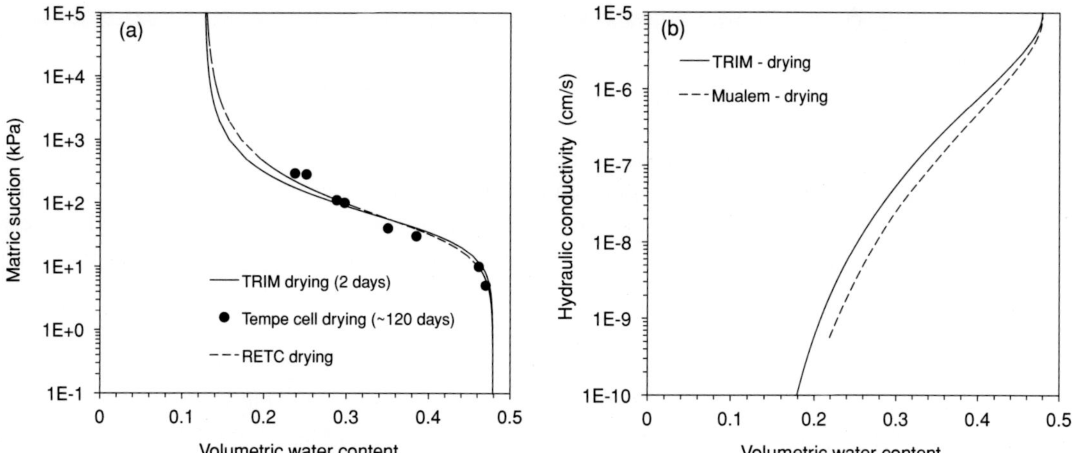

Figure 8.11 Comparison of the SWRC and HCF obtained for a remolded silty clay with the TRIM method and with a Tempe cell under drying conditions: (a) SWRC and (b) HCF (from Wayllace and Lu, 2012).

some differences are evident in the functions for the wetting loop that result from the 28% difference in the n parameter estimates.

In a similar way, results obtained for remolded silty clay tested under drying conditions are shown in Figure 8.11. For this soil, the SWRC and HCF measured with both methods are similar. The hydrologic parameters determined with the TRIM method and Tempe cell data are reported in Table 8.2. The differences in values for the van Genuchten parameters α and n are 18% and 8%, respectively.

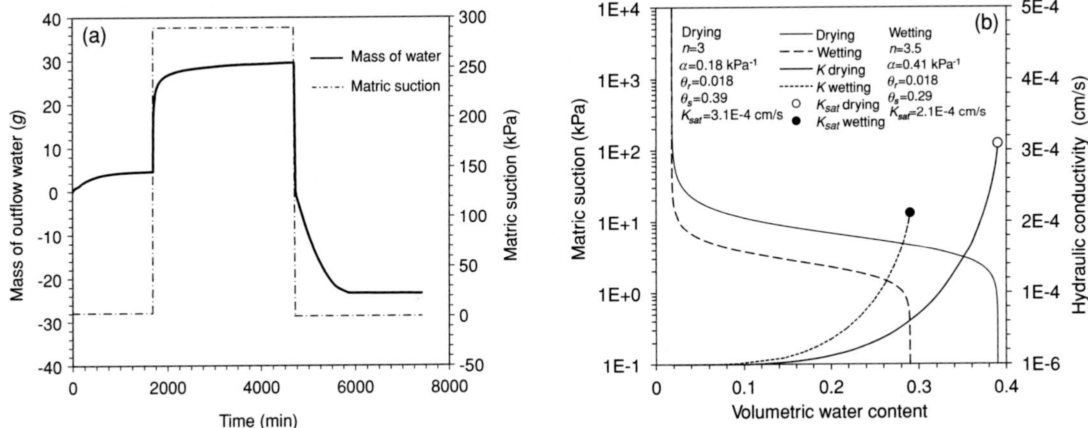

Figure 8.12 Results obtained for remolded poorly graded sand: (a) transient outflow results, (b) the SWRC and HCF for wetting and drying (from Wayllace and Lu, 2012).

8.5 Application of the TRIM to different soils

8.5.1 TRIM test on sandy soil

The SWRC and HCF were obtained for the three soils characterized in Table 8.2, they are a remolded poorly graded sand, an undisturbed silty clay, and a remolded silty clay. The first material is a sandy colluvium obtained from Vashon Advance Outwash Sand collected from a coastal bluff near Edmonds, Washington (Minard, 1983). The specimen tested was remolded and compacted to a porosity of 0.39; a saturated hydraulic conductivity of 2.1×10^{-4} cm/s was measured using the constant head method. After the specimen was saturated, a 3 kPa suction increment was applied for 28 hours followed by a suction increment of 290 kPa. Data for wetting conditions were obtained by applying matric suction equal to zero, which was accomplished by reducing applied air pressure to 0 kPa and maintaining the water container on the balance and the ceramic disk in the flow cell at the same level (Figure 8.12a). The results obtained for this soil are presented in Figure 8.12b.

8.5.2 TRIM test on undisturbed silty clay soil

An undisturbed specimen of a silty clay obtained from a landslide-prone hillside near the San Francisco Bay region, California, was collected in a 2.5-inch diameter tube. The porosity and K_s measured were 0.44 and 1.1×10^{-5} cm/s, respectively. The specimen was placed directly into the flow cell and was saturated for about 12 hours. A small suction increment of 5 kPa was applied for 8 hours, followed by a larger suction increment of 290 kPa. Hydrologic properties for wetting conditions were measured by applying a positive

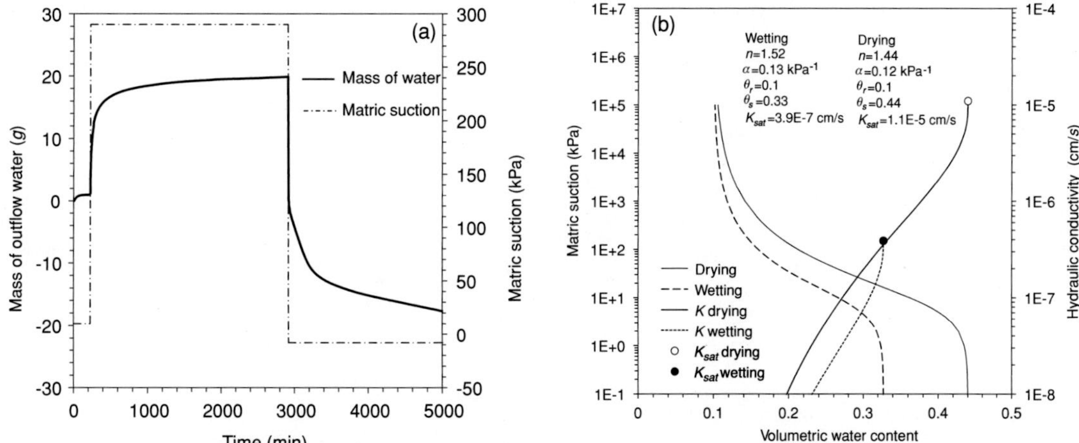

Figure 8.13 Results obtained for undisturbed silty clay: (a) transient outflow results, (b) the SWRC and HCF for wetting and drying (from Wayllace and Lu, 2012).

pressure head of 0.8 kPa at the base of the sample. This was accomplished by decreasing the applied air pressure to 0 kPa while maintaining the ceramic disk in the flow cell at a position 8 cm lower than the water level in the container (Figure 8.13a). Figure 8.13b shows the results obtained for the undisturbed CL-ML soil sample.

8.5.3 TRIM test on remolded silty clay soil

The third soil tested was remolded silty clay obtained from the same location as the second specimen described above. The soil was air dried and then compacted directly into the flow cell to a porosity of 0.48. A constant head test was performed to measure a K_s equal to 1.0×10^{-5} cm/s. Saturation of the sample was achieved in approximately 14 hours, then 10 kPa of suction was applied for 4 hours followed by a larger increment in suction of 290 kPa. Similar to the undisturbed CL-ML sample, the wetting state was measured by applying a positive pressure head of 0.8 kPa at the base of the sample (Figure 8.14a). Measured SWRC and HCF for both wetting and drying conditions are shown in Figure 8.14b.

Results obtained for the three different soils illustrate the range of applicability of the TRIM method. Soils ranging from clean sands to fine-grained samples were tested, as well as both undisturbed and remolded soil samples.

8.6 Quantification of SSCC using TRIM

Because quantitative relationships between SWRC and SSCC have been established (Equations (6.13)–(6.15)), test results from SWRC can be directly used to quantify SSCC, or vice

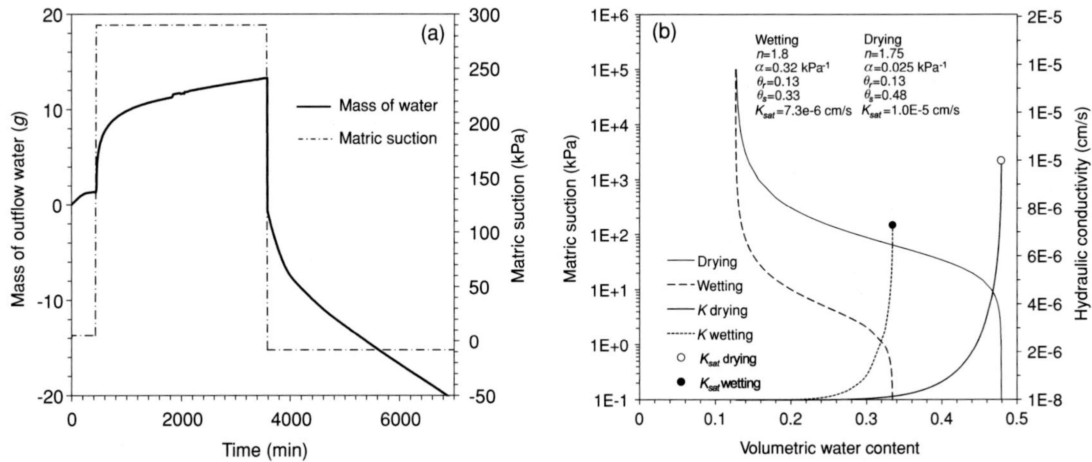

Figure 8.14 Results obtained for undisturbed silty clay: (a) transient outflow results, (b) SWRC and HCF for wetting and drying (from Wayllace and Lu, 2012).

versa. Thus, the results obtained for the three different soils in the previous sections by the TRIM tests are used to quantify the corresponding SSCC. Figure 6.15 shows the SSCC for these three soils under both wetting and drying conditions. As predicted by the unified effective stress with the suction stress concept for parameter $n > 2.0$, suction stress for the poorly graded sand (Figure 8.12 for SWRC and HCF and Figure 8.15a for SSCC) varies with the water content in a non-monotonic fashion. Large hysteresis behavior is evident. For example, the saturated water content for drying is 0.39 and for wetting is 0.29; a 25% difference. The minimum suction stress in this sandy soil is -3.16 kPa and occurs during drying at a water content of 0.23. The minimum suction stress during the wetting process is -1.52 kPa and occurs at a water content of 0.21. The difference in the suction stress hysteresis is 1.62 kPa (-1.52 kPa $+ 3.16$ kPa). This amount is significant enough to determining the stability of sandy hillslopes under rainfall conditions, as will be demonstrated in Sections 9.5 and 10.4.

For the two silty clay soils, SSCC varies monotonically with the water content as shown in Figure 8.15b and c. The hysteresis behaviors for the two silty soils are quite different. The maximum suction stress for the undisturbed silty clay soil shown in Figure 8.15b is $\sim -1{,}215$ kPa and occurs at a water content of 10.7%, which is close to the residual water content of 10%, whereas the remolded silty clay soil shown in Figure 8.15c has maximum suction stress -198 kPa and occurs at the water content of 12.6%. The air-entry pressure ($1/\alpha$) for the undisturbed silty clay soil (Figures 8.13 and 8.15b) under drying ($1/0.12 =$ 8.3 kPa) is similar to the air-entry pressure under wetting ($1/0.13 = 7.7$ kPa), whereas the air-entry pressure for the remolded silty clay soil (Figures 8.14 and 8.15c) under drying is 40 kPa, but is 3 kPa under wetting. Thus, the difference in suction stress due to hysteresis in this soil could be up to 37 kPa. This amount of difference in suction stress (effective stress) is sufficient to trigger deep landslides in silty and clayey soils, as will be illustrated through a case study in Sections 9.6 and 10.5. In hillslope environments, hysteresis in

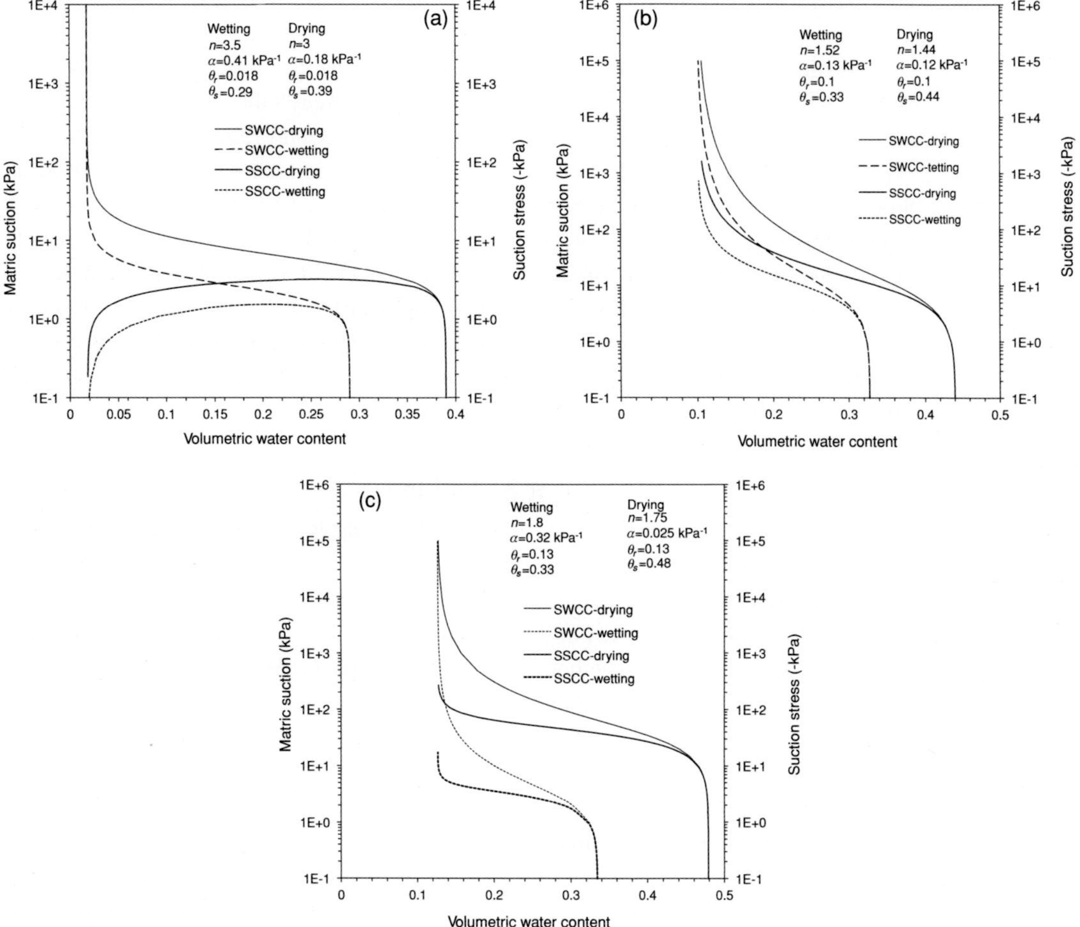

Figure 8.15 Results obtained for three soils under both drying and wetting TRIM tests: (a) SSCC for a poorly graded sand, (b) SSCC for an undisturbed silty clay soil, and (c) SSCC for a remolded silty clay soil.

SWRC, HCF, and SSCC can greatly affect the dynamics of soil water distribution, water movement, effective stress distribution, and ultimately stability of slopes.

8.7 Summary

The main advantages of the TRIM method are (1) only one soil specimen is needed to obtain all hydrologic properties, namely, SWRC, HCF, and SSCC, (2) the apparatus accommodates either remolded or undisturbed soil samples, (3) the capability to determine the hydrologic properties of specimens under wetting conditions, a quality that most standard tests do not have, and (4) the testing time required to obtain all hydrologic functions is approximately one week, which is a significant improvement over other methods that may require several months or longer to obtain a SWRC.

Validation of the technique for a variety of soils was demonstrated by (1) verifying that the parameters obtained through inverse modeling are reliable and independent of the initial values assigned to them, (2) ensuring the experimental results are repeatable, and (3) comparing results with data obtained using other traditional methods. Recommended procedures and performance of the transient water retention technique are illustrated using test results for three soils: a remolded sample of poorly graded sand, an undisturbed sample of silty clay, and a remolded sample of silty clay. The TRIM method provides a potentially fast, accurate, and simple testing tool for obtaining SWRC, HCF, and even SSCC of various types of soils under both wetting and drying states with a high range of matric suction several orders of magnitude above the air-entry pressure of the ceramic stone used in the experimental setup.

8.8 Problems

1. What are the common methods for measuring soil suction? What are the advantages and disadvantages of each of these methods?
2. What are the common methods for measuring SWRC? What are the advantages and limitations of each of these methods?
3. What are the common methods for measuring HCF? What are the advantages and limitations of each of these methods?
4. What is the principle of the TRIM test?
5. How long will it take to complete both the wetting and drying tests for a soil specimen using TRIM?
6. Describe the three types of tests that are used to examine the validity of the TRIM method.
7. Is the hydraulic conductivity always greater under wetting conditions compared to the hydraulic conductivity measured under drying conditions?
8. What are the main advantages of TRIM over other methods?
9. How would you obtain a SSCC from a TRIM test?
10. The hydraulic and mechanical properties of a silty loam specimen collected in the San Francisco Bay region have been measured under both wetting and drying conditions from TRIM tests and are represented by van Genuchten's (1980) model for SWRC, Mualem's (1976) model for HCF, and Lu *et al.*'s (2010) model for SSCC with the following parameters:

α^d(kPa^{-1})	α^w(kPa^{-1})	n^d	n^w	θ_s^d	θ_s^w	θ_r^d	θ_r^w	K_s^d(cm/s)	K_s^w (cm/s)
0.33	0.51	1.30	1.40	0.39	0.36	0.02	0.02	4.8E-05	2.2E-06

Calculate and plot the matric suction head as a function of volumetric water content for both wetting and drying conditions. Calculate and plot the specific moisture capacity as a function of water content for both wetting and drying conditions. Calculate and plot the hydraulic conductivity as a function of water content for both wetting and drying conditions. Calculate and plot suction stress as a function of water content for both wetting and drying conditions. For moisture content varying between 0.1 and 0.35, estimate the maximum differences in suction, hydraulic conductivity, and suction stress due to hysteresis.

PART V

HILLSLOPE STABILITY

FACTOR OF SAFETY ANALYSIS

9 Failure surface based stability analysis

9.1 Classical methods of slope stability analysis

9.1.1 Factor of safety for slope stability

The classical methods of slope stability analysis were developed in the disciplines of soil mechanics and foundation engineering to assess the failure potential of excavations, highway and road embankments, earth dams and levees, and natural and engineered hillsides. The classical methods are based on the concept of "limit equilibrium," which defines the limiting state when the shear stress in a slope is in just-stable mechanical equilibrium with the shear strength of the slope material (e.g., Fellenius, 1936; Morgenstern and Price, 1965; Duncan and Wright, 2005). Distributions of shear stress and shear strength along a potential failure surface are used to establish limiting equilibrium. Consider the finite slope shown in Figure 9.1. Shear stress develops in the slope due to gravity and topographic relief. The spatial distribution of shear stress can be quantified using methods such as linear elastostatics theory described in Chapter 5. The pattern of shear stress often develops preferentially along certain planes or surfaces, such as the potential failure surface ABC. The ability of hillslope materials to resist this shear stress along the potential failure plane can be quantified using the shear strength of the materials, which is described in Chapter 6. Thus, the stability of hillslopes can be assessed quantitatively by determining the ratio of shear strength of the soil τ_f to shear stress developed for mechanical equilibrium τ_d. This ratio is called factor of safety FS, i.e.,

$$FS_s = \frac{\tau_f}{\tau_d} \tag{9.1}$$

where the subscript s denotes shear stress/strength based analysis. In limit-equilibrium analysis, both force and moment equilibrium principles should be satisfied when explicit equations for the factor of safety defined by Equation (9.1) are established. In practice, most theories employ only one or partially both principles.

As described in Chapter 7, determination of the shear strength of hillslope materials is subject to great uncertainty and many theories and testing methods have been developed to address the uncertainty in failure behavior, heterogeneity of shear properties, and the presence of plant roots. An additional challenge involved in quantifying the shear strength of hillslope materials is the effective normal stress, which requires not only the spatial and temporal distribution of total stress (Chapter 5), but also the distribution pore-water pressure or suction stress (Chapters 4, 5, and 6). It is important to note that effective stress

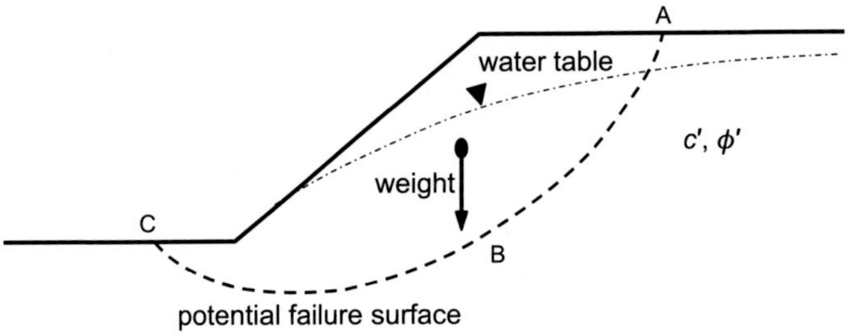

Figure 9.1 Illustration of a potential failure surface in a hillslope.

in surficial hillslope materials varies over a broad range due to the transient nature of pore-water conditions in this environment driven by infiltration and evapo-transpiration.

The shear strength behavior of hillslope materials under variably saturated conditions can be first approximated using the Mohr–Coulomb failure criterion and the effective stress principle:

$$\tau_f = c' + \sigma' \tan\phi' \tag{9.2}$$

where the generalized effective stress for variably saturated materials is (Lu and Likos, 2006; see details in Chapter 6)

$$\sigma' = \sigma - u_a - \sigma^s \tag{9.3}$$

with suction stress σ^s equal to pore-water pressure u_w below the water table and equal to a characteristic function of soil suction above the water table. The Mohr–Coulomb criterion (9.2) leads to a material's shear strength that is described by two components: cohesion c' and the friction angle ϕ'. The shear strength due to the frictional component requires explicit knowledge of effective stress defined in Equation (9.3).

Substituting Equation (9.2) into Equation (9.1) yields

$$\tau_d = \frac{c'}{\text{FS}} + \sigma' \frac{\tan\phi'}{\text{FS}} = c'_d + \sigma' \tan\phi'_d \tag{9.4}$$

In classical soil mechanics, the first term on the right hand side of Equation (9.4) is called the mobilized or developed cohesion c'_d, and the parameter ϕ'_d in the second term, the mobilized friction angle. Equation (9.4) also implies that

$$\text{FS} = \frac{\tau_f}{\tau_d} = \text{FS}_{c'} = \frac{c'}{c'_d} = \text{FS}_{\phi'} = \frac{\tan\phi'}{\tan\phi'_d} \tag{9.5}$$

In applying the concept of the factor of safety to an entire slope, classical soil mechanics assumes that FS is the same at all points along the potential failure surface. Thus, this approach to stability analysis requires an a-priori assumption of the geometry and location of the potential failure surface. Despite abundant field evidence that failure surfaces are often curved or complex surfaces, most classical theories for slope stability analysis assume the failure surfaces can be approximated by simple geometric forms. A variety of geometric forms have been incorporated into stability analyses, such as failure surfaces with planar, circular, and logarithmic spiral geometries. Infinite slope and Culmann (1875) methods are

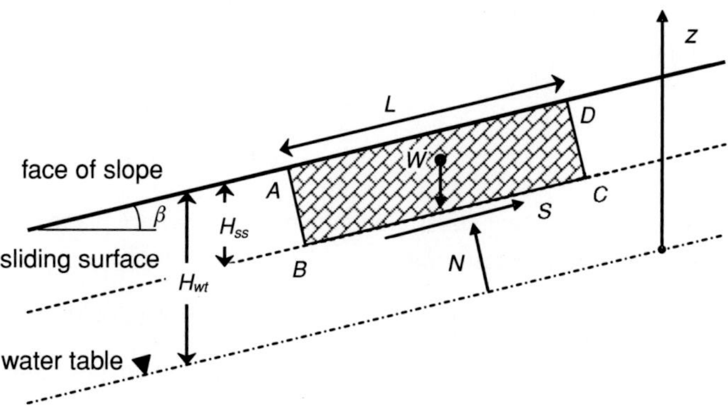

Figure 9.2 Representation of shallow slope stability by infinite-slope model.

two simple models that use a plane as the geometry of the potential failure surface. With these simplifications, the stability of the entire slope can be assessed by a single quantity FS, defined by Equation (9.1) or (9.5).

9.1.2 Infinite-slope stability model

For many rainfall-induced landslides, the failure surfaces are often shallow (upper few meters) and parallel to the slope surface. Under such conditions, stress concentration is ignored and a one-dimensional limit-equilibrium model called the "infinite-slope stability model" is frequently used. The infinite-slope model serves as an excellent illustration for slope stability analysis of translational hillslope failure and will be discussed here under various hydrologic conditions: dry, saturated, and variably saturated. Under variably saturated conditions, theories for the equilibrium state of infinite slopes under both steady-state infiltration and evaporation, and transient infiltration will be described.

Consider the hillslope shown in Figure 9.2. The mechanical equilibrium conditions of the near surface can be established by incorporating all the forces involved in a representative block ABCD. The body force W due to gravity can be considered as acting at the center of the block. For symmetry, forces along AB and CD have the same magnitude but are opposite in sign, leading to them canceling each other. The weight of the block with length L and depth H_{ss} is

$$W = \gamma L H_{ss} \cos\beta \tag{9.6}$$

where γ is the unit weight of hillslope material and β is the slope inclination angle. By mechanical equilibrium, the total forces along the potential failure surface S and normal to the potential failure surface N due to the weight are

$$S = W \sin\beta = \gamma L H_{ss} \sin\beta \cos\beta \tag{9.7}$$

$$N = W \cos\beta = \gamma L H_{ss} \cos^2\beta \tag{9.8}$$

Normalizing the above forces by the area of the potential slip plane $L \times 1$ leads to the total shear τ_d and normal stress σ.

$$\tau_d = \gamma H_{ss} \sin\beta \cos\beta \tag{9.9}$$

$$\sigma = W \cos\beta = \gamma H_{ss} \cos^2\beta \tag{9.10}$$

The resisting shear stress along the potential failure plane can be assessed using the Mohr–Coulomb failure criterion:

$$\tau_f = c + \sigma \tan\phi \tag{9.11}$$

Substituting the normal stress σ (Equation (9.10)) into the above equation leads to

$$\tau_f = c + \gamma H_{ss} \cos^2\beta \tan\phi \tag{9.12}$$

At the limit-equilibrium state, the shear resistance should be equal to the prevailing shear stress along the potential failure plane. Thus, the state of the stability of the slope can be assessed by the ratio of the prevailing shear resistance to the shear stress and it is defined as factor of safety:

$$\mathrm{FS}_s = \frac{\tau_f}{\tau_d} \tag{9.13}$$

Substituting the prevailing shear stress (Equation (9.9)) and the shear strength (Equation (9.12)) into the above equation yields

$$\mathrm{FS} = \frac{c + \gamma H_{ss} \cos^2\beta \tan\phi}{\gamma H_{ss} \sin\beta \cos\beta} = \frac{\tan\phi}{\tan\beta} + \frac{2c}{\gamma H_{ss} \sin 2\beta} \tag{9.14}$$

Equation (9.14) is the infinite-slope stability model for dry slopes. It is useful to point out that the first term on the right hand side of Equation (9.14) quantifies the contribution of the friction angle of the hillslope material to the stability of the slope, and the second term quantifies the contribution of cohesion of the hillslope material to stability. The slope is stable if $\mathrm{FS} > 1.0$ and unstable if $\mathrm{FS} < 1.0$; for $\mathrm{FS} = 1.0$, the slope is in a state of limiting equilibrium. A critical thickness of the hillslope material can be identified by setting $\mathrm{FS} = 1.0$ in Equation (9.14):

$$H_{ss}^{cr} = \frac{2c}{\gamma \sin 2\beta} \frac{\tan\beta}{(\tan\beta - \tan\phi)} \tag{9.15}$$

For cohesionless materials such as dry sand, the second term in Equation (9.14) vanishes, and the critical thickness at which $\mathrm{FS} = 1.0$ leads to $\beta = \phi$, i.e., the slope fails when the inclined angle is equal to the friction angle.

If the water table in a hillslope is above the potential sliding surface, i.e., $H_{wt} < H_{ss}$, pore-water pressure u_w must be considered. A unified approach to analyzing the saturated stability under both hydrostatic and seepage conditions employs Terzaghi's effective stress σ' in lieu of total stress for both normal stress and shear strength, i.e.,

$$\sigma' = \sigma - u_w \tag{9.16}$$

$$\tau_f = c' + \sigma' \tan\phi' \tag{9.17}$$

For effective stress conditions a different set of material properties for cohesion (drained c') and friction angle (drained ϕ') are appropriate to describe the shear strength behavior under saturated conditions. Substituting Equations (9.16) and (9.17) into Equation (9.12) leads to the factor of safety under saturated conditions:

$$\mathrm{FS} = \frac{c' + \left(\gamma H_{ss}\cos^2\beta - u_w\right)\tan\phi'}{\gamma H_{ss}\sin\beta\cos\beta} = \frac{\tan\phi'}{\tan\beta} + \frac{2c'}{\gamma H_{ss}\sin 2\beta} - \frac{u_w}{\gamma H_{ss}\sin 2\beta}\tan\phi' \tag{9.18}$$

As shown in the above equation, the negative sign in front of the third term indicates that an increase in pore-water pressure tends to have a destabilizing effect on hillslopes. With the aid of a trigonometric relation, the above equation is also commonly written in the following form:

$$\mathrm{FS} = \frac{\tan\phi'}{\tan\beta} + \frac{2c'}{\gamma H_{ss}\sin 2\beta} - \frac{u_w}{\gamma H_{ss}}(\tan\beta + \cot\beta)\tan\phi' \tag{9.19}$$

or

$$\mathrm{FS} = \frac{\tan\phi'}{\tan\beta} + \frac{2c'}{\gamma H_{ss}\sin 2\beta} - r_u(\tan\beta + \cot\beta)\tan\phi' \tag{9.20}$$

with the pore-water pressure parameter r_u:

$$r_u = \frac{u_w}{\gamma H_{ss}} \tag{9.21}$$

9.1.3 Culmann's finite-slope stability model

All slopes are of finite dimensions, and many failure surfaces are not strictly parallel to the slope surface. Culmann (1875) developed a theory for finite slopes with a failure plane that is not parallel to the slope surface. Culmann's method gives fair results for steeply inclined or nearly vertical slopes, but poor results for less-steep slopes. The planar failure surface assumption is violated in the middle part of slopes and beneath the toe, where the failure surfaces are commonly curved. Extensive field study in the early and middle parts of the last century led to replacing the assumption of planar failures by other geometries, such as circular (Fellenius, 1927, 1936; Taylor, 1937) and logarithmic spirals (e.g., Michalowski, 2002). Nonetheless, Culmann's method provides an excellent illustration of the concepts of limit-equilibrium analysis and factor of safety calculation with respect to the friction and cohesion properties of hillslope materials.

For a given steep finite slope (Figure 9.3) with a friction angle ϕ', slope height H, unit weight γ, and slope angle β, Culmann's method can be used to identify the critical angle θ_{cr} when FS $= 1.0$. The total weight of the slope wedge W forming the slope surface OAB and the potential failure plane OB can be calculated as

$$W = \frac{1}{2}\gamma(AB)H(1) = \frac{\gamma H}{2}(H\cot\theta - H\cot\beta) = \frac{\gamma H^2}{2}(\cot\theta - \cot\beta) \tag{9.22}$$

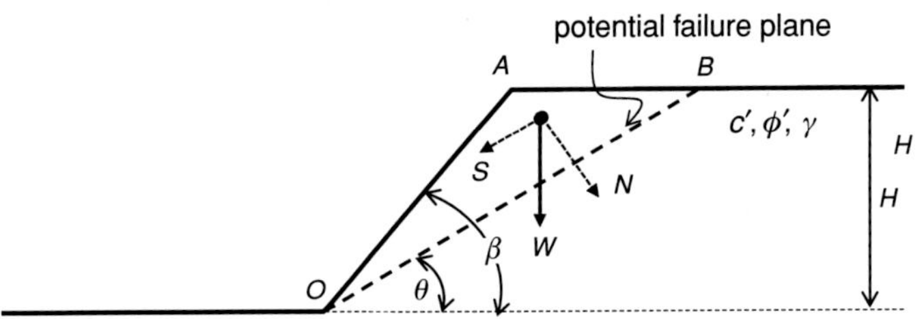

Figure 9.3 Illustration of Culmann's limit-equilibrium method for finite slopes.

The components of the weight W along the potential failure plane S and normal to the failure plane N are respectively:

$$S = W\sin\theta = \frac{\gamma H^2}{2}(\cot\theta - \cot\beta)\sin\theta \tag{9.23}$$

$$N = W\cos\theta = \frac{\gamma H^2}{2}(\cot\theta - \cot\beta)\cos\theta \tag{9.24}$$

The corresponding shear and normal stresses on the potential failure plane OB are then:

$$\tau = \frac{S}{OB} = \frac{\frac{\gamma H^2}{2}(\cot\theta - \cot\beta)\sin\theta}{\frac{H}{\sin\theta}} = \frac{\gamma H}{2}(\cot\theta - \cot\beta)\sin^2\theta \tag{9.25}$$

$$\sigma = \frac{N}{OB} = \frac{\frac{\gamma H^2}{2}(\cot\theta - \cot\beta)\cos\theta}{\frac{H}{\sin\theta}} = \frac{\gamma H}{2}(\cot\theta - \cot\beta)\cos\theta\sin\theta \tag{9.26}$$

The mobilized shear stress along the potential failure plane OB, following the Mohr–Coulomb failure criterion (9.17) and Equation (9.4), can be written with the normal stress in Equation (9.26):

$$\tau_d = c_d + \sigma\tan\phi_d = c_d + \frac{\gamma H}{4}(\cot\theta - \cot\beta)\sin 2\theta\tan\phi_d \tag{9.27}$$

The mobilized shear stress thus depends on the shear strength parameters of the soil and the angle θ that defines the potential failure plane. This stress is the reaction of the shear stress along the potential failure plane τ in Equation (9.25):

$$\frac{\gamma H}{2}(\cot\theta - \cot\beta)\sin^2\theta = c_d + \sigma\tan\phi_d = c_d + \frac{\gamma H}{4}(\cot\theta - \cot\beta)\sin 2\theta\tan\phi_d \tag{9.28}$$

Rearranging Equation (9.28) leads to an expression of the mobilized cohesion c_d in terms of the shear strength parameters of the soil and the angle θ:

$$c_d = \frac{\gamma H}{4}(\cot\theta - \cot\beta)\left[2\sin^2\theta - \sin 2\theta\tan\phi_d\right] \tag{9.29}$$

The maximum mobilized cohesion is solely a function of the angle θ, as the other parameters γ, H, and ϕ are constants for a given finite slope. By taking the derivative of the above equation with respect to θ and setting the result equal to zero,

$$\frac{\partial c_d}{\partial \theta} = \frac{\partial}{\partial \theta}\left\{\frac{\gamma H}{4}(\cot\theta - \cot\beta)\left[2\sin^2\theta - \sin 2\theta \tan\phi_d\right]\right\} = 0 \tag{9.30}$$

yields the critical θ_{cr} for the mobilized cohesion c_d as

$$\theta_{cr} = \frac{\beta + \phi_d}{2} \tag{9.31}$$

$$c_d = \frac{\gamma H}{4}\frac{1 - \cos(\beta - \phi_d)}{\sin\beta \cos\phi_d} \tag{9.32}$$

From Equation (9.5), when FS $= 1.0$, $\text{FS}_c = \text{FS}_\phi = 1.0$, which leads to $c = c_{d\max}$, and $\phi_d = \phi$, and the maximum height $H_{\max}$ from Equation (9.32):

$$H_{\max} = \frac{4c}{\gamma}\frac{\sin\beta\cos\phi}{1 - \cos(\beta - \phi)} \tag{9.33}$$

For a slope angle $\beta = 90°$ (vertical cut slope) the maximum height of the slope can be obtained from Equation (9.33):

$$H_{\max} = \frac{4c}{\gamma}\frac{\sin 90°\cos\phi}{1 - \cos(90° - \phi)} = \frac{4c}{\gamma}\frac{\cos\phi}{1 - \sin\phi} \tag{9.34}$$

For example, if a cut slope has the shear strength parameters of $c = 10$ kPa and $\phi = 30°$, and unit weight γ of 20 kN/m^3, the maximum height $H_{\max}$ of the cut slope is

$$H_{\max} = \frac{4c}{\gamma}\frac{\cos\phi}{1 - \sin\phi} = \frac{4(10)(\text{kN/m}^2)}{(20)(\text{kN/m}^3)}\frac{\cos 30°}{1 - \sin 30°} = 3.46\,(\text{m})$$

For the same slope material, if FS $= 2.0$ is imposed, the height of the cut slope H can also be calculated as follows:

$$\therefore \text{FS} = 2.0 = \text{FS}_c = \frac{c}{c_d}$$

$$\therefore c_d = \frac{c}{FS_c} = \frac{10(\text{kPa})}{2} = 5(\text{kPa})$$

$$\therefore \text{FS} = 2.0 = \text{FS}_\phi = \frac{\tan\phi}{\tan\phi_d}$$

$$\therefore \tan\phi_d = \frac{\tan\phi}{\mathrm{FS}_\phi} = \frac{\tan 30^\circ}{2}$$

$$\therefore \phi_d = \tan^{-1}\left(\frac{\tan 30^\circ}{2}\right) = 16.13^\circ$$

Substituting the mobilized friction $\phi_d = 16.13^\circ$ and the mobilized cohesion $c_d = 5$ kPa into Equation (9.32) and setting $\beta = 90^\circ$ yields

$$H = \frac{4c_d}{\gamma}\frac{\sin\beta\cos\phi_d}{1-\cos(\beta-\phi_d)} = \frac{4(5)(\mathrm{kN/m^2})}{20(\mathrm{kN/m^3})}\frac{\sin 90^\circ \cos 16.13^\circ}{1-\cos(90^\circ - 16.13^\circ)} = 1.33\,(\mathrm{m})$$

For saturated slopes of poorly drained materials such as clay, the shear strength is often considered under undrained conditions, and $\phi' = 0$, and $c = c_u$, leading to the maximum vertical height of cut slopes

$$H_{\max} = \frac{4c_u}{\gamma} \tag{9.35}$$

For steep finite slopes under unsaturated conditions, Culmann's finite slope theory can be reformulated using the unified effective stress equation (6.4) described in Chapter 6. Replacing the total stress σ in Equation (9.28) by the effective stress equation (9.3), i.e., $\sigma' = \sigma - \sigma^s$ for $u_a = 0$, and following the above similar derivation, Equation (9.33) can be generalized to

$$H_{\max} = \frac{4\,(c' - \sigma^s\tan\phi')}{\gamma}\frac{\sin\beta\cos\phi}{1-\cos(\beta-\phi)} \tag{9.36}$$

For moist cohesionless materials such as beach sand, $c' = 0$ and Equation (9.36) becomes

$$H_{\max} = -\frac{4\sigma^s\tan\phi'}{\gamma}\frac{\sin\beta\cos\phi'}{1-\cos(\beta-\phi')} \tag{9.37}$$

As an example of application, the above equation is used to estimate the maximum height of sand castles. Consider a sand with a minimum suction stress of -1.2 kPa (see Chapter 6), a friction angle of 42°, a unit weight of $\gamma = 17$ kN/m^3, and a maximum angle β for sand castle slopes of 70°. For these conditions and parameters the maximum height of sand castles $H_{\max}$ built out of this sand is 2.17 m.

Other analytical methods that better represent the geometry of failure surfaces and impose both force and moment mechanical equilibrium in finite slopes were developed in the last century. For example, the logarithmic spiral method (Frohlich, 1953; Leshchinsky and Volk, 1985) satisfies both moment and force equilibrium principles (static equilibrium or statically determinant). The assumption of logarithmic-shaped failure surfaces is considered to be superior over circular-shaped failure surfaces as both force and moment mechanical equilibria are considered and the results for factor of safety calculation are shown to be more accurate (Leshchinsky and San, 1994).

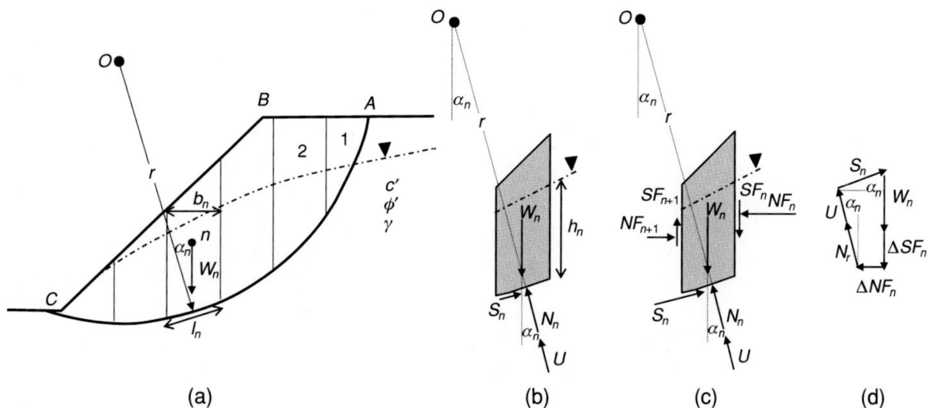

Figure 9.4 Illustration of the method of slices for limit-equilibrium analysis: (a) conceptualization of slope failure surface and discretization of slope into slices, (b) free-body force diagram for the ordinary method, (c) free-body force diagram for Bishop's method, and (d) force vector equilibrium diagram for Bishop's method of slices.

9.2 Method of slices for calculating factors of safety

9.2.1 Ordinary method of slices

Analytical solutions for calculating factors of safety have been developed for failure surface geometries other than planar, such as circular and logarithmic spirals. Trial search procedures are often involved, with the aid of slope stability charts (e.g., Fellenius, 1927; Taylor, 1937; Terzaghi and Peck, 1967; Michalowski, 2002, among many others). Alternatively, computer codes have been developed that use the method of slices, and they are widely used in geotechnical engineering practice.

As shown in Figure 9.4a, slopes can be divided into a number of slices separated by vertical boundaries. The geometry of slopes is predetermined, while the location of the potential failure surface having the minimum factor of safety can be found through an iterative process. A circular shape of the potential failure surface AC is commonly assumed. Limit equilibrium for all slices is used to establish the factor of safety. In the method devised by Fellenius (1927), commonly referred to as the ordinary method of slices, only forces on the potential failure surface and body forces are considered in the moment equilibrium, as shown in Figure 9.4. This greatly simplifies the theory and provides an excellent means to illustrate the concepts used by the other refined methods of slices. Other methods are more rigorous in dealing with inter-slice forces for force equilibrium, and can incorporate more complicated seepage conditions and the complex geometry of the failure surface, as illustrated in Figures 9.4c and 9.4d and demonstrated later in this chapter.

The factor of safety for the ordinary method of slices can be established by considering the free body diagram shown in Figure 9.4b for the nth slice in Figure 9.4a. The inter-slice

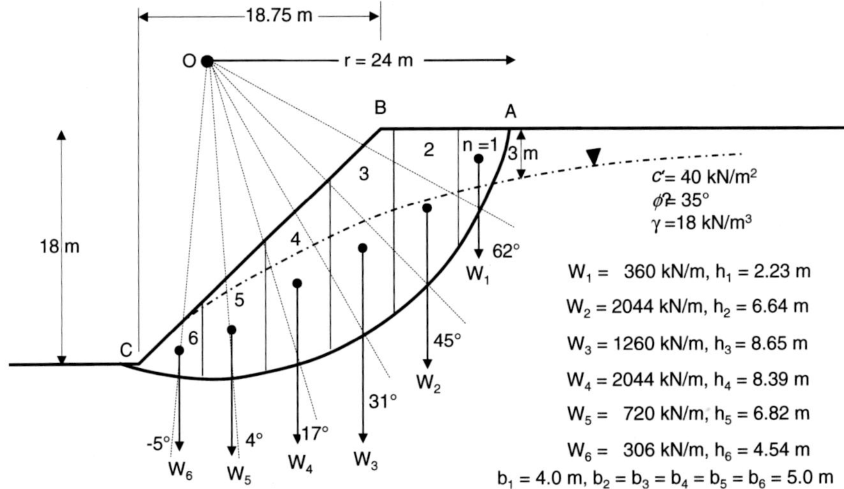

Figure 9.5 Illustration of ordinary and Bishop's simplified methods of slices for a finite slope composed of silt. The 1/m in the weight of slices reflects the unit depth in the direction perpendicular to the slope cross section.

forces are completely ignored. The normal reacting force N_n is obtained by considering the force equilibrium with the weight of slice W_n:

$$N_n = W_n \cos \alpha_n \tag{9.38}$$

In light of the concept of factor of safety defined in Equation (9.4), the mobilized shear resistance along the nth segment of the potential failure surface L_n is

$$S_n = \tau_d l_n = \frac{\tau_f l_n}{\mathrm{FS}_s} = \frac{l_n}{\mathrm{FS}_s}\left(c' + (\sigma - u_w)\tan\phi'\right) \tag{9.39}$$

The total stress σ can be obtained by dividing the normal force N_n by the length l_n, and the pore-water pressures u_w can be approximated by the distance from the failure surface to the water table h_n times the unit weight of water γ_w (9.8 kN/m^3). Considering the driving moment from the weight of the slope (Equation (9.38)) and the resisting moment from the mobilized shear resistance (Equation (9.39)), and applying the moment equilibrium of the n slices with respect to the point O gives

$$\sum_{n=1}^{m} W_n r \sin\alpha_n = \sum_{n=1}^{m} \tau_d l_n r = \sum_{n=1}^{m} \frac{l_n}{\mathrm{FS}_s}\left(c' + (\sigma - u_w)\tan\phi'\right) r$$

$$\sum_{n=1}^{m} W_n r \sin\alpha_n = \sum_{n=1}^{m} \tau_d l_n r = \sum_{n=1}^{m} \frac{l_n}{\mathrm{FS}_s}\left[c' + \left(\frac{W_n \cos\alpha_n}{l_n} - \gamma_w h_n\right)\tan\phi'\right] r$$

$$\mathrm{FS}_s = \frac{\sum_{n=1}^{m}\left[c' l_n + (W_n \cos\alpha_n - \gamma_w h_n l_n)\tan\phi'\right]}{\sum_{n=1}^{m} W_n \sin\alpha_n} \tag{9.40}$$

Table 9.1 Example of computational procedures for the ordinary method of slices

1	2	3	4	5	6	7	8	9	10	11	12	Driving force	Resistance
Slice	W_n	h_n	u_n	α_n	$\sin\alpha_n$	$\cos\alpha_n$	b_n	l_n	$c'l_n$	u_nl_n	$W_n\cos\alpha_n$	$W_n\sin\alpha_n$	10 + (12 − 11)
Unit	(kN/m)	(m)	(kPa/m)	(deg)			(m)	(m)	(kN/m)	(kN/m)	(kN/m)	(kN/m)	$\tan\phi'$ (kN/m)
1	360	2.23	21.85	62	0.88	0.47	4.0	8.52	340.8	186.2	169.0	317.8	328.7
2	2044	6.64	65.07	45	0.70	0.71	5.0	7.07	282.8	460.1	1445.3	1445.3	972.6
3	1260	8.65	84.77	31	0.51	0.86	5.0	5.83	233.3	494.4	1080.0	648.9	643.3
4	2044	8.39	82.22	17	0.29	0.96	5.0	5.22	209.1	429.8	1954.6	597.6	1276.8
5	720	6.82	66.84	4	0.07	1.00	5.0	5.01	200.4	335.0	718.2	50.2	468.8
6	306	4.54	44.49	−5	0.08	1.00	5.0	5.01	200.7	223.3	304.8	−26.7	257.8
Σ												3033.30	3948.2
FS_s													1.30

A finite slope composed of consolidated silt inclined at 43.83° shown in Figure 9.5 is used to illustrate the computational procedure. The slope is first divided into six slices with a horizontal spacing $b_n = 5$ m, except the first slice, which is 4 m. The area of each slice is calculated and given a unit weight of 18 kN/m^3, the weight of each slice W_n is determined (the column 2 in Table 9.1). The water table location is near the slope surface and the distance from the potential failure surface to the water table h_n for each slice is listed in column 3, and the pore pressure u_n calculation in the column 4. The inclination angle of the potential failure surface of each slice α_n and its cosine and sine values are shown in columns 5, 6, and 7. The horizontal width and the width along the potential failure surface of each slice are shown in columns 8 and 9. Three resistance forces contributing to the resistance moment for each slice (numerator in Equation (9.40)) are calculated (columns 10, 11, and 12) and summed in the last column. The driving force (denominator in Equation (9.40)) is computed and shown in column 13. Summation of all the resistance forces (3948.29 kN/m) divided by the summation of all the driving forces (3033.30 kN/m) gives the factor of safety of the slope $FS_s = 1.30$. In the above example, the location of the potential failure surface (center O) and number of slices are set arbitrarily. To use this method in practice, one should iteratively adjust the location of the failure surface and number of slices to find the minimum factor of safety. This iterative process can be done with manual calculations or computer programs.

The treatment of pore-water pressure as the height of a water table above the failure surface acting as a normal force in the ordinary method contributes to the low factors of safety that are obtained with this method compared to other methods (Turnbull and Hvorslev, 1967) discussed in the following sections.

9.2.2 Bishop's simplified method of slices

To improve the accuracy of the ordinary method of slices, Bishop (1955) considered the inter-slice forces to calculate factors of safety. Consider the nth slice shown in Figure 9.4c, Force equilibrium requires that the summation of all forces in the horizontal direction (x) and vertical direction (z) on the nth slice should be zero. Mathematically, the force equilibrium in two-dimensional space requires $\Sigma F_{xi} = 0$, and $\Sigma F_{zi} = 0$. Graphically, it can be illustrated in the vector diagram shown in Figure 9.4d, where all the forces involved should complete a closed polygon. In Bishop's simplified method of slices, only the vertical equilibrium is imposed. Consider the mobilized shear force along the nth segment of the potential failure surface (Equation (9.4)):

$$S_n = c'_d l_n + N_n \tan\phi'_d = \frac{c' l_n}{\mathrm{FS}_s} + N_n \frac{\tan\phi'}{\mathrm{FS}_s} \tag{9.41}$$

Imposing the force equilibrium condition $\Sigma F_{zi} = 0$ on the nth slice leads to

$$W_n + \Delta SF_n - U_z = N_n \cos\alpha_n + S_n \sin\alpha_n \tag{9.42}$$

Recognizing that the force due to pore-water pressure is $u_n b_n$ and the mobilized shear force in Equation (9.41), Equation (9.42) becomes

$$W_n + \Delta SF_n - u_n b_n = N_n \cos\alpha_n + \left(\frac{c' l_n}{\mathrm{FS}_s} + N_n \frac{\tan\phi'}{\mathrm{FS}_s}\right)\sin\alpha_n \tag{9.43}$$

or

$$N_n = \frac{W_n + \Delta SF_n - u_n b_n - \dfrac{c' l_n}{\mathrm{FS}_s}\sin\alpha_n}{\cos\alpha_n + \dfrac{\tan\phi'}{\mathrm{FS}_s}\sin\alpha_n} = \frac{W_n + \Delta SF_n - u_n b_n - \dfrac{c' l_n}{\mathrm{FS}_s}\sin\alpha_n}{I(\alpha_n, \phi', \mathrm{FS}_s)} \tag{9.44}$$

where $I(\alpha_n, \phi', \mathrm{FS}_s)$ is

$$I = \cos\alpha_n + \frac{\tan\phi'}{\mathrm{FS}_s}\sin\alpha_n \tag{9.45}$$

The moment equilibrium principle is imposed by taking the driving force W_n and resistant force S_n with respect to point O at the center of the potential failure arc AC, for all m slices, i.e., $\Sigma M_i = 0$,

$$\sum_{n=1}^{m} W_n r \sin\alpha_n = \sum_{n=1}^{m} S_n r \tag{9.46}$$

Substituting Equation (9.44) into Equation (9.42) to eliminate N_n, and substituting the resulting equation into Equation (9.46) gives

$$\mathrm{FS}_s = \frac{\sum\limits_{n=1}^{m} \left(c' b_n + W_n \tan\phi' - u_n b_n \tan\phi + \Delta SF_n \tan\phi'\right)/I}{\sum\limits_{n=1}^{m} W_n \sin\alpha_n} \tag{9.47}$$

To assess ΔSF_n in the above equation requires additional information. In Bishop's simplified method, ΔSF_n is set to be zero, or the shear forces on each side of the slice are equal in magnitude but opposite in direction and Equation (9.47) becomes

$$\mathrm{FS}_s = \frac{\sum\limits_{n=1}^{m} \left(c' b_n + W_n \tan\phi' - u_n b_n \tan\phi\right)/I(\alpha_n, \phi', \mathrm{FS}_s)}{\sum\limits_{n=1}^{m} W_n \sin\alpha_n} \tag{9.48}$$

Table 9.2 Example of computational procedures for the Bishop's simplified method of slices

1	2	3	4	5	6	7	8	9	10	11	Driving force	13	Resistance
Slice	W_n	h_n	u_n	α_n	$\sin\alpha_n$	b_n	$u_n b_n$	$c' b_n$	$u_n b_n \tan\phi'$	$W_n \tan\phi'$	$W_n \sin\alpha_n$	I	$(9+11-10)/13$
Unit	(kN/m)	(m)	(kPa/m)	(deg)		(m)	(kN/m)	(kN/m)	(kN/m)	(kN/m)	(kN/m)	(Eq. 9.45)	(kN/m)
1	360	2.23	21.85	62	0.88	4.0	87.4	160.0	61.2	252.0	317.8	0.90	391.6
2	2044	6.64	65.07	45	0.70	5.0	325.3	200.0	227.8	1431.2	1445.3	1.05	1338.4
3	1260	8.65	84.77	31	0.51	5.0	423.8	200.0	296.7	882.2	648.9	1.11	710.2
4	2044	8.39	82.22	17	0.29	5.0	411.1	200.0	287.8	1431.2	597.6	1.10	1224.0
5	720	6.82	66.84	4	0.06	5.0	334.1	200.0	234.0	504.1	50.2	1.03	455.9
6	306	4.54	44.49	−5	0.08	5.0	222.4	200.0	155.8	214.2	−26.6	0.95	270.9
Σ											3033.3		4391.2
FS_s													1.45

Equation (9.48) is routinely used and requires an iterative procedure to determine FS_s as it appears on both side of the equation. Fortunately, for most problems, unique FS_s can be reached upon several iterations, as shown in the following example. Consider the same silt slope defined in Figure 9.5. To begin, $FS_s = 1.30$ is obtained from one trial of the ordinary method of slices. Columns 2–12 shown in Table 9.2 are computed only once based on the problem defined in Figure 9.5. Initially $FS_s = 1.30$ is used to calculate I for column 13 by Equation (9.45), leading to the resistance for each slice (column 14), and the total resistance and $FS_s = 1.41$ after the first iteration. $FS_s = 1.41$ is used for the second iteration to update new I for column 13, the resistances in column 14 and a new $FS_s = 1.44$ is obtained. After four iterations, a unique value of $FS_s = 1.45$ is obtained. It can also be shown that the convergent value for FS_s (1.45) is insensitive to the initial value.

This section provided an overview of two limit-equilibrium methods of slices used for slope stability analysis. In the ordinary method only forces on the potential failure surface and body forces are considered in the moment equilibrium, whereas in Bishop's simplified method the inter-slice forces are considered. The next section extends each of these methods to partially saturated conditions using the suction stress concept and examines the effect of steady infiltration on stability using an infinite-slope analysis.

9.3 Landslides under steady infiltration

9.3.1 Extension of classical methods to unsaturated conditions

For shallow landslide problems, the infinite-slope stability model and Culmann's method can be readily expanded to account for variably saturated conditions with a straightforward application of the unified effective stress concept described in Chapter 6. In the same manner, all the limit-equilibrium theories such as the ordinary method and Bishop's simplified method of slices can be readily established for deeply seated landslide problems using the unified effective stress principle.

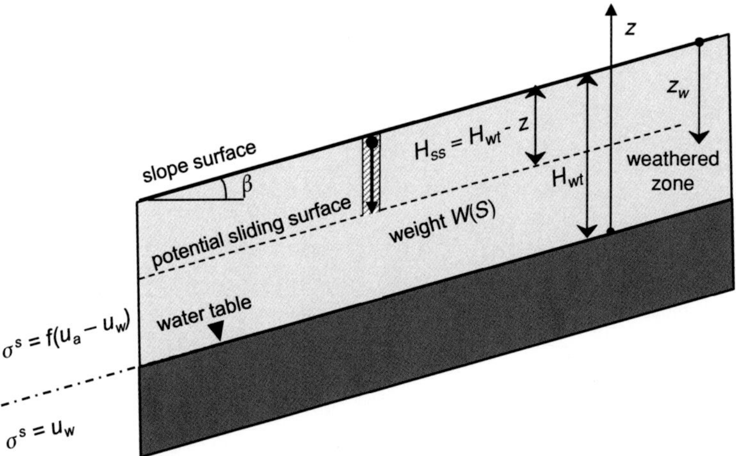

Figure 9.6 Conceptual illustration of an infinite-slope model under variably saturated conditions.

For the infinite slope shown in Figure 9.6, two quantities, the moist unit weight γ and suction stress σ^s are dependent on the soil moisture content or saturation and are rigorously considered here. The unified effective stress described in Chapter 6 can be directly used (Lu and Likos, 2006), i.e.,

$$\sigma' = \sigma - u_a - \sigma^s \tag{9.49}$$

To consider the variation in unit weight that results from changes in moisture content in an infinite-slope analysis, integration of the moisture content profile above the potential slip surface is needed. The weight of soil column per unit cross section area W_v shown in Figure 9.6 can be obtained by integration of the unit weight γ in the vertical direction z from the potential sliding surface $z = H_{wt} - H_{ss}$ to the slope surface $z = H_{wt}$ as

$$W_v = \int_{H_{wt}-H_{ss}}^{H_{wt}} \gamma \, dz \tag{9.50}$$

Substituting W_v for $\gamma\ H_{ss}$ and suction stress σ^s for u_w in Equation (9.19) yields the general equation for factor of safety for infinite slopes under variably saturated conditions:

$$\mathrm{FS} = \frac{\tan\phi'}{\tan\beta} + \frac{2c'}{W_v \sin 2\beta} - r_u\,(\tan\beta + \cot\beta)\tan\phi' \tag{9.51}$$

$$r_u = \frac{\sigma^s}{W_v} \tag{9.52}$$

Suction stress σ^s can be defined in terms of matric suction $(u_a - u_w)$ for variably saturated materials (Lu *et al.*, 2010):

$$\sigma^s = -(u_a - u_w) = u_w \qquad u_a - u_w \le 0 \tag{9.53}$$

$$\sigma^s = -\frac{(u_a - u_w)}{(1 + [\alpha(u_a - u_w)]^n)^{(n-1)/n}} \qquad u_a - u_w > 0 \tag{9.54}$$

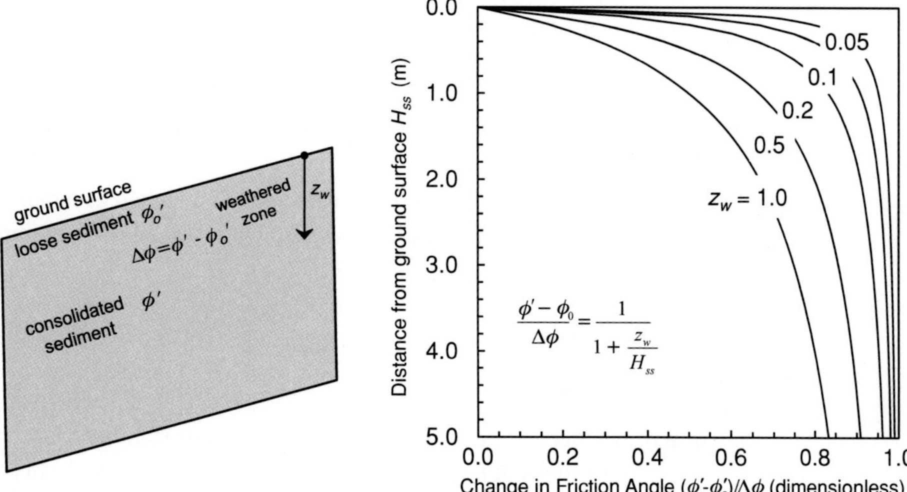

Figure 9.7 Graphic illustration of relative change in friction angle as a function of depth for various thicknesses of the weathered zone parameter z_w (from Lu and Godt, 2008).

The unit weight γ can be expressed using the specific gravity G_s, the unit weight of water γ_w, void ratio e, and the degree of saturation S:

$$\gamma = \gamma_w \frac{G_s + eS}{1+e} \tag{9.55}$$

The degree of saturation S can be analytically expressed using a soil-water retention function. If van Genuchten's model (1980) is used, the degree of saturation S is expressed as a function of matric suction $(u_a - u_w)$ with the unsaturated hydrologic parameters S_r, α, and n:

$$S = S_r + (1 - S_r)\left\{\frac{1}{1+[\alpha(u_a - u_w)]^n}\right\}^{1-1/n} \tag{9.56}$$

Furthermore, under a steady unsaturated vertical infiltration rate q, the profile of matric suction in Equation (9.56) can be expressed using the unsaturated hydrologic properties k_s, α, n, and S_r (Lu and Likos, 2004a; also in Chapter 3):

$$(u_a - u_w) = \frac{-1}{\alpha} ln\left[(1+q/k_s)e^{-\gamma\alpha z} - q/k_s\right] \tag{9.57}$$

As shown in Chapter 6, the friction angle of hillslope materials may vary with depth due to the decrease in porosity with depth. This phenomenon could be important in shallow environments, as illustrated in Figure 9.7. Following the model of Lu and Godt (2008), a mathematical expression describing the friction dependence on the soil depth H_{ss} is

$$\phi' = \phi_o' + \frac{\Delta\phi}{1+\frac{z_w}{H_{ss}}} \tag{9.58}$$

where ϕ'_o is the friction angle at the slope surface, $\Delta\phi$ is the total change in friction angle, and z_w is the thickness of the weathered zone. If soil samples from different depths are tested to determine ϕ'_o, $\Delta\phi$, and z_w, Equation (9.58) can be used to quantify the shear strength increase with depth in a hillslope environment. Figure 9.7 illustrates the relative change in friction angle as a function of soil depth for various weathering zone parameters z_w.

With the above conceptualization of a weathered soil mantle, a generalized factor of safety equation for the infinite-slope model under variably saturated soil conditions becomes (Lu and Godt, 2008)

$$FS(z) = \frac{\tan\phi'(z)}{\tan\beta} + \frac{2c'}{W_v \sin 2\beta} - \frac{\sigma^s}{W_v}(\tan\beta + \cot\beta)\tan\phi'(z) \tag{9.59}$$

Equation (9.59), together with Equations (9.53)–(9.58), completely defines a generalized analytical hydro-mechanical framework for infinite-slope stability under variably saturated conditions. Illustrations of how the generalized infinite-slope model can be used for steady infiltration conditions will be described in Section 9.3.2 and for transient infiltration conditions in Section 9.4.

The extension of Culmann's method to variably saturated conditions was described in Section 9.2 (Equations (9.36) and (9.37)).

Generalization of the method of slices for variably saturated conditions is illustrated here for the ordinary and Bishop's routine methods. For the ordinary method, replacing pore-water pressure $\gamma_w h_n$ by suction stress σ^s in Equation (9.40) gives

$$FS_s = \frac{\sum_{n=1}^{m} [c' l_n + (W_n \cos\alpha_n - \sigma^s l_n)\tan\phi']}{\sum_{n=1}^{m} W_n \sin\alpha_n} \tag{9.60}$$

And for Bishop's simplified method or Bishop's routine method, replacing pore-water pressure u_n by suction stress σ^s in Equation (9.48) gives

$$FS_s = \frac{\sum_{n=1}^{m} \left(c' b_n + W_n \tan\phi' - \sigma^s_n b_n \tan\phi'\right) / I(\alpha_n, \phi', FS_s)}{\sum_{n=1}^{m} W_n \sin\alpha_n} \tag{9.61}$$

Because suction stress σ^s is negative, inclusion of it in Equations (9.60) and (9.61) will lead to more rigorous and generally greater factor of safety values. A quantitative illustration of application of the expanded ordinary and Bishop's routine methods is shown here for the same silt slope described in Figure 9.5, except the location of the water table is changed and the unit weight of soil is reduced accordingly, as shown in Figure 9.8.

The water table decline shown in Figure 9.8 may reflect the situation after a long period of drought or the installation of horizontal wells to de-water the slope. Two methods of incorporating suction stress above the water table are used for comparison. In the first case, zero pore-water pressure or suction stress above the water table is assumed and hydrostatic conditions below the water table are assumed. In the second case, a constant suction stress

Table 9.3 Example of the ordinary method of slices for case 1: classical treatment of pore pressure

1	2	3	4	5	6	7	8	9	10	11	12	Driving force	Resistance
Slice	W_n	h_n	u_n	α_n	$\sin\alpha_n$	$\cos\alpha_n$	b_n	l_n	$c'l_n$	$u_n l_n$	$W_n\cos\alpha_n$	$W_n\sin\alpha_n$	$10 + (12 - 11)\tan\phi'$
Unit	(kN/m)	(m)	(kPa/m)	(deg)			(m)	(m)	(kN/m)	(kN/m)	(kN/m)	(kN/m)	(kN/m)
1	320	0	0.00	62	0.88	0.47	4.0	8.52	340.81	0.00	150.23	282.54	446.0
2	1817	0	0.00	45	0.71	0.71	5.0	7.07	282.84	0.00	1284.81	1284.81	1182.4
3	1120	0	0.00	31	0.51	0.86	5.0	5.83	233.33	0.00	960.03	576.84	905.5
4	1817	2.27	22.25	17	0.29	0.96	5.0	5.23	209.14	116.31	1737.61	531.24	1344.3
5	640	2.82	27.64	4	0.07	1.00	5.0	5.01	200.49	138.52	638.44	44.64	550.5
6	272	2.31	22.64	−5	−0.09	1.00	5.0	5.02	200.76	113.62	270.96	−23.71	310.9
Σ												2696.38	4739.8
FS_s													1.76

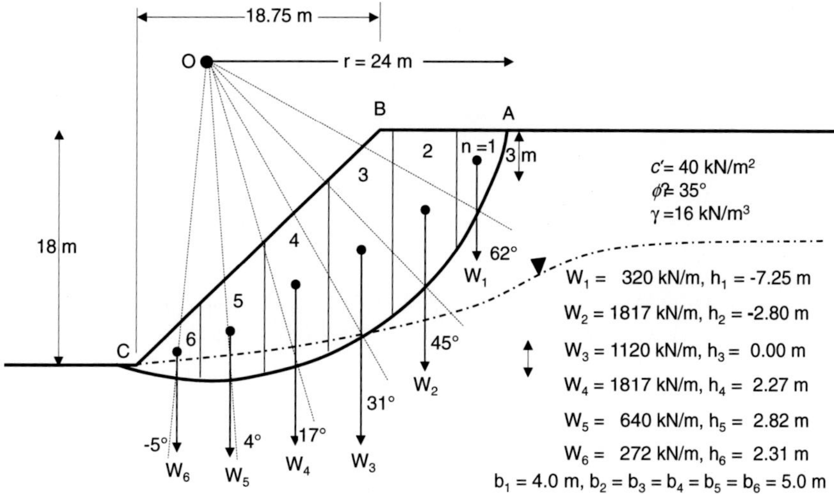

Figure 9.8 Illustration of ordinary and Bishop's simplified methods of slices for a finite silt slope for various saturation conditions.

of −49 kPa above the water table and hydrostatic conditions below the water table are assumed. More rigorous treatment of suction variation with moisture content variation above the water table can be found later in this section and in Chapter 6. The computational procedures for both ordinary and Bishop's simplified methods for the first case are shown in Tables 9.3 and 9.4.

Compared to the situation before de-watering, the factor of safety of the slope increases from 1.30 to 1.76 for results from the ordinary method, and from 1.45 to 1.90 for Bishop's simplified method. Both methods show that de-watering increases the factor of safety of the slope by about 30%.

The computational procedures for both ordinary and Bishop's simplified methods for the second case are shown in Tables 9.5 and 9.6.

Table 9.4 Example of Bishop's simplified method of slices for case 1: classical treatment of pore pressure

1	2	3	4	5	6	7	8	9	10	11	Driving force	13	Resistance
Slice	W_n	h_n	u_n	α_n	$\sin\alpha_n$	b_n	$u_n b_n$	$c' b_n$	$u_n b_n \tan\phi'$	$W_n \tan\phi'$	$W_n \sin\alpha_n$	I	(9 + 11 – 10)/13
Unit	(kN/m)	(m)	(kPa/m)	(deg)		(m)	(kN/m)	(kN/m)	(kN/m)	(kN/m)	(kN/m)	(Eq. 9.43)	(kN/m)
1	320	0	0.00	62	0.88	4.0	0.0	160.0	0.00	224.1	282.54	0.79	483.2
2	1817	0	0.00	45	0.71	5.0	0.0	200.0	0.00	1272.3	1284.81	0.97	1521.4
3	1120	0	0.00	31	0.51	5.0	0.0	200.0	0.00	784.2	576.84	1.05	940.1
4	1817	2.27	22.25	17	0.29	5.0	111.2	200.0	77.88	1272.3	531.24	1.06	1310.4
5	640	2.82	27.64	4	0.07	5.0	138.2	200.0	96.75	448.1	44.64	1.02	538.8
6	272	2.31	22.64	−5	0.09	5.0	113.2	200.0	79.26	190.5	−23.71	0.96	322.8
Σ											2696.38		5116.7
FS_s													1.90

Table 9.5 Example of the ordinary method of slices for case 2: expanded theory for unsaturated conditions

1	2	3	4	5	6	7	8	9	10	11	12	Driving force	Resistance
Slice	W_n	h_n	u_n	α_n	$\sin\alpha_n$	$\cos\alpha_n$	b_n	l_n	$c' l_n$	$u_n l_n$	$W_n \cos\alpha_n$	$W_n \sin\alpha_n$	10 + (12 – 11) $\tan\phi'$
Unit	(kN/m)	(m)	(kPa/m)	(dec)			(m)	(m)	(kN/m)	(kN/m)	(kN/m)	(kN/m)	(kN/m)
1	320	−5	−49.00	62	0.88	0.47	4.00	8.52	340.81	−417.49	150.23	282.54	738.33
2	1817	−5	−49.00	45	0.71	0.71	5.00	7.07	282.84	−346.48	1284.81	1284.81	1425.09
3	1120	0	0.00	31	0.52	0.86	5.00	5.83	233.33	0.00	960.03	576.84	905.55
4	1817	2.27	22.25	17	0.29	0.96	5.00	5.23	209.14	116.31	1737.61	531.24	1344.38
5	640	2.82	27.64	4	0.07	1.00	5.00	5.01	200.49	138.52	638.44	44.64	550.54
6	272	2.31	22.64	−5	−0.1	1.00	5.00	5.02	200.76	113.62	270.96	−23.71	310.94
Σ												2696.38	5274.82
FS_S													1.96

Table 9.6 Example of Bishop's simplified method of slices for case 2: expanded theory for unsaturated conditions

1	2	3	4	5	6	7	8	9	10	11	Driving force	13	Resistance
Slice	W_n	hn	u_n	α_n	$\sin\alpha_n$	b_n	$u_n b_n$	$c' b_n$	$u_n b_n \tan\phi'$	$W_n \tan\phi'$	$W_n \sin\alpha_n$	I	(9 + 11 – 10)/13
Unit	(kN/m)	(m)	(kPa/m)	(dec)		(m)	(kN/m)	(kN/m)	(kN/m)	(kN/m)	(kN/m)	(Eq. 9.43)	(kN/m)
1	320	−5	−49.00	62	0.88	4.00	−196.00	160.00	−137.24	224.07	282.54	0.77	677.38
2	1817	−5	−49.00	45	0.71	5.00	−245.00	200.00	−171.55	1272.28	1284.81	0.95	1734.99
3	1120	0	0.00	31	0.52	5.00	0.00	200.00	0.00	784.23	576.84	1.03	953.50
4	1817	2.27	22.25	17	0.29	5.00	111.23	200.00	77.88	1272.28	531.24	1.06	1320.84
5	640	2.82	27.64	4	0.07	5.00	138.18	200.00	96.75	448.13	44.64	1.02	539.89
6	272	2.31	22.64	−5	−0.1	5.00	113.19	200.00	79.26	190.46	−23.71	0.97	321.96
Σ											2696.38		5548.57
FS_S													2.06

Table 9.7 Summary of computed factor of safety by different methods and saturation conditions

Computed FS	Classical ordinary	Bishop's simplified	Unsaturated ordinary	Unsaturated Bishop's simplified
Before de-watering	1.30	1.45	1.30	1.45
After de-watering	1.76	1.90	1.96	2.06

Compared to the situation before de-watering, the factor of safety of the slope increases from 1.36 to 1.96 for the ordinary method and from 1.45 to 2.06 for Bishop's simplified method. Part of the difference results from the treatment of pore-water pressure as a normal force in the ordinary method rather than an effective stress as is done in Bishop's simplified method. Both methods show that de-watering increases the factor of safety of the slope by about 42–44%. Under the de-watering situation, the factor of safety of the slope increases from 1.76 by the classical method to 1.96 by the expanded theory for unsaturated conditions for the ordinary method, and the factor of safety increases from 1.90 by the classical to 2.06 by the expanded theory for unsaturated conditions for Bishop's simplified method. Both methods show that the expanded theory for unsaturated conditions increases the factor of safety calculation by about 8–13%. The ranges of the differences in factor of safety demonstrated here are significant for determining the stability of hillslopes.

The computed values of factor of safety for all the cases are summarized in Table 9.7.

9.3.2 Impact of infiltration rate on slope stability

The expanded infinite-slope stability model for variably saturated soil conditions described in Section 9.3.1 is used in the following analysis as an illustration. A generalized factor of safety equation without explicit consideration of the impact of moisture content on unit weight of the soil such as given by Equation (9.59) can be expressed as

$$\mathrm{FS}(z) = \frac{\tan\phi'(z)}{\tan\beta} + \frac{2c'}{\gamma\,(H_{wt}-z)\sin 2\beta} - \frac{\sigma^s}{\gamma\,(H_{wt}-z)}(\tan\beta+\cot\beta)\tan\phi'(z) \quad (9.62)$$

Under steady, one-dimensional (vertical) flux at the ground surface q (m/s), profiles of suction stress (developed in Chapter 6) can be calculated as (Lu and Godt, 2008)

$$\sigma^s = -\frac{1}{\alpha}\frac{\ln\left[(1+q/K_s)e^{-\gamma_w\alpha z} - q/K_s\right]}{\left(1+\{-\ln[(1+q/K_s)e^{-\gamma_w\alpha z} - q/K_s]\}^n\right)^{(n-1)/n}} \qquad u_w \le 0 \quad (9.63)$$

where K_s is saturated hydraulic conductivity (m/s), q is the steady flux rate (m/s), which is negative for infiltration and positive for evaporation, and α and n are unsaturated hydro-mechanical properties.

Equations (9.62) and (9.63) provide a theoretical basis to assess the stability of infinite slopes composed of different materials under various steady infiltration/evaporation conditions. The following example illustrates an application to assess the changes in stability of 45° sandy hillslopes composed of three different sands: coarse, medium, and fine. For these sandy soils, cohesion $c' = 0$, averaged moist unit weight $\gamma = 18$ kN/m^3 and the other hydro-mechanical parameters for the hilllslopes are given in Table 9.8.

Table 9.8 Hydro-mechanical properties of sands used in the example analysis

Soil type	n(unitless)	α(kPa^{-1})	k_s(m/s)	ϕ_o (degree)	$\Delta\phi$(degree)	β(degree)	z_w(m)
Coarse sand	7.5	0.45	–	40	6	45	0.5
Medium sand	5.5	0.14	–	40	6	45	0.5
Fine sand	4.75	0.08	1.0×10^{-6}	40	6	45	0.5

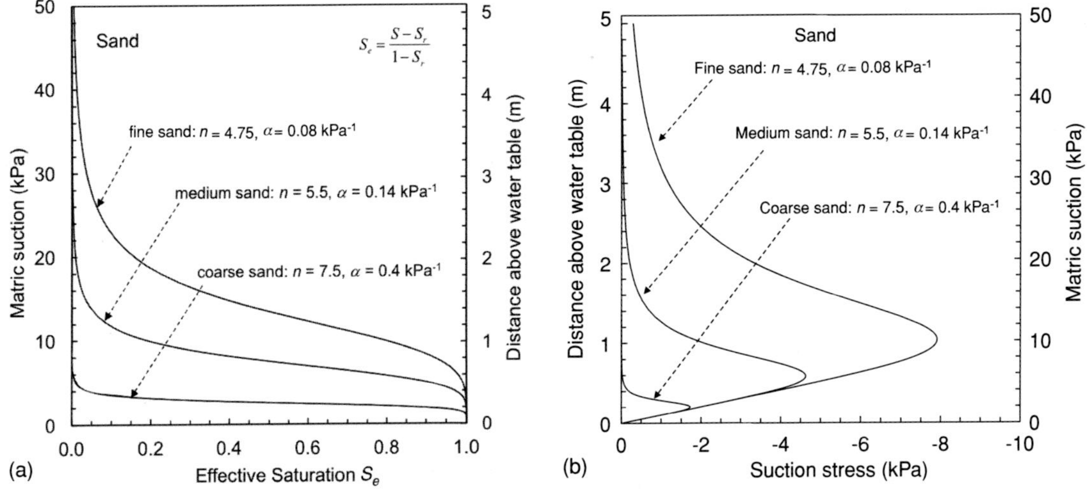

Figure 9.9 Hydrostatic profiles of (a) effective degree of saturation, and (b) suction stress with a thickness of 5 m (from Lu and Godt, 2008).

Profiles of soil moisture content, suction stress, and factor of safety for the sandy hillslopes under zero infiltration or hydrostatic conditions are examined first. The profiles of soil moisture content and soil suction for the sandy soils under hydrostatic equilibrium conditions can be assessed using Equations (3.44)–(3.46) and the unsaturated parameters shown in Table 9.8. The one-dimensional soil moisture and soil suction profiles under steady infiltration/evaporation conditions (also see Lu and Godt, 2008) in terms of the effective degree of saturation and matric suction can be written as

$$(u_a - u_w) = \frac{-1}{\alpha} \ln\left[(1 + q/K_s)e^{-\gamma\alpha z} - q/K_s\right] \tag{9.57}$$

$$S_e = \frac{S - S_r}{1 - S_r} = \left\{\frac{1}{1 + [\alpha(u_a - u_w)]^n}\right\}^{1-1/n} \tag{9.56}$$

For hydrostatic conditions ($q = 0$) the potential gradient is zero and matric suction increases linearly with height above the water table. Profiles of the effective degree of saturation for three hypothetical sands for these conditions are shown in Figure 9.9a. For the coarse sand, the effective degree of saturation varies dramatically near the water table from 100% saturation at the water table to the residual state about 0.6 m (or at 6 kPa of soil suction) above the water table. For the medium sand, the zone of saturation greater than the

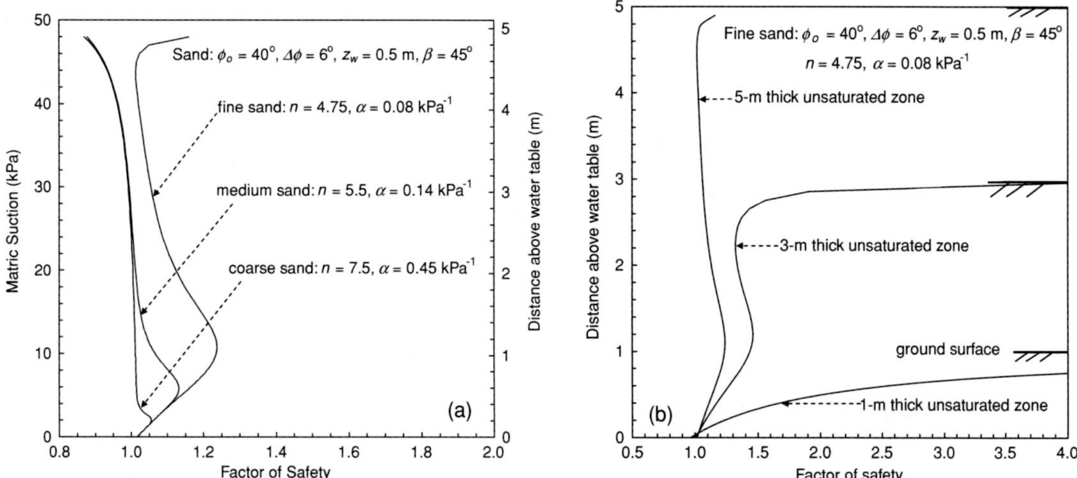

Figure 9.10 Hydrostatic profiles of (a) factor of safety in various sandy slopes with a thickness of 5 m, and (b) factor of safety in fine sand slopes of various thickness (from Lu and Godt, 2008).

residual state extends to about 2 m (or 20 kPa of soil suction) above the water table. For the fine sand, this zone reaches more than 5 m (or 50 kPa of soil suction) above the water table.

Suction stress profiles for the three hypothetical sands can be predicted using Equation (9.63) for $q = 0$ and are shown in Figure 9.9b. In general, suction stress for the sandy soils is sensitive to soil suction over a range from 0 to less than several kPa and diminishes to nearly zero for soil suctions greater than several hundred kPa. Within several hundred kPa of soil suction, suction stress in sands decreases from zero at zero soil suction (100% saturation) to a minimum between 8 and 10 kPa of soil suction, and then increases towards zero again as soil suction increases. The minimum suction stress for the hypothetical sandy soils shown in Figure 9.9b is -1.75 kPa for the coarse sand, -4.25 kPa for the medium sand, and -8.0 kPa for the fine sand. Figure 9.9b also shows that the distance above the water table at which the minimum suction stress occurs will increase as the soil becomes finer. For example, for the fine sand, the minimum suction stress of -8.0 kPa occurs at about 1.0 m above the water table.

The effect of suction stress above the water table on the factor of safety of the sandy slope can be assessed by Equation (9.62), and is shown in Figure 9.10a. For illustration, the inclination of the slope is assumed to be 45°. The friction angle is 40° at the ground surface and increases to 46° at a depth of 5 m with the weathering zone parameter z_w set as 0.5 m. Under the zero infiltration condition with water tables 5 m vertically below the ground surface and assuming steady, slope-parallel flow below the water table (Figure 9.10a), the factor of safety is 1.03 at the water table for all three sands. The maximum factor of safety nearly coincides with the location of the minimum suction stress (refer to Figure 9.9b) in all three sands. For slopes of coarse, medium, and fine sand, the maximum factors of safety are about 1.06, 1.13, and 1.24 respectively. The potential increases in suction stress by several tens of percent in sandy soils could be important for the stability of many natural slopes, where environmental changes such as the passage of a transient wetting front could overcome suction stress and potentially trigger shallow landslides.

Above the location of the maximum factor of safety in sandy hillslopes, the factor of safety varies greatly depending on the type of sands (Figure 9.10a). For the coarse and medium sands, the profile of the factor of safety shows three distinct horizons: a zone of sharp reduction right above the location of the maximum factor of safety, above which lie zones of gradual reduction and a zone of increasingly sharp reduction in the factor of safety near the ground surface. As an example, for the slope of medium sand (Figure 9.10a) the zone of sharp reduction is located between 0.6 m and 1.5 m above the water table, the gradual zone of reduction is between 1.5 m and 3.5 m above the water table, and the zone of increasingly sharp reduction is within 1.5 m of the ground surface. For slopes of both coarse and medium sands, the factor of safety reaches 1.0 at about 2.4 m above the water table, implying that under hydrostatic conditions slopes inclined at 45° of these two sands will not be stable if the unsaturated zone is thicker than 2.4 m. For these slopes to remain stable, additional factors contributing to stability, such as the support by vegetation roots, must be invoked.

The focus of the following example is on the variation in the profiles of effective saturation, suction stress, and factors of safety in a slope of fine sand because the saturated hydraulic conductivity of this material is approximately equal to realistic rainfall rates. Steady rainfall rates much less than the saturated hydraulic conductivity produce less pronounced changes over hydrostatic conditions. For the slope of fine sand, the variation of the factor of safety shows two distinct horizons: a zone of reduction right above the location of the maximum factor of safety, and a sharp increase in the factor of safety near the ground surface (Figure 9.10a). The minimum factor of safety ($\sim$1.01) occurs at the transition of these two zones about 0.5 m below the ground surface. The existence of suction stresses of -8.0 kPa in the fine sand soil under hydrostatic conditions implies that the factor of safety in the unsaturated zone could be reduced to less than 1.0 under infiltration conditions. These scenarios will be explored in a later section. It is important to point out that the sharp variation of the factor of safety with depth cannot be predicted by the classical infinite-slope stability model. For the classical model, the factor of safety is a constant over the entire layer of cohesionless sand. The occurrence of the minimum factor of safety near the ground surface depends on the thickness of the unsaturated zone (Figure 9.10b). When the unsaturated zone is relatively thin, such as 1 m shown in Figure 9.10b, the minimum disappears and the factor of safety increases monotonically with increasing distance away from the water table.

Under steady infiltration conditions, the impact of suction stress on the potential of the factor of safety to be less than 1.0, or predicted slope failure in the unsaturated zone, can be explored by applying various steady infiltration conditions in Equation (9.63). The fine sand slope with the same hydro-mechanical properties and geometry given for the previous sandy soil examples is used for illustration. A series of infiltration rates of -1.5×10^{-7} m/s, -2.5×10^{-7} m/s, and -4.9×10^{-7} m/s are used. A steady infiltration rate of -1.5×10^{-7} m/s for a period of one week is equivalent to a total rainfall of 91 mm, a moderate rainfall event for many hillslope environments. A steady infiltration rate

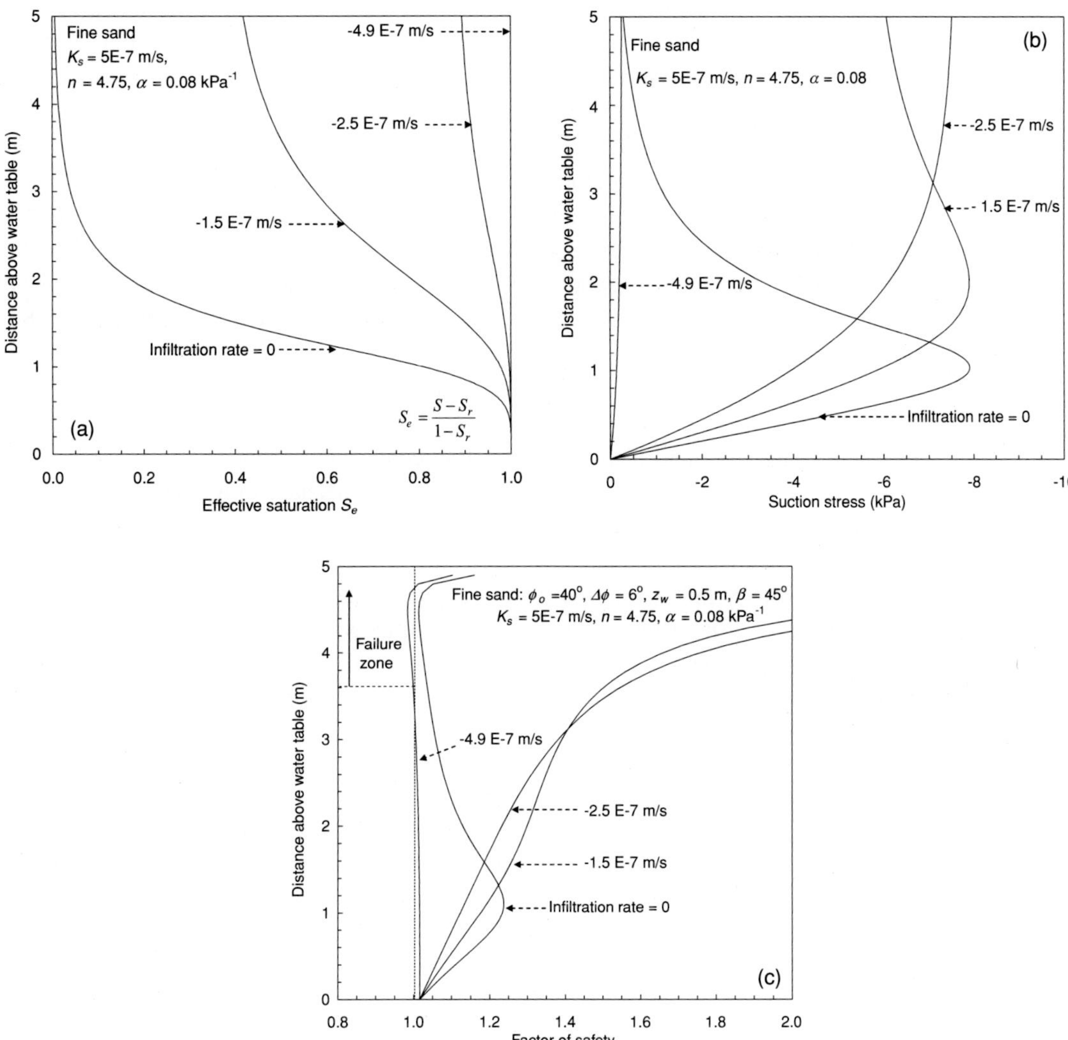

Figure 9.11 Profiles of (a) effective degree of saturation, (b) suction stress, and (c) factor of safety in a 5 m deep fine sand slope under various steady flux conditions (from Lu and Godt, 2008).

of -4.9×10^{-7} m/s for a week is equivalent to a total rainfall of 296 mm, a heavy rainfall event. For a sandy soil, the wetting front can reach a depth of several meters within a month following rainfall events of these magnitudes, so that the steady model used here yields an approximate range of the changes in soil suction, suction stress, and the factor of safety that can be expected. The steady profiles of the effective degree of saturation and suction stress under the assumed infiltration rates as well as under hydrostatic conditions are predicted using Equations (9.56), (9.57), and (9.63), as shown in Figure 9.11a and Figure 9.11b, respectively.

The profiles of the effective degree of saturation (Figure 9.11a) are sensitive to the applied infiltration rates, varying from nearly zero saturation under hydrostatic conditions to nearly full saturation under the infiltration rate of -4.9×10^{-7} m/s. The profile of suction stress also varies with the variation in infiltration rate. The magnitude of the minimum suction stress remains about the same as the infiltration rate decreases from 0 to -2.5×10^{-7} m/s, but its location moves toward the ground surface. Changing the infiltration rate from -2.5×10^{-7} m/s to -4.9×10^{-7} m/s causes a drastic increase in suction stress to nearly zero throughout the soil profile. Correspondingly, a zone where the factor of safety is less than 1.0 is observed between 0.3 m and 1.4 m below the ground surface, indicating that shallow landslides may occur on this slope under heavy precipitation conditions (Figure 9.11c).

9.3.3 Impact of moisture variation on slope stability

In recent years, the rapid increase in moisture content in hillslope materials that results from heavy rainfall has been considered as a possible mechanism for destabilizing slopes as it provides an increase in the driving force along a potential sliding surface. On the other hand, as illustrated in this chapter, increasing the soil weight can also increase the mobilized shear resistance by mobilizing the friction angle. A rigorous analysis examining each of the two counteracting mechanisms through an infinite-slope stability analysis (Equation (9.62)) is provided here.

The factor of safety for the infinite-slope stability model can be expressed as (Equation (9.62) and Figure 9.6)

$$FS = \frac{\tan\phi'}{\tan\beta} + \frac{2c'}{\gamma H_{ss}\sin 2\beta} - \frac{\sigma^s}{\gamma H_{ss}}(\tan\beta + \cot\beta)\tan\phi' \tag{9.62}$$

where an averaged unit weight γ above the potential sliding surface is used. The variation or change in factor of safety with respect to changes in unit weight γ, according to Equation (9.63), is

$$\Delta FS = -\frac{2c'\Delta\gamma}{\gamma^2 H_{ss}\sin 2\beta} + \frac{\sigma^s\Delta\gamma}{\gamma^2 H_{ss}}(\tan\beta + \cot\beta)\tan\phi'$$

or

$$\Delta FS = -\frac{\Delta\gamma}{\gamma}\left(\frac{2c'}{\gamma H_{ss}\sin 2\beta} - \frac{\sigma^s}{\gamma H_{ss}}(\tan\beta + \cot\beta)\tan\phi' + \frac{\tan\phi'}{\tan\beta} - \frac{\tan\phi'}{\tan\beta}\right) \tag{9.64}$$

Substituting Equation (9.63) into Equation (9.64) gives

$$\Delta FS = -\frac{\Delta\gamma}{\gamma}\left(FS - \frac{\tan\phi'}{\tan\beta}\right) \tag{9.65}$$

or

$$\frac{\Delta FS}{FS} = -\frac{\Delta\gamma}{\gamma} + \frac{\frac{\tan\phi'}{\tan\beta}}{FS}\frac{\Delta\gamma}{\gamma} = -\frac{\Delta\gamma}{\gamma} + \lambda\frac{\Delta\gamma}{\gamma} \tag{9.66}$$

$$\lambda = \frac{\frac{\tan\phi'}{\tan\beta}}{FS} \tag{9.67}$$

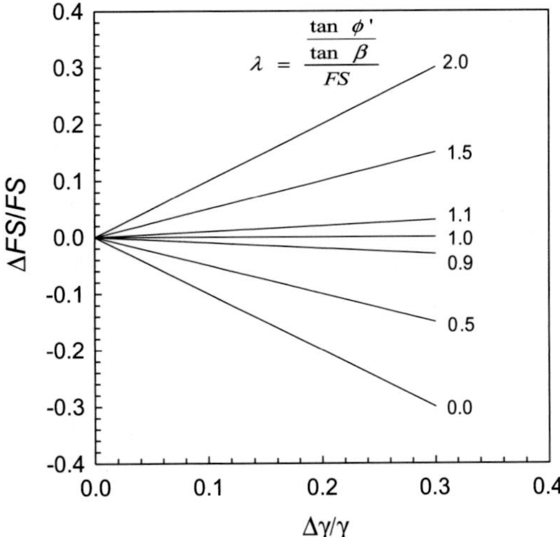

Figure 9.12 Impact of soil weight on slope stability in an infinite-slope theory.

The dimensionless number λ can be called the slope stability number for soil unit weight. The physical implications of changes in the unit weight of soil on the factor of safety now can be clearly interpreted from Equation (9.66). The first term on the right hand side of Equation (9.66) indicates that increases in the unit weight lead to decreases in factor of safety or the stability of slopes at the same rate. The second term indicates that increases in the unit weight will also lead to increases in the factor of safety by a factor λ that depends on the friction angle of materials, slope angle, and the prevailing factor of safety. If λ is less than unity, an increase in the soil weight will result in a decrease in slope stability. If λ is equal to unity, a change in soil weight will have no effect on slope stability. If λ is greater than unity, an increase in the soil weight will result in an increase in slope stability. The net effect on the relative changes in the factor of safety as a function of relative change in the soil unit weight is shown in Figure 9.12 for various values of λ.

If λ is less than unity, Equation (9.67) gives

$$\frac{\tan\phi'}{\tan\beta} < \mathrm{FS} \tag{9.68}$$

Substituting Equation (9.63) into Inequality (9.68) gives

$$\frac{\tan\phi'}{\tan\beta} < \mathrm{FS} = \frac{\tan\phi'}{\tan\beta} + \frac{2c'}{\gamma H_{ss}\sin 2\beta} - \frac{\sigma^s}{\gamma H_{ss}}(\tan\beta + \cot\beta)\tan\phi' \tag{9.69}$$

Recognizing $\sin 2\beta = 2/(\tan\beta + \cot\beta)$, Inequality (9.69) becomes

$$c' > \sigma^s \tan\phi' \tag{9.70}$$

for destabilizing condition. On the other hand, if λ is greater than unity, Equation (9.67) gives

$$\frac{\tan\phi'}{\tan\beta} > \mathrm{FS} \tag{9.71}$$

Substituting Equation (9.63) into Inequality (9.6) gives

$$\frac{\tan\phi'}{\tan\beta} > \text{FS} = \frac{\tan\phi'}{\tan\beta} + \frac{2c'}{\gamma H_{ss}\sin 2\beta} - \frac{\sigma^s}{\gamma H_{ss}}(\tan\beta + \cot\beta)\tan\phi' \tag{9.72}$$

or

$$c' < \sigma^s \tan\phi' \tag{9.73}$$

for stabilizing conditions. For neutral conditions, it can be shown that

$$c' = \sigma^s \tan\phi' \tag{9.74}$$

Since, above the water table, suction stress is always negative, the Inequality (9.70) holds. Below the water table, suction stress σ^s is equal to the pore-water pressure u_w. Inequalities (9.70) or (9.73) may or may not be satisfied, depending on pore-water pressure u_w, cohesion c' and friction angle ϕ'. Depending on seepage conditions, pore-water pressures below the water table in hillslopes generally increase with depth below the water table. Thus, a critical location where Equation (9.74) holds will define if increasing soil weight will destabilize or stabilize the potential sliding surface. Above this location, increasing soil weight will tend to destabilize the slope and vice versa. Therefore, in a qualitative sense an increase in soil weight will tend to destabilize slopes if the potential sliding surface is above the water table. If the potential sliding surface is below the water table but above some critical depth, increasing the soil weight will still tend to destabilize slopes. Below this critical depth however, an increase in soil weight will stabilize slopes. Quantitative assessments of changes in soil weight on slope stability require explicit assessment of the factor of safety and can be accomplished using the generalized infinite-slope framework established early in this section for slopes under various steady infiltration conditions.

A slope composed of silty sand with geometry and hydro-mechanical properties shown in Figure 9.13 will be used to examine quantitatively the impact of changes in soil unit weight on slope stability. The generalized infinite-slope model fully defined by Equations (9.50) to (9.59) will be used under four steady vertical unsaturated seepage conditions: hydrostatic ($q = 0$), infiltration rate $q = -5 \times 10^{-7}$ m/s, $q = -8 \times 10^{-7}$ m/s, and $q = -9.9 \times 10^{-7}$ m/s. The soil unit weight profile from the slope surface to the water table under the hydrostatic condition is first calculated by Equations (9.55)–(9.57) and plotted in Figure 9.14a. It can be seen that soil unit weight varies from its saturated value at $\sim$19 kN/m^3 to a relatively dry state of $\sim$15 kN/m^3. Both profiles of matric suction and degree of saturation can also be examined using Equations (9.57) and (9.56) but are not shown here. The corresponding suction stress profile is calculated using Equation (9.63) and shown in Figure 9.14b. For this material, the profile of suction stress is non-linear under hydrostatic conditions. The minimum suction stress is -12.75 kPa, which occurs at 1.63 m above the water table. This substantial amount of suction stress stabilizes the slope and such an effect can be seen in the profile of factor of safety computed from Equation (9.59) and plotted in Figure 9.14c. The factor of safety reaches a maximum of 1.41 at 1.81 m above the water table. The minimum factor of safety of 1.01 occurs at the water table. Assuming the hillslope material is homogeneous, the slope is stable under hydrostatic conditions.

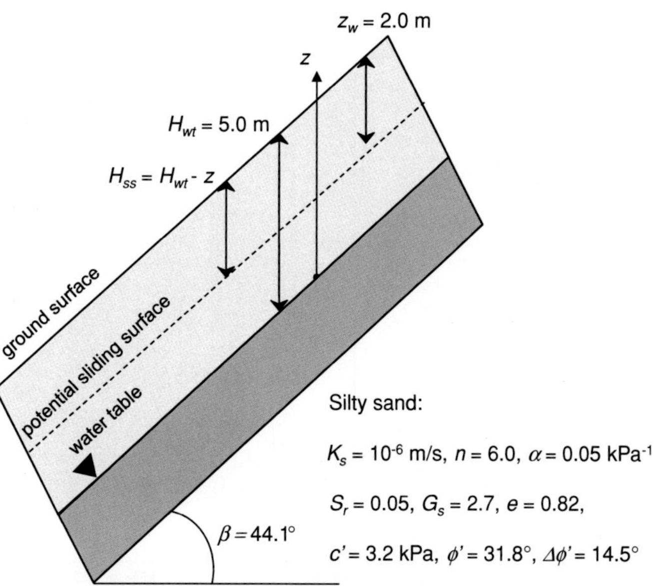

Figure 9.13 Illustration of a 45° unsaturated silty sand slope for factor of safety analysis by the generalized infinite-slope model.

If a steady infiltration rate of $q = -5 \times 10^{-7}$ m/s, or 43.2 mm/day is imposed, the soil unit weight increases roughly constantly with height above the water table and ranges from its saturated value of 19.0 kN/m^3 at the water table to 18.7 kN/m^3 at the ground surface (Figure 9.14a). The corresponding suction stress increases in the lower part of the soil layer near the water table, but decreases in the upper part near the slope surface (Figure 9.14b). The minimum value of suction stress of −11.4 kPa is near the slope surface, leading to a significant increase in factor of safety in the upper part of the soil layer (Figure 9.14c). The profiles of factor of safety plotted as dashed lines are computed using the average value of soil unit weight over the entire 5 m layer under hydrostatic conditions, whereas the solid lines are computed by using the integrated weight of soil column from the slope surface to the point of interest, as defined by Equations (9.50) and (9.55)–(9.57). The differences in the factor of safety values between the average soil weight at the hydrostatic and the variably saturated soil weight vary from zero at the water table to 0.1 at the depth about 1.0 m beneath the slope surface, or about 5% difference.

If a steady infiltration rate of $q = -8.0 \times 10^{-7}$ m/s, or 69.1 mm/day is imposed, the soil unit weight increases to its saturated value of 19.0 kN/m^3 throughout the entire soil layer (Figure 9.14a). The corresponding suction stress profile decreases to less than half of that under the previous infiltration rate (Figure 9.14b). The profile of the factor of safety is significantly smaller than that under the previous infiltration rate (Figure 9.14c). Persistent differences of about 5% between the average soil weight under hydrostatic conditions and the variably saturated soil weight can be seen throughout most of the soil layer.

If a steady infiltration rate of $q = -9.9 \times 10^{-7}$ m/s, or nearly equal to its saturated hydraulic conductivity, is imposed, the soil unit weight retains its saturated value of 19.0 kN/m^3 (Figure 9.14a), but suction stress further increases to near zero throughout

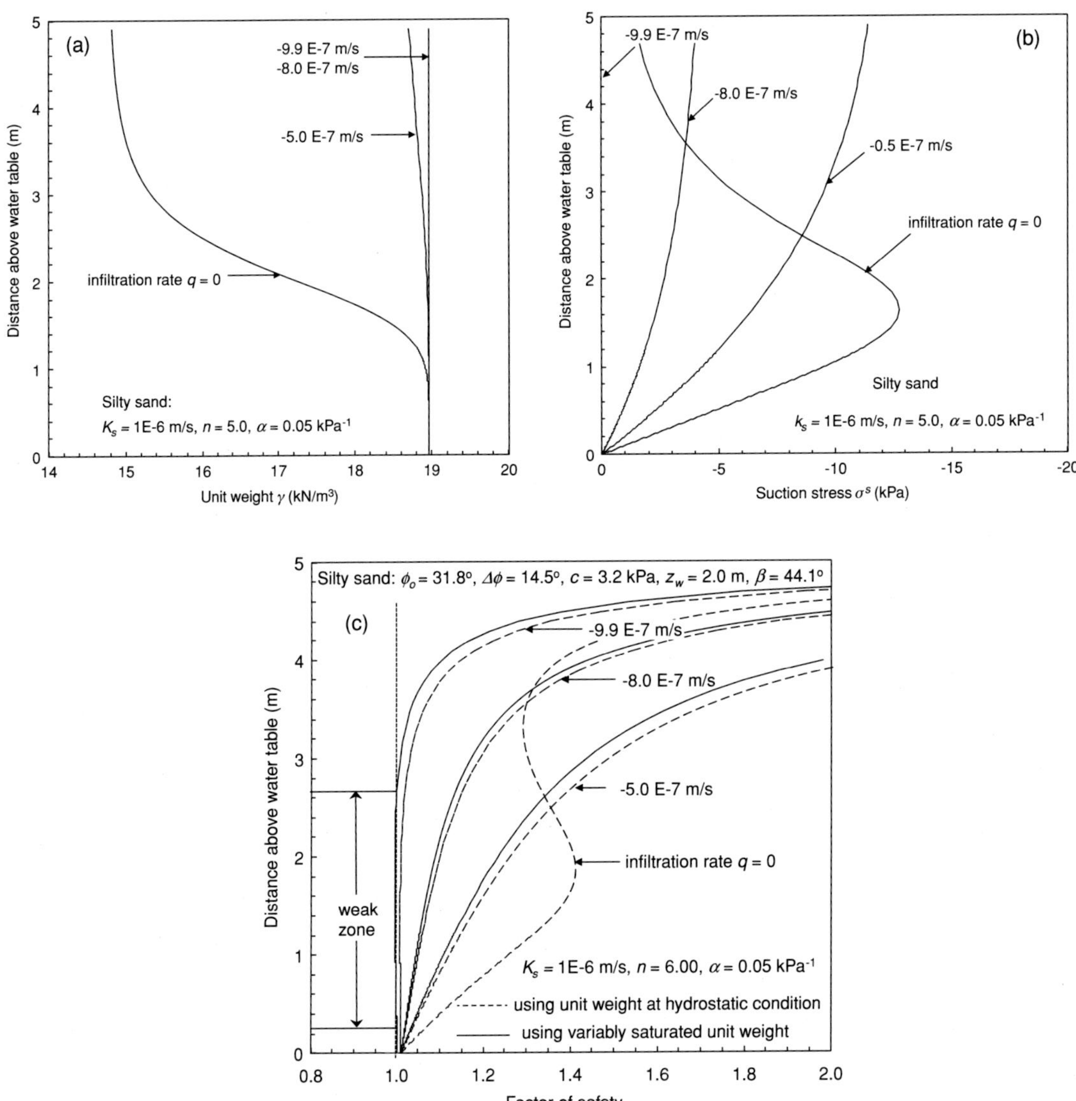

Figure 9.14 Profiles of (a) soil unit weight, (b) suction stress, and (c) factor of safety in a 5 m unsaturated silty sand slope under various steady infiltration rate conditions.

the entire layer (Figure 9.14c). Factor of safety is very close to 1.0 for depths 2 m below the slope surface. A persistent difference of about 3–4% between the average soil weight under hydrostatic conditions and the variably saturated soil weight can be seen throughout the entire soil layer. The further reduction in factor of safety that occurs without any change in the soil weight from the infiltration rate of $q = -9.9 \times 10^{-7}$ m/s to the infiltration rate of $q = -8.0 \times 10^{-7}$ m/s results from the coupled effect of soil weight and suction stress, as shown in the last term in Equation (9.59). Under these conditions, the profile of factor of safety predicted using the variably saturated soil weight indicates that a 2.7 m thick zone above the water table is susceptible to failure as its factor of safety is less than 1.0.

The above example illustrates that increases in soil weight due to wetting or increased infiltration can reduce the factor of safety by a small percentage, which may be sufficient to initiate landslides. The examples illustrate that the generalized infinite-slope theory defined by Equations (9.50)–(9.59) can quantitatively describe the evolution of stability in shallow hillslope environments.

This section described the extension of classical methods of slope stability analysis to partially saturated conditions using the suction stress concept. A comparison of results from both the ordinary method of slices and Bishop's simplified method shows that factors are larger by about 10% if unsaturated conditions are considered. The impact of steady infiltration on the stability of partially saturated infinite slopes was also examined. When the steady infiltration rate approaches the saturated hydraulic conductivity of the hillside materials, instability above the water table can occur. Finally, analysis of the counteracting effect of soil water content on soil unit weight on slope stability shows that increases in soil unit weight that result from infiltration can reduce the calculated factors of safety by a small percentage, which may be sufficient to induce slope movement in some settings.

9.4 Shallow landslides induced by transient infiltration

The close link between rainfall and shallow landslide initiation is well documented (e.g., Caine, 1980; Godt *et al.*, 2006). The quantitative understanding of this link is described in the following sections by analyzing the transient pore-water response to infiltration and consequent changes in the state of stress and slope stability (e.g., Haneberg, 1991; Reid, 1994; Iverson, 2000). Steady flow models, such as that described in Section 9.3, provide insight into the pore-water conditions at the time of slope failure, but little information on the rainfall needed to reach that point. In partially saturated soils, changes in moisture content θ resulting from rainfall and infiltration are accompanied by changes in suction $(u_a - u_w)$ or pressure head h_m, which is described by the soil water characteristic curve $\theta(h_m)$. In terms of pressure head, the slope of this curve, or specific moisture capacity $C(h_m) = \partial\theta/\partial h_m$ describes the available storage in partially saturated soil. Soil water diffusivity $D(h_m) = K(h_m)/C(h_m)$ is the relative amount of storage available as a function of the pressure-dependent hydraulic conductivity, and is described in Chapter 4.

Iverson (2000) showed that for the limiting case of nearly saturated hillslopes, the dominant direction of strong pore-pressure transmission resulting from rainfall can be assessed using two time scales. Assuming a reference soil water diffusivity, D_0, in an isotropic, homogeneous hillslope, the time scale for strong pore-pressure diffusion normal to the slope is H^2/D_0 where H is the slope normal depth. The time scale for strong pore-pressure diffusion parallel to the slope to some point located below a catchment with contributing area A is A/D_0. For hillslopes where the ratio of these two time scales, $\varepsilon = H/\sqrt{A} = 1$, the long- and short-term pressure-head responses to rainfall can be adequately described by one-dimensional linear and quasi-linear approximations to the

Richards equation. Typically, for locations near the failure surface of existing or potential landslides, where contributing area A may range from $\sim$100 m^2 to 10^4 m^2, slope normal depths of a few to tens of meters, and soil water diffusivities from 10^{-2} to 10^{-5} m^2/s, slope-normal response times range from tens of minutes to hundreds of days. In contrast, time scales for slope-parallel pressure diffusion typically range from several hours to years.

The one-dimensional analytical solution of the Richards equation by Srivastava and Yeh (1991) described in Chapter 4 combined with the infinite-slope stability analysis as implemented in the TRIGRS model (Savage *et al.*, 2004; Baum *et al.*, 2008, 2010) provides a means to examine the timing of potential instability considering the effects of the unsaturated zone. The following sections describe the changes in stability above the water table for a steep (45°) hillslope composed of four hypothetical materials. The analytical solution (Equations (4.39)–(4.44)) isolates the effects of material properties on the timing and depth of potential instability. Figure 9.15 shows the soil water and suction stress characteristics of coarse, medium, and fine sand, and silt. The soil water characteristics were calculated using the Gardner (1958) model for the constitutive relations between pressure head h_m and hydraulic conductivity K and volumetric moisture content θ

$$K(h_m) = K_s \exp(\alpha h_m) \tag{9.75}$$

$$\theta = \theta_r + (\theta_s - \theta_r) \exp(\alpha h_m) \tag{9.76}$$

where α is the inverse of the air-entry pressure head or height of the capillary fringe. The suction stress characteristic curve (Lu and Likos, 2004b, 2006) as a function of matric suction $(u_a - u_w)$ is

$$\sigma^s = -S_e(u_a - u_w) = -\frac{\theta - \theta_r}{\theta_s - \theta_r}(u_a - u_w) \tag{9.77}$$

Table 9.9 gives the material hydrologic and strength properties for the hypothetical examples. They cover the range typical for soils that mantle steep hillsides and are prone to shallow landsliding under partially saturated conditions.

The Gardner (1958) exponential model is used to calculate the suction stress characteristic curve and thus its shape is controlled by the α parameter (Figure 9.15c). For all of the soils, suction stress is nearly zero at small suctions, decreases to a minimum at several meters of suction, and then increases with increasing suction. Minimum suction stresses vary amongst the soils from about -1.0 kPa at a suction head of about 0.3 m for the coarse sand, -3.5 kPa at a suction head of about 1.0 m for the medium sand, -5.1 kPa at a suction head of about 1.4 m for the fine sand, and -7.2 kPa at a suction head of about 2.0 m for the silt. The pressure head response to an infiltration rate I_Z equivalent to the saturated hydraulic conductivity K_s was computed for a 2.0 m thick unsaturated zone above a stationary boundary with a constant pressure head $h_m = 0$ in the sand examples and a 5.0 m thick unsaturated zone in the silt example. Initial conditions were prescribed as hydrostatic above a water table at the lower boundary for all examples.

Infinite-slope stability analyses are statically determinate and therefore convenient for examining the role of infiltration and consequent transient changes in the distribution of soil

Table 9.9 Hydrologic and shear strength properties for hypothetical soils (Gardner, 1958)

Soil type	$\alpha(\mathrm{m}^{-1})$	$K_s(\mathrm{ms}^{-1})$	θ_s	θ_r	c'(kPa)	$\phi'(^\circ)$	$\Delta\phi'(^\circ)$	$\beta(^\circ)$	Z_W(m)
Coarse sand	3.5	1.0×10^{-5}	0.41	0.05	0.0	40	7	45	0.5
Medium sand	1.0	7.5×10^{-6}	0.41	0.05	0.0	40	7	45	0.5
Fine sand	0.7	5.0×10^{-6}	0.41	0.05	0.0	40	7	45	0.5
Silt	0.5	9.0×10^{-7}	0.45	0.10	1.7	30	15	45	1.5

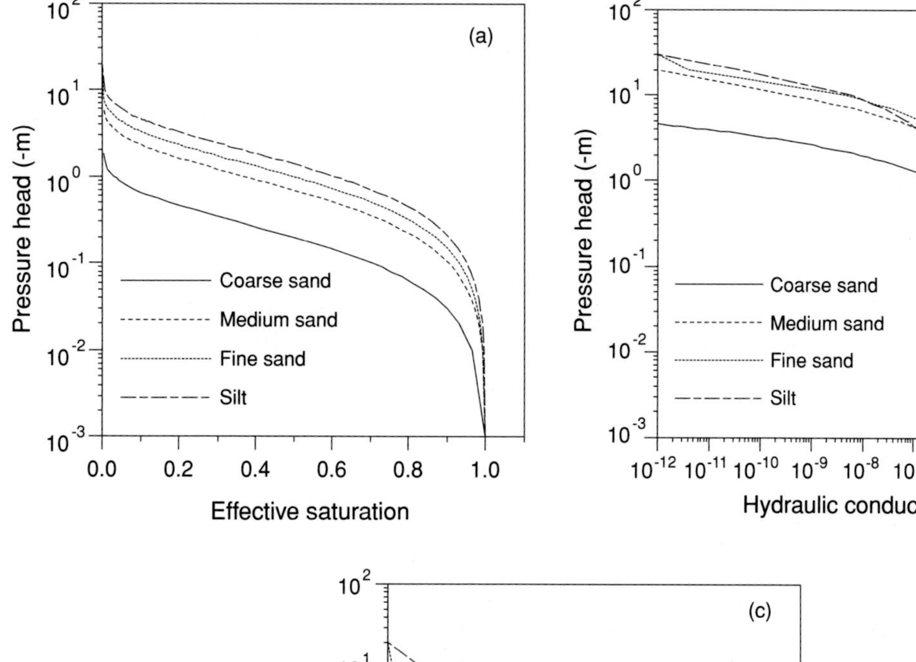

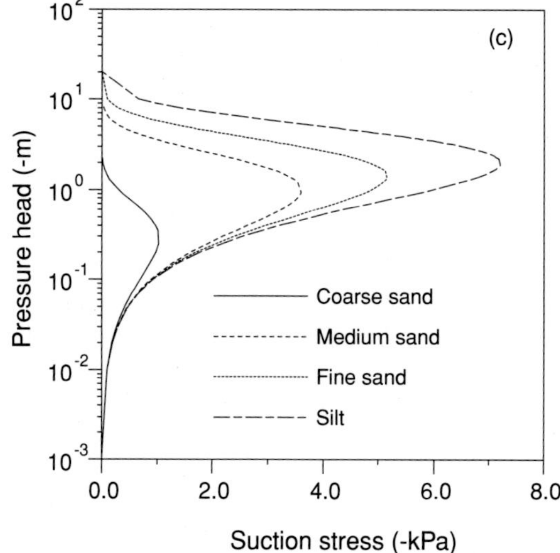

Figure 9.15 (a) Effective saturation, (b) hydraulic conductivity, and (c) suction stress as a function of soil suction for four hypothetical soils (from Godt *et al.*, 2012).

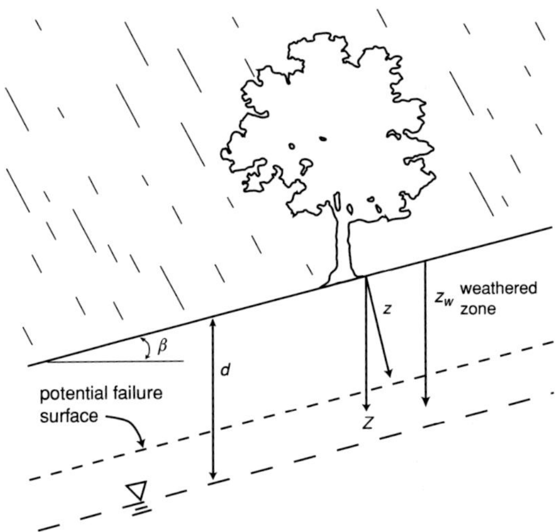

Figure 9.16 Schematic cross section and definitions for a variably saturated infinite slope (from Godt *et al.*, 2012).

moisture on stability (e.g., Iverson, 2000). For transient infiltration conditions, the factor of safety at depth Z below the ground surface at time t is given by (Baum *et al.*, 2008, 2010)

$$\mathrm{FS}\,(Z,t) = \frac{\tan\phi'(Z)}{\tan\beta} + \frac{c' + \sigma^s\,(Z,t)\tan\phi'\,(Z)}{\gamma_s Z \sin\beta\cos\beta} \tag{9.78}$$

where $\phi'(Z)$ is a depth-dependent angle of internal friction and c' is the soil cohesion, γ_s the soil unit weight, and β the slope angle (Figure 9.16). The depth dependence of the friction angle as a function of the depth of the weathered zone z_w is described in Section 9.3.1 and shown in Figure 9.7.

9.4.1 Stability of a coarse sand hillslope

Figure 9.17 shows the changes in pressure head, degree of saturation, suction stress, and stability as a function of time that result from infiltration equivalent to the saturated hydraulic conductivity for the coarse sand example. Under initially hydrostatic conditions with a water table located 2.0 m below the ground surface (Figure 9.17a) the effective saturation (Figure 9.17b) is reduced to zero at about 1.5 m above the water table. Under hydrostatic conditions suction stress has a minimum of −1.0 kPa at 0.3 m above the water table, is zero at the water table, and diminishes to −0.2 kPa at 1.2 m above the water table (Figure 9.17c). The changes in suction stress with height above the water table under hydrostatic conditions lead to factors of safety of less than unity for these initial conditions (Figure 9.17d) given the slope geometry, material strength, and hydrologic properties selected for this example.

Although unstable initial conditions are not realistic, results illustrate the minimum behavior of suction stress and its effect on stability in coarse-grained soils. Under drying conditions resulting from seasonal weather patterns or following wildfire, the dramatic reduction of suction stress near the ground surface may lead to potential instability in the form of shallow landslides and dry ravel processes (e.g., Gabet, 2003).

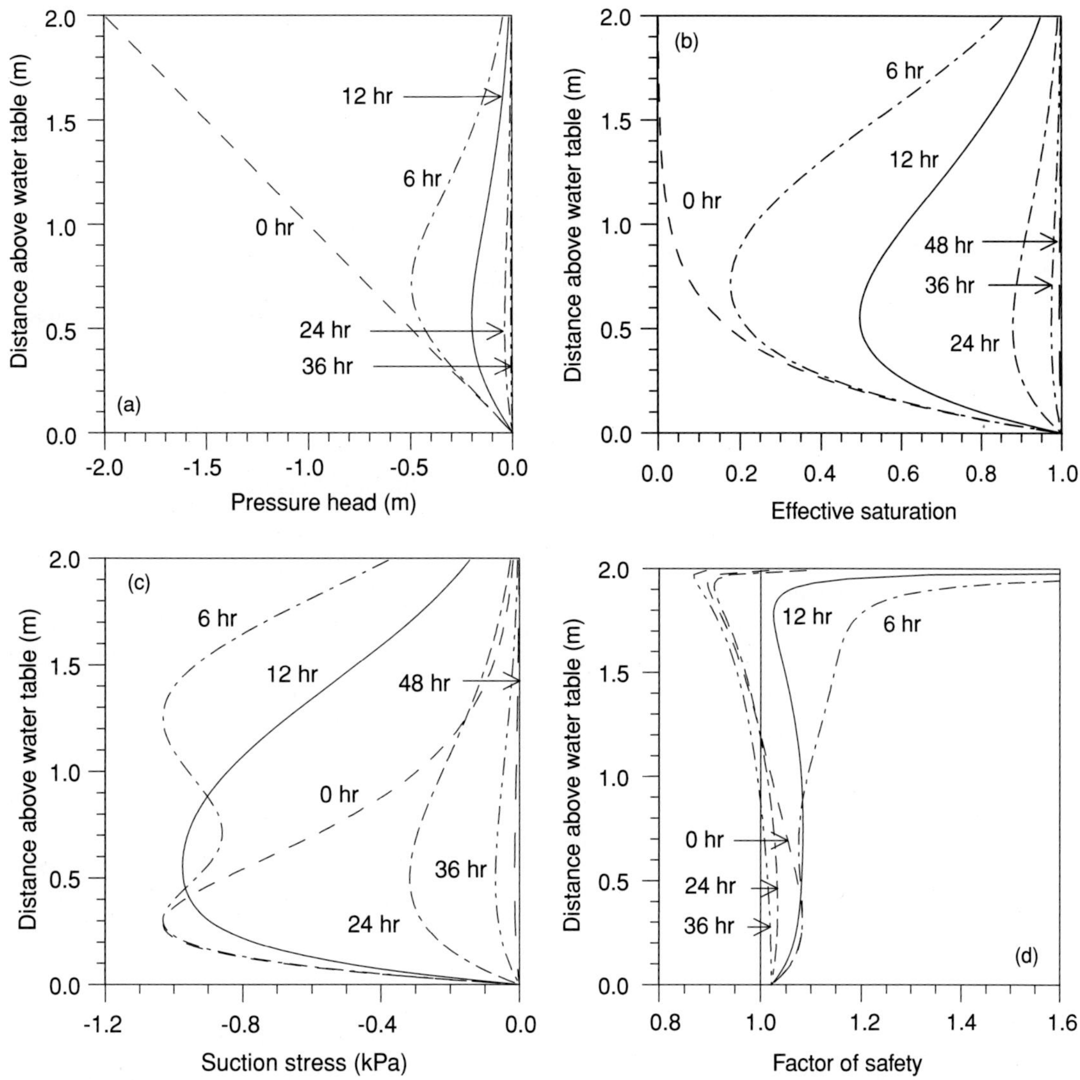

Figure 9.17 Profiles of (a) pressure head, (b) effective saturation, (c) suction stress, and (d) factor of safety for coarse sand with $\alpha = 3.5$, saturated hydraulic conductivity $K_s = 1 \times 10^{-5}$ m/s and material strength properties in Table 9.9. Prescribed flux at the ground surface is equivalent to K_s (from Godt *et al.*, 2012).

After 6 hours of rainfall, pressure heads increase to near zero at the ground surface with a consequent increase in effective saturation. The effect of infiltration on the profile of suction stress is complex, with a large relative decrease in suction stress to a minimum of about -1.0 kPa over the upper 1.0 m from the hydrostatic initial condition after 6 hours of rainfall. This decrease in suction stress (more negative) leads to an increase in the factor of safety and a shift to relatively stable conditions compared to the initial state. After 12 hours of infiltration, pressure heads increase to more than -0.2 m throughout the profile and relative saturations are at a minimum of 0.5 at about 0.6 m above the water table. The profile is stable throughout at 12 hours, but a zone of factors of safety approaching unity develops at about 0.2 m below the ground surface at this time. At later times (24 and 48 hours in Figures 9.17a, b), pressure heads are nearly zero throughout the profile and effective saturation is almost 1.0. Suction stress approaches zero throughout the profile and the upper 0.8 m is potentially unstable.

Figure 9.17 shows that, for this limiting hypothetical case in which a 2.0 m thick coarse sand layer overlies a water table of constant depth, landslides a few tens of centimeters thick may result from infiltration equivalent to the saturated conductivity that persists for more than about 12 hours. Less thick deposits, accreting water tables, and wetter initial conditions will all tend to reduce the time needed to achieve potential instability under a given flux of water at the ground surface.

9.4.2 Stability of a medium sand hillslope

This example describes the transient effects of infiltration on the stability of a hypothetical hillslope with a surficial cover of medium sand. Under initially hydrostatic conditions with a water table located 2.0 m below the ground surface (Figure 9.18a) the effective saturation decreases from 1.0 at the water table to about 0.1 at the ground surface (Figure 9.18b). The profile of suction stress for the medium sand (Figure 9.18c) has a similar form to that for the coarse sand example (Figure 9.17c) with a minimum in the lower part of the profile increasing towards the ground surface. Suction stress is zero at the water table, reaches a minimum at about −3.6 kPa at 1.0 m above the water table and increases to about −2.8 kPa at the ground surface (Figure 9.18c). In contrast to the coarse sand, the medium sand is stable under the same initially hydrostatic conditions (Figure 9.18d). Factors of safety increase monotonically above the water table for the medium sand due to the contribution of suction stress to stability.

After 6 hours of infiltration, pressure heads increase to −0.3 m near the ground surface with a minimum (−0.6 m) 1.1 m above the water table. The degree of effective saturation follows a similar pattern to that of the pressure head with a minimum of 0.55 at 1.1 m above the water table. The corresponding change in suction stress after 6 hours is small, with a maximum increase of about 0.4 kPa over that of the initial hydrostatic case at 0.9 m above the water table. This small change in suction stress results in a similarly small change in the factor of safety. Following 12 hours of infiltration, pressure heads increase to greater than −0.3 m and effective saturations to greater than 0.7 throughout the profile. Suction stresses increase to more than −2.2 kPa with a broad peak between about 0.8 and 1.4 m above the water table and a consequent reduction in the factor of safety. Pressure heads, effective saturation, and suction stress continue to increase as infiltration progresses to 24 hours. Although factors of safety decrease, no part of the profile is potentially unstable. After 36 hours of infiltration, pressure heads are nearly zero and effective saturation is close to 1.0 throughout the profile. At this time suction stress peaks at −0.2 kPa at 1.0 m above the water table and the profile of the factor of safety indicates potential instability in a zone that extends from 0.1 to 0.5 m below the ground surface. Pressure head, effective saturation,

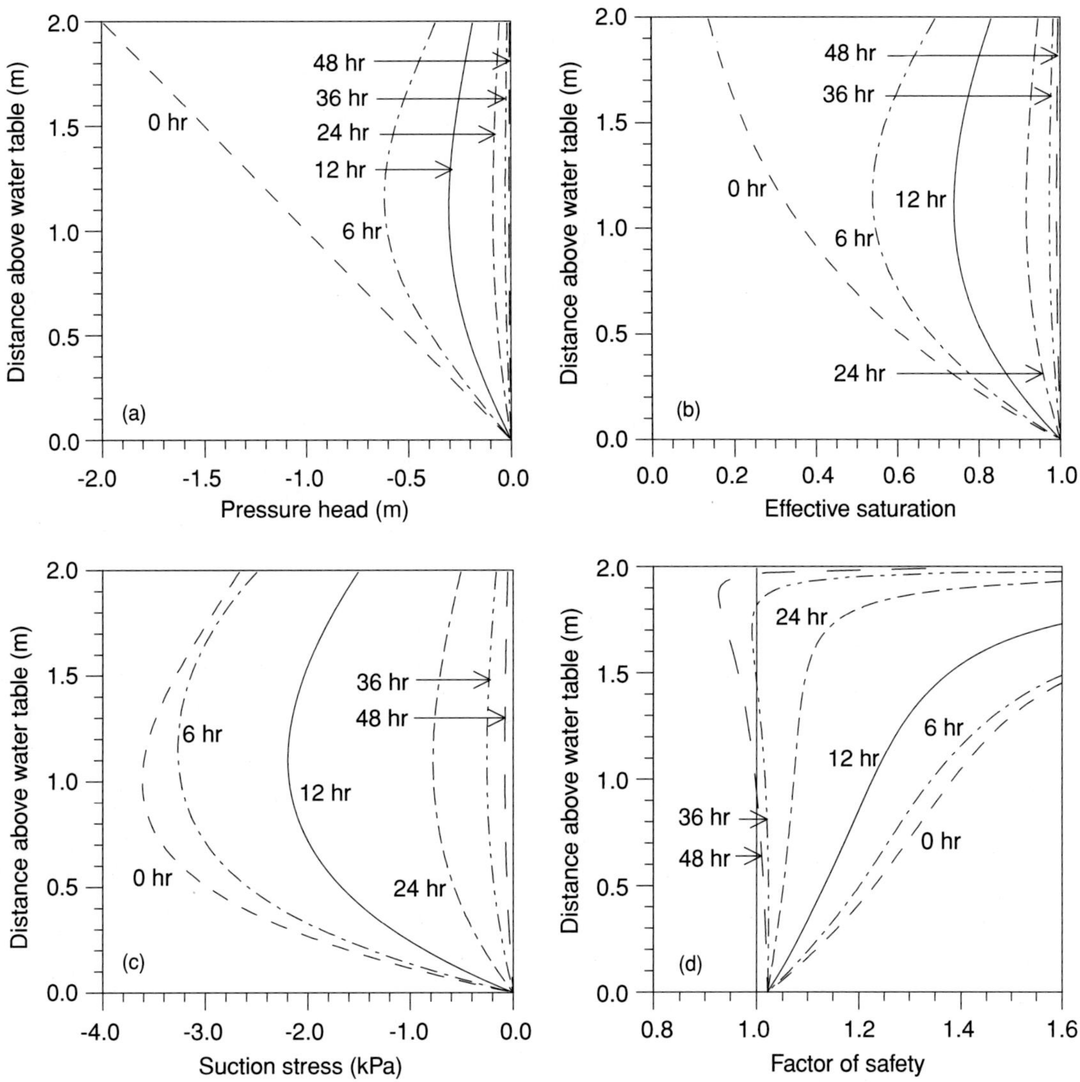

Figure 9.18 Profiles of (a) pressure head, (b) effective saturation, (c) suction stress, and (d) factor of safety for medium sand with $\alpha = 1.0$, saturated hydraulic conductivity $K_s = 7.5 \times 10^{-6}$ m/s and material strength properties in Table 9.9. Prescribed flux at the ground surface is equivalent to K_s (from Godt *et al.*, 2012).

and suction stress continue to increase with continued infiltration and at 48 hours the upper half of the profile is potentially unstable.

9.4.3 Stability of a fine sand hillslope

This hypothetical example examines the effect of transient infiltration on the potential for landslide occurrence in a fine sand hillslope. The pattern of pressure head, effective

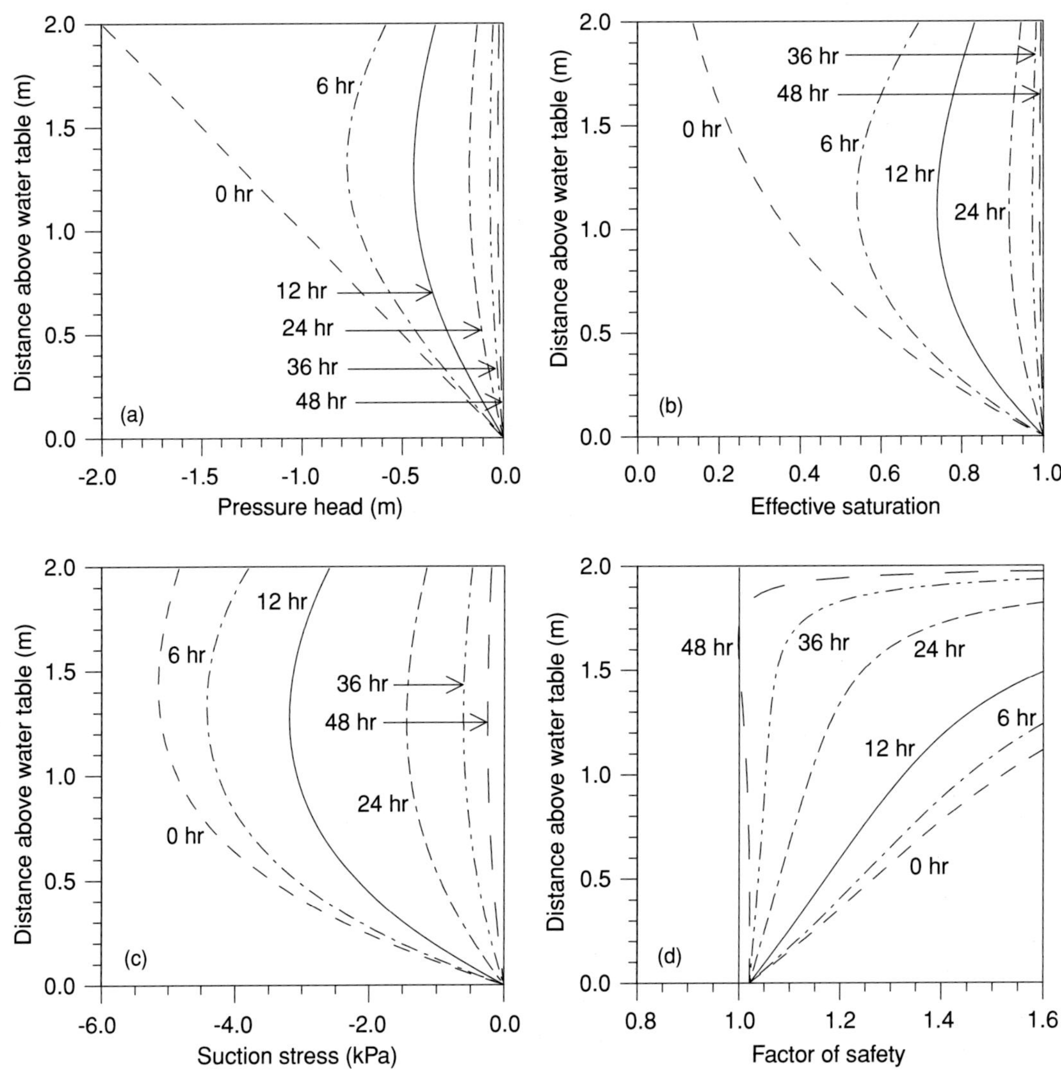

Figure 9.19 Profiles of (a) pressure head, (b) effective saturation, (c) suction stress, and (d) factor of safety for fine sand with $\alpha = 0.7$, saturated hydraulic conductivity $K_s = 5 \times 10^{-6}$ m/s and material strength properties in Table 9.9 Prescribed flux at the ground surface is equivalent to K_s (from Godt *et al.*, 2012).

saturation, and suction stress increase in the fine sand slope that results from infiltration (Figure 9.19) is very similar to that of the medium sand example (Figure 9.18). Under initially hydrostatic conditions with a water table located 2.0 m below the ground surface (Figure 9.19a) effective saturation decreases from 1.0 at the water table to about 0.15 at the ground surface (Figure 9.19b). Suction stress decreases from zero at the water table to a broad minimum of about −5.2 kPa in a zone between 1.3 and 1.6 m above the water table (Figure 9.19c). Under initially hydrostatic conditions, factors of safety increase monotonically above the water table (Figure 9.19d) similar to the medium sand example described in the previous subsection.

After 6 hours of infiltration, pressure heads increase to −0.6 m at the ground surface and the profile has a minimum of −0.8 m at 1.3 m above the water table. The profile of the relative degree of saturation has a similar pattern and is 0.7 at the ground surface with a minimum of about 0.55 in a zone extending from 1.0 to 1.4 m above the water table. Suction stress is about −3.8 kPa at the ground surface with a peak of about −4.5 kPa between 1.2 and 1.5 m above the water table. Because suction stress only diminishes by less than 1.0 kPa after 6 hours of infiltration, factors of safety remain much more than unity throughout the profile. After 12 hours of infiltration, pressure heads are more than −0.4 m and saturations are greater than about 0.75 throughout the profile. However, suction stress is less than −2.5 kPa in the upper 1.3 m of the profile and the slope remains stable. After 24 hours of infiltration, pressure heads increase to more than −0.2 m in the upper 1.5 m of the profile and saturations are greater than 0.9. Suction stress is less than −1.0 kPa in this same zone and stability holds. At 36 hours, pressure heads are nearly zero and the profile is almost saturated, but suction stress of about −0.5 kPa persists and although factors of safety approach unity, the slope is still stable. Not until 48 hours are pressure heads and saturation sufficiently great to diminish suction stress, leading to factors of safety less than unity and potential instability of the upper 0.5 m of the hillslope.

9.4.4 Stability of a silt hillslope

This example describes the transient changes resulting from infiltration into a 5 m thick unsaturated layer of silt above a stationary boundary with a constant pressure head $h_m = 0$. Initial conditions are hydrostatic and in equilibrium with the lower boundary condition (Figure 9.20a). The upper boundary is a prescribed flux I_Z at the ground surface equivalent to the saturated hydraulic conductivity of the silt (Table 9.9).

Under hydrostatic initial conditions (Figure 9.20a) the effective saturation decreases from 1.0 at the water table to 0.08 at the ground surface (Figure 9.20b). Suction stress decreases with distance above the water table to about −7.2 kPa at about 2.0 m above the water table and increases to −4.0 kPa at the ground surface (Figure 9.20c), and the soil profile is initially stable (Figure 9.20d). Following 4 days of infiltration, pressure heads are about −0.8 m at the ground surface and increase above the initial hydrostatic values in the upper 4.0 m of the soil profile with a minimum of about −1.8 m at 2.7 m above the water table. Effective saturation increases to about 0.75 at the ground surface with a minimum of 0.4 about 2.8 m above the water table. Suction stress decreases from the initially hydrostatic conditions in a zone extending from 2.3 m above the water table to the ground surface, which leads to a consequent increase in the factor of safety over the same region. After 8 days of infiltration, pressure heads increase to more than −1.0 m and effective saturations increase to more than 0.6 throughout the profile. The broad zone of suction stress less than about −7.0 kPa is reduced by about half its thickness after 8 days and suction stress increases to −4.0 kPa at the ground surface. Factors of safety only decrease slightly. Pressure heads,

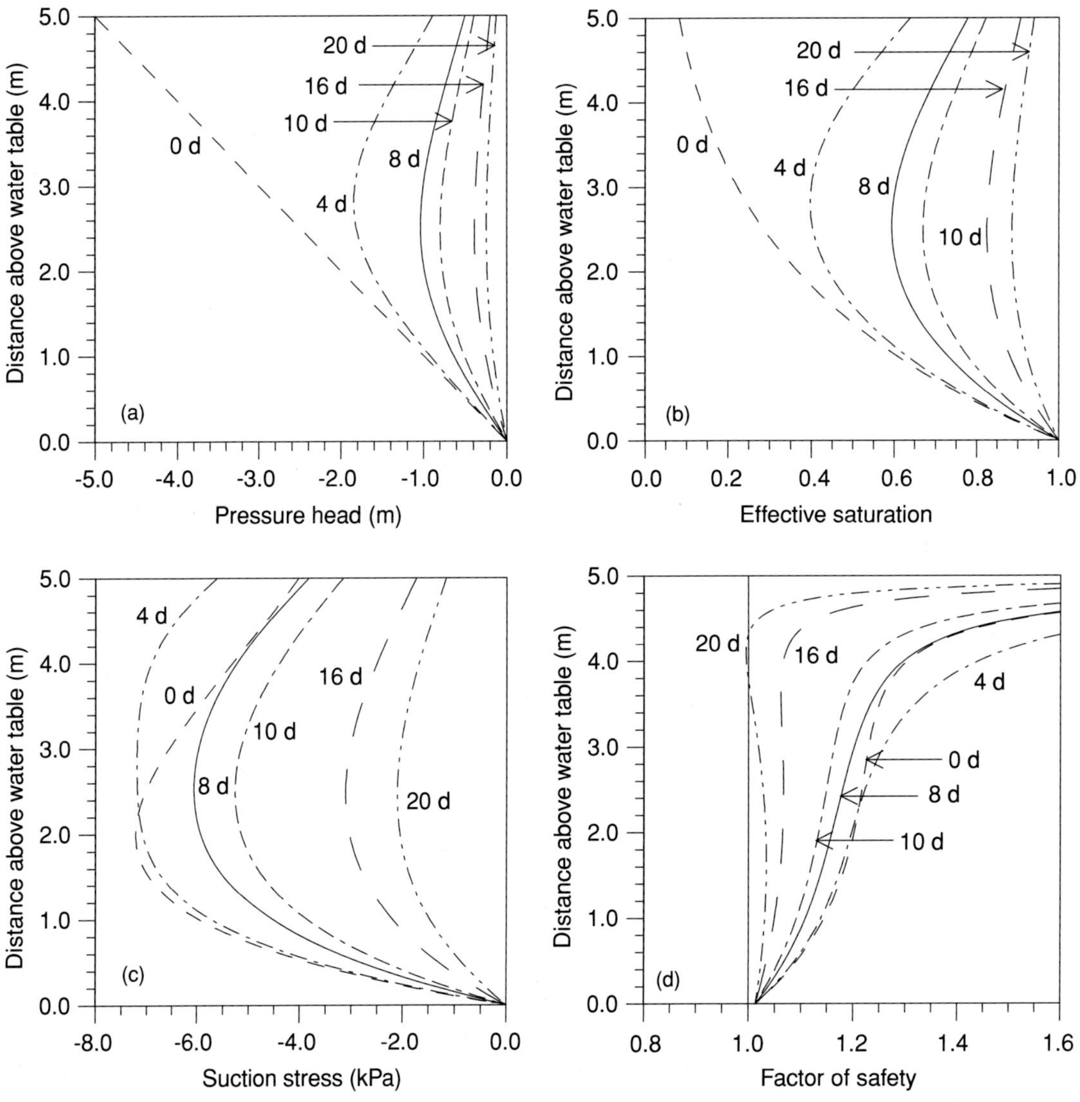

Figure 9.20 Profiles of (a) pressure head, (b) effective saturation, (c) suction stress, and (d) factor of safety for silt with $\alpha = 0.5$, saturated hydraulic conductivity $K_s = 9 \times 10^{-7}$ m/s and material strength properties in Table 9.9 Prescribed flux at the ground surface is equivalent to K_s (from Godt *et al.*, 2012).

effective saturation, and suction stress continue to increase and factors of safety continue to decrease with infiltration after 10 and 16 days. After 20 days of infiltration, pressure heads are more than -0.2 m and effective saturation is greater than about 0.9 throughout the profile. Suction stress increases to -1.1 kPa at the ground surface, with a minimum of about -2.0 kPa at 2.5 m above the water table. This change in suction stress in response to infiltration leads to potential instability in a zone near the ground surface, extending from 3.6 to 4.4 m above the water table.

9.4.5 Summary of model results

This section has provided several examples of the effects of transient infiltration on the timing and depth of slope failure above the water table. In general, failure depth increases for the finer-grained soils. This results in part from the influence of the α parameter (i.e., inverse of the air-entry head) in the Gardner (1958) formulation of the SWCC on both the hydrologic and mechanical properties of the materials. The maximum contribution to stability from suction stress is relatively small for the coarse sand, which has a large α compared to that for the fine sand. Because the α parameter also controls the hydraulic conductivity function, it influences the simulated shape of the wetting front during infiltration. The simulated degree of saturation from the coarse sand example shows a more abrupt change with depth compared to that for the fine sand example, which in turn leads to a greater reduction in suction stress and stability near the ground surface. Because the applied rainfall flux at the ground surface was equivalent to the saturated conductivity K_s of the materials, the variation in the timing of failure for the hypothetical examples is largely a function of the variation in hydraulic conductivity among the different materials. For example, the saturated hydraulic conductivity of the coarse sand is about five times that of the fine sand, and the time to failure is longer by about the same amount. Variation in the time to failure also results from differences in the hydraulic conductivity function at pressure heads less than zero and the failure and overall profile depths. Rainfall flux at rates much less than the saturated conductivity of the soil would not tend to induce instability for the boundary conditions used in the modeling examples. The following section provides two case studies of landslides induced by precipitation: one by rainfall and the other by melting snow.

9.5 Case study: Rainfall-induced shallow landslide

9.5.1 Site geology, geomorphology, and monitoring program

This section describes an application of the suction stress concept to assess the potential for instability at an instrumented hillslope along the Puget Sound near Edmonds, Washington, about 15 km north of Seattle (Figure 9.21a) where a shallow landslide occurred in the apparent absence of positive pore-water pressures under partially saturated soil conditions (Godt *et al.*, 2009). Steep hillslopes in this area are subject to frequent shallow landslides during the winter wet season, when extended rainy periods last several days. Pleistocene-age glaciation, shoreline wave attack, and mass movement processes have formed steep (>30°) 50 to 100 m high coastal bluffs above many areas along the Puget Sound shoreline. In this area, shallow landslides typically involve the loose, sandy, colluvial deposits derived from the glacial and non-glacial sediments that form the bluffs (Galster and Laprade, 1991).

The hillslope is a steep (45°) coastal bluff covered by a thin, sandy (<2.0 m) colluvium (Figure 9.21b) and was instrumented with water content probes and tensiometers in two

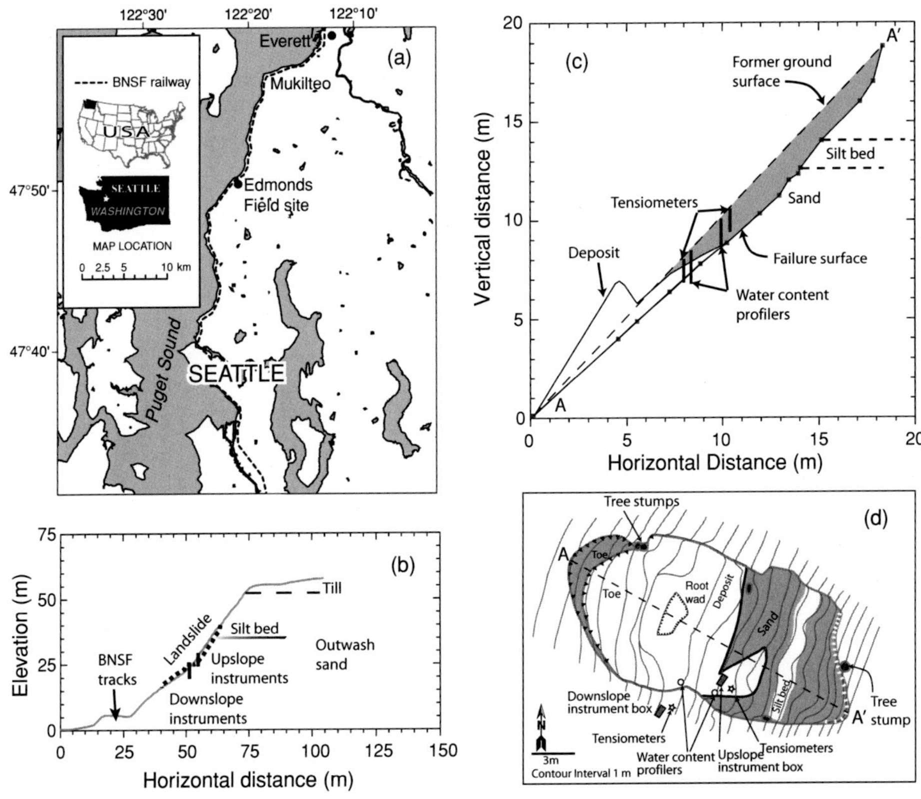

Figure 9.21 (a) Location of monitoring array and landslide at the Edmonds field site, near Seattle, Washington, (b) hillslope cross section, (c) detailed cross section, and (d) map showing the location of the instrument array and the shallow landslide. Two rain gauges were located at the toe of the bluff about 300 m north of the slide (after Godt *et al.*, 2009).

profiles. During the 16-month period this instrument array was operational, no positive pore-water pressures were observed. On 14 January 2006 a 25 m long by 11 m wide shallow landslide initiated in colluvium near the instrument array. The depth of the failure surface was between 1.0 and 2.0 m below the original ground surface. The failure surface was apparently coincident with the contact between the loose sandy colluvium and the consolidated glacial outwash sand (Figure 9.21c). The slide exposed a thin (<2 m) silt bed near the headscarp and a shallow zone (<0.8 m below the ground surface) of small diameter (<2 mm) blackberry and grass roots. Only a few larger diameter alder tree roots penetrated the failure surface.

9.5.2 Numerical modeling of transient flow

A numerical, one-dimensional, finite-element solution to the Richards equation (HYDRUS 1-D, Šimůnek *et al.*, 2008) was applied to simulate the pore-water response to rainfall at the Edmonds site. The relation between pressure head h_m and water content θ is given by the van Genuchten (1980) formulation

$$\theta(h_m) = \theta_r + \frac{\theta_s - \theta_r}{(1 + |\alpha h_m|^n)^\omega}, \qquad \omega = 1 - \frac{1}{n} \tag{9.79}$$

where θ_r and θ_s are the residual and saturated moisture contents respectively, α, n, and ω are curve-fitting parameters. The parameter α is considered to be the approximate inverse of the air-entry pressure head. The pressure-head-dependent hydraulic conductivity $K(h_m)$ is predicted using the statistical pore size distribution model of Mualem (1976)

$$K(h_m) = K_s S_e^{0.5} \left[1 - \left(1 - S_e^{1/\omega}\right)^\omega\right]^2, \qquad \mathrm{S}_e = \frac{\theta - \theta_r}{\theta_s - \theta_r} \tag{9.80}$$

The numerical solution allows the use of an inverse-modeling approach described in detail in Chapter 8 to estimate the soil water characteristics from field monitoring information. The model domain was a one-dimensional 3.76 m deep homogeneous profile above a no-flow boundary. The flux at the upper boundary was taken from measured rainfall (Figure 9.22a). Initial conditions were assumed to be hydrostatic and in equilibrium with a water table coincident with the lower boundary and consistent with the measured pressure-head profile in the middle of September 2005 (Figure 9.22c).

Slope stability is calculated from both the observed and modeled water contents using Equation (9.78), where suction stress σ^s is a function of the effective degree of saturation S_e:

$$\sigma^s = -\frac{S_e}{\alpha}\left(S_e^{\frac{n}{1-n}} - 1\right)^{\frac{1}{n}} \tag{9.81}$$

or matric suction

$$\sigma^s = -\frac{(u_a - u_w)}{(1 + [\alpha(u_a - u_w)]^n)^{(n-1)/n}} \tag{9.82}$$

Hillside material properties are given in Table 9.10.

Measured rainfall (Figure 9.22a)), water contents (Figure 9.22b), and pressure heads (Figure 9.22c) at a depth of 1.5 m for the period from 28 October 2005 to 21 November 2005 were used to inversely estimate the saturated hydraulic conductivity, saturated and residual moisture contents, and the fitting parameters for Equations (9.79) and (9.80) and are listed in Table 9.10.

9.5.3 Comparison of model results with observations

Figure 9.22b compares observed and modeled water contents. The agreement between observations and model results between 28 October 2005 and 21 November 2005 is best at the 1.5 m depth, as this was the target for the inverse modeling procedure. At both the 1.0 and 1.5 m depths, the timing of changes in water content match quite well, although in general, modeled water contents decrease more rapidly following the cessation of rainfall. This may result because any lateral downslope flow or hysteresis in the soil water characteristics was ignored in the one-dimensional model, both of which would tend to

Table 9.10 Soil water characteristics from the inverse modeling procedure and material strength parameters

$K_s(\mathrm{ms}^{-1})$	θ_s	θ_r	$\alpha(\mathrm{m}^{-1})$	n	ω	$\phi'(°)$	$c'(\mathrm{kPa})$
4.18×10^{-7}	0.40	0.05	3.75	1.16	0.14	36	1.1

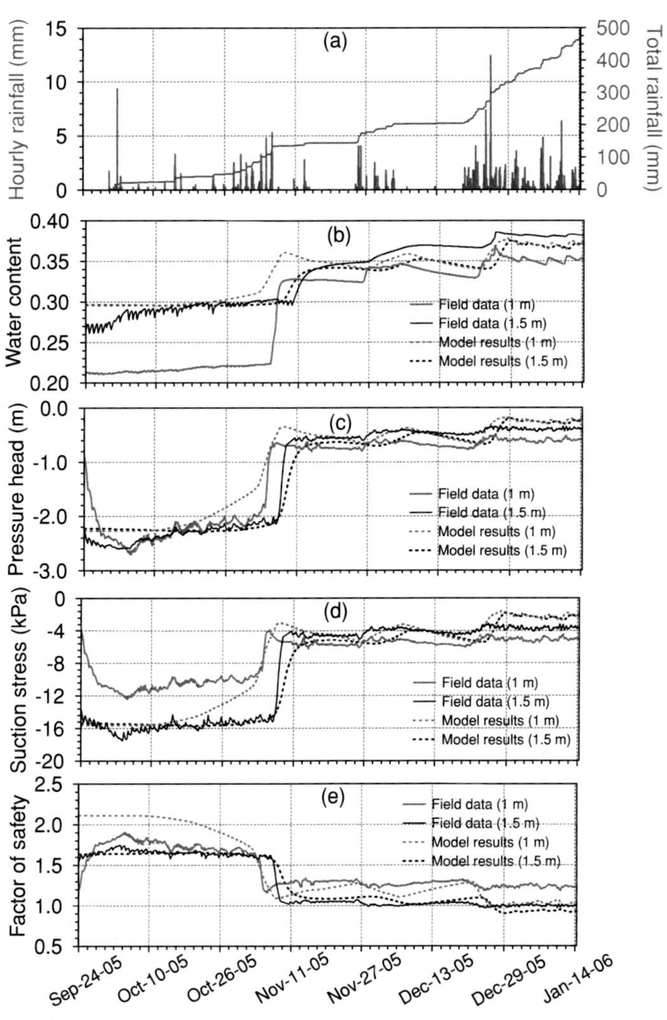

Figure 9.22 (a) Hourly and cumulative rainfall, (b) modeled and observed soil water content, (c) observed and modeled pressure head, (d) modeled and calculated suction stress, and (e) modeled and calculated factor of safety for the period 24 September 2005 to 14 January 2006 at 1.0 and 1.5 m depths. The landslide at the site occurred on 14 January 2006 (from Godt *et al.*, 2012). See also color plate section.

attenuate any decrease in moisture content during drainage. Quantitative discussion of the hysteresis and lateral flow can be found in Chapters 8 and 4, respectively. Compared to the 1.5 m depth, the agreement between the modeled and observed water contents is not as close at 1.0 m. Initially the simulated water content is about 0.08 too large compared to the observations. This apparent bias towards higher water contents at the 1.0 m depth is evident throughout the simulation (Figure 9.22b). This discrepancy likely results from minor differences in the hillside materials and thus differences in soil water characteristics at the two depths.

The timing and magnitude of the modeled and observed pressure-head variation are in general agreement at both the 1.0 and 1.5 m depths (Figure 9.22c). The arrival of the wetting front at the 1.0 m depth at the end of October 2005 is well captured by the simulation. The simulated arrival of the wetting front at 1.5 m depth is less abrupt and delayed compared to the observation (Figure 9.22c). However, differences between observed and modeled pressure head decrease as the soil becomes wetter. Modeled pressure heads are greater than the observation at the 1.0 depth after the beginning of November and greater at both depths during the wet period leading to the landslide at the site beginning in the middle of December 2005 (Figure 9.22c). Modeled pressure heads also show a greater response to rainfall than observations during the period from the end of December to 14 January. This may result from limitations of pressure transducers and tensiometers to resolve pressure heads near zero.

Suction stress was calculated using Equation (9.77) and is shown for both the modeled and observed water contents and pressure heads assuming a saturated water content of 0.4 (Figure 9.22d). At the 1.0 m depth, suction stress calculated from the monitoring data increases from about −12 kPa to −4 kPa after the rainy period in late October 2005 and fluctuates around −5 kPa for the remaining period (Figures 9.22a and 9.22d). At 1.5 m the calculated suction stress has a similar pattern to that calculated for the 1.0 m depth, but the increase from about −18 kPa to about −4 kPa follows the increase at 1.0 m by 2.6 days as the wetting front moves through the soil (Figure 9.22c) and remains at about −4 kPa.

The differences in timing between modeled changes in suction stress compared to those calculated from the monitoring data mimic the differences between the modeled and observed pressure heads (Figures 9.22d and 9.22c). The differences between the modeled and calculated magnitudes are generally less than a few kPa, but have a substantial impact on modeled factors of safety.

Factors of safety from Equation (9.78), calculated from the monitoring data, are greater than 1.5 at both 1.0 and 1.5 m depth prior to the onset of rainfall and consequent increases in water content, pressure head, and suction stress that occur at the beginning of November 2005 (Figure 9.22). Factors of safety from the monitoring data at 1.5 m decrease to near unity (indicating the potential for slope failure) at this time and remain close to 1.0 for the remainder of the record. At the 1.0 m depth, factors of safety reach a minimum of about 1.2 during several periods at the beginning of November, December, and at the end of December. On 14 January 2006, factors of safety calculated from the monitoring data were less than unity at the 1.5 m depth for about 50 hours prior to the occurrence of the landslide (Godt *et al.*, 2009).

Factors of safety from Equation (9.78) calculated from modeled water contents and pressure heads are greater than unity at both the 1.0 and 1.5 m depths in September and much of October 2005. They decrease at both depths in early November in response to the rainfall, increases in water content, pressure head, and suction stress (Figure 9.22). However, unlike factors of safety calculated from the monitoring data, factors of safety from the model results do not approach unity at the 1.5 m depth on 27 December. Late on 13 January, about 24 hours before the occurrence of the landslide at the site and following a rainfall of 19.6 mm, factors of safety calculated from the modeling results at both the 1.0 and 1.5 m depths are very close to unity, indicating potential instability.

9.6 Case study: Snowmelt-induced deeply seated landslide

9.6.1 Site geology, morphology, and hydrology

This case study describes the recurring instability of a landslide that impacts a heavily traveled section of Interstate 70 (I-70) in the Rocky Mountains of Colorado. Infiltration at this location is driven largely by the melting of a deep snowpack that accumulates in the winter months. The landslide is located about 1.5 miles (2.5 km) west of the Eisenhower Tunnel (2.7 km long) in Summit County, Colorado. The tunnel opened in 1979 and carries traffic on I-70 under the Continental Divide at an elevation of 3401 m above sea level. It is one of the highest vehicular tunnels in the world. A geologic cross section cutting through east–west aligned I-70 at the site is shown in Figure 9.23a. The average snow depth that accumulates during the months from November to April is 1.07 m and the depth varies from 0.50 to 2.76 m. This snowpack melts in late spring and early summer (May to June) to yield an average of 0.36 m of water with substantial variability from 0.19 to 0.9 m. Summer season precipitation is similarly variable, and heavy rainfall can trigger debris flows and shallow landslides (Godt and Coe, 2007). Because the annual snowmelt process takes from several days to weeks, much of the water infiltrates into the slope and embankment.

Site geologic investigation (Gallagher, 1997) indicates that four distinct materials underlie the highway: colluvium and fill from spoil or waste from the boring of the tunnel, highly fractured decomposed gneiss, intact gneiss, and a thin layer of alluvium near the toe of the slope (Figure 9.23a). Measurements taken a single time from sampling boreholes indicate that the water table is located at depths of 15 m to 25 m beneath the highway and adjacent downslope embankment, and surface drainage conditions indicate that the water table is probably roughly parallel to the slope. The depth of the water table likely varies greatly from season to season and year to year, controlled by episodic infiltration events. Bedrock composed of intact gneiss located at depths from 12 m to 15 m below the highway surface presumably acts as a hydrologic barrier described in Section 3.6 of Chapter 3.

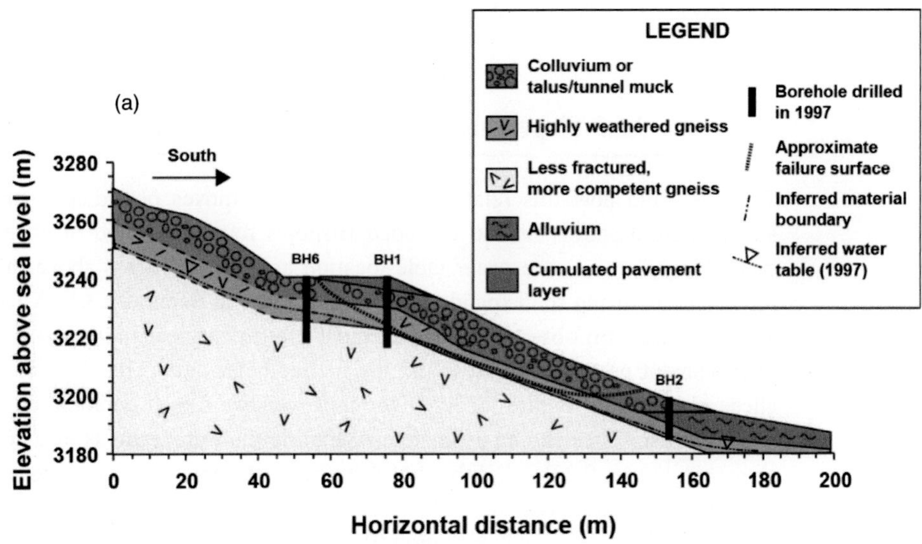

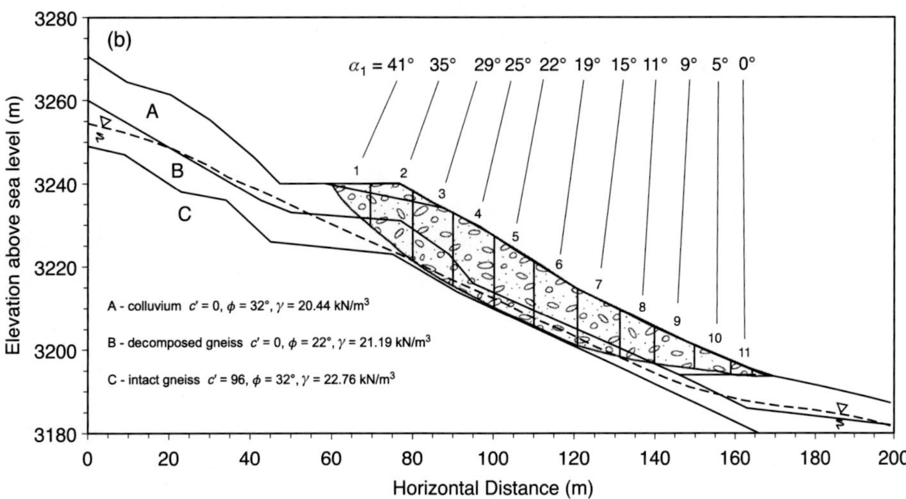

Figure 9.23 Case study at the I-70 milepost 212: (a) geologic cross section showing site stratigraphy, topography, and the inferred sliding surface; and (b) cross section showing material properties, and slice discretization used in the stability analysis.

Since the opening of the tunnel in 1979, records over the past four decades indicate that the highway pavement surface moves downward episodically at an average rate of 3 cm/year, leading to about 0.6 m of pavement settlement that has accumulated on the eastbound lanes (see the pavement wedge shown in Figure 9.23a). In recent years, borehole inclinometers have been installed at depths of up to 35 m in the embankment (Wayllace *et al.*, 2012); monitoring data indicate that the failure surface is located beneath the highway and between the interface of the intact bedrock and the well-fractured, decomposed gneiss.

9.6.2 Slope stability analysis with and without suction stress

To understand how this relatively deep landslide moves periodically under infiltration supplied from snowmelt, the extended Bishop's routine method of slices is used. Two controlling factors – the water table location and suction stress above the water table – are examined using the expanded theory described in Section 9.3.1. A failure surface is assumed based on observation of pavement movement near the highway divide, surface displacement near the toe area, and the inclinometer data. The slope is divided into 11 slices as shown in Figure 9.23b. If the unit weight of each geologic unit is known, the weight of each slice can be estimated. Assuming the water table is located at the position shown in Figure 9.23b, slightly above the bedrock, pore pressure below the water table is hydrostatic, and above the water table is zero, the calculation of each term by Bishop's simplified method shown in Equation (9.48) leads to a factor of safety of 0.96 after several iterations. The computational procedures and steps leading to the factor of safety are shown in Table 9.11. Thus, considering zero pore-water pressure in the vadose zone leads to the failure of the slope embankment, which could be the case when the melted snow infiltrates towards the water table.

If the water table is assumed to be at the same location as before, but suction stress above the water table is non-zero, say -20 kPa, the extended Bishop's routine method (Equation (9.61)) can be used to assess the factor of safety. This provides an approximation of the late summer or early fall conditions after the infiltrated water drains downslope. The computational procedures and steps leading to the factor of safety are shown in Table 9.12. The factor of safety calculated in this manner converges to 1.01, after several iterations, leading to the suspension of movement.

9.6.3 Slope stability analysis with water table rise

If the water table rises 1.00 m after infiltration reaches the saturated zone, results following the same procedure as described above lead to a decrease in the factor of safety to 0.99, as shown in Table 9.13. Thus, a rising water table alone could also lead to movement of the embankment.

The above analysis shows that for hillslopes of marginal stability, variations in either the water table location or suction stress in the vadose zone can initiate and arrest movement. By explicitly considering the impact of moisture changes on effective stress, the coupled hydro-mechanical framework provides a means to describe the physical mechanisms that lead to perennial and episodic landslide movement driven by snowmelt at this site. Knowledge of the dynamic fields of moisture content, soil suction, and suction stress at the site, and consequently the time-dependent instability analysis, requires more rigorous transient analysis, which will be described in Chapter 10.

Table 9.11 Computational procedures for calculating the factor of safety by Bishop's simplified method with the slope properties of $\gamma = 21.19$ kN/m^3, $c' = 0$, and $\phi_r = 22°$. Initial value of the factor of safety is set to be 1.10

1	2	3	4	5	6	7	8	9	10	11	Driving force	13	Resistance
Slice	W_n	h_n	u_n	α_n	$\sin\alpha_n$	b_n	$u_n b_n$	$c' b_n$	$u_n b_n \tan\phi'$	$W_n \tan\phi'$	$W_n \sin\alpha_n$	I	$(9 + 11 - 10)/13$
Unit	(kN/m)	(m)	(kPa/m)	(deg)		(m)	(kN/m)	(kN/m)	(kN/m)	(kN/m)	(kN/m)	(Eq. 9.48)	(kN/m)
1	1666	0	0.00	42	0.67	10.00	0.00	0.00	0.00	673.11	1114.77	1.02	656.85
2	3136.12	0	0.00	35	0.57	10.00	0.00	0.00	0.00	1267.07	1798.80	1.06	1194.74
3	3877.77	1.2	11.76	29	0.49	10.00	117.60	0.00	47.51	1566.72	1879.98	1.08	1408.42
4	3941.34	3.5	34.30	25	0.43	10.00	343.00	0.00	138.58	1592.40	1665.68	1.08	1340.95
5	3708.25	4.6	45.08	22	0.38	10.00	450.80	0.00	182.14	1498.23	1389.13	1.08	1213.17
6	3199.69	3.4	33.32	19	0.32	10.00	333.20	0.00	134.62	1292.76	1041.72	1.08	1069.84
7	3072.55	1.2	11.76	15	0.26	10.00	117.60	0.00	47.51	1241.39	795.23	1.07	1110.74
8	2436.85	0	0.00	11	0.19	10.00	0.00	0.00	0.00	984.55	464.97	1.06	927.13
9	1674.01	0	0.00	9	0.15	10.00	0.00	0.00	0.00	676.34	261.87	1.05	641.98
10	911.17	0	0.00	5	0.09	10.00	0.00	0.00	0.00	368.14	79.41	1.03	356.42
11	317.85	0	0.00	0	0	10.00	0.00	0.00	0.00	128.42	0.00	1.00	128.42
Σ											10491.58		10048.66
FS_s													0.96

Table 9.12 Computational procedures for calculating the factor of safety by the extended Bishop's simplified method with the slope properties of $\gamma = 21.19$ kN/m^3, $c' = 0$, and $\phi_r = 22°$. Initial value of the factor of safety is set to be 1.10. Shaded areas are changes from Table 9.11

1	2	3	4	5	6	7	8	9	10	11	Driving force	13	Resistance
Slice	W_n	h_n	σ^s	α_n	$\sin\alpha_n$	b_n	$u_n b_n$	$c' b_n$	$u_n b_n \tan\phi'$	$W_n \tan\phi'$	$W_n \sin\alpha_n$	I	$(9 + 11 - 10)/13$
Unit	(kN/m)	(m)	(kPa/m)	(deg)		(m)	(kN/m)	(kN/m)	(kN/m)	(kN/m)	(kN/m)	(in Eq. 9.61)	(kN/m)
1	1666	−4.37	−20.00	42	0.67	10.00	−200.00	0.00	−80.81	673.11	1114.77	1.01	745.85
2	3136.12	−2.1	−20.00	35	0.57	10.00	−200.00	0.00	−80.81	1267.07	1798.80	1.05	1285.41
3	3877.77	1.2	11.76	29	0.48	10.00	117.60	0.00	47.51	1566.72	1879.98	1.07	1421.74
4	3941.34	3.5	34.30	25	0.42	10.00	343.00	0.00	138.58	1592.40	1665.68	1.08	1351.93
5	3708.25	4.6	45.08	22	0.37	10.00	450.80	0.00	182.14	1498.23	1389.13	1.08	1221.96
6	3199.69	3.4	33.32	19	0.33	10.00	333.20	0.00	134.62	1292.76	1041.72	1.08	1076.58
7	3072.55	1.2	11.76	15	0.26	10.00	117.60	0.00	47.51	1241.39	795.23	1.07	1116.34
8	2436.85	−1.1	−10.78	11	0.19	10.00	−107.80	0.00	−43.55	984.55	464.97	1.06	971.78
9	1674.01	−4.83	−20.00	9	0.16	10.00	−200.00	0.00	−80.81	676.34	261.87	1.05	720.91
10	911.17	−5.43	−20.00	5	0.09	10.00	−200.00	0.00	−80.81	368.14	79.41	1.03	435.42
11	317.85	−6.52	−20.00	0	0	10.00	−200.00	0.00	−80.81	128.42	0.00	1.00	209.22
Σ											10491.58		10557.15
FS_s													1.01

Table 9.13 Computational procedures for calculating the factor of safety by the extended Bishop's simplified method with the slope properties of $\gamma = 21.19$ kN/m^3, $c' = 0$, and $\phi_r = 22°$, and a 1 m water table rise. Initial value of the factor of safety is set to be 1.10. Shaded areas are changes from Table 9.12

1	2	3	4	5	6	7	8	9	10	11	Driving force	13	Resistance
Slice	W_n	h_n	σ^s	α_n	$\sin\alpha_n$	b_n	$\sigma^s{}_n b_n$	$c' b_n$	$\sigma^s{}_n b_n \tan\phi'$	$W_n \tan\phi'$	$W_n \sin\alpha_n$	I	(9 + 11 − 10)/13
Unit	(kN/m)	(m)	(kPa/m)	(deg)		(m)	(kN/m)	(kN/m)	(kN/m)	(kN/m)	(kN/m)	(in Eq. 9.61)	(kN/m)
1	1666	−3.37	−20.00	42	0.67	10.00	−200.00	0.00	−80.81	673.11	1114.77	1.02	739.85
2	3136.12	−1.1	−20.00	35	0.57	10.00	−200.00	0.00	−80.81	1267.07	1798.80	1.06	1276.86
3	3877.77	2.2	21.56	29	0.48	10.00	215.60	0.00	87.11	1566.72	1879.98	1.07	1377.03
4	3941.34	4.5	44.10	25	0.42	10.00	441.00	0.00	178.18	1592.40	1665.68	1.08	1308.82
5	3708.25	5.6	54.88	22	0.37	10.00	548.80	0.00	221.73	1498.23	1389.13	1.08	1180.17
6	3199.69	4.4	43.12	19	0.33	10.00	431.20	0.00	174.22	1292.76	1041.72	1.08	1035.94
7	3072.55	2.2	21.56	15	0.26	10.00	215.60	0.00	87.11	1241.39	795.23	1.07	1076.12
8	2436.85	−0.1	−0.98	11	0.19	10.00	−9.80	0.00	−3.96	984.55	464.97	1.06	932.30
9	1674.01	−3.83	−20.00	9	0.16	10.00	−200.00	0.00	−80.81	676.34	261.87	1.05	719.60
10	911.17	−4.43	−20.00	5	0.09	10.00	−200.00	0.00	−80.81	368.14	79.41	1.03	434.97
11	317.85	−5.52	−20.00	0	0	10.00	−200.00	0.00	−80.81	128.42	0.00	1.00	209.22
Σ											10491.58		10290.88
FS_s													0.98

9.7 Problems

1. What is the conventional way of assessing the factor of safety of a hillslope?
2. In classical slope stability assessments, does the location of the failure surface have to be known or assumed prior to the analysis?
3. In classical slope stability assessments, is failure assumed to occur simultaneously at each point along the rupture surface?
4. What is the factor of safety for an infinite slope composed of material with zero cohesion?
5. What is the maximum angle a slope can be inclined if its materials have zero cohesion?
6. Does cohesion increase or decrease the stability of an infinite slope? Will an increase in pore-water pressure tend to decrease or increase the stability of the slope?
7. What is the maximum height of a cut slope using Culmann's finite-slope stability theory? Assume a vertical cut slope of material with cohesion c.
8. In Fellenius' ordinary method of slices, are inter-slice forces considered in the assessment of the stability of slope? Is Fellenius' method based on force and/or moment equilibrium?
9. In Bishop's simplified method of slices, are inter-slice forces considered in the assessment of the stability of slope? Is Bishop's simplified method based on force and/or moment equilibrium?
10. Is the factor of safety in Bishop's simplified method expressed in an implicit or explicit form? How does this affect the calculation of the factor of safety?
11. What are the major differences between infinite-slope stability models under saturated conditions and unsaturated conditions?
12. When suction stress increases (becomes less negative), will an infinite slope become less or more stable? Why?
13. Why does Bishop's simplified method generally produce a greater factor of safety than that calculated using the ordinary method of slices?
14. Where are landslides most likely to occur on an unsaturated hillslope of sandy soil?
15. Describe the general role of steady infiltration on the stability of a hillslope.
16. What is the relative percentage change of soil weight above the water table from hydrostatic conditions to full saturation?
17. Will an increase in soil weight above the water table always promote instability?
18. What is the physical meaning of the positive and negative signs on the right hand side of Equation (9.66)?
19. If the parameter λ is less than one, will an increase in soil weight above the water table lead to a decrease in slope stability?
20. Consider a hillslope composed of sandy soil. In general, how much change (as a percentage) would you expect to occur in the factor of safety with a change in pore-water conditions from hydrostatic to nearly saturated?
21. What are the major differences between steady-state and transient analysis in terms of (a) soil suction, (b) soil moisture content, and (c) stresses?

22 Why is there a time delay or diffusion in infiltration processes in hillslopes?

23 In sandy hillslopes under heavy rainfall, what are the ranges of variation in (a) soil suction, (b) effective saturation, (c) suction stress, and (d), factor of safety?

24 In silt hillslopes under heavy rainfall, what are the ranges of variation in (a) soil suction, (b) effective saturation, (c) suction stress, and (d) factor of safety?

25 For the case study of rainfall-induced shallow landslides in the Seattle area, at which depth are comparisons between the observed data and model results better in terms of (a) water content, (b) suction head, (c) suction stress, and (d) factor of safety?

26 For the case study of rainfall-induced shallow landslides in the Seattle area, how much stress change from the beginning of the rainy season is needed to trigger landslides? Correspondingly, how much change in the factor of safety occurred over this same period?

27 For the case study of landsliding induced by snowmelt, approximately how deep is the failure surface?

28 For the case study of landsliding induced by snowmelt, what is the difference in the factor of safety estimates if suction stress is considered or not?

29 By how much is the factor of safety reduced for the case study of landsliding induced by 1 m of snowmelt?

30 From the case analyses presented in this chapter, are you convinced or not that it is necessary to consider suction stress? Why?

31 For the hillslope with configuration and material properties shown in Figure 9.5, except that the water table is located at the slope surface and under hydrostatic conditions, calculate the factor of safety by (a) the classical ordinary method, (b) Bishop's routine method.

32 For the hillslope shown in Figure 9.5, except that the water table is horizontal and located at the bottom of slice 5, calculate the factor of safety by (a) the classical ordinary method, (b) Bishop's routine method, (c) the extended method (uniform suction stress of -20 kPa above the water table), and (d) the extended Bishop's routine method (uniform suction stress of -20 kPa above the water table). From Problem 31 and this problem, draw your conclusions regarding the importance of considering variably saturated hillslopes.

33 A long, linear sandy hillslope is considered as an infinite slope with the following material properties: $\phi_o = 38°$, $\Delta\phi = 6°$, $z_w = 0.8$ m, $\beta = 42°$, $k_s = 6 \times 10^{-7}$ m/s, $n = 4.5$, $\alpha = 0.1$ kPa^{-1}, $\gamma = 20$ kN/m^3. Assess and plot the profiles of suction, effective saturation, suction stress, and factor of safety when the steady infiltration rate q is 5.8×10^{-7} m/s.

10 Stress field based stability analysis

10.1 Hydro-mechanical framework

10.1.1 Failure modes in hillslopes

Slope stability analyses based on known or assumed rupture geometry prevail in the civil design of foundations and other earth structures. However, several inherent limitations impede the application of these approaches for prediction of rainfall-induced landslides in natural settings. Importantly, further theoretical development to assess landslide initiation and progression is also constrained by these limitations. The major limitations are: simplification of the geometry and location of the landslide failure surface, and simplification of the treatment of variably saturated flow and stress conditions. In recent years, some efforts have been made in overcoming these limitations, yet frameworks that are ready for practical applications remain to be established. This chapter attempts to provide a framework to move beyond both limitations.

Simplification of the geometry and location of failure surfaces was an effective, powerful, and necessary strategy for slope stability analysis prior to the widespread availability of computational tools that provide a means to quantitatively simulate the stress distribution in hillslopes. The assumption that all points along a pre-defined failure surface simultaneously reach a limiting equilibrium state greatly simplifies computational procedures and leads to single, consistent stability indicator or factor of safety (FS) for the entire slope. In reality, the state of stress along a pre-defined failure surface or any surface in a hillslope varies from point to point, and catastrophic failure is likely the result of progressive failure within a hillslope. As described in Chapters 5 and 7, the failure modes of hillslope materials are primarily shear and tensile. For example, consider the total stress field that results from gravity and geologic history shown in Figure 10.1. At point A, near the crest of the slope, the distribution of total stress promotes tensile failure, whereas for the middle of the slope (shown as point B) and near the toe (shown as point C), the distribution of total stress favors shear failure. It is unlikely that all these points along a potential rupture surface will fail simultaneously. Depending on the geologic and geomorphic factors that control the stress history of the hillslope, a failure that initiates in some part of the slope and then propagates is a likely scenario. While some studies have demonstrated that failure can initiate either near the crest and/or the toe of the slope (e.g., Eberhardt *et al.*, 2004; Duncan and Wright, 2005), others have suggested that failure could initiate internally or due to strain localization (e.g., Andrade and Borja, 2006; Borja *et al.*, 2006; Borja and White, 2010; Lu *et al.*, 2012).

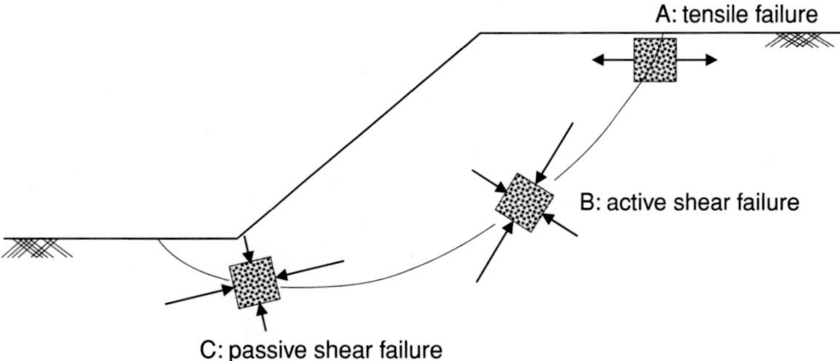

Figure 10.1 Illustration of potential failure modes in a hillslope under the distribution of total stress.

Because of the inaccuracy of current numerical models in capturing stress concentrations of the irregular topography of slopes and lack of physical experimental validations, realistic representations of stress and strain fields by numerical models remain challenging.

Because the geomorphic history, current state of stress, and the strength of materials (both tensile and shear) at each point completely define the failure state, the initiation and progression of failure are site specific. However, it is possible to define failure patterns for typical geologic settings, geomorphic history, strengths of materials, and additional triggering mechanisms such as earthquakes and rainfall. Seeking the connections among these controls and the consequent pattern of failure initiation and progression demands integration of knowledge from different disciplines and should be a high priority in landslide research.

A quantitative illustration of the strength and limitations of the limit-equilibrium methodology is provided here using a stability analysis of a 60° cut slope under gravitational forces. Several common limit-equilibrium methods, namely Culmann, ordinary, Bishop, Janbu, and Morgenstern–Price are used. Results in terms of the calculated FS along with the least-stable failure surfaces defined using a search algorithm are shown in Figure 10.2a. For the given shear strength parameters, all of the limit-equilibrium methods yield factors of safety within 8%, ranging from 0.97 for Janbu's method to 1.05 for Culmann's method. Regions with factors of safety less than 1.0 are also shaded. Other than Culmann's method, all the other limit-equilibrium methods predict failure of the slope with the same failure surface geometry.

Next, the stress field under gravity for linear elastic material is computed (see theory in Chapter 5) using a finite-element model (FEM) (e.g., Dawson *et al.*, 1999; Griffiths and Lane, 1999) and results are shown in Figure 10.3. The vertical stress field σ_{zz} (Figure 10.3b) is controlled by the self-weight of the slope and thus increases with the depth from the two horizontal free stress surfaces. The exception is near the toe of the slope, where stress concentration results from the geometry of the free surface. The horizontal stress field σ_{xx} (Figure 10.3a) is controlled by the field of vertical stress and the free stress boundary along the slope surface and the corner at the toe, which leads to a region of concentrated stress beneath the toe. The shear stress field τ_{xz} (Figure 10.3c) is characterized by a region of concentrated stress near the toe. Using a limit-equilibrium method to calculate a FS from the stress field simulated with the FEM yields a global FS of 1.03 and a failure surface shown in Figure 10.2a. The failure surface and FS calculated in this manner differ from the other limit-equilibrium methods and lie somewhere between results of the Culmann

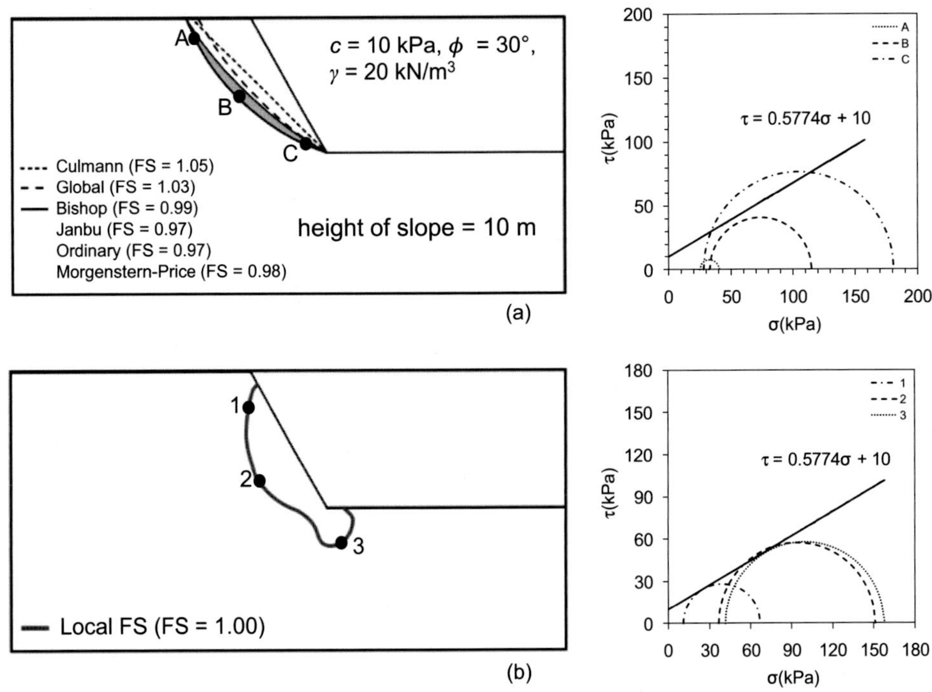

Figure 10.2 Illustration of differences between failure surface based and stress field based slope stability analyses: (a) a 60° slope analyzed by failure surface based limit-equilibrium methods, where the state of stress along the presumed failure surface may not reach failure state all together; and (b) the same slope analyzed by a stress field based method where the state of stress along the identified failure surface is all at failure state.

and the other methods. The state of stress along the failure surface for the other limit-equilibrium methods are shown in Figure 10.2a at three points A, B, and C as Mohr circles along with the failure envelope of the slope materials. Point A near the crest of the slope is far from failure, point B inside the slope is still stable, and point C near the toe exceeds the failure envelope. The calculated state of stress at point C is not physical, as it cannot exceed the strength of the materials; it can only approach this limit. As a result of the simplification that a single FS value is determined for the entire slope, regions where the state of stress approaches the strength of the slope materials are not realistically identified using conventional limit-equilibrium methods.

Because FEM analysis fields of stress are computed, methods for defining a field of FS or stability at each point within a slope can be developed (e.g., Iverson and Reid, 1992). Such methods take advantage of detailed information provided in the stress field, and thus provide a means to analyze spatial and temporal variation of hillslope stability. A method to quantify a scalar field of FS called local factor of safety or LFS will be introduced and illustrated in Section 10.2. The field of LFS obtained from the stress fields shown in Figure 10.3 and the Mohr–Coulomb shear strength parameters shown in Figure 10.2a for the contour of LFS = 1.0 is shown in Figure 10.2b. The states of stress at points 1, 2, and 3 along this contour are shown in the form of Mohr circles in Figure 10.2b. In spite of

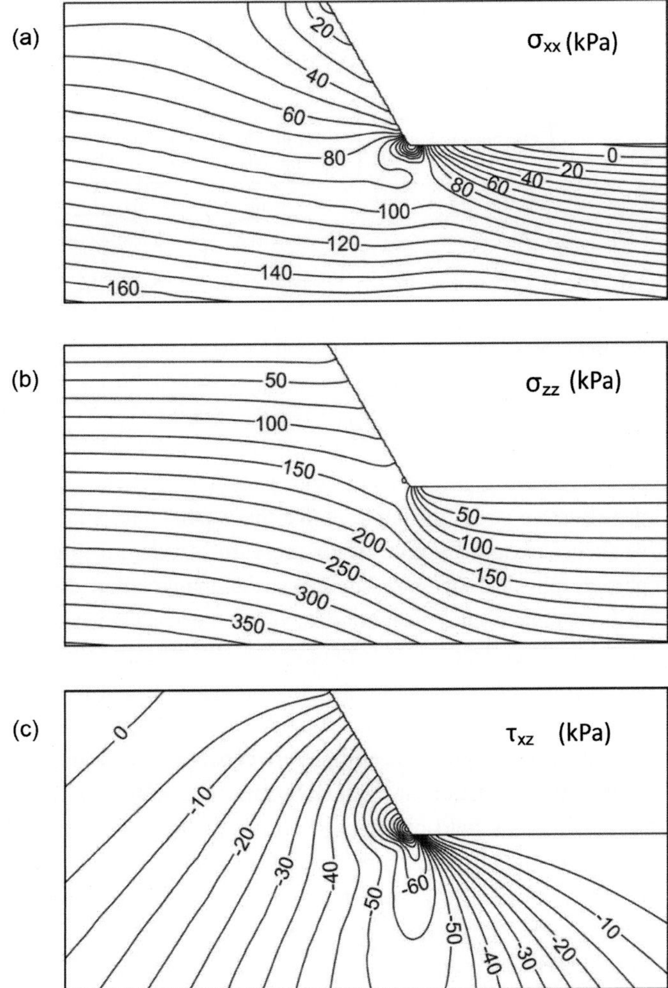

Figure 10.3 Stress fields from FEM analysis of a cut slope shown in Figure 10.2.

differences in the magnitude in the principal stresses, all three points reach failure state or the Mohr–Coulomb failure envelope. Zones of initial failure defined in such manner satisfy limiting equilibrium along the entire contoured region and thus should provide a rational basis to define the failure surface and its progression. Slope stability analysis based on stress fields can also accommodate other mechanisms that influence the state of stress, such as excavation, earthquake loading, and rainfall-induced infiltration. This last mechanism will be described in this chapter.

10.1.2 Unified effective stress principle

A second limitation in conventional slope stability analysis is the lack of a unified and consistent theory for the treatment of pore-water pressure in variably saturated porous

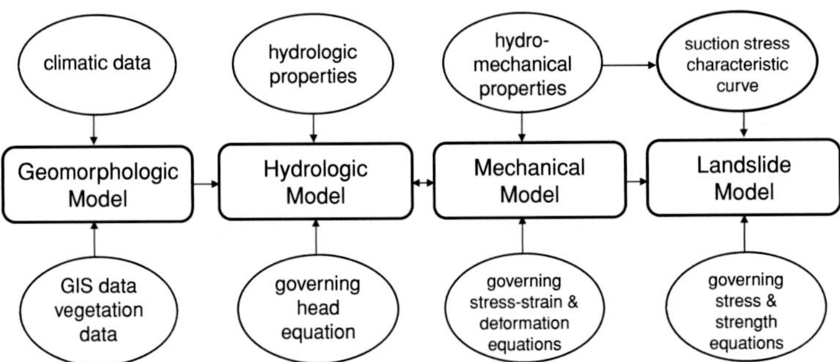

Figure 10.4 One-way coupled hydro-mechanical framework for stability analysis of variably saturated hillslopes.

media. Most conventional slope stability methods seek simplifications or ignore the effects of pore-water pressure above the water table. Some recent works completely abandon the effective stress principle above the water table, and rather focus on the modification of shear strength by pore-water pressure. As demonstrated in Chapter 6, it is the inter-particle physical stress or suction stress changes that drive changes in strength behavior in both tensile and shear modes. It has been shown (Lu, 2008) that using soil suction as the independent stress variable to modify shear strength criteria is physically unsound and theoretically inconsistent with Terzaghi's principle of effective stress established for saturated porous media. The effective stress principle for variably saturated porous media described in Chapter 6 is completely consistent with Terzaghi's effective stress principle. Application of the unified effective stress equation (6.14) or (6.15) allows for an accurate and continuous field of effective stress in variably saturated hillslopes to be obtained, as shown in this chapter. Under the effective stress principle, the failure criterion or shear strength criterion (e.g., Mohr–Coulomb) remains the same for all degrees of saturation. Furthermore, shear strength variation due to cementation and plant roots can be explicitly considered (Chapter 7).

10.1.3 Hydro-mechanical framework

The unified effective stress principle is also consistent with the continuous fields of soil suction and saturation established in hillslope hydrology covered in Chapters 3 and 4. For variably saturated hillslopes under transient rainfall conditions, coupling between soil suction and effective stress should be considered. A framework that explicitly accounts for the coupling among hillslope geomorphology, unsaturated hydrology, and mechanical stress and deformation would provide a rigorous and rational basis for development of physically based slope stability analyses. The logical relations among these aspects are illustrated in Figure 10.4. It is emphasized here that while the geomorphologic and hydrologic models provide necessary information for any sound mechanical model, it is the mechanical model that provides the ultimate step towards a realistic physical description of the state of stability of any hillslope. The power of such an approach will be illustrated in the following sections.

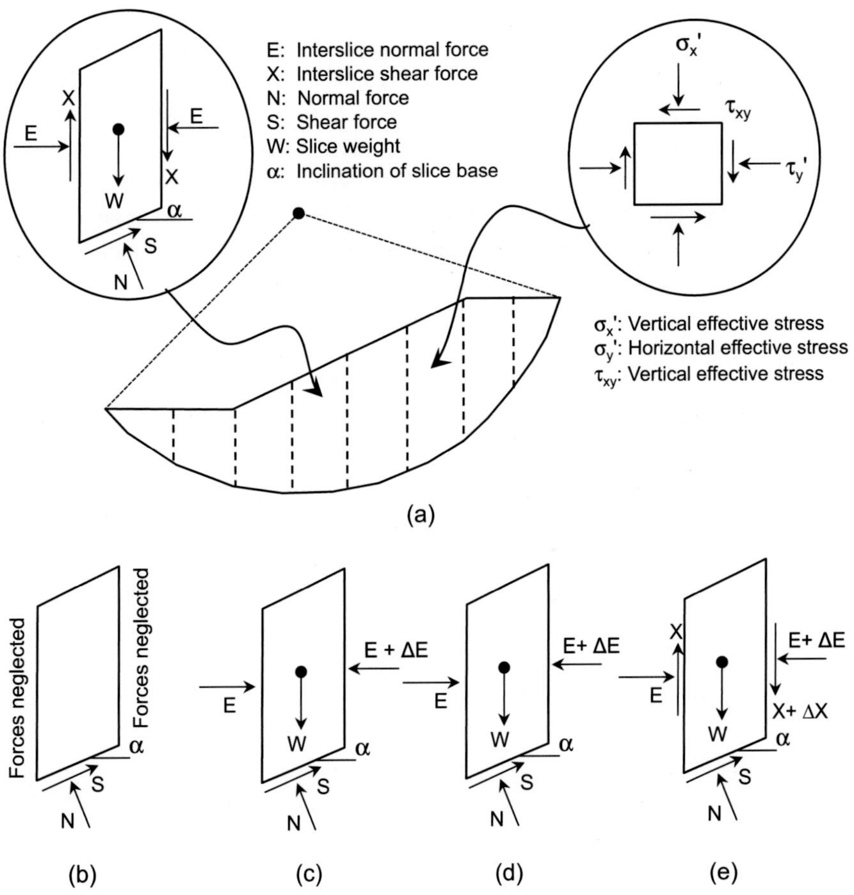

Figure 10.5 (a) Slice discretization of a potential sliding mass and forces acting on a typical slice for the classical limit-equilibrium methodology. (b)–(e) Illustration of the commonality and differences among the classical methods of slices for: (b) ordinary (Fellenius) method, (c) Bishop's simplified method, (d) Janbu's simplified method, and (e) Morgenstern–Price method (from Lu *et al.*, 2012).

10.2 Scalar field of factor of safety

10.2.1 Rationale for scalar field of factor of safety

Slope stability analyses have largely been performed using limit-equilibrium methods because of their proven effectiveness and reliability (e.g., Fellenius, 1936; Bishop, 1955; Morgenstern and Price, 1965; Janbu, 1973; Lade, 1992; Lam and Fredlund, 1993; Michalowski, 2002; Duncan and Wright, 2005). In the classical methods, the geometry and location of a potential failure surface are approximated and/or predetermined. All two-dimensional limit-equilibrium methods consist of discretizing the mass of a potential failure slope into small vertical slices and treating each individual slice as a unique sliding block, as shown in Figure 10.5a.

By applying the principles of force and/or moment equilibrium, all classical methods seek a single slope stability indicator, the factor of safety, which is typically defined as the ratio of the available shear strength to the shear stress required for equilibrium along the prescribed failure surface. Several variations of limit-equilibrium methods, such as the ordinary Fellenius (1936), Bishop (1955), Janbu (1973), and Morgenstern–Price (1965) methods, among many others, have been developed and differ primarily in how the inter-slice forces and the equilibrium principles are handled. For later comparison with the new method proposed here, the commonalities and differences among the classical methods are illustrated in Figure 10.5 and highlighted below.

The ordinary method (Figure 10.5b, Fellenius, 1936) is the simplest and the oldest method of slices. It ignores all the inter-slice forces because of the assumption that the forces are parallel to the base of each slice. The method satisfies only moment equilibrium and assumes a circular slip surface. Due to its simplicity, it is possible to compute FS using hand calculations. The ordinary method is generally less accurate compared to the other methods of slices. Some studies (e.g., Duncan and Wright, 2005; Fredlund and Krahn, 1977; Abramson *et al.*, 1996) have shown that the accuracy is decreased for effective stress analyses and under high pore-water pressure conditions.

Bishop's simplified method (Figure 10.5c, Bishop, 1955) satisfies vertical force equilibrium for each slice and overall moment equilibrium at the center of the circular trial surface (Figure 10.5a). This method also ignores inter-slice shear forces (X). Bishop's simplified method is considered to be more accurate than the ordinary method, especially for effective stress analyses with high pore-water pressures (e.g., Duncan and Wright, 2005; Abramson *et al.*, 1996).

The Janbu simplified method (Figure 10.5d) satisfies only overall horizontal force equilibrium and also assumes that there are no inter-slice shear forces (X). The assumption that all the inter-slice forces are horizontal often leads to a smaller FS when compared to more rigorous methods that satisfy complete force and moment equilibrium. To account for the effect of shear forces, a correction factor related to soil cohesion, angle of friction, and the shape of the failure surface was presented by Janbu (e.g., Duncan and Wright, 2005; Abramson *et al.*, 1996).

The Morgenstern and Price method (Figure 10.5e; 1965) considers both shear (X) and normal (E) inter-slice forces and satisfies both moment and force equilibrium. This method assumes an arbitrary mathematical function to describe the direction of the inter-slice forces (e.g., Fredlund and Krahn, 1977).

Some recent developments of the FS include more accurate computation of the inter-slice stress distribution (e.g., Duncan, 1996; Yu *et al.*, 1998; Swan and Seo, 1999) and numerical algorithms for shear strength reduction analysis from finite-element methods under a continuum mechanics framework (e.g., Matsui and San, 1992; Smith and Griffiths, 2004).

Because the classical use of the FS seeks a single indicator for the entire slope, it cannot describe where the failure begins or how the failure progresses. To overcome these limitations and fully take advantage of modern computational capabilities for numerical solutions of stress fields in slopes, a stress invariant based scalar field of FS is proposed in the next section.

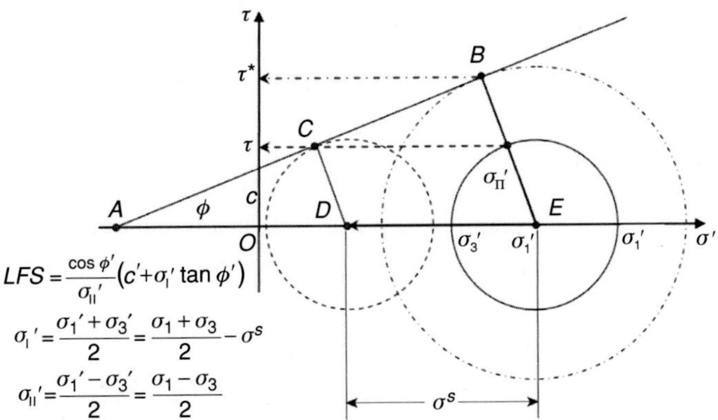

Figure 10.6 Illustration of the concept of the scalar field of LFS (from Lu *et al.*, 2012).

10.2.2 Definition of scalar field (or local) of factor of safety

The basic idea is to identify a scalar quantity at each point in a hillslope that is indicative of the stability of that point. As illustrated in two-dimensional stress space in Figure 10.6, if the current state of stress at a point of interest is the solid circle, the stress quantity that provides a measure of the failure state is Coulomb stress τ. Coulomb stress can be uniquely defined by shifting the Mohr circle leftward until it touches the Mohr–Coulomb failure envelope. In the literature (Iverson and Reid, 1992; Reid and Iverson, 1992; King *et al.*, 1994; Toda *et al.*, 1998), the shear stress on the Mohr circle along the direction perpendicular to the Mohr–Coulomb failure criterion is called Coulomb stress. In the study of the stress state of the earth's crust, scalar fields of Coulomb stress calculated in this manner around faults have been widely used to understand the occurrence of earthquakes. The stress paths in slopes follow the leftward shifting pattern under variably saturated soil conditions because the size of the Mohr circle is determined by the difference in principal total stresses. The principal total stresses are mainly affected by the geometry of the slope and the soil's self-weight, however, the former does not change for a given slope and changes in the latter are relatively small compared with changes in pore-water pressure (or suction stress above the water table) during the seepage process. On the other hand, infiltration and seepage greatly affect the suction stress (above the water table) or pore-water pressure (below the water table), leading to the leftward shift of the Mohr circle. This process likely occurs throughout the entire slope.

The pattern of stress path shift implies that the direction of failure under variably saturated soil conditions would be similar to the direction of Coulomb stress. If the reference point for soil strength at the current state of stress (solid Mohr circle) can be estimated by the intercept of the extension of the Coulomb stress with the Mohr–Coulomb envelope shown as point *B* in Figure 10.6, a FS at each point within the entire slope can be defined as:

$$\mathrm{LFS} = \frac{\tau^*}{\tau} \tag{10.1}$$

By the similarity of triangles *ACD* and *ABE*, the LFS can be expressed conveniently in terms of the ratio of the adjusted first stress invariant of the current state of stress to the adjusted first stress invariant of the potential failure state under the Mohr–Coulomb criterion. The adjustment here comes from the fact that in hillslopes tensile stress could occur in the area near the slope crest. To include such a situation, a stress value of *AD*, in lieu of the stress invariant *OD* (Figure 10.6) is used. Similarly, a stress value of *AE* in lieu of the stress invariant *OE*, is used. Graphically, the LFS defined by Equation (10.1) is equal to *AE* over *AD*. If the Mohr–Coulomb failure criterion is used, it can be shown that

$$\mathrm{LFS} = \frac{\cos\phi}{\sigma'_{II}}\left(c' + \sigma'_I \tan\phi'\right) \tag{10.2}$$

where c' is the drained cohesion of the slope material, ϕ' is the drained friction angle of the slope material, σ'_I and σ'_{II} are the mean and deviatoric effective stress in two-dimensional space defined by

$$\sigma'_I = \frac{\sigma'_1 + \sigma'_3}{2} = \frac{\sigma_1 + \sigma_3}{2} - \sigma^s \tag{10.3a}$$

$$\sigma'_{II} = \frac{\sigma'_1 - \sigma'_3}{2} = \frac{\sigma_1 - \sigma_3}{2} \tag{10.3b}$$

with the generalized effective stresses for variably saturated porous media defined by suction stress σ^s defined in Equation (6.5) (Lu and Likos, 2004b, 2006) as

$$\sigma' = \sigma - u_a - \sigma^s \tag{10.4}$$

where u_a is the prevailing air pressure. Suction stress σ^s is a characteristic function of a soil and varies with soil suction (Equation (6.14)) or saturation (Equation (6.15)) for all soils (Lu and Likos, 2004b; Lu and Griffiths, 2004; Lu *et al.*, 2010) by using the same set of parameters to describe the soil-water characteristic curve in van Genuchten's (1980) model (Equation (6.13)), i.e.,

$$\sigma^s = -(u_a - u_w) \quad \text{for} \quad u_a - u_w \le 0 \tag{10.5a}$$

$$\sigma^s = -\frac{(u_a - u_w)}{\left\{1 + [\alpha(u_a - u_w)]^n\right\}^{(n-1)/n}} \quad \text{for} \quad u_a - u_w \ge 0 \tag{10.5b}$$

where α and n are unsaturated soil parameters identical to those in van Genuchten's (1980) soil-water characteristic curve model that relates the equivalent degree of saturation S_e (equal to the degree of saturation S normalized by the residual degree of saturation S_r) to soil suction $(u_a - u_w)$ (Equation (6.13)):

$$S_e = \frac{S - S_r}{1 - S_r} = \left\{\frac{1}{1 + [\alpha(u_a - u_w)]^{\,n}}\right\}^{1-1/n} \tag{10.6}$$

Thus, in hillslopes the LFS defined by Equations (10.1)–(10.5) is a scalar field quantity that varies spatially and temporally. Under the classical limit-equilibrium concept, within a slope for points or regions where the LFS values are greater than unity, the states of stress are less than their limit states. Whereas for those with a FS less than or equal to unity, the states of stress have reached their limit states or local failures have already occurred. Thereafter, LFS has potential to delineate zones of stability or failure in slopes.

The following sections compare the LFS with several classical methods using examples of slopes with various configurations under either dry or saturated fully drained conditions. Applications of the LFS to slope stability analysis under variably saturated conditions are then illustrated for a slope under transient rainfall conditions with comparisons to some recent developments of infinite-slope stability theory under transient infiltration conditions.

10.2.3 Comparisons with the classical factor of safety methodologies

In the following examples a finite-element code (HILLSLOPE FS^2) was used to compute fields of stress, soil suction, and saturation, and the LFS for a two-dimensional slope. HILLSLOPE FS^2 simulates unsaturated flow and stress problems. It combines two existing FEM codes (Lu *et al.*, 2010). The first is based on FEM2D (Reddy, 1985) and solves the governing partial differential equations for stress and displacement described by the linear elasticity theory in Section 5.4. The second part is a solution of the Richards equation (Equation (4.18)) for unsaturated flow and is based on the U.S. Department of Agriculture model SWMS_2D (Šimůnek *et al.*, 1994). SLOPE/W (GeoStudio, 2007) was used to compute a series of limit-equilibrium analyses for comparison. The failure surfaces in these analyses were identified using an iterative procedure to identify the least stable surface. The field of LFS is computed using Equation (10.1). Since linear elasticity with no failure is assumed, a value of LFS at a point greater than one indicates that the computed shear stress is less than the presumed shear strength. Conversely, the computed zones or areas when LFS values are less than or equal to one only indicate regions of potential failure. These regions could differ in extent and geometry from post-failure zones computed by more sophisticated elasto-plastic models in which LSF less than unity is not physical or permissible.

Factor of safety and LFS in vertical cuts

First, the stress field in a vertical cut of either dry or saturated, but fully drained soil, under gravity for four different stability numbers m (0.05, 0.10, 0.15, and 0.20) was computed. The stability number is defined as $m = c'/\gamma\ H$ with γ being the unit weight of the slope material, and H the height of the slope (Taylor, 1937). A small m value indicates that the slope is generally less stable and vice versa. The field of LSF is computed by Equation (10.2) and plotted in Figure 10.7 (left column) for the computed stress field and strength parameters of the slope material. Three weak or potentially unstable zones are present in a vertical cut slope located near the vertical face, beneath the toe, and far field from the slope at the same elevation as the toe on horizontal ground. The low LFS zone behind the vertical face is characterized by the principal maximum stress in the vertical direction, whereas the low LFS zones beneath the toe and far field from the slope are characterized by the principal maximum stress in the horizontal direction. For all the m values examined here, the LFS in the far-field area away from the slope is never less than 1.0, whereas the LFS in the other two zones could be less than 1.0 and the area expands as the m value decreases (left column in Figure 10.7). The most pronounced weak zone is behind the vertical face with a progressive up-left trend along approximately 60° (or $\frac{\pi}{4} + \frac{\phi'}{2} = 60°$) from horizontal as the m value decreases.

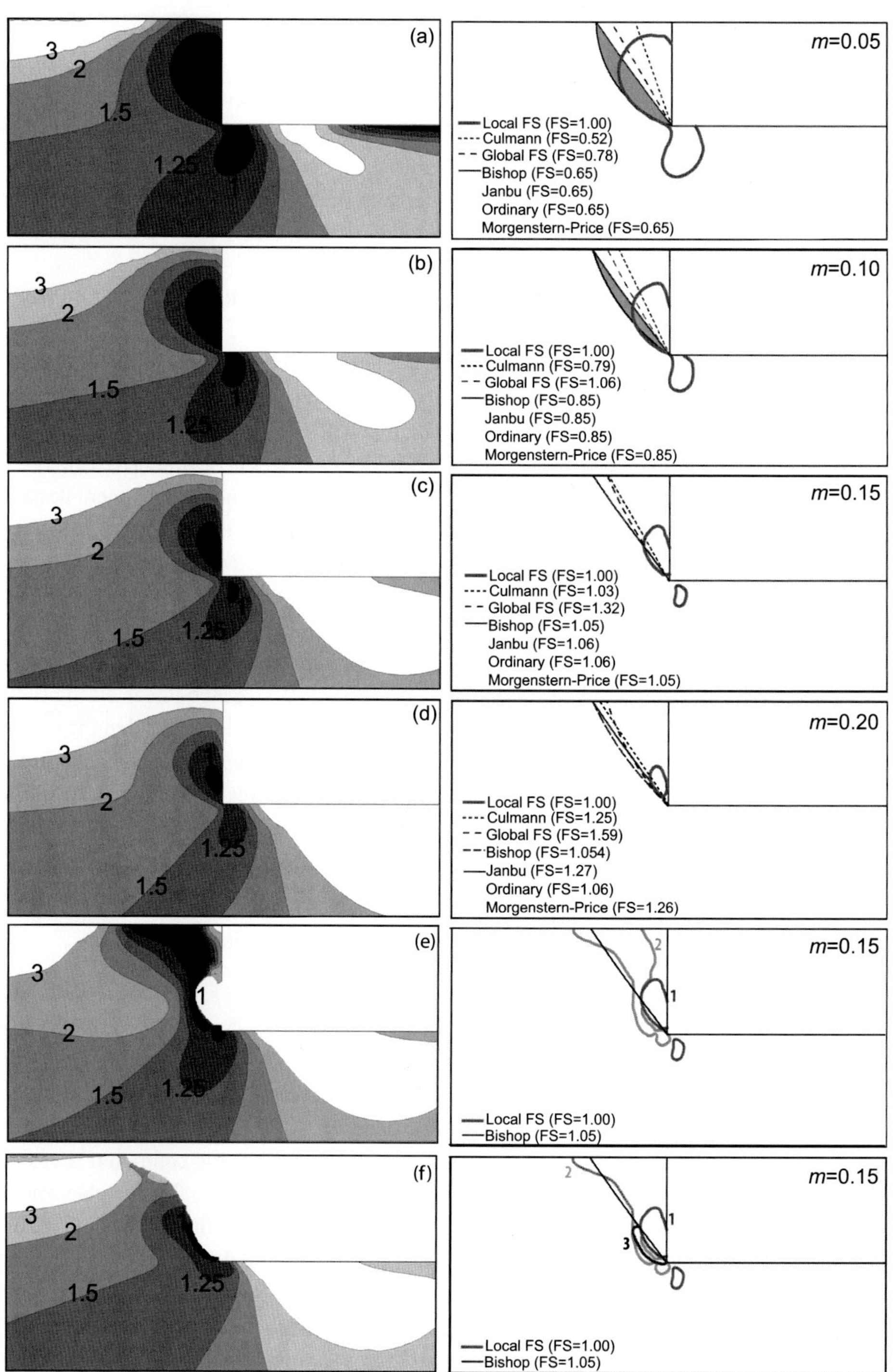

Figure 10.7 Distribution of the LFS with different stability numbers $m = c'/\gamma H$ calculated by HILLSLOPE (FS)[2] (Lu *et al.*, 2010) in vertical cuts (left). The comparison of LFS with FS by various limit-equilibrium methods (right). The black zones in the figures in the left column indicate the regions where all the slip surfaces have FS less than 1.0. (a) $m = 0.05$, (b) $m = 0.10$, (c) $m = 0.15$, (d) $m = 0.20$, and (e and f) $m = 0.15$ ($\gamma = 20$ kN/m^2, $\phi' = 30°$). The zone with a LFS less than 1.0 is removed from (e) and subsequent iterative calculation of the LFS (f) illustrates progressive development of the failure surface (after Lu *et al.*, 2012).

To better illustrate this trend, contours of LFS equal to 1.0 are plotted in the right column in Figure 10.7 (thick solid lines) together with other potential failure surfaces identified by four classical limit-equilibrium stability methods. To broaden the comparison and assessment, potential failure surfaces predicted by the Culmann method (e.g., Lohnes and Handy, 1968) and by a hybrid method (Global FS – SLOPE/W), in which finite elements are used to compute the stress distribution, and the FS of a least stable surface is identified using an iterative method of slices, are also plotted in the right column in Figure 10.7.

For a relatively strong vertical cut ($m = 0.20$ and shown in Figure 10.7d), all four limit-equilibrium methods (ordinary, Bishop, Janbu, and Morgenstern–Price), as well as the Culmann and the hybrid global methods, predict that the vertical cut is stable or the FS is greater than 1.0, whereas the LSF method delineates a failure zone behind the vertical face and near the toe. The potential failure surfaces for Janbu, ordinary, and Morgenstern–Price are identical (thin solid curve), and very close to that given by the global FS in the lower portion of the cut, but different than that by Bishop (long dashed curve) in the middle portion of the cut. The potential failure surface predicted by the Culmann method is quite different than the rest of the limit-equilibrium methods. It predicts the failure of the cut (short dashed line with FS = 1.0). Note that the lower part of the LFS = 1.0 contour follows closely the potential failure surfaces predicted by the other methods. Overall, Bishop's method yields the lowest FS value (1.054) and is close to the FS given by the ordinary method (1.06). Both the Janbu and Morgenstern–Price methods yield similar FS values (1.27 vs. 1.26), much higher than the FS by Bishop and ordinary methods, yet much smaller than the global FS method (1.59).

For vertical cuts with a stability number $m = 0.15$, all the classical limit-equilibrium methods predict the same potential failure surface with nearly identical FS values (1.05–1.06). Compared to the case where $m = 0.20$, the global FS method yields a much higher FS value (1.32) and the potential failure surface is farther away from that given by the four limit-equilibrium methods (Figure 10.7d). Again, the lower part of the LFS = 1.0 contour follows closely the potential failure surfaces predicted by the other methods. Furthermore, the LFS method identifies an additional failure zone beneath the toe where the LFS is less than unity. As the stability number m decreases (shown respectively for m values of 0.10 and 0.05 in Figures 10.7a and 10.7b, all four classical limit-equilibrium methods predict instability of the vertical cut with an identical FS value (0.85 for $m = 0.10$ and 0.65 for $m = 0.05$) and similar failure surface. For limit-equilibrium analysis, zones where the FS is less than unity can also be identified and are shown as shaded areas in Figures 10.7c and 10.7d. Once again it can be observed that the lower part of the LFS = 1.0 contour follows very closely with the potential failure surfaces predicted by the classical limit-equilibrium methods. Compared to the potential failure surfaces with lower m values (Figures 10.7c and 10.7d), both the global FS and Culmann methods predict nearly identical failure surfaces, but away from those by either the classical limit-equilibrium methods or the LFS method.

In light of the above analysis for the stability of vertical cuts it appears that the failure surface initiates from the toe region behind the vertical face and progresses up-and-left toward the top of the cuts. This trend can only be predicted by a LFS-based methodology as the limit-equilibrium methodology explicitly excludes detailed information on the stress distribution and prescribes the failure surface along the general direction of the Rankine active failure planes, i.e., $\frac{\pi}{4}+\frac{\phi'}{2} = 60°$. A natural question here is: does the limit-equilibrium methodology predict accurate or realistic failure surfaces? Answers to this question can be gained by applying the LFS methodology progressively to obtain the final stable configuration of the slope as follows.

For a vertical cut with a stability number $m = 0.15$ (Figure 10.7c), all four limit-equilibrium methods show the slope is about to fail (FS = ~1.05–1.06). To illustrate the potential of the LFS to predict the progressive process of failure in slopes, we assume that the material of the slope is brittle, it will lose all its cohesion strength after failure, and by the force of gravity the failed mass will be separated from the cut and have no further stabilizing or destabilizing effect. Following these assumptions, the zone above the toe where LFS is less than unity is removed and then the LFS distribution is computed for the modified domain (Figure 10.7e). This process is then repeated until all LFS are greater than unity (through Figures 10.7e and 10.7f). As shown in Figure 10.7f, the final surface configuration produced in such manner is similar to those failure surfaces predicted by the limit-equilibrium methodology. The final slope configuration predicted by the LSF methodology would be more rigorous if elasto-plastic models were to be incorporated in the slope stability analysis, thus providing additional understanding of the formation and evolution of slope surfaces.

FS and LFS in 60° slopes

The distributions of LFS are shown in the left column in Figure 10.8 for an example 60° slope. The limit-equilibrium methods produce similar (potential) failure surfaces except for the case of stability number $m = 0.05$ (Figure 10.8a), where differences near the top of the slope between the classical limit-equilibrium and the hybrid global FS methods appear. Only one zone with LFS less than unity is identified for all m values. This zone occurs near, but above the toe and propagates upward throughout the entire face of the slope as the stability number m decreases. For stability number $m = 0.20$ (Figure 10.8d), a zone of LFS less than unity coincides with the potential failure plane predicted by the Culmann method, but is located above the potential failure surface predicted by all other limit-equilibrium methods. For stability number $m = 0.20$ (Figure 10.8d), $m = 0.15$ (Figure 10.8c), and $m = 0.10$ (Figure 10.8b), all the limit-equilibrium methods predict that the slope is stable with similar FS given by the four classical limit-equilibrium methods. The FS (1.97) is slightly higher for the Morgenstern–Price method for the stability number $m = 0.20$ case and is closer to that given by the global FS method (FS = 2.02). For the stability number $m = 0.05$ (Figure 10.8a), the FS given by all the limit-equilibrium methods are similar and predict that the slope is unstable (FS < 1.0). However, the global FS method predicts stability (FS = 1.03). The contour of unity for the LFS follows the potential failure surfaces given by the limit-equilibrium methods near the toe for stability numbers $m = 0.15$ and $m = 0.10$, but is somewhat different for stability number $m = 0.05$. Moving towards the crest of the slope, the contour of unity for the LFS differs significantly from the failure surfaces predicted by all the limit-equilibrium methods. The LFS unity contour exits the

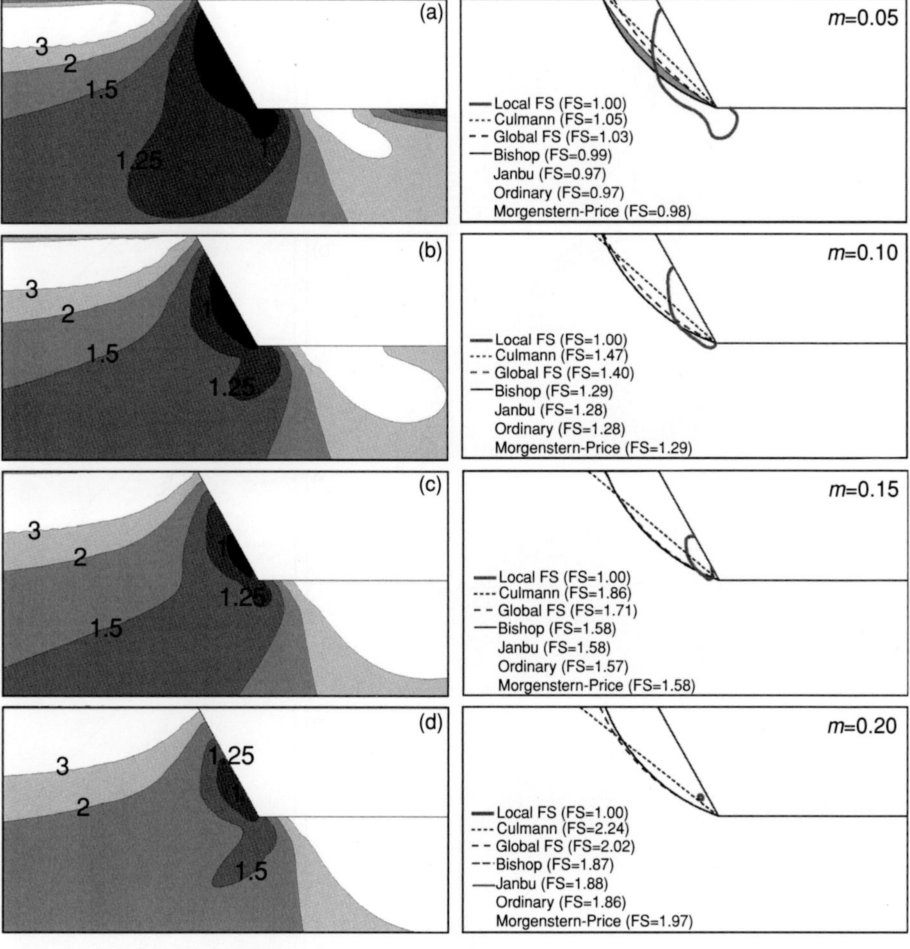

Figure 10.8 Distribution of the LFS for hillslopes with different stability numbers $m = c'/\gamma H$ calculated by HILLSLOPE (FS)^2 in 60° slopes (left). The comparison of LFS with FS given by various limit-equilibrium methods (right). The black zones in the left column indicate the regions where all the slip surfaces with FS less than 1.0 are located from the limit-equilibrium analyses. (a) $m = 0.05$, (b) $m = 0.10$, (c) $m = 0.15$, and (d) $m = 0.20$ ($\gamma = 20$ kN/m^2, $\phi' = 30°$).

domain on the slope face, whereas the equilibrium methods always predict that the failure surface will intercept the ground surface behind the slope crest. The LFS method predicts failure under conditions where the other methods predict stability, providing additional insight into the shape of the failure surface and sensitivity to the distribution of stress. The LFS method predicts small unstable zones on the slope face for the $m = 0.15$ and $m = 0.10$ cases, whereas the limit-equilibrium methods indicate the slope is stable. This implies that the LFS provides an increase in precision and sensitivity that produces a more conservative prediction of the failure geometry over that given by the limit-equilibrium methods.

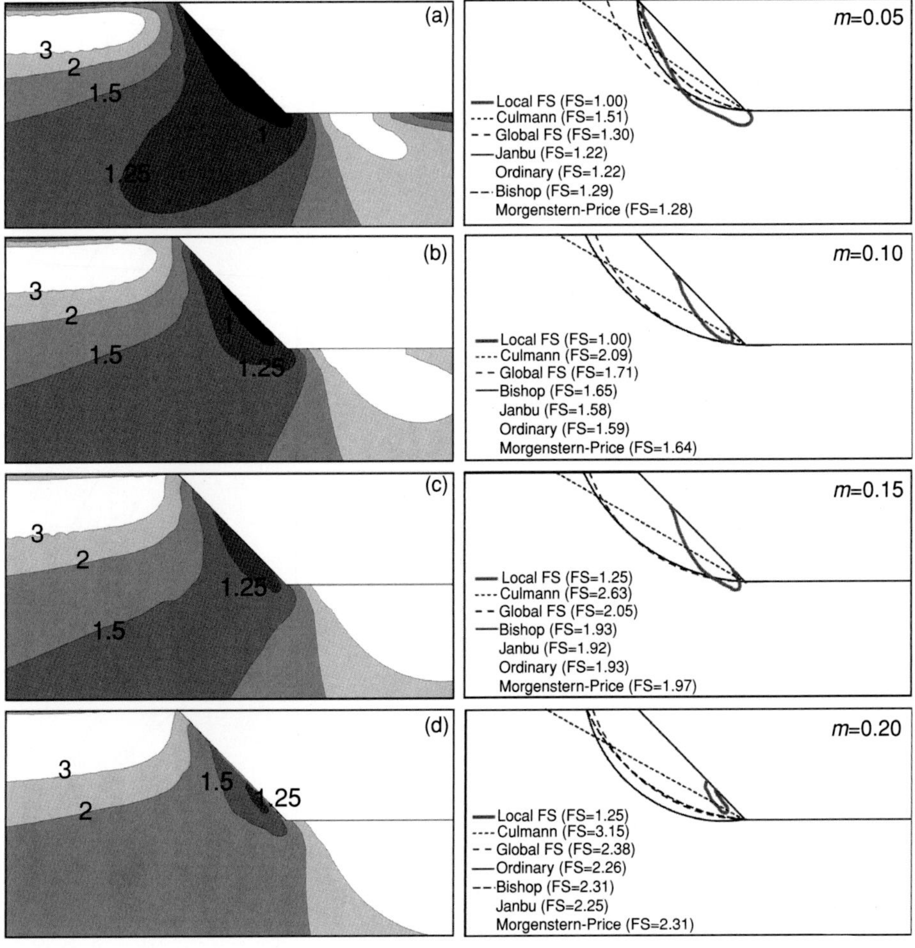

Figure 10.9 Distribution of the LFS for hillslopes with different stability numbers $m = c'/\gamma H$ calculated by HILLSLOPE (FS)^2 in 45° slopes (left). The comparison of LFS with FS given by various limit-equilibrium methods (right). The black zones in the left column indicate the regions where all the slip surfaces with FS less than 1.0 are located from the limit-equilibrium analyses. (a) $m = 0.05$, (b) $m = 0.10$, (c) $m = 0.15$, and (d) $m = 0.20$ ($\gamma = 20$ kN/m^2, $\phi' = 30°$).

FS and LFS in 45° slopes

More gently inclined slopes are generally more stable (Figure 10.9). All the equilibrium methods assess the slopes as stable (FS $>$ 1.0) for all stability numbers analyzed (*m* from 0.05 to 0.20) with the calculated FS values within 10% among the methods. However, the geometry of the potential failure surfaces predicted by different methods is quite different. For example, for stability number $m = 0.05$ (Figure 10.9a), the global FS method and Culmann method predict that the failure surfaces daylight at the top and at the toe of the slope at almost the same locations, but the surfaces are quite different elsewhere within the slope.

All four limit-equilibrium methods consistently predict that the failure surface would exit at the crest of the slope (Figure 10.9a). No failure is predicted for the LFS method for stability numbers $m = 0.15$ and 0.20 (Figures 10.9c and 10.9d), and failure zones are only observed for stability numbers $m = 0.05$ and 0.10 (Figures 10.9a and 10.9b). As the stability numbers decrease, the failure begins from the lower part of the slope near the toe and progresses upward toward the crest (Figures 10.9a and 10.9b). The contour for LFS = 1.0 agrees with the potential failure surfaces predicted by the four limit-equilibrium methods (Figure 10.9a), except in the region beneath the toe of the slope. Assuming that linear elasticity theory and the Mohr–Coulomb failure criterion adequately describe the stress distribution and failure conditions, the 45° slope with the stability numbers $m = 0.10$ and 0.05 will be unstable, as predicted by the new LFS methodology. On the other hand, the limit-equilibrium methods give stable results with a FS ranging from 1.58 to 1.71 for stability number $m = 0.10$ and from 1.22 to 1.30 for stability number $m = 0.05$ for potential failure surfaces that are very similar to the LFS contour of 1.0 for the stability number $m = 0.05$ (Figure 10.9a). Comparison of the LFS contours between the 45° and 60° slopes with the same stability number (Figure 10.8a versus Figure 10.9a, and Figure 10.8b versus Figure 10.9b) shows that the LFS contour of 1.0 or failure zone becomes shallower, indicating qualitatively that the radius of curvature of the LFS = 1.0 becomes larger as the slope becomes gentler. This trend of failure zone evolution with the slope angle is also predicted in the analytical solution of the distribution of gravitational stress with Coulomb failure criteria in finite slopes by Savage (1994).

FS and LFS in 30° slopes

The trend of increasing radii of curvature of the failure surface predicted by the LFS methodology with decreasing slope gradient can be further examined in slopes inclined at 30° (Figure 10.10). The LFS methodology predicts a low FS zone of shallow depth nearly parallel to the slope surface for the slopes with stability numbers of $m = 0.10$ and $m = 0.15$ (Figures 10.10b and 10.10c). As the stability number decreases (from Figures 10.10d to 10.10a), this zone progressively elongates and eventually a quasi-translational failure zone with LFS < 1.0 covers nearly the entire slope surface (Figure 10.10a). A similar failure zone is predicted by Savage's (1994) exact solutions of stresses in finite slopes. In contrast, none of the limit-equilibrium methods predict this shape of the failure surface or the potential instability for a slope with a stability number $m = 0.05$. The smallest FS calculated by the limit-equilibrium methods is 1.64 (Janbu) and all potential failure surfaces are curved, in contrast to the nearly translational surface predicted by the LFS method. Shallow landslides, typically translational slope failures a few meters thick of unlithified soil mantle or regolith, may dominate mass-movement processes in hillslope environments (Cruden and Varnes, 1996; Trustrum *et al.*, 1999; Sidle and Ochiai, 2006). None of the limit-equilibrium methods described here can adequately capture the shape of such failures, whereas the LFS methodology can predict the initiation, evolution, and geometry of such failures.

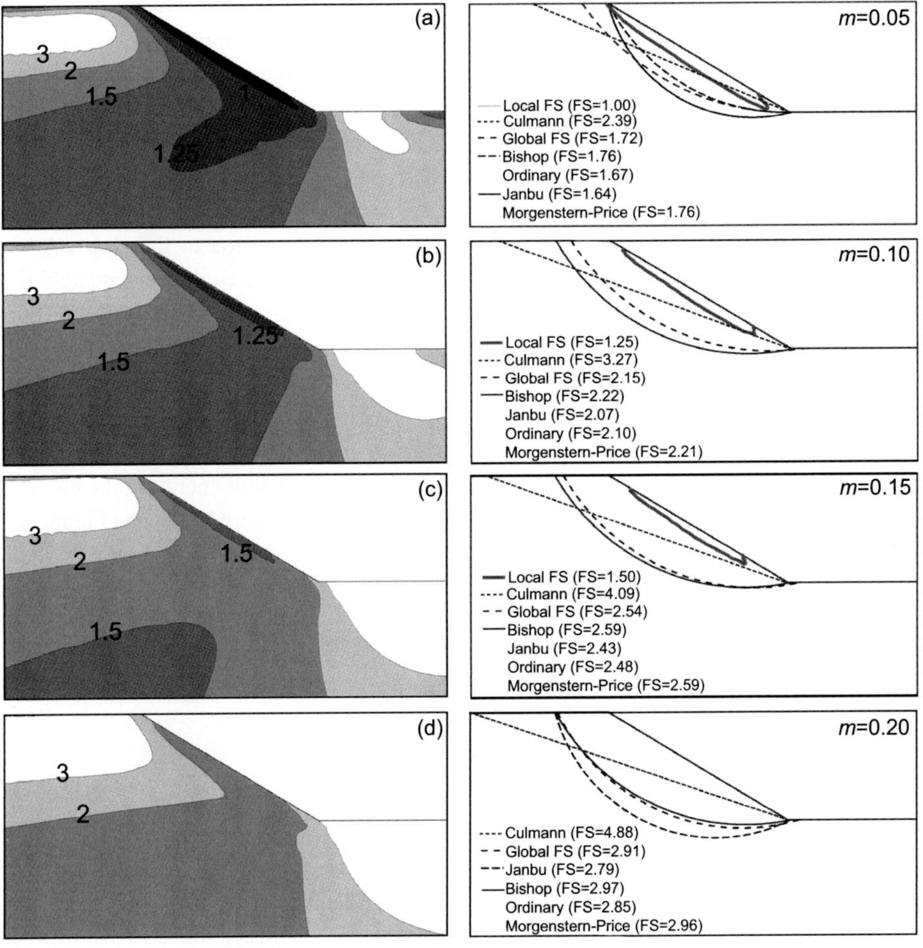

Figure 10.10 Distribution of the LFS with different stability numbers $m = c'/\gamma H$ calculated by HILLSLOPE (FS)^2 in 30° slopes (left). The comparison of LFS with FS by various limit-equilibrium methods (right). The black zones in the left column indicate the regions where all the slip surfaces with FS less than 1.0 are located from the limit-equilibrium analyses. (a) $m = 0.05$, (b) $m = 0.10$, (c) $m = 0.15$, and (d) $m = 0.20$ ($\gamma = 20$ kN/m^2, $\phi' = 30°$) (after Lu *et al.*, 2012).

Slope stability analysis of translational landslides has typically relied on the infinite-slope model (e.g., Dietrich *et al.*, 1995; Wu and Sidle, 1995; Baum *et al.*, 2010), which requires several strict assumptions, including slope parallel failure surfaces, saturated materials, a sharp contrast between the properties of overlying materials and substrate, and that failure occurs simultaneously at given depths. The assumptions of saturated materials, sharp contrast between layered materials, and fixed failure depth have been removed using variably saturated theory for effective stress analysis (Lu and Godt, 2008). Comparison of the LFS methodology with the infinite-slope stability models is presented in the next section for slopes under infiltration conditions.

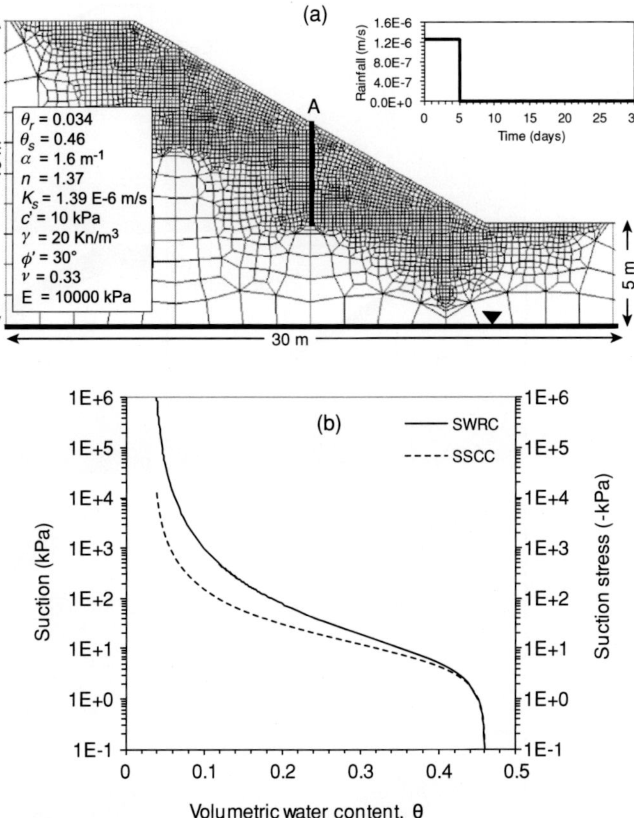

Figure 10.11 (a) The geometry, material properties, FEM mesh, and rainfall scenario used in the silty slope simulation, and (b) soil water characteristic curve (SWCC) and suction stress characteristic curve (SSCC) of the silty soil hillslope used in the simulations (after Lu *et al.*, 2012).

10.3 Transient hillslope stability analysis

A 30° slope under variably saturated conditions is used as an example here and shown in Figure 10.11a. The water table is set at 5 m below the toe of the slope and forms the lower boundary. An initial hydrostatic condition above the water table is imposed and results in soil suction varying linearly with distance from the water table (not shown) and volumetric soil moisture varying non-linearly above the water table (shown in Figure 10.12a). The slope material is a silty soil with hydro-mechanical properties defined in Figure 10.11b. Both left and right sides of the domain are assumed to be no-flow boundaries and the top boundary is subject to a 5-day rainfall flux with a rate of 4.5 mm/hr (or 90% of the saturated hydraulic conductivity 1.39×10^{-6} m/s) (Figure 10.11a). A total simulation time of 5 years is considered.

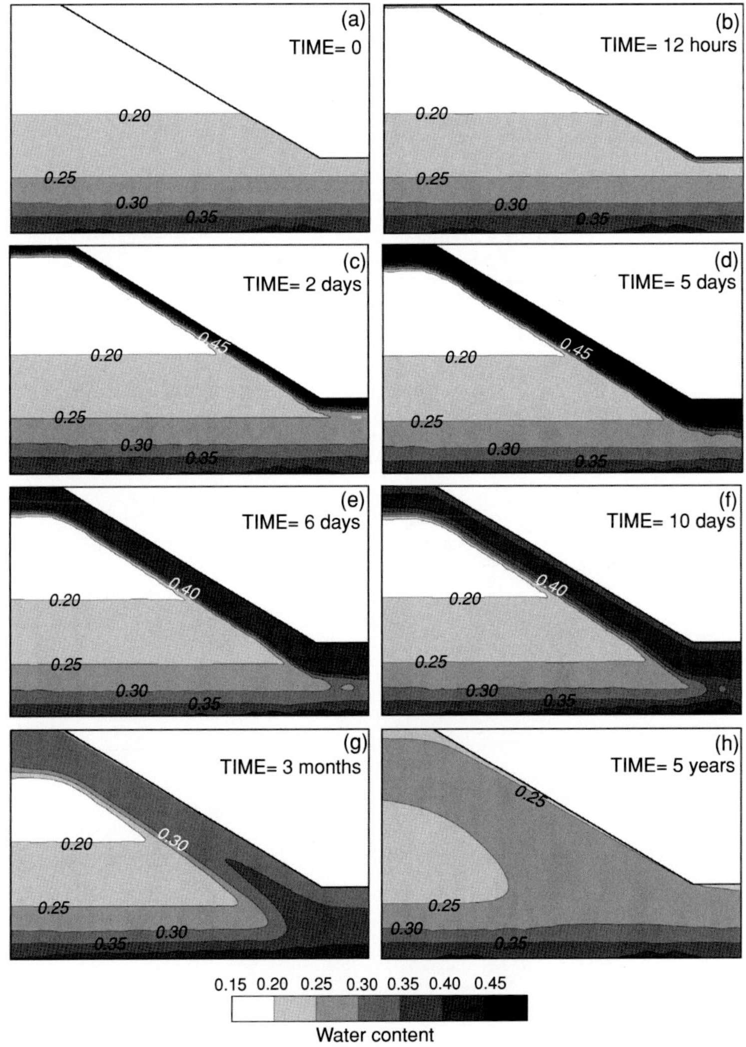

Figure 10.12 Simulated contours of moisture content at different times for a silty slope under rainfall conditions (a) at time = 0, (b) at time = 12 hours, (c) at time = 2 days, (d) at time = 5 days, (e) at time = 6 days, (f) at time = 10 days, (g) at time = 3 months, and (h) at time = 5 years.

The variation in the soil moisture content θ resulting from the 5-day rainfall is shown in Figure 10.12. Because soil near the slope surface is initially quite dry ($\theta < 20\%$) under hydrostatic conditions, and the prescribed flux is nearly the same as the saturated hydraulic conductivity, a nearly saturated zone preceded by a sharp wetting front develops during the 5-day rainfall period (Figures 10.12b–d). The progress of the wetting front is also illustrated in Figure 10.13 as profiles of moisture content at the middle of the slope (location A in Figure 10.11a) at different times from the beginning of the rainfall.

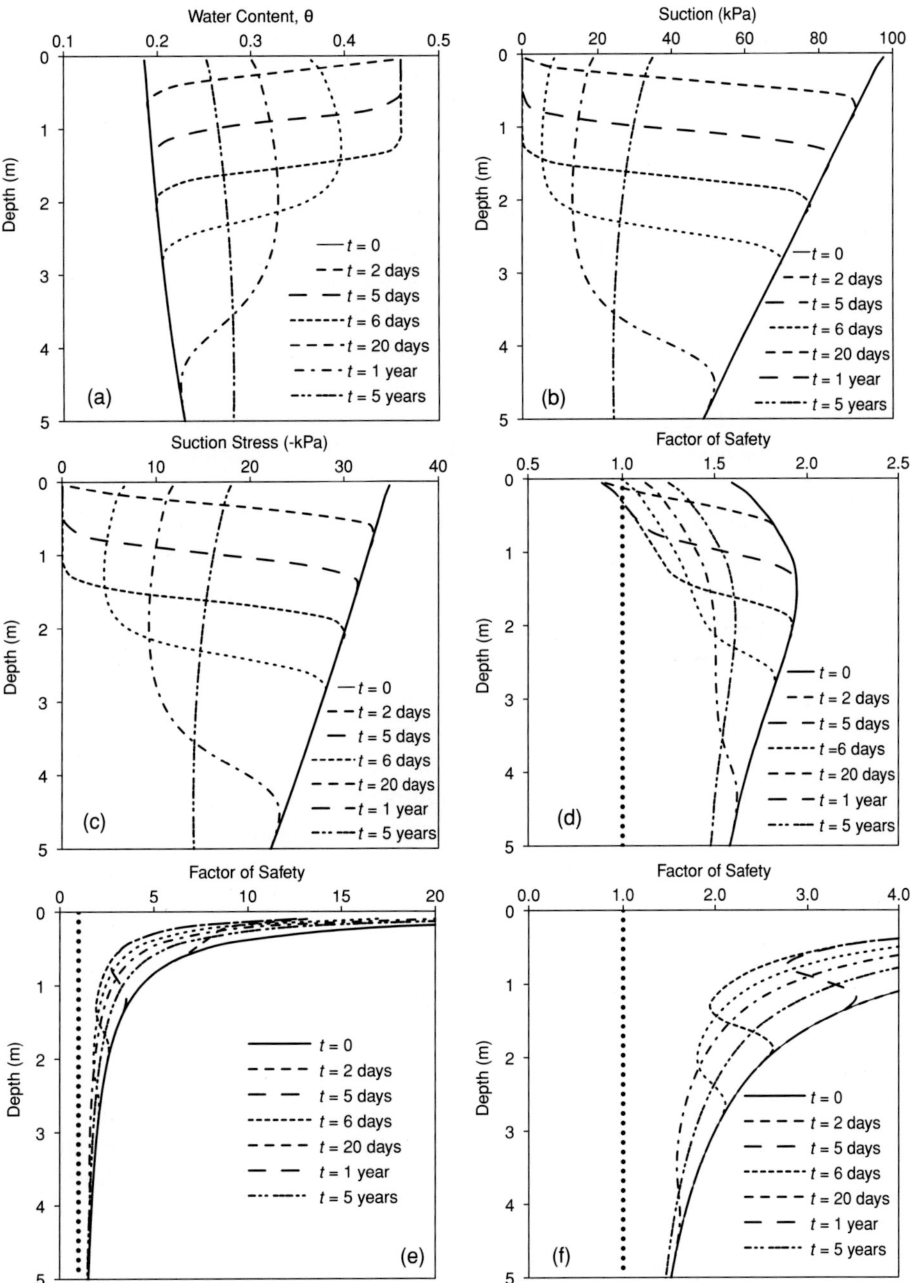

Figure 10.13 Profiles of (a) moisture content, (b) suction, (c) suction stress, (d) LFS, (e) and (f) FS calculated by infinite-slope stability analysis in the middle of the silty slope at different times.

The wetting front becomes more diffuse as time elapses. Soil suction in the slope progresses in a similar pattern to that of the moisture content, as illustrated by the suction profiles at different times in Figure 10.13b. After the cessation of rainfall, a drying zone develops near the slope surface behind the wetting front (Figures 10.12e, 10.12f, 10.13a, and 10.13b).

The 5-day rainfall episode results in 540 mm water infiltrating into the slope and causes moisture content redistribution throughout the entire slope. The infiltrated water reaches the water table below the toe area in about 30 days, as shown in Figures 10.12f, 10.12g, 10.13a, and 10.13b. After 5 years, the moisture content field is still far from the initial hydrostatic conditions (Figures 10.12h, 10.13a, and 10.13b). Thus, given the imposed boundary conditions, it will take years for the moisture content and suction to recover to the pre-rainfall hydrostatic conditions.

The dynamics of the mechanical response of the slope to the 5-day rainfall episode can be quantitatively investigated by examining the suction stress field. Under the framework of the unified effective-stress principle defined in Equations (10.4) and (10.5), the field of suction stress in the slope can be determined once either the moisture content or soil suction field is known. Figure 10.14 shows the suction stress field in the slope at different times. Before the rainfall, as shown in Figure 10.13c, suction stress varies nearly linearly with depth below the ground surface and has much smaller absolute values than soil suction (shown in Figure 10.13b) at the same location. For example, on the slope surface at location A, soil suction is −98.0 kPa, whereas the corresponding suction stress is −34.2 kPa (see Figure 10.11b). After the rainfall begins, the wetting front moves into the soil, increasing water contents and increasing soil suction and suction stress as shown in Figures 10.13c and 10.14b–10.14d. At the end of the 5-day rainfall, a zone of zero suction stress with a thickness of about 1.2 m forms, as shown in Figures 10.13c and 10.14d. After the cessation of rainfall, suction stress continues to increase in the zone in front of the wetting front, but decreases in the drying zone near the slope surface (Figures 10.13c and 10.14e–10.14h). At the end of the 5-year simulation period, suction stress is still much greater than that of the pre-rainfall hydrostatic condition indicating smaller effective stresses than that under hydrostatic conditions.

The dynamics of slope stability now can be quantified using the LFS field in light of varying effective stress in the slope. The field of LFS is calculated simply by using Equations (10.2)–(10.4) with the total gravitational stress field from the FEM simulations. The field of LFS in the slope at different simulation times is shown in Figure 10.15. Before the rainfall episode, the slope is stable (Figure 10.15a) and the LFS near the slope surface is between 1.50 and 2.00. After 12 hours of rainfall, a thin zone nearly parallel to the slope surface with LFS between 1.25 and 1.50 develops. After 2 days, a thin zone nearly parallel to the slope surface with LFS less than 1.0 develops. This zone extends from the toe to an area near the crest of the slope (Figure 10.15c) and varies in thickness with a maximum thickness of about 20 cm near the middle of the slope (Figures 10.15c and 10.13d). At the end of the 5-day rainfall episode, this zone remains nearly the same shape, but the depth doubles to a maximum thickness of about 40 cm (Figures 10.15d and 10.13d). This zone of LFS < 1.0 spatially and temporally coincides with the large increase in suction stress shown in Figure 10.14d. Within this zone, as a result of increases in suction stress

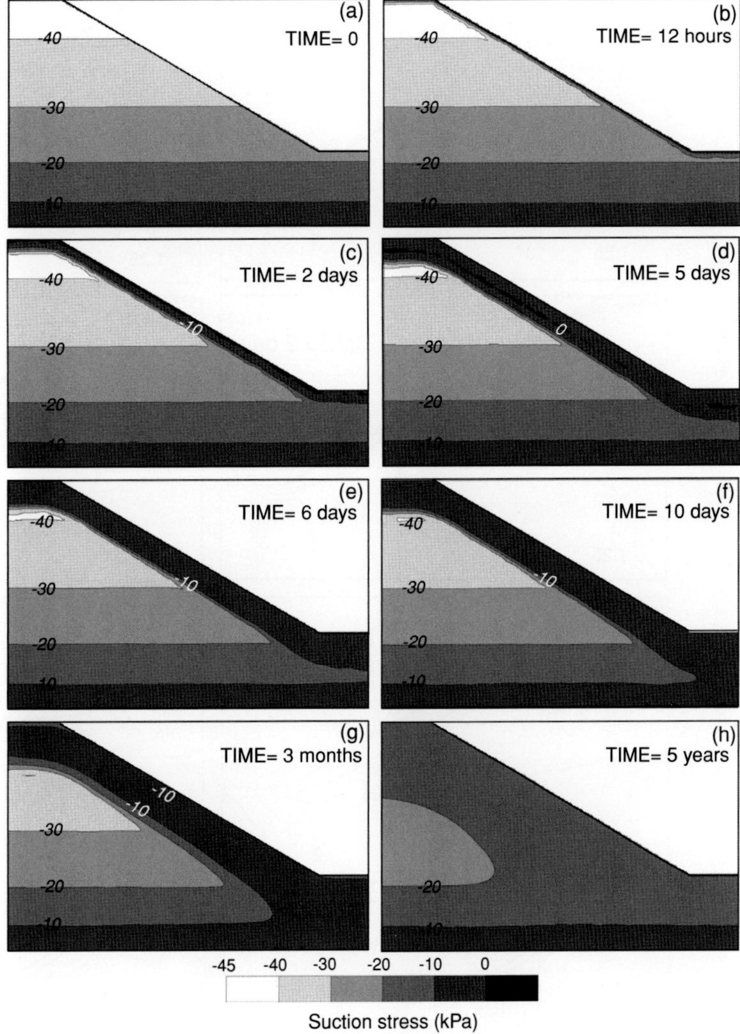

Figure 10.14 Simulated contours of suction stress at different times in a silty slope (as in Figure 10.12) under rainfall conditions (a) at time = 0, (b) at time = 12 hours, (c) at time = 2 days, (d) at time = 5 days, (e) at time = 6 days, (f) at time = 10 days, (g) at time = 3 months, and (h) at time = 5 years.

σ^s or decreases in effective stress (Equation (10.4)), the states of stress shift leftward in Mohr diagram space (as shown in Figure 10.6, moving from the solid Mohr circle to the dashed Mohr circle) leading to the failure of the soil. After the cessation of rainfall, the zone of decreasing LFS progresses downward roughly coincident with the evolution of moisture content and suction stress (Figures 10.12 and 10.14). The LFS zone of less than 1.0 gradually diminishes (Figures 10.15e and 10.15f) and eventually disappears (Figures 10.15g and 10.15h). At the end of the 5-year simulation, the low LFS zone between 1.25 and 1.50 still persists (Figures 10.13d and 10.15h). It is interesting to note that while the

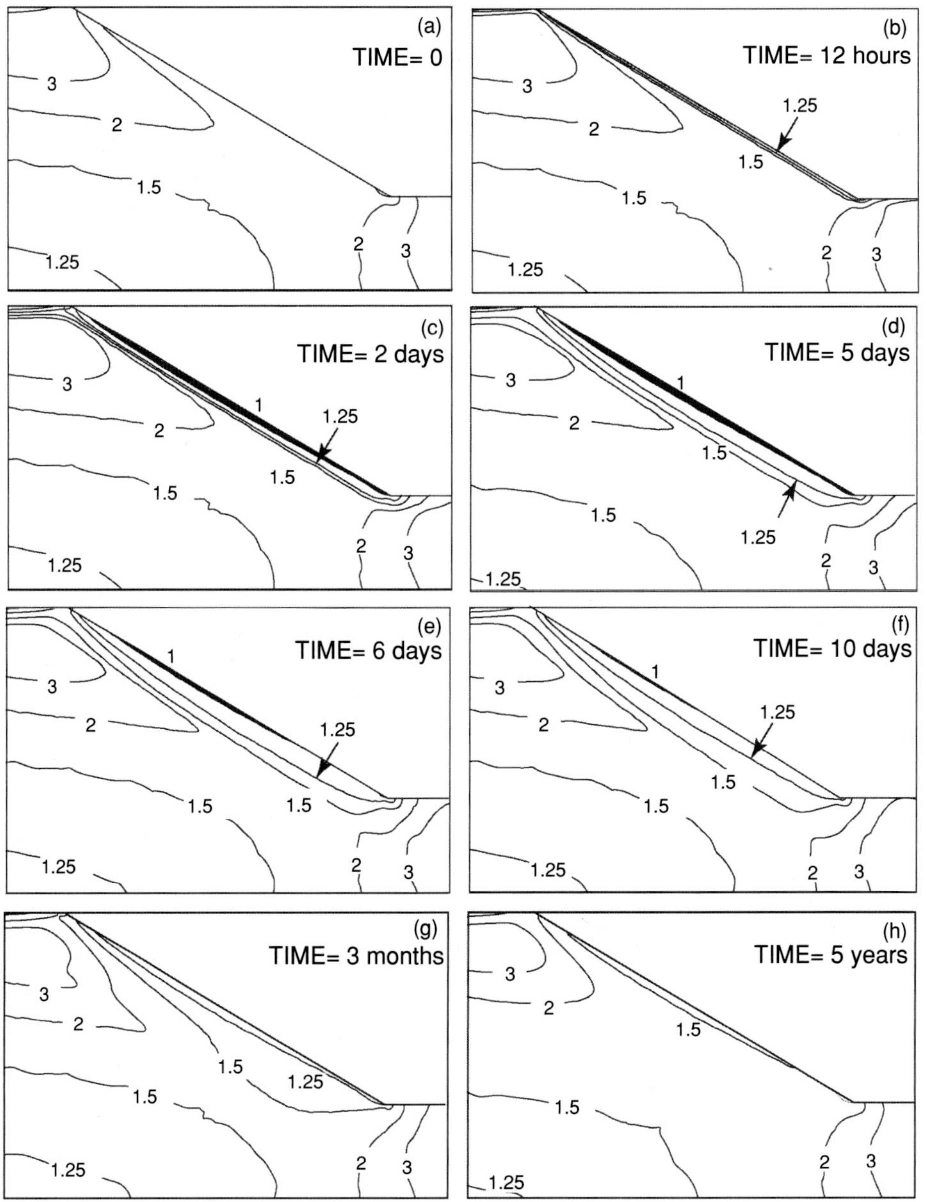

Figure 10.15 Simulated contours of LFS at different times in a silty slope (as in Figures 10.12 and 10.14) under rainfall conditions (a) at time = 0, (b) at time = 12 hours, (c) at time = 2 days, (d) at time = 5 days, (e) at time = 6 days, (f) at time = 10 days, (g) at time = 3 months, and (h) at time = 5 years.

geometry of the moisture content and suction stress contours are more or less parallel to the slope surface (Figures 10.12 and 10.14), the contours of LFS near the slope surface have a lens shape that is consistent with field observations of shallow landslides. The LFS, as illustrated here, provides rich quantitative information regarding initial failure location and timing, geometry, and progression.

The LFS method also provides a quantitative way to examine the suitability of the widely used infinite-slope stability model for shallow landslides. In a recent work by Lu and Godt (2008), the classical infinite-slope stability model is expanded to accommodate variably saturated conditions by using the unified effective stress equation (10.4) in lieu of Terzaghi's effective stress equation for saturated soil. The FS is (Lu and Godt, 2008)

$$F(z,t) = \frac{\tan\phi'}{\tan\beta} + \frac{2c'}{\gamma\, z \sin 2\beta} - \frac{\sigma^s(z,t)}{\gamma\, z}(\tan\beta + \cot\beta)\tan\phi' \qquad (10.7)$$

where β is the slope angle, γ is the moist unit weight of soil within the distance z from the slope surface, and t is time (see Chapter 9). Equation (10.7) is used to calculate FS profiles at location A shown in Figure 10.11a, using the suction stress profiles at different times shown in Figure 10.13c. The calculated FS profiles are shown in Figures 10.13e and 10.13f (same as Figure 10.13e but with different scales in FS values). The FS for hydrostatic conditions ($t = 0$) calculated from the infinite-slope stability model increases monotonically from about 1.5 at 5 m below the slope surface to more than 20 near the slope surface (Figure 10.13e). The monotonic variation pattern is also obtained for cases in which the soil is either dry or saturated using the infinite-slope model. Under unsaturated seepage conditions, the FS profile is no longer monotonic and the location of minimum FS can occur within the unsaturated zone, as illustrated in Figure 10.13f. Lu and Godt (2008) and Godt *et al.* (2009) show that FS could be less than 1.0 in partially saturated sandy soils on steep slopes under heavy rainfall conditions (see Chapter 9). For the silty slope after the 5-day rainfall episode, the FS can be reduced to less than 2.0 but never approaches 1.0, implying that this slope is stable if infinite-slope theory is used. Comparing Figures 10.13d and 10.13f, both LFS and the infinite-slope models produce about the same values of FS at relatively large distances from the slope surface, say $z > 4.0$ m. However, using the effective stress calculation from the FEM (in Figures 10.13d and 10.15), the LFS is shown to be less than 1.0 near the slope surface for a certain time period. The overestimation of stability near the slope surface by the infinite-slope model compared to the LFS method results from the appearance of distance z in the denominator in Equation (10.7), and is overcome by the LFS method.

10.4 Case study: Rainfall-induced landslide

This section provides a synthesis of many of the topics that are the subject of this book by describing an application of a finite-element model to simulate the hydrologic and mechanical conditions leading up to the occurrence of a shallow landslide. The slide was located at a USGS monitoring site near Seattle, Washington, and is described in more detail in Chapter 9. The transient pore-water response to rainfall infiltration was simulated using a two-dimensional solution of the Richards equation (Equation (4.18)), described in detail in Chapter 4. The pressure-head field was then combined with a static field of total gravitational-induced stress using the unified effective stress concept, described in Chapter 6, to produce a FS at each point in the model domain, as described in the previous section.

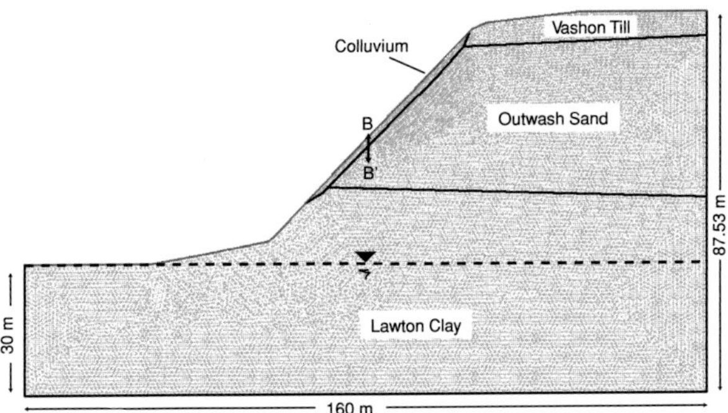

Figure 10.16 Geometry, finite-element mesh, initial conditions, and simplified geology used in the numerical simulations of a rainfall-induced shallow landslide at the USGS monitoring site near Seattle, Washington.

The hydrologic properties of the materials were determined using the TRIM (Chapter 8) and inverse modeling techniques (Chapter 9).

10.4.1 Two-dimensional numerical model

A simplified geologic cross section of the hillslope is shown in Figure 10.16. The hillslope is a steep (>40°) coastal bluff along the Puget Sound north of Seattle, Washington. The bluff is composed of a glacial sedimentary sequence of fine-grained lacustrine silt (Lawton Clay) overlain by an advance outwash sand, which is capped by a till. The bluff face is mantled by a thin (<2 m thick) sandy colluvium. The unsaturated hydrologic response of the hillslope to rainfall was monitored by the USGS for 16 months prior to the occurrence of a shallow landslide at the site on 14 January 2006 (Godt *et al.*, 2009). The failure surface of the slide was located about 1.5 m below the former ground surface at the contact between the colluvium and better-consolidated outwash sand.

The model domain comprised 13,075 finite-element nodes with finer mesh spacing near the ground surface. The initial pressure-head distribution was assumed to vary linearly above the water table located at the approximate elevation of Puget Sound, consistent with the monitoring observations at the beginning of the simulated period. The lateral boundaries of the model domain were specified as no-flow and the lower boundary was assigned a prescribed head of 30 m. A time-varying flux was applied to the upper boundary consistent with the 478 mm of rainfall measured at the site over the 3.5 months leading up to the landslide (Figure 10.17). Periods of no rainfall were specified as zero flux, neglecting any flow of water out of the domain that might result from evaporation or transpiration from vegetation.

Material hydrologic and mechanical properties for each of the geologic units in the model are summarized in Table 10.1, and soil water, hydraulic conductivity, and suction stress characteristic curves are shown in Figure 10.18. The soil water characteristics of the advance outwash sand were determined using the TRIM method described in Chapter 8. Hydrologic properties of the colluvium were estimated using monitoring data of water content and soil

Table 10.1 Hydrologic and mechanical properties used in the simulation of pore-water response and slope stability

	Hydrologic properties					Mechanical properties				
Unit	α (m^{-1})	n	K_s ($\times 10^{-6}$ m/s)	θ_s	θ_r	c' (kPa)	ϕ' (°)	γ (kN/m^3)	σ	E (kPa)
Colluvium	3.75	1.16	7.51	0.40	0.05	32	40	19	0.25	20,000
Advance outwash	1.77	3.00	2.10	0.39	0.02	30	40	20	0.40	20,000
Vashon Till	3.75	1.16	7.51	0.40	0.05	96	36	20	0.40	20,000
Lawton Clay	1.52	1.39	0.418	0.39	0.14	75	36	20	0.40	20,000

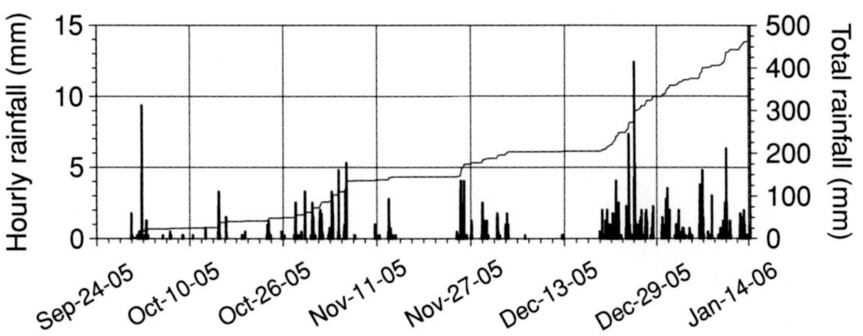

Figure 10.17 Hourly and cumulative rainfall recorded at the USGS monitoring site near Seattle, Washington.

suctions at two depths (1.0 and 1.5 m) and an inverse modeling procedure described in Section 9.5. No laboratory or monitoring data are available for the Lawton Clay or till; therefore, representative soil water characteristics were assigned to these units based on published values. The mechanical properties of the materials were also selected from previously published values (Debray and Savage, 2001; Harp *et al.*, 2008). The material strengths were increased over those measured in the laboratory tests to maintain the stability of the slope (i.e., factors of safety greater than unity) at the beginning of the simulation. The friction angle of the colluvium was increased by about 4 degrees and the cohesion by about 31 kPa.

10.4.2 Simulated hydrologic response to rainfall

Figure 10.19 shows the evolution of the simulated soil water content. The initial moisture content ranged between 0.27 and 0.32 in the colluvium, between 0.03 and 0.09 in the advance outwash, and between 0.36 and 0.38 in the Lawton Clay, reflecting the differences in the soil water characteristics of the various materials. On 17 October 2005, following a period of intermittent rainfall beginning on 29 September during which about 40 mm of rainfall accumulated, moisture contents near the ground surface increase as the wetting front progresses roughly parallel to the slope face. Figure 10.20 shows the coincident increase in pore-water pressures. By 9 November, following another 95 mm of rainfall, a zone of nearly zero pressure head begins to form near the toe of the slope in the Lawton Clay due to

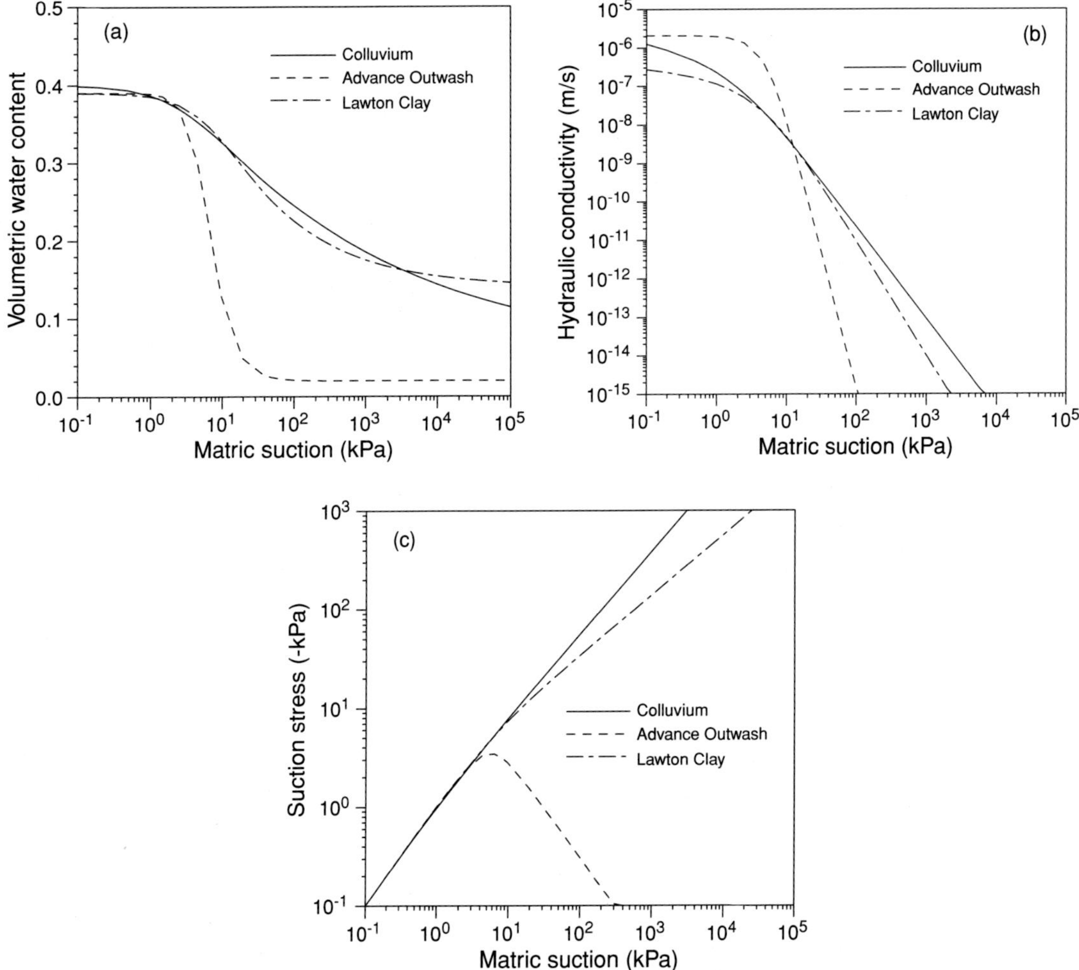

Figure 10.18 Hydro-mechanical properties for the geologic materials used in the simulations at the USGS monitoring site near Seattle, Washington: (a) soil water characteristic curves, (b) hydraulic conductivity functions, and (c) suction stress characteristic curves.

the proximity of the water table and the relatively wet initial conditions there. The zone of nearly zero pressure head expands as time progresses; by 25 December an additional 167 mm of rainfall has fallen and the colluvium becomes nearly saturated and the wetting front moves into the underlying advance outwash. This condition of near saturation persists until the time of the landslide at the site (Figure 10.20).

Figures 10.21a and 10.21b show the progress of the wetting front with time and the consequent changes in pressure head and moisture content near the slope surface along a 5 m deep profile B–B′ shown in Figure 10.16. The profile is located near the center of the slide and spans the contact between colluvium and the underlying advance outwash sand, which is located about 2 m below the slope surface. The contact between the two materials

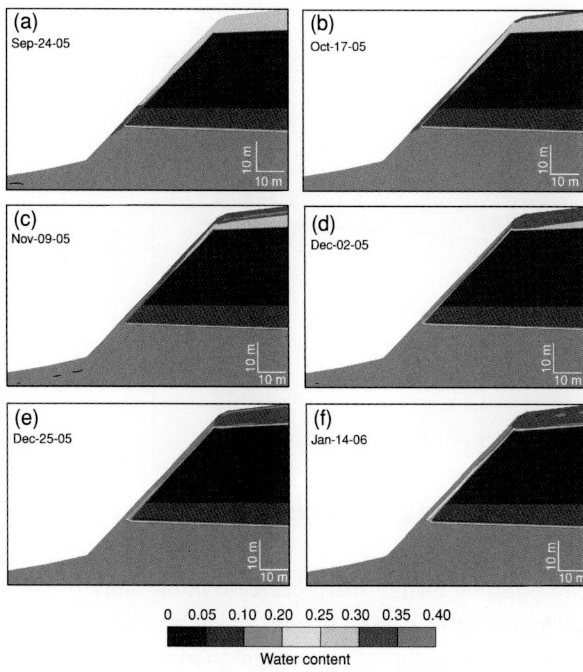

Figure 10.19 Evolution of simulated soil water content resulting from the rainfall shown in Figure 10.17. See also color plate section.

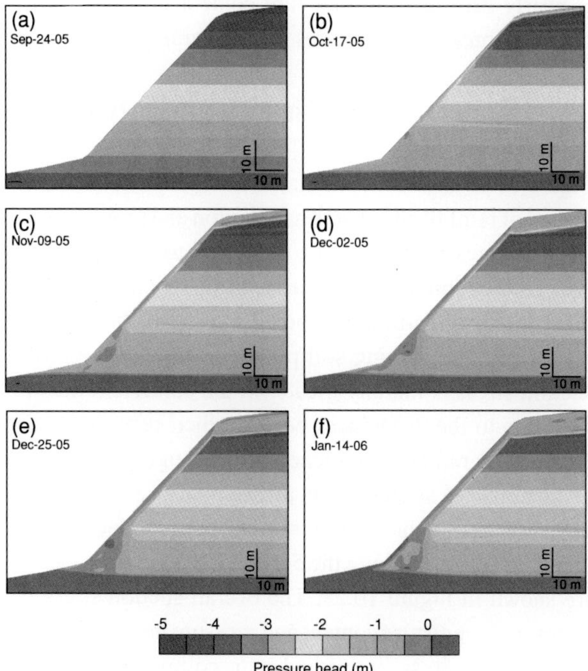

Figure 10.20 Evolution of simulated pressure head resulting from the rainfall shown in Figure 10.17. See also color plate section.

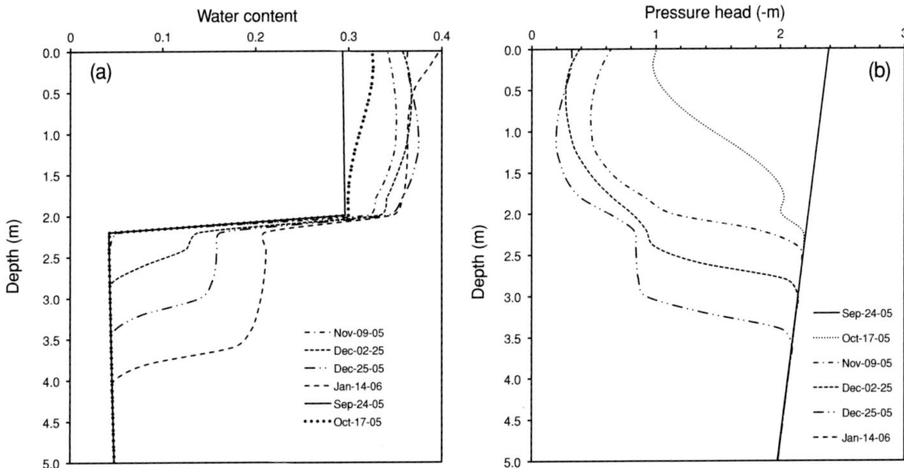

Figure 10.21 Simulated evolution of (a) water content, and (b) pressure head at the B–B′ profile.

is responsible for the sharp decrease in moisture content in this part of the profile. With the beginning of rainfall in late September and early October, moisture contents near the surface of the colluvial layer begin to increase. By early November the wetting front has progressed down to the sand layer and reaches about 4 m below the surface by the middle of January. On 14 January when the shallow landslide occurs, the simulated pressure head is nearly zero and the moisture content is about 0.4, or almost saturation. However, no positive pore-water pressures were obtained in the 5 m deep profile.

10.4.3 Simulated changes in stress and stability

Figure 10.22 shows the simulated changes in the suction stress field resulting from the infiltration of rainfall and changes in suction and moisture content. Suction stress was calculated from the pressure head field, converted to pressure units, using the closed form Equation (10.5b). The initial field of suction stress generally varies along with the initial hydrostatic pressure-head distribution moving up from the water table (Figure 10.21b). Differences result from the varying soil properties among the layers. For example, the magnitude of suction stress is initially greater in the colluvium than in the underlying advance outwash sand due to the differences between their respective suction stress characteristics (Figure 10.18c). As rainfall commences, the changes in the field of suction stress follow the same general pattern as changes in pressure head, and by 25 December suction stresses diminish to nearly zero in the colluvium near the slope face.

Suction stresses along the 5 m deep B–B′ profile (see Figure 10.16) for the selected times are shown in Figure 10.23. The overall suction stress response to rainfall and coincident increases in moisture content and pressure head is of opposite sign for the colluvium and the sand; suction stress diminished in the colluvium and increased in magnitude for the outwash sand. Simulated suction stress in the colluvium increased from a minimum of −16.3 kPa initially to a maximum of −0.22 kPa near the surface on 14 January. In the outwash sand it

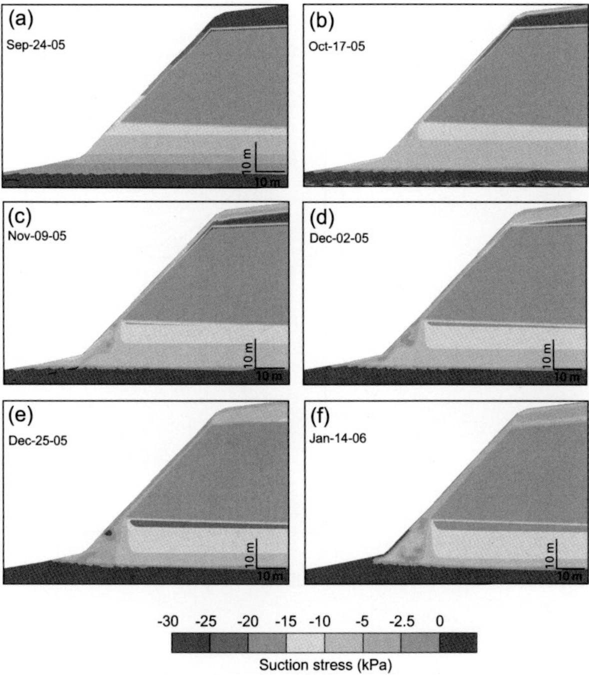

Figure 10.22 Evolution of simulated suction stress resulting from the rainfall shown in Figure 10.17. See also color plate section.

decreased from −1.40 kPa to −3.1 kPa for the same simulated period. This results primarily from the differences in the soil water and suction stress characteristics of the two materials (Figure 10.18). The simulated moisture contents along the B–B′ profile increased from 0.29 to 0.39 in the colluvium, and from 0.042 to 0.21 in the outwash sand (Figure 10.21a). For these ranges of water content, suction stress increases for the colluvium. However, for the sand, suction stress decreases for this range of water contents (Figure 10.17).

The stability of the slope was evaluated using the LFS concept with effective stresses described in Sections 10.1–10.3. The variations in field of FS with time are shown in Figure 10.24. Before the rainfall began, the slope was stable and the LFS was greater than unity at the slope surface. Stress concentrations near the toe of the slope result in LFS less than unity. On 17 October, after a period of intermittent rainfall during which about 40 mm accumulated (Figure 10.17), a potentially unstable region near the ground surface low on the slope begins to develop (Figure 10.24b). As rainfall continues throughout the simulated period, the potentially unstable areas evolve, growing in size and moving upslope parallel to the surface. On 14 January, when the landslide occurs at the site, a slope-parallel zone of instability between about 0.3 m and 1.5 m thick is simulated (Figure 10.24f). This zone is roughly consistent with field observations of the depth and extent of the rupture surface of the landslide at the site (Godt *et al.*, 2009).

Figure 10.25 shows profiles of the simulated FS along the B–B′ profile. Factors of safety increase with depth in the colluvium to almost 1.2 at 2 m below the ground surface. Factors of safety decrease sharply at the contact with the advance outwash sand, coincident with

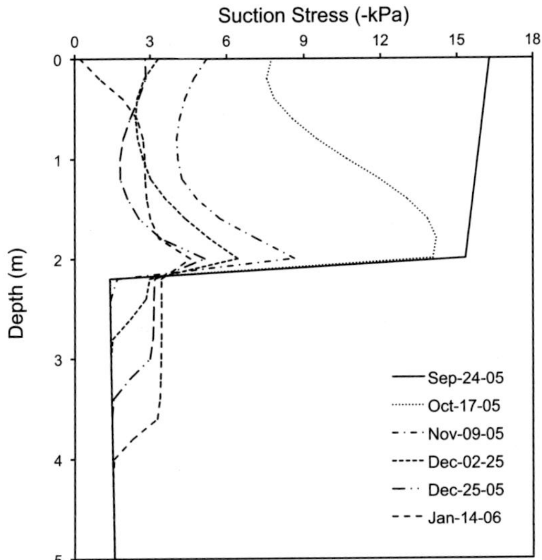

Figure 10.23 Simulated evolution of suction stress at the B–B′ profile.

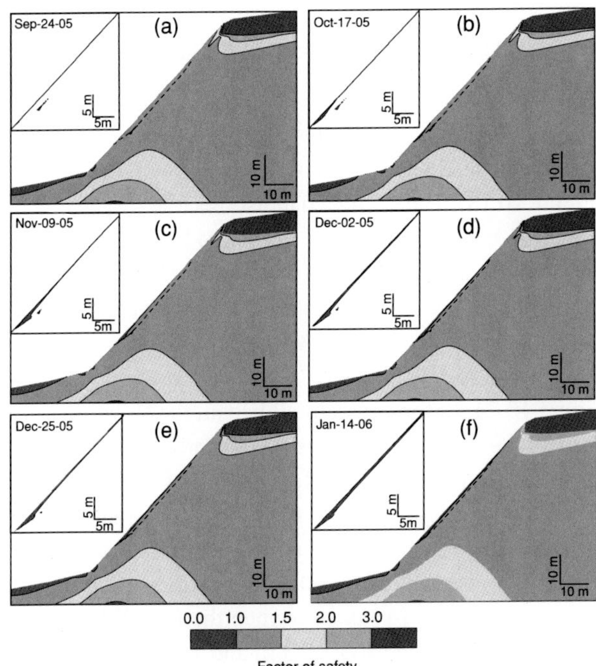

Figure 10.24 Evolution of FS for the rainfall shown in Figure 10.17. The inset in the upper left corner of each figure shows the area where the landslide occurred. The black dashed line shows the appoximate location of the failure surface from Godt *et al.* (2009). See also color plate section.

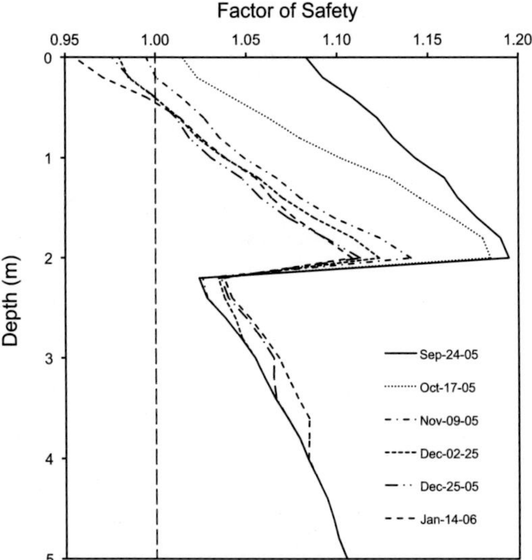

Figure 10.25 Simulated evolution of LFS at the B–B′ profile.

the increase in suction stress at this location (Figure 10.23). With time, infiltating rainfall induced changes in pore-water conditions and suction stress leading to simulated instability in the upper few tens of centimeters of the slope surface along this profile. On January 14 the area of potential instability extends over a large part of the slope face (Figure 10.24) and the depth of factors of safety less than unity extends to about 0.2 m below the ground surface (Figure 10.25).

Several questions remain to be resolved with this example application. Among these is the need to increase the material strengths of the surficial materials to achieve LFS > 1.0 at the beginning of the simulation. Some additional strength, over that measured in the laboratory, may arise from the contribution of vegetation roots. Vegetation at this location consists mainly of deciduous alder trees, blackberry vines, and grasses. However, the contribution to material shear strength was estimated to be only a few kPa (Godt *et al.*, 2009). Also, the need for additional strength may be a function of the overestimation of shear stresses near the slope face by the finite-element solution. Finite-element solutions are very sensitive to boundary geometry, particularly for slopes inclined at such steep angles. Future research should focus on resolving these issues.

10.5 Case study: Snowmelt-induced deeply seated landslide

10.5.1 Site hydrology and displacement monitoring

This example was used previously in Chapter 9 to illustrate the method of slices extended to variably saturated conditions. Although the earlier analysis showed that both suction stress

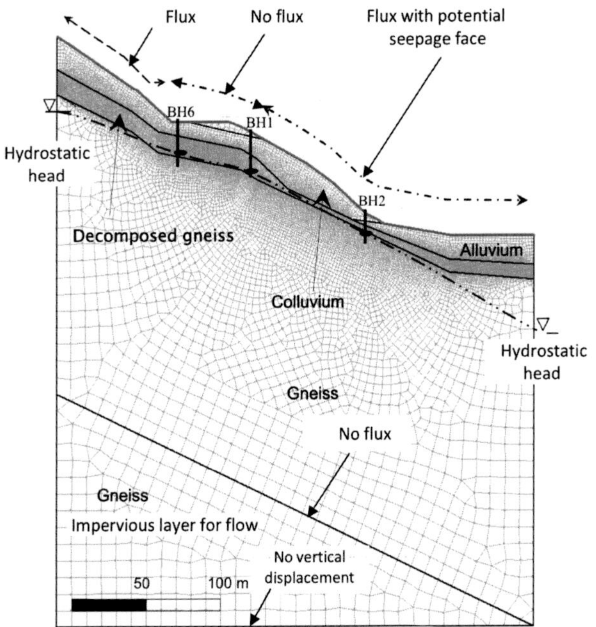

Figure 10.26 Borehole locations and the water table position from the I-70 site investigation. Also shown here is the finite-element mesh used to discretize the domain and the boundary conditions for the numerical model.

and changes in the water table position could trigger landslide movement at the (I-70) site, it does not provide an assessment of the timing and duration of instability over the annual cycle. With additional information on the subsurface hydrogeology and monitoring data on displacement, a better understanding of both the spatial and temporal characteristics of movement at the site can be gained. In what follows, a stress field based analysis is presented. The basic geologic, geomorphologic, and hydrologic conditions of the site are described in Section 9.6 and are illustrated in Figure 9.23 in the form of a cross-section diagram perpendicular to Interstate Highway 70. Field observations indicate that the head of the slide is located between the lanes of I-70 between boreholes BH6 and BH1 at the north end and the toe of the slide is located near borehole BH2 (Figure 10.26). To identify the position of the water table and the sliding surface, three boreholes (BH1, BH2, and BH6) were advanced in July 1996 and data from inclinometers were collected between the summers of 2007 and 2008. The borehole (vertical bar) and the water table (solid ellipse) locations are shown in Figure 10.26.

Displacement monitoring results from the inclinometers are shown in Figure 10.27 and indicate that boreholes BH1 and BH6 penetrated the sliding surface. Borehole BH2 is located near the toe of the slide. The depths of the water table at BH6, BH1, and BH2 are 25.3 m, 12.2 m, and 9.8 m, respectively. The slide apparently moves year round, but accelerates during the summer season. The displacement record at BH2 shows upslope movement, probably in response to the complex rotation of the slide in this area.

Based on this site investigation, a two-dimensional finite-element model shown in Figure 10.26 was constructed with appropriate hydrologic and mechanical boundary

Table 10.2 Hydro-mechanical properties used in the two-dimensional stress field based slope stability analysis

	Unsaturated flow				Total stress			Stability		
Layer	θ_s	θ_r	u_b (kPa)	n (m/sec)	K_s (kPa)	E (kN/m^3)	ν (kPa)	γ (Ì)	c'	ϕ'
Pavement layer	0.45[a]	0[a]	2.24[a]	1.12[a]	1.0E−5[a]	1E+5[a]	0.33[a]	19.62[c]	0[c]	32[a]
Colluvium	0.45[b]	0[a]	2.24[b]	1.12[b]	1.0E−5[b]	1E+5[a]	0.33[a]	20.41[c]	0[c]	34[c]
Decomposed gneiss	0.45[a]	0[a]	2.24[a]	1.12[a]	1.7E−6[a]	1E+5[a]	0.33[a]	21.19[c]	5[c]	23.3[c]
Intact gneiss	0.45[a]	0[a]	2.24[a]	1.12[a]	9.8E−7[a]	2E+5[a]	0.33[a]	22.76[c]	95.7[c]	34[c]
Alluvium	0.45[a]	0[a]	2.24[a]	1.12[a]	1.7E−6[a]	1E+5[a]	0.33[a]	20.41[c]	0[c]	32[c]

[a] Assumed
[b] Measured
[c] Provided by Colorado Department of Transportation

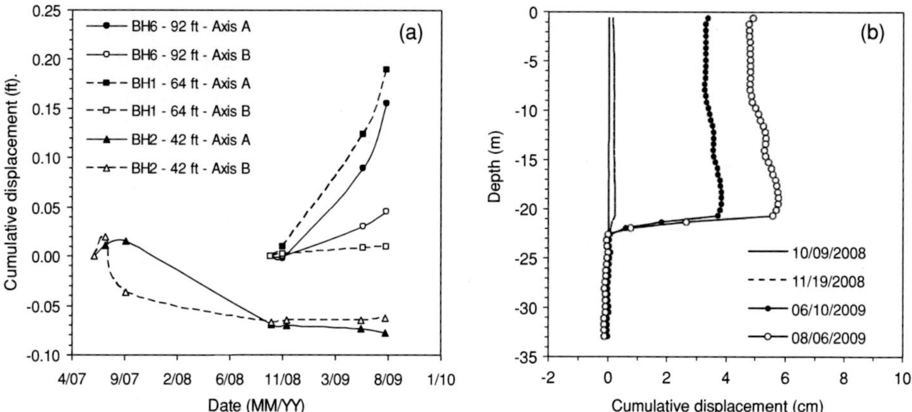

Figure 10.27 Displacement data from the inclinometers installed at boreholes BH1, BH2, and BH6: (a) cumulative displacements in the downslope (Axis A) and highway (Axis B) directions as functions of time, and (b) profile of downslope displacements in BH1 at different times.

conditions. Five materials were defined for the slope domain; their hydro-mechanical properties, based on laboratory testing of the samples and similar rocks from the literature, are summarized in Table 10.2.

The annual cycle of infiltration, over a 5-year period, was simulated with the transient flux boundary conditions shown in Figure 10.28. Initial steady conditions were established by imposing a constant flux equivalent to 20 cm/year. The sensitivity of the stability of the slope to snowmelt amount was examined by using three different infiltration rates of 15 cm/month, 30 cm/month, and 50 cm/month applied for a 2-month period beginning in April. These amounts are roughly consistent with measured snow–water equivalent depths near the site. Thus, the total infiltration amounts resulting from snowmelt for the low, medium, and high scenarios were 30 cm, 60 cm, and 100 cm, respectively.

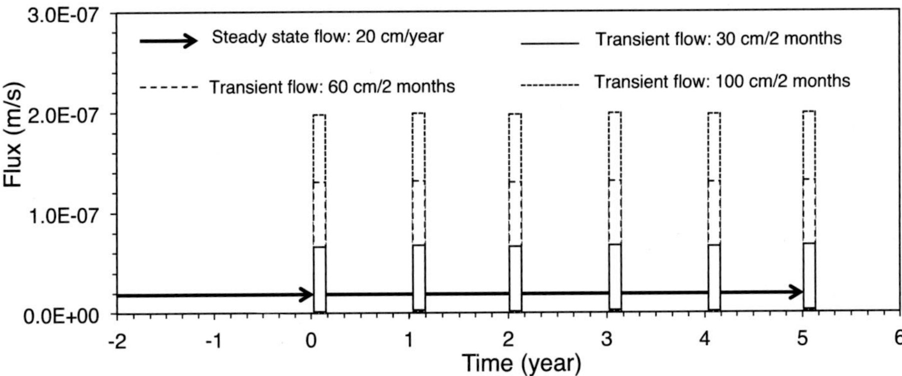

Figure 10.28 The 5-year cyclic surface infiltration boundary conditions for the FE simulations.

The finite-element results were obtained using GeoStudio 2007. The SIGMA/W module was used to calculate the total stress distribution in the slope. This module solves the governing force equilibrium equations for plane-strain elastic stress-displacement described in Section 5.4. The SEEP/W module was used to simulate the hydrologic response of the slope and solve the Richards equation described in Section 4.2, and the SLOPE/W module was used to calculate the FS using a method of slices described in Section 9.2. The hydro-mechanical framework described in the early part of this chapter was also used to calculate the factors of safety.

In the total stress and deformation analysis, the only load considered is the self-weight of the hillslope materials. Boundary conditions on the sides and bottom of the domain were set to zero-x and zero-y displacement, respectively.

The hydrologic modeling was performed in two stages. The first was used to establish initial conditions and the second to simulate the transient response to rainfall. The model boundary conditions consist of a flux boundary with possible seepage at the ground surface, and no flow at the lower, right, and left boundaries. Initial conditions were established by applying a steady uniform flux of 6.34×10^{-9} m/s (equivalent to 20 cm/year) to the upper boundary. The highway pavement was considered impermeable. The water table level was maintained at 18 m (59 ft) below the surface. Then, transient seepage conditions were simulated over a 6-year period using the three infiltration scenarios shown in Figure 10.28 and described above.

Slope stability was evaluated using a finite-element, stress-based method of slices which consists of eight steps: (1) evaluate the state of stress at each node shown in Figure 10.26; (2) for each slice, identify the element that is located at the middle of the bottom of the slice; (3) compute the normal and shear stress at this location,; (4) using the calculated normal stress, compute the shear strength for the slice; (5) convert stress into forces by multiplying the values by the length of the slice; (6) repeat the process for all the slices; (7) integrate the forces over the length of the slip surface; and (8) calculate the FS defined as the ratio of the available shear resistance to the mobilized shear along the assumed slip surface. For these simulations, generalized effective stresses (e.g., Equations (9.49) and (9.54)) were used. Pore-water pressure, volumetric water content, and suction stress values are reported at three simulation observation points that were located at boreholes BH1, BH2, and BH6 shown as solid ellipses in Figure 10.26.

10.5.2 Simulated transient suction and suction stress fields

The initial distribution of simulated pore-water pressure prior to the onset of snowmelt (1 April) resulting from the steady background flux of 20 cm/year is shown in Figure 10.29a. The failure surface is shown as a dashed curve. The water table, shown as the zero-pressure head contour, closely follows the contact between the intact and decomposed gneiss. Pore-water pressure values near the ground surface are about –150 kPa near the crest of the slope and –50 kPa in the toe area. The corresponding initial values of suction stress are –100 kPa near the crest increasing to –25 kPa near the toe (Figure 10.29b). The distribution of pore-water pressure and suction stress that results after 6 years are shown in Figures 10.29c and 10.29d. Over the 6-year period, 60 cm of water from melting snow was allowed to infiltrate during the two months of April and May, in each year. Under this scenario, the simulated water table moves towards the ground surface, and both pore-water pressure and suction stress in the vadose zone increase. Changes in suction stress contribute directly to changes in the field of effective stress as shown by Equation (10.4).

As described in Sections 9.2 and 10.1, quantitative assessment of the stability of a hillslope can be accomplished by examining the distribution and evolution of total and effective stress. The gravitational field mainly controls the field of total stress and the effective stress distribution is mainly controlled by the distribution of total stress and pore-water pressure. The simulated distribution of mean effective stress for the 100 cm infiltration scenario is shown in Figure 10.30a at the beginning of the 6-year infiltration cycle (1 April). Figure 10.30b shows simulation results on 1 June of the sixth year after 100 cm of snowmelt infiltrates in each April and May. In general, effective stress is reduced as a result of infiltration. The amount that the effective stress is reduced (Figure 10.30c) can be obtained by subtracting the mean effective stress distribution shown in Figure 10.30b from that shown in Figure 10.30a. Most of the reduction in effective stress occurs above the failure surface; as much as 45 kPa beneath the slope surface south of I-70. The contours of effective stress are nearly parallel to the slope surface and the reduction in effective stress diminishes with increasing depth. Along the sliding surface, the reduction in effective stress is about 5 kPa.

The pore-water pressures simulated at the locations where the water table was observed in boreholes BH1, BH2, and BH6 in 1996 for the 60 cm infiltration scenario are shown in Figure 10.31a. Borehole BH1 is located south of I-70; the simulation results are shown at the interface between the decomposed gneiss and the gneiss layers. The initial pressure head is −1.5 m; it increases during the first 2 years and then cycles between −0.25 m to 0.2 m during the last three years (Figure 10.31a). The change in the effective degree of saturation corresponds to the changes in pressure head, with values ranging from

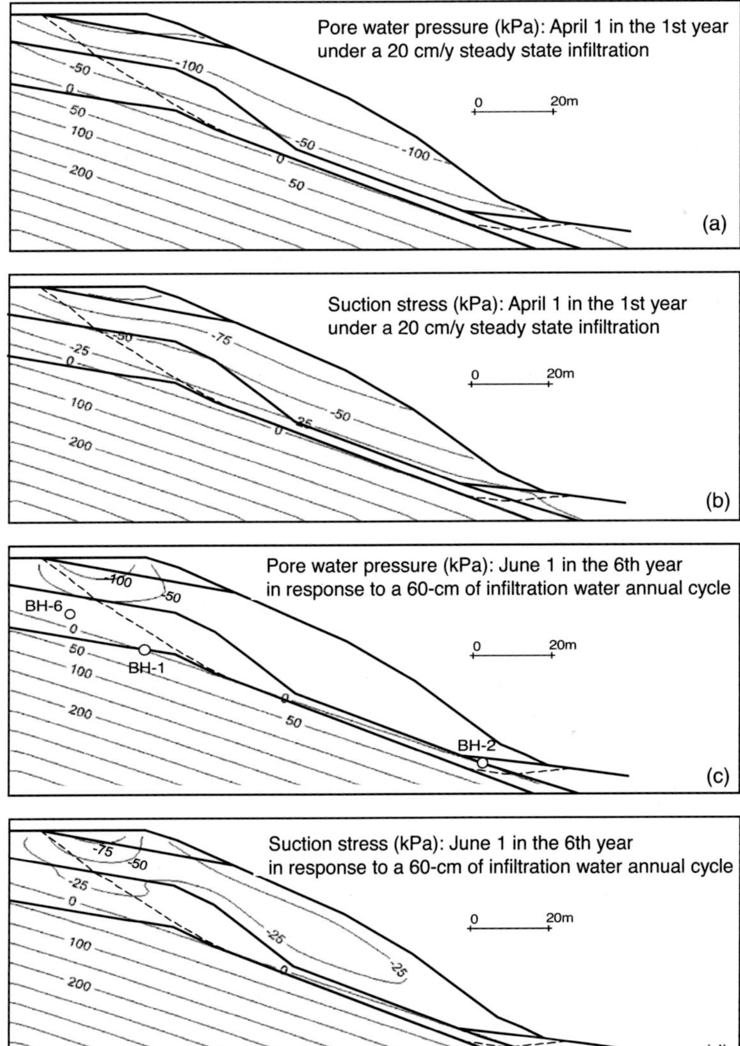

Figure 10.29 Simulation results showing (a) pore-water pressure distribution at time $t = 0$, (b) pore-water pressure distribution at $t = 6$ years (after six annual cycles), (c) suction stress distribution at $t = 0$, and (d) suction stress distribution at $t = 6$ years (after six annual cycles). The dashed line indicates the approximate location of the failure surface.

0.94 to 1.0 (Figure 10.31b). The corresponding suction stress varies inversely in magnitude with respect to the degree of saturation and cycles between −2.3 kPa and 1.8 kPa (Figure 10.31c). The peak values for pressure head, saturation, and suction stress lag the end of the infiltration period by about 4 months, indicating that infiltration travel time from the surface to the water table is about 4 months. After three annual cycles, the behavior of pore-water pressure, saturation, and suction stress reaches a quasi-steady periodicity.

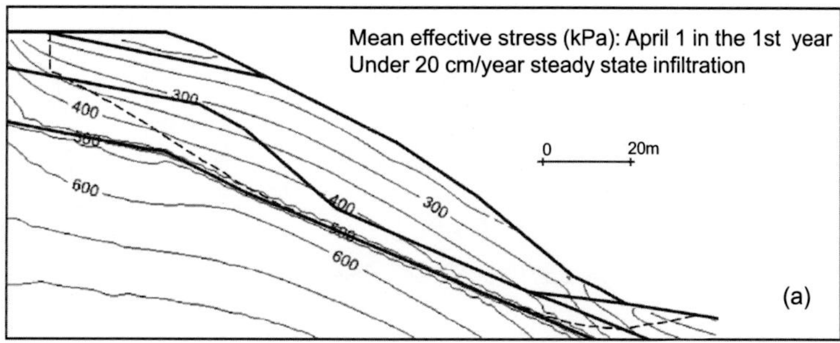

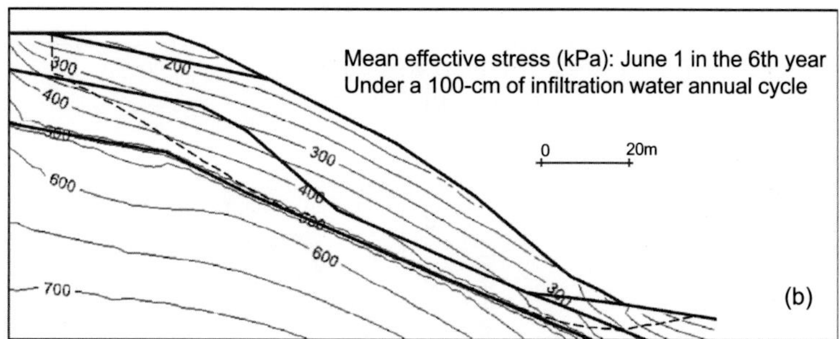

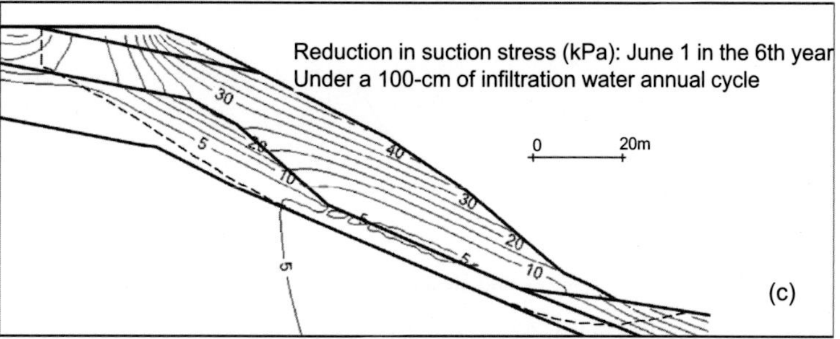

Figure 10.30 Simulation results showing (a) mean effective stress distribution at $t = 0$, (b) mean effective stress distribution at $t = 6$ years (in the sixth annual cycle) for the 100 cm/2 months infiltration rate annually, and (c) changes in mean effective stress (or changes in suction stress distribution) at $t = 6$ years (in the sixth annual cycle). The water table is above the slip surface. The changes in suction stress occur above the water table.

Borehole BH6 is located north of I-70; simulation results are shown at 10 m below the ground surface. The simulated patterns of pressure head, effective degree of saturation, and suction stress are similar to BH1. Because the soil at BH6 is initially drier, pressure heads are more negative (−2.7 m) and suction stress values are smaller (−5.0 kPa) than at the observation point at BH1. Finally, borehole BH2 is located close to the toe of the slope. Although initial conditions here are similar to BH1, the variations in pressure head, effective saturation, and suction stress are greater. After the quasi-steady periodic condition

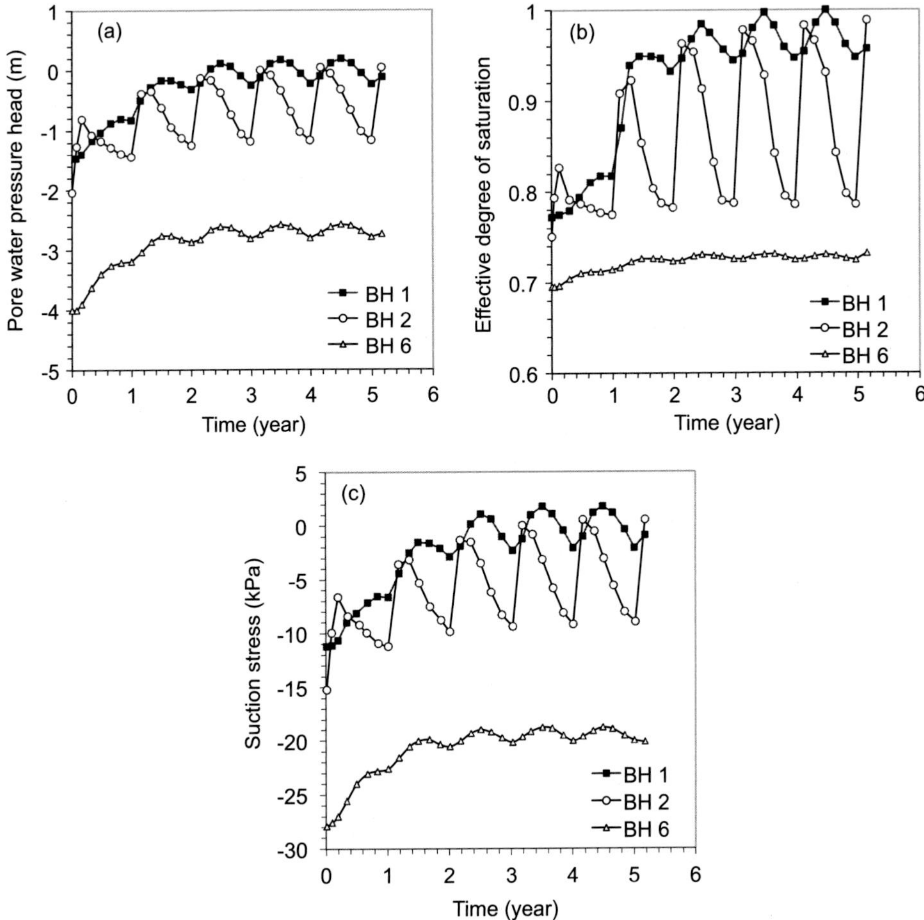

Figure 10.31 Simulation results showing time-dependent variations at the three water table locations measured in the boreholes in 1996 for (a) pore-water pressure, (b) the effective degree of saturation, and (c) suction stress.

is reached (3 years), the annual fluctuations of pore-pressure head at this location are between −1.2 m and 0 m, the effective degree of saturation varies between 0.8 and 0.98, and the suction stress varies between −9.4 kPa and 0.5 kPa.

10.5.3 Simulated transient slope stability conditions

The simulated suction stress distributions along the sliding surface under the three different infiltration scenarios are shown in Figure 10.32a. The FS of the slope as a function of time is calculated for the three infiltration conditions (Figure 10.32b). In all three cases, FS varies cyclically so that the slope is more stable when the soil is drier (Figure 10.32b). Failure was reached during the 100 cm infiltration scenario, and the simulated cyclic behavior of FS is consistent with the inclinometer measurements shown in Figure 10.27.

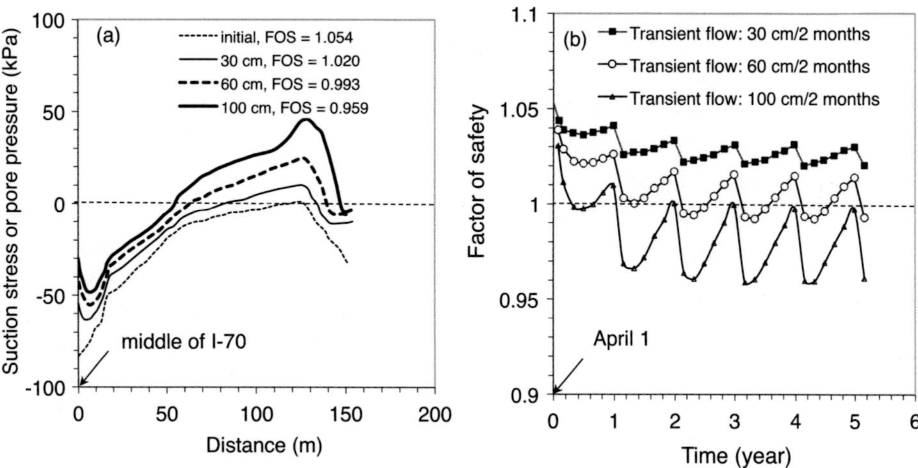

Figure 10.32 Simulation results showing (a) suction stress distribution along the failure surface on June 1 of the sixth year, and (b) global factor of safety as a function of time and infiltration scenarios.

As defined in the unified effective stress principle for variably saturated materials, the change in suction stress is a direct indicator of changes in effective stress. Initially, the sliding surface is above the water table and suction stresses are all negative, resulting in a FS of 1.054 (Figure 10.32a–b). As the total amount of infiltrated water in the three scenarios increases, the lower-middle part of the slope (distance > 60 m) becomes saturated, leading to positive suction stress or pore-water pressure and a reduction in effective stress. The peak suction stress (or pore-water pressure) occurs around 125 m from the north boundary of the sliding surface or 25 m from the toe at the south end. For the 100 cm infiltration scenario, the peak pore-water pressure at this location reaches 45 kPa and the FS reduces to 0.959 (Figure 10.32b). Thus, it appears that reducing pore-water pressures near the toe by enhancing drainage could effectively stabilize the slope, increasing the FS by as much as 9.5%.

From this stress field based slope stability analysis, it appears that suction stress variation above the water table, pore-water pressure variation below the water table due to infiltration, and the corresponding cyclic water-table fluctuations are responsible for the perennial episodes of landslide movement. The annual variation of suction stress and pore-water pressure along the failure surface are on the order of 10 kPa for the 60 cm infiltration scenario, resulting in a FS that fluctuates between 0.99 and 1.02 (Figure 10.32b) with movement assumed when FS is less than 1.00. Snowmelt to the amount of 60 cm is sufficient to initiate movement a month or so after melt begins and could persist for 6 months. This case study illustrates that a suction stress field based slope stability analysis framework may be capable of predicting the state of hillslope stability within a few percentage change in the FS.

As methods and technologies applied to surface and subsurface mapping continue to advance, stress field based stability analysis will increasingly become routine for site-specific and regional assessment of hillslope stability. The hydro-mechanical framework described in this chapter provides a theoretical and physical basis for simulating transient moisture and suction distributions and the consequent distribution of effective stress under rainfall conditions. Comparison of stability results obtained using the local factor of safety method with those from limit-equilibrium and stress-based methods shows that the LFS method has the potential to overcome some of the inherent drawbacks in using a single factor of safety. The LFS method provides a means to effectively and accurately quantify slope failure initiation and progression. This is of particular importance in the study of rainfall-induced landslides. The case study presented for the sandy hillslope demonstrates that the hydro-mechanical framework, combined with the LFS analysis, provides a means to predict the spatial and temporal occurrence of landslides where instability is generated by a reduction of a few kPa of effective stress. The case study of the deeply seated landslide induced by melting snow demonstrates that suction stress can be used in lieu of pore-water pressure to simulate episodic landsliding under variably saturated conditions.

10.6 Problems

1 What are the major differences between the failure surface based slope stability analysis and the stress field based slope stability analysis?
2 If shear stress field is the determining factor in a slope's stability, where is failure most likely to initiate?
3 In the unified effective stress approach, does the shear strength criterion change between saturated and unsaturated hillslope materials?
4 In the hydro-mechanical framework described in this book, what is the link between hydrologic and mechanical processes?
5 What are the controlling parameters and variables in the definition of LFS?
6 In the region where the LFS value is equal to 1, what does it mean?
7 From the comparisons between the classical limit analysis for the FS and the LFS, what are the major differences?
8 From the comparisons of the failure regions between hillslopes with different inclination angles, what are the major differences in terms of the geometry of the failure regions?
9 What is the purpose of conducting a transient analysis of slope stability?
10 Describe the character and dynamics of the wetting front in the simulations shown in Figures 10.19–10.21. How would these change if the slope was initially wetter? What if the rainfall intensity was greater or less?
11 What influences would these changes in the wetting front have on stability?
12 What occurs near the slope surface after the cessation of rainfall in Figures 10.19–10.21? What effect does this have on stability? Would you expect the same effects if the hillslope materials were sand?
13 In the simulated case shown in Section 10.3, why doesn't the hillslope return to the initial hydrostatic state after the 5-year period?

14 Why is the magnitude of suction stress at a given point in the hillslope less than that of soil suction?

15 What are the differences between the stability estimated with the LFS method and that from an infinite-slope stability analysis? What factors contribute to this difference?

16 Consider a 45° homogeneous hillslope of silty sand with a height of 20 m, average unit weight of 20 kN/m^3, drained cohesion of 40 kPa, drained internal friction angle of 30°, unsaturated hydro-mechanical properties of $n = 5.0$ and $\alpha = 0.05$ kPa^{-1}, and saturated hydraulic conductivity of $K_s = 10^{-6}$ m/s. Under a steady-state infiltration rate of $q = -0.5 \times 10^{-7}$ m/s and the assumption that flow is in the vertical direction within the entire hillslope, assess the stability of the hillslope by drawing the profiles of the LFS at the toe, middle, and crest of the slope. What are your major conclusions from this analysis?

(Hint: the LFS (Equations (10.2) and (10.3)) shown in this chapter can be divided into two terms: total stress and suction stress. The profiles of the total stress at each slope location can be determined by the linear elastic theory and tables in Chapter 5 (e.g., Figure 5.15 or Table 5.6), and the profiles of suction stress (Equation (9.63)) can be quantified using the theory in Chapter 9).

References

Abdul, A. S., and Gillham, R. W. (1984). Laboratory studies of the effects of the capillary fringe on streamflow generation, *Water Resources Research*, 20, 691–698.

Abramento, M., and Carvalho, C. S. (1989). Geotechnical parameters for the study of natural slope instabilization at Serra do Mar, Brazil, in *Proceedings of the 12th International Conference on Soil Mechanics and Foundation Engineering*, Rio de Janeiro.

Abramson, L. W., Lee, T. S., Sharma, S., and Boyce, G. M. (1996). *Slope Stability and Stabilization Methods*, New York: John Wiley & Sons.

Adams, D. K., and Comrie, A. C. (1997). The North American Monsoon, *Bulletin of the American Meteorological Society*, 78, 2197–2213.

Aldridge, R., and Jackson, R. J. (1968). Interception of rainfall by manuka (*Leptospermum scoparium*) at Taita, New Zealand, *New Zealand Journal of Science*, 11, 301–317.

Allen, M. R., and Ingram, W. J. (2002). Constraints on future changes in climate and the hydrological cycle, *Nature*, 419, 224–232.

Anderson, M. G., and Burt, T. P. (1978). The role of topography in controlling throughflow generation, *Earth Surface Processes*, 3, 331–344.

Anderson, M. G., and Howes, S. (1985). Development and application of a combined soil water-slope stability model, *Quarterly Journal of Engineering Geology*, 18, 225–236.

Anderson, M. G., Kemp, M. J., and Lloyd, D. M. (1988). Applications of soil water finite difference models to slope stability problems, in *Proceedings of the 5th International Symposium on Landslides*, 1, 525–530, Lausanne, 10–15 July 1988.

Andrade, J. E., and Borja, R. I. (2006). Capturing strain localization in dense sands with random density, *International Journal for Numerical Methods in Engineering*, 67(11), 1531–1564.

ASTM (1985). *Classification of Soils for Engineering Purposes*, Annual Book of ASTM Standard, D2487–83, 04.08, American Society for Testing and Materials, 395–408.

Baum, R. L., Messerich, J., and Fleming, R. W. (1998). Surface deformation as a guide to kinematics and three-dimensional shape of slow-moving, clay-rich landslides, Honolulu, Hawaii, *Environmental and Engineering Geoscience*, 4(3), 283–306.

Baum, R. L., McKenna, J. P., Godt, J. W., Harp, E. L., and McMullen, S. R. (2005). *Hydrological Monitoring of Landslide-Prone Coastal Bluffs near Edmonds and Everett, Washington, 2001–2004*, U.S. Geological Survey Open-File Report 2005-1063, p. 42.

Baum, R. L., Savage, W. Z., and Godt, J. W. (2008). *TRIGRS: A FORTRAN Program for Transient Rainfall Infiltration and Grid-Based Regional Slope-Stability Analysis, version 2.0*, U.S. Geological Survey Open-File Report 2008-1159.

Baum, R. L., Godt, J. W., and Savage, W. Z. (2010). Estimating the timing and location of shallow rainfall-induced landslides using a model for transient, unsaturated infiltration, *Journal of Geophysical Research*, 115, F03013, doi:10.1029/2009JF001321.

Benson, C. H. and Gribb, M. (1997). Measuring unsaturated hydraulic conductivity in the laboratory and field, in S. Houston and D. G. Fredlund, eds., *Unsaturated Soil Engineering Practice*, ASCE Special Technical Publication No. 68, pp. 113–168.

Berdichevsky, V. (2009). *Variational Principles of Continuum Mechanics*, New York: Springer.

Bergeron, T. (1960). Operation and results of 'Project Pluvius', in *Physics of Precipitation*, American Geophysical Union Monograph no. 5, pp. 152–157.

Betson, R. P. (1964). What is watershed runoff? *Journal of Geophysical Research*, 69, 1541–1552.

Beven, K. J. (1996). Equifinality and uncertainty in geomorphological modeling, in B. L. Rhoades and C. E. Thorne, eds., *The Scientific Nature of Geomorphology*, Chichester: Wiley, pp. 289–313.

Beven, K. (2004). Robert E. Horton's perceptual model of infiltration processes, *Hydrological Processes*, 18, 3447–3460.

Bishop, A. W. (1954). The use of pore water coefficients in practice, *Géotechnique*, 4(4), 148–152.

Bishop, A. W. (1955). The use of the slip circle in the stability analysis of slopes, *Géotechnique*, 5, 7–17.

Bishop, A. W. (1959). The principle of effective stress, *Teknisk Ukeblad I Samarbeide Med Teknikk*, Oslo, Norway, 106(39), 859–863.

Bishop, A. W., and Garga, V. K. (1969). Drained tests on London clay, *Geotechnique*, 19(2), 309–312.

Bocking, K. A., and Fredlund, D. G. (1980). Limitations of the axis translation technique, in *Proceedings 4th International Conference on Expansive Soils*, Denver, Colo., pp. 117–135.

Borga, M., Fontana, G. D., Ros, D. D. and Marchi, L. (1998). Shallow landslide hazard assessment using a physically based model and digital elevation data, *Environmental Geology*, 35, 81–88.

Borja, R. I., and White, J. A. (2010). Continuum deformation and stability analyses of a steep hillside slope under rainfall infiltration, *Acta Geotechnica*, 5(1), 1–14.

Borja, R. I., Oettl, G., Ebel, B. A., and Loague, K. (2006). Hydrologically driven slope failure initiation in variably saturated porous media, in *Modern Trends in Geomechanics*, Springer Proceedings in Physics, 106, Part IV, pp. 303–311.

Bouwer, H. (1966). Rapid field measurement of air entry value and hydraulic conductivity of soil as significant parameters in flow system analysis, *Water Resources Research*, 2, 729–738.

Brand, E. W. (1981). Some thoughts on rain-induced slope failures, in *Proceedings of the International Conference on Soil Mechanics and Foundation Engineering*, 10(3), pp. 373–376.

Brand, E. W., Dale, M. J., and Nash, J. M. (1986). Soil pipes and slope stability in Hong Kong, *Quarterly Journal of Engineering Geology and Hydrogeology*, 19(3), 301–303.

Brooks, R. H., and Corey, A. T. (1964). *Hydraulic Properties of Porous Media*, Colorado State University, Hydrology Paper No. 3, March.

Brooks, S. M., and Richards, K. S. (1994). The significance of rainstorm variations to shallow translational hillslope failure, *Earth Surface Processes and Landforms*, 19(1), 85–94.

Brooks, S. M., Anderson, M. G., and Collison, A. J. C. (1995). Modeling the role of climate, vegetation and pedogenesis in shallow translational hillslope failure, *Earth Surface Processes and Landforms*, 20, 231–242.

Brooks, S. M., Crozier, M. J., Preston, N. J., and Anderson, M. G. (2002). Regolith stripping and the control of shallow translational hillslope failure: application of a two-dimensional coupled soil hydrology-slope stability model, Hawke's Bay, New Zealand, *Geomorphology*, 45, 165–179.

Bucknam, R. C., Coe, J. A., Chavarria, M. M., *et al.* (2001). *Landslides Triggered by Hurricane Mitch in Guatemala: Inventory and Discussion*, U.S. Geological Survey Open-File Report 01-443.

Buol, S. W., Southard, R. J., Graham, R. C., and McDaniel, P. A. (2003). *Soil Genesis and Classification*, 5th Edition, Ames, Iowa: Iowa State Press, Blackwell.

Burns, D. A., Hooper, R. P., and McDonnell, J. J. (1998). Base cation concentrations in subsurface flow from a forested hillslope: the role of flushing frequency, *Water Resources Research*, 34, 3535–3544.

Cai, F., Ugai, K., Wakai, A., and Li, Q. (1998). Effects of horizontal drains on slope stability under rainfall by three-dimensional finite element analysis, *Computers and Geotechnics*, 23, 255–275.

Caine, N. (1980). The rainfall intensity–duration control of shallow landslides and debris flows, *Geogrfiska Annaler A*, 62, 23–27.

Cannon, S. H., and Ellen, S. D. (1988). Rainfall that resulted in abundant debris-flow activity during the storm, in S. D. Ellen and G. F. Wieczorek, eds., *Landslides, Floods, and Marine Effects of the Storm of January 3–5, 1982, in the San Francisco Bay Region, California*, U.S. Geological Survey Professional Paper 1434, pp. 27–34.

Cardinali, M., Ardizzone, F., Galli, M., Guzzetti, F., and Richenbach, P. (2000). Landslides triggered by rapid snow melting: the December 1996 – January 1997 event in Central Italy, in P. Claps, and F. Siccardi, eds., *Proceedings of the 1st Plinius Conference on Mediterranean Storms*, Bios, Cosenza, pp. 439–448.

Casagrande, A. (1936). Characteristics of cohesionless soils affecting the stability of slopes and earth sills, in *Contributions to Soil Mechanics, 1925–1940*, Boston Society of Civil Engineers.

Chagnon, S. A. (ed.) (2000). *El Niño 1997–1998: The Climate Event of the Century*, Oxford University Press.

Chen, W. F. (1975). *Limit Analysis and Soil Plasticity*, Amsterdam: Elsevier.

Chen, L., and Young, M. H. (2006). Green–Ampt infiltration model for sloping surface, *Water Resources Research*, 42, W07420, doi:10.1029/2005WR004468.

Chiu, T. F. and Shackelford, C. D. (1998). Unsaturated hydraulic conductivity of compacted sand–kaolin mixtures, *Journal of Geotechnical and Geoenvironmental Engineering*, 124(2), 160–170.

Cho, S. E., and Lee, S. R. (2001). Instability of unsaturated soil slopes due to infiltration, *Computers and Geotechnics*, 28, 185–208.

Chorley, R. J. (1978). The hillslope hydrological cycle, in, M.J. Kirkby, ed., *Hillslope Hydrology*, Chichester: John Wiley & Sons, pp. 1 – 42.

Chu, S. T. (1978). Infiltration during an unsteady rain, *Water Resources Research*, 14(3), 461–466.

Clark, C. R., and Nevels, J. B. (2007). Triggering mechanisms for the Garvin landslide, in *Proceedings of the 86th Transportation Research Board (TRB) Annual Meeting*, Paper 07-3148.

Coe, J. A., and Godt, J. W. (2001). *Debris Flows Triggered by the El Niño Rainstorm of February 2–3, 1998, Walpert Ridge and Vicinity, Alameda County, California*, U.S. Geological Survey Miscellaneous Field Studies Map MF-2384, 3 sheets, scale 1:24,000.

Coe, J. A., Godt, J. W., and Tachker, P. (2004). *Map Showing Recent (1997–98 El Niño) and Historical Landslides, Crow Creek and Vicinity, Alameda and Contra Costa Counties, California*, U.S. Geological Survey Scientific Investigations Map SIM-2859, scale 1:18,000.

Coe, J. A., McKenna, J. P., Godt, J. W., and Baum, R. L. (2009). Basal-topographic control of stationary ponds on a continuously moving landslide, *Earth Surface Processes and Landforms*, 34, 264–279.

Cohen, D., Schwarz, M., and Or, D. (2011). An analytical fiber bundle model for pullout mechanics of root bundles, *Journal of Geophysical Research*, 116, F03010, doi:10.1029/2010JF001886.

Coleman, J. D. (1962). Stress /strain relations for partly saturated soils, *Géotechnique*, 12(4), 348–350.

Collins, B. D., and Sitar, N. (2008). Processes of coastal bluff erosion in weakly lithified sands, Pacifica, California, USA, *Geomorphology*, 97, 483–501.

Collins, B. D., and Znidarcic, D. (2004). Stability analyses of rainfall induced landslides, *Journal of Geotechnical and Geoenvironmental Engineering*, 13(4), 362–372.

Collison, A. J. C., Anderson, M. G., and Lloyd, D. M. (1995). Impact of vegetation on slope stability in a humid tropical environment: a modeling approach, *Proceedings of the Institute of Civil Engineers: Water and Maritime Engineering*, 112, 168–175.

Corey, A. T. (1957). Measurement of water and air permeability in unsaturated soil, *Proceedings of Soil Science Society of America*, 21(1), 7–10.

Cornforth, D. H. (1964). Some experiments on the influence of strain conditions on the strength of sand, *Geotechnique*. 14(2), 143–167.

Cornforth, D. H. (2005). *Landslides in Practice: Investigation, Analysis, Remedial/Preventive Options in Soils*, New York: John Wiley and Sons.

Crozier, M. J (1986). *Landslides: Causes, Consequences, and Environment*, London: Croom Helm.

Crozier, M. J. (1999). Prediction of rainfall-triggered landslides: a test of the antecedent water status model, *Earth Surface Processes and Landforms*, 24, 825–833.

Cruden, D. M., and Varnes, D. J. (1996). Landslide types and processes, in A. K. Turner and R. L. Schuster eds., *Landslides: Investigation and Mitigation*, Transportation Research Board Special Report, 247, Washington, D.C.: National Academy Press, pp. 36–75.

Cui, Y. J., and Delage, P. (1996). Yielding and plastic behaviour of an unsaturated compacted silt, *Géotechnique*, 46, 291–311.

Culmann, C. (1875). *Die Graphische Statik*, Zurich: Meyer and Zeller.

Dai, F. C., Xu, C., Yao, X., Xu, L., Tu, X. B., and Gong, Q. M. (2011). Spatial distribution of landslides triggered by the 2008 Ms 8.0 Wenchuan earthquake, China, *Journal of Asian Earth Sciences*, 40, 883–895.

Daniel, D. (1983). Permeability test for unsaturated soil, *Geotechnical Testing Journal*, 6(2), 81–86.

Dawson, E. M., Roth, W.H., and Drescher, A. (1999). Slope stability by strength reduction, *Géotechnique*, 49, 835–840.

Day, R. W., and Axten, G. W. (1989). Surficial stability of compacted clay slopes, *Journal of Geotechnical and Geoenvironmental Engineering*, 115(4), 577–580.

Debray, S., and Savage, W. Z. (2001). *A Preliminary Finite-Element Analysis of a Shallow Landslide in the Alki Area of Seattle, Washington*, U.S. Geological Survey Open-File Report 01-0357.

Dettinger, M. D., Ralph, F. M., Das, T., Neiman, P. J., and Cayan, D. R. (2011). Atmospheric rivers, floods, and the water resources of California, *Water*, 3, 445–478.

Dietrich, W. E., and Sitar, N. (1997). Geoscience and geotechnical engineering aspects of debris-flow hazard assessment, in C. Chen, ed., *Debris-Flow Hazards Mitigation: Mechanics, Prediction, and Assessment*, ASCE.

Dietrich, W. E., Reiss, R., Hsu, M. L., and Montgomery, D. R. (1995). A process-based model for colluvial soil depth and shallow landsliding using digital elevation data, *Hydrological Processes*, 9(3–4), 383–400, doi:10.1002/hyp.3360090311.

Dirksen, C. (1978). Transient and steady flow from subsurface line sources at constant hydraulic head in anisotropic soil, *Transactions of the ASAE*, 21(2), 913–919.

Douglas, I. (1976). Erosion rates and climate: geomorphological implications, in E. Derbyshire, ed., *Geomorphology and Climate*, London: John Wiley & Sons, pp. 269–287.

Duncan, J. M. (1996). State of the art: limit equilibrium and finite-element analysis of slopes, *Journal of Geotechnical and Geoenvironmental Engineering*, 122(7), 577–596.

Duncan, J. M., and Wright, S. G. (2005). *Soil Strength and Slope Stability*, Hoboken, N.J.: John Wiley & Sons.

Dunne, T., and Black, R. D. (1970). An experimental investigation of runoff production in permeable soils, *Water Resources Research*, 6(2), 478–490.

Dunne, T., Moore, T. R., and Taylor, C. H. (1975). Recognitions and prediction of runoff-producing zones in humid regions, *Hydrological Sciences Bulletin*, 20, 305–327.

Durner, W., and Iden, S. C. (2011). Extended multistep outflow for the accurate determination of soil hydraulic properties near water saturation, *Water Resources Research*, 47, W08526, doi:10.1029/2011WR010632.

Ebel, B. A., Loague, K., Vanderkwaak, J. E., *et al.* (2007). Near-surface hydrologic response for a steep, unchanneled catchment near Coos Bay, Oregon: 2. Physics-based simulations, *American Journal of Science*, 307, 709–749.

Ebel, B. A., Loague, K., and Borja, R. I. (2010). The impacts of hysteresis on variably saturated hydrologic response and slope failure, *Environmental Earth Science*, doi:10.1007/s12665-009-0445-2.

Eberhardt, E., Stead, D., and Coggan, J. S. (2004). Numerical analysis of initiation and progressive failure in natural rock slopes: the 1991 Randa rockslide, *International Journal of Rock Mechanics and Mining Sciences*, 41, 69–87.

Eckel, E. B., 1958, Introduction, in, E. B. Eckel, ed., *Landslides and Engineering Practice*, Highway Research Board Special Report 29, National Academy of Sciences, National Research Council, Washington, D.C., 1–5.

Ellen, S. D., Cannon, S. H., and Reneau, S. L. (1988). Distribution of debris flows in Marin County, in S. D. Ellen and G. F. Wieczorek, eds., *Landslides, Floods, and Marine Effects of the Storm of January 3–5, 1982, in the San Francisco Bay Region, California*, U.S. Geological Survey Professional Paper 1434, pp. 113–132.

Emmanuel, K. (2005). Increasing destructiveness of tropical cyclones over the past 30 years, *Nature*, 436, 686–688.

Eswaran, H., Rice, T., Ahrens, R., and Stewart, B. A. (eds) (2002). *Soil Classification: A Global Reference*, Boca Raton, Fla.,: CRC Press.

Fannin, R. J., and Jaakkola, J. (1999). Hydrological response of hillslope soils above a debris-slide headscarp, *Canadian Geotechnical Journal*, 36, 1111–1122.

Fannin, R. J., Eliadorani, A., and Wilkinson, J. M. T. (2005). Shear strength of cohesionless soils at low stress, *Geotechnique*, 55(6), 467–478.

Fellenius, W. (1927). *Erdstatische Berchnungen*, Revised edition, Berlin: W. Ernst u. Sons.

Fellenius, W. (1936). Calculation of the stability of earth dams, in *Transactions of the 2nd Congress on Large Dams*, Washington, D.C., 4, 445–463.

Fleming, R. W., and Taylor, F. A. (1980). *Estimating the Costs of Landslide Damage in the United States*. U.S. Geological Survey Circular 832.

Fourie, A. B., Rowe, D., and Blight, G. E. (1999). The effect of infiltration on the stability of the slopes of a dry ash dump, *Géotechnique*, 49(1), 1–13.

Fox, G. A., and Wilson, G. V. (2010). The role of subsurface flow in hillslope and streambank erosion: a review, *Soil Science Society of America Journal*, 74, 717–733.

François, B., Tacher, L., Bonnard, Ch., Laloui, L., and Triguero, V. (2007). Numerical modeling of the hydrogeological and geomechanical behavior of a large slope movement: the Triesenberg landslide (Liechtenstein), *Canadian Geotechnical Journal*, 44, 840–857.

Fredlund, D. G., and Krahn, J. (1977). Comparison of slope stability methods of analysis, *Canadian Geotechnical Journal*, 14(3): 429–439.

Fredlund, D. G., and Morgenstern, N. R. (1977). Stress state variables for unsaturated soils, *Journal of Geotechnical Engineering Division*, 103, 447–466.

Fredlund, D. G., Vanapalli, S. K., Xing, A., and Pufahl, D. E. (1995). Predicting the shear strength for unsaturated soils using the soil water characteristic curve, in *Proceedings of the 1st International Conference on Unsaturated Soils*, Paris, pp. 63–69.

Freer, J., McDonnell, J. J., Beven, K. J., *et al.* (2002). The role of bedrock topography on subsurface storm flow, *Water Resources Research*, 38, 1269, doi:10.1029/2001WR000872.

Freeze, R. A. (1969). The mechanism of natural ground-water recharge and discharge 1. One-dimensional, vertical, unsteady, unsaturated flow above a recharging or discharging ground-water flow system, *Water Resources Research*, 5, 153–171.

Freeze, R.A. (1972). Role of subsurface flow in generating surface runoff. 1. Base flow contributions to channel flow, *Water Resources Research*, 8(5), 609–623.

Freeze, R. A., and Cherry, J. A. (1979). *Groundwater*, Englewood Cliffs, N.J.: Prentice Hall.

Fritz, P., Cherry, J. A., Weyer, K. V., and Sklash, M. G. (1976). Runoff analyses using environmental isotopes and major ions, in *Interpretation of Environmental Isotopes and Hydrochemical Data in Groundwater Hydrology*, Vienna: International Atomic Energy Agency, pp.111–130.

Frohlich, O. K. (1953). The factor of safety with respect to sliding of a mass of soil along the arc of a logarithmic spiral, in *Proceedings of the 3rd International Conference on Soil Mechanics and Foundation Engineering, Switzerland*, 2, 230–233.

Gabet, E. J. (2003). Sediment transport by dry ravel, *Journal of Geophysical Research*, 108(B1), 2049, doi:10.1029/2001JB001686.

Gabet, E. J., and Dunne, T. (2002). Landslides on coastal sage-scrub and grassland hillslopes in a severe El Niño winter: the effects of vegetation conversion on sediment delivery, *Geological Society of America Bulletin*, 114(8), 983–990.

Gale, M. R., and Grigal, D. F. (1987). Vertical root distributions of northern tree species in relation to successional status, *Canadian Journal of Forest Research*, 17, 829–834.

Gallagher, M. D. (1997). *Progress Report of Geological and Geotechnical Data Acquisition, Embankment Slope Instability Study, Straight Creek, Interstate 70 West of Eisenhower Tunnel, Summit County, Colorado*, Kumar and Associates, Inc.

Galster, R. W. and Laprade, W. T. (1991). Geology of Seattle, Washington, United States of America, *Bulletin of the Association of Engineering Geologists*, 18, 235–302.

Gardner, W. R. (1956). Calculation of capillary conductivity from pressure plate outflow data. *Soil Science Society of America Proceedings*, 20, 317–320.

Gardner, W. R. (1958). Some steady-state solutions of unsaturated moisture flow equation with application to evaporation from a water table, *Soil Science*, 85, 228–232.

Gee, G. W., Campbell, M. D., Campbell, G. S., and Campbell J. H. (1992). Rapid measurement of low soil-water potentials using a water activity meter, *Soil Science Society of America Journal*, 56 (4), 1068–1070.

GeoStudio (2007). Geoslope International, Inc.

Gillham, R. W. (1984). The capillary fringe and its effect on water-table response, *Journal of Hydrology*, 67, 307–324.

Gillham, R. W., and Abdul, A. S. (1986). Reply on Comment on "Laboratory studies of the effects of the capillary fringe on streamflow generation" by E. Zaltsburg, *Water Resources Research*, 22, 839.

Godt, J. W., and Coe, J. A. (2007). Alpine debris flows triggered by a 28 July 1999 thunderstorm in the central Front Range, Colorado, *Geomorphology*, 84, 80–97.

Godt, J. W., and Savage, W. Z. (1999). El Niño 1997–98: direct costs of damaging landslides in the San Francisco Bay region, in J. S. Griffiths, M. R. Stokes, and R. G. Thomas, eds., *Proceedings of the 9th International Conference and Field Trip on Landslides*, Rotterdam: A. A. Balkema, pp. 47–55.

Godt, J. W., Baum, R. L., and Chleborad, A. F. (2006). Rainfall characteristics for shallow landsliding in Seattle, Washington, USA, *Earth Surface Processes and Landforms*, 31, 97–110.

Godt, J. W., Baum, R. L, and Lu, N. (2009). Landsliding in partially saturated materials, *Geophysical Research Letters*, 36, L02403.

Godt, J. W., Şener-Kaya, B., Lu, N., and Baum, R. L. (2012). Infinite-slope stability under transient partially saturated conditions, *Water Resources Research*, 48, W05505, doi:10.1029/2011WR011408.

Goswami, B. N., Venugopal, V., Sengupta, D., Madhusoodanan, M. S., and Xavier, P. K. (2006). Increasing trend of extreme rain events over India in a warming environment, *Science*, 314, 1442–1445.

Gray, D. H., and Ohashi, H. (1983). Mechanics of fiber reinforcement in sand, *Journal of Geotechnical Engineering*, 109(3), 335–353.

Grayson, R. B., Western, A. W., and Chiew, F. H.–S. (1997). Preferred states in spatial soil moisture patterns: local and nonlocal controls, *Water Resources Research*, 33, 2897–2908.

Green, W. H., and Ampt, G. A. (1911). Studies on soil physics, Part I.The flow of air and water through soils, *Journal of Agricultural Science*, 4, 1–23.

Gribb, M. (1996). Parameter estimation for determining hydraulic properties of a fine sand from transient flow measurements, *Water Resources Research*, 32 (7), 1965–1974.

Griffiths, D. V. and Lane, P. A. (1999). Slope stability analysis by finite elements, *Géotechnique*, 49, 387–403.

Griffiths, P. G., Magirl, C. S., Webb, R. H., *et al.* (2009). Spatial distribution and frequency of precipitation during an extreme event: July 2006 mesoscale convective complexes and floods in southeastern Arizona, *Water Resources Research*, 45, W07419.

Groisman, P. Y., Knight, R. W., Easterling, D. R., *et al.* (2005). Trends in intense precipitation in the climate record, *Journal of Climate*, 18, 1326–1350.

Gupta, S. C., Farrell, D. A., and Larson, W. E. (1974). Determining effective soil water diffusivities from one-step outflow experiments, *Soil Science Society of America Proceedings*, 38, 710–716.

Guzzetti, F. (2000). Landslide fatalities and the evaluation of landslide risk in Italy, *Engineering Geology*, 58, 89–107.

Hack, J. T., and Goodlett, J. C. (1960). *Geomorphology and Forest Ecology of a Mountain Region in the Central Appalachians*, U.S. Geological Survey Professional Paper 347.

Hafiz, M. A. A. (1950). Strength characteristics of sands and gravels in direct shear, Ph.D. dissertation, University of London.

Hamilton, J. M., Daniel, D. E., and Olson, R. E. (1981). Measurement of hydraulic conductivity of partially saturated soils, in T. F. Zimmie and C. O.Riggs, eds., *Permeability and Groundwater Contaminant Transport*, ASTM STP 746, pp. 182–196.

Haneberg, W. C. (1991). Pore pressure diffusion and the hydrologic response of nearly saturated, thin landslide deposits to rainfall, *Journal of Geology*, 99(6), 886–892.

Hansen, W. R. (1965). *Effects of the Earthquake of March 27, 1964, at Anchorage, Alaska*, U.S. Geological Survey Professional Paper 542-A.

Hardy, B. (1992). Two-pressure calibration on the factory floor, *Sensors*, July, 15–19.

Harp, E. L., and Jibson, R. L. (1995). *Inventory of Landslides Triggered by the 1994 Northridge, California Earthquake*, U.S. Geological Survey Open-File Report, 95-213.

Harp, E. L., and Jibson, R. W. (1996). Landslides triggered by the 1994 Northridge, California, earthquake, *Bulletin of the Seismological Society of America*, 86(1B), S319–332.

Harp, E. L., Wells, W. G., and Sarmiento, J. G. (1990). Pore pressure response during failure in soils, *Bulletin of the Geological Society of America*, 102, 428–438.

Harp, E. L., Hagaman, K. W., and McKenna, J. P. (2002). *Digital Inventory of Landslides and Related Deposits in Honduras Triggered by Hurricane Mitch*, U.S. Geological Survey Open-File Report, 02-61.

Harp, E. L., Michael, J. A., and Laprade, W. T. (2008). Shallow landslide hazard map of Seattle, Washington, in R. L. Baum, J. W. Godt, and L. M. Highland, eds., *Landslides and Engineering Geology of the Seattle, Washington, Area*, Geological Society of America, Reviews in Engineering Geology, 20, pp. 67–82.

Harp, E. L., Keefer, D. K., Sato, H. P., and Yagi, H. (2010). Landslide inventories: the essential part of seismic landslide hazard analyses, *Engineering Geology*, doi:10.1016/j.engeo.2010.06.013.

Harr, R. D. (1977). Water flux in soil and subsoil on a steep forested slope, *Journal of Hydrology*, 33, 37–58.

Hassanizadeh, S. M., and Gray, W. G. (1987). High velocity flow in porous media, *Transport in Porous Media*, 2(6), 521–531.

Healy, R. W. (2008). Simulating water, solute, and heat transport in the subsurface with the VS2DI software package, *Vadose Zone Journal*, 7(2), 632–639.

Hewlett, J. D., and Hibbert, A. R. (1963). Moisture and energy conditions within a sloping soil mass during drainage, *Journal of Geophysical Research*, 68(2), 1081–1087.

Hewlett, J. D., and Hibbert, A. R. (1967). Factors affecting the response of small watersheds to precipitation in humid areas, in *Proceedings of the International Symposium on Forest Hydrology*, Pennsylvania State University, Pergamon, pp. 275–290.

Hilf, J. W. (1956). *An Investigation of Pore Water Pressure in Compacted Cohesive Soils*, Technical Memorandum No. 654, U.S. Department of the Interior, Bureau of Reclamation, Design and Construction Division, Denver, Colo.

Hillel, D. (1982). *Introduction to Soil Physics*, New York: Academic.

Hoek, E., and Bray, J. W. (1981). *Rock Slope Engineering*, Institution of Mining and Metallurgy, London.

Hogentogler, C. A., and Terzaghi, K. (1929). Interrelationship of load, road and subgrade, *Public Roads*, 37–64.

Holland, G. J., and Webster, P. J. (2007). Heightened tropical cyclone activity in the North Atlantic: natural variability or climatic trend, *Philosophical Transactions of the Royal Society A*, 365, 2695–2716.

Holzer, T. L., Hanks, T. C., and Youd, T. L. (1989). Dynamics of liquefaction during the 1987 Superstition Hills, California earthquake, *Science*, 244, 56–59.

Hong, Y., Adler, R., and Huffman, G. (2006). Evaluation of the potential of NASA multi-satellite precipitation analysis in global landslide hazard assessment, *Geophysical Research Letters*, 33, L22402.

Horton, R. E. (1933). The role of infiltration in the hydrologic cycle, *Transactions of American Geophysical Union*, 14, 446–460.

Horton, R. E. (1939). Analysis of runoff-plate experiments with varying infiltration capacity, *Transactions of American Geophysical Union*, 20, 693–711.

Horton, R. E. (1945). Erosional development of streams and their drainage basins: hydrophysical approach to quantitative morphology, *Geological Society of America Bulletin*, 56, 275–370.

Houston, S. L., Houston, W. N., and Wagner, A. (1994). Laboratory filter paper suction measurements, *Geotechnical Testing Journal*, 17(2), 185–194.

Hovius, N., Stark, C. P., and Allen, P. A. (1997). Sediment flux from a mountain belt derived by landslide mapping, *Geology*, 25, 231–234.

Hsieh, P. A., Wingle, W., and Healy, R. W. (1999). VS2DI: *A Graphical Software Package for Simulating Fluid Flow and Solute or Energy Transport in Variably Saturated Porous Media*. U.S. Geological Survey Water-Resource Investigation Report 99-4130, Reston, Va.

Hungr, O., Evans, S. G., Bovis, M. J., and Hutchinson, J. N. (2001). A review of the classification of landslides of the flow type, *Environmental and Engineering Geosciences*, 7(3), 221–238.

Hursh, C. R. (1936). Storm-water and absorption, in Discussion on list of terms with definitions: report to the Committee on Absorption and Transpiration, *Transactions of the American Geophysical Union*, 17, 301–302.

Hutchinson, J. N. (1968). Mass movement, in R. W. Fairbridge, ed., *Encyclopedia of Geomorphology*, New York: Reinhold, pp. 688–695.

Hutchinson, J. N. (1988). General report: morphological and geotechnical parameters of landslides in relation to geology and hydrogeology, in C. Bonnard, ed., *Proceedings of the 5th International Symposium on Landslides*, Rotterdam: A. A. Balkema, pp. 3–36.

Hutchinson, J. N., and Bhandari, R. K. (1971). Undrained loading, a fundamental mechanism of mudflows and other mass movements, *Géotechnique*, 21, 353–358.

Ingles, O. G. (1962). A theory of tensile strength for stabilized and naturally coherent soils, in *Proceedings of 1st Conference of the Australian Road Research Board*, 1, pp. 1025–1047.

Ijjasz-Vasquez, E. J. and Bras, R. L. (1995). Scaling regimes of local slope versus contributing area in digital elevation models, *Geomorphology*, 12, 299–311.

Israelachvili, J. (1992). *Intermolecular and Surface Forces*, 2nd Edition, San Diego: Academic.

Iverson, R. M. (1997). The physics of debris flows, *Reviews of Geophysics*, 35(3), 245–296.

Iverson, R. M. (2000). Landslide triggering by rain infiltration, *Water Resources Research*, 36, 1897–1910.

Iverson, R. M., and Major, J. S. (1987). Rainfall, ground-water flow, and seasonal movement at Minor Creek landslide, northwestern California: physical interpretations and empirical relations, *Geological Society of America Bulletin*, 99, 579–594.

Iverson, R. M. and Reid, M. E. (1992). Gravity-driven groundwater flow and slope failure potential 1. Elastic effective-stress model, *Water Resources Research*, 28(3), 925–938, doi:10.1029/91WR02694.

Iverson, R. M., Reid, M. E., and LaHusen, R. G. (1997). Debris-flow mobilization from landslides, *Annual Review of Earth and Planetary Sciences*, 25, 85–138.

Jackson, C. R. (1992). Hillslope infiltration and lateral downslope unsaturated flow, *Water Resources Research*, 28(9), 2533–2539.

Jackson, C. R. (1993). Reply, *Water Resources Research*, 29(12), 4169–4170.

Jackson, R. B., Canadell, J., Ehleringer, J. R., *et al.* (1996). A global analysis of root distributions for terrestrial biomes, *Oecologia*, 108(3), 389–411.

Janbu, N. (1973). Slope stability computations, in R. C. Hirschfeld and S. J. Poulos, eds., *Embankment-Dam Engineering: Casgrande Volume*, New York: John Wiley & Sons, pp. 49–86.

Jayatilaka, C. J., and Gillham, R. W. (1996). A deterministic-empirical model of the effect of the capillary fringe on near-stream area runoff: 1. Description of the model, *Journal of Hydrology*, 184, 299–315.

Jennings, J. E. B., and Burland, J. B. (1962). Limitation to the use of effective stresses in unsaturated soils, *Géotechnique*, 12, 125–144.

Jewell, R. A. (1980). Some factors which influence the shear strength of reinforced sand, *CUED/D-Soils/TR85*, Cambridge University Engineering Department, Cambridge.

Johnson, K. A. and Sitar, N. (1990). Hydrologic conditions leading to debris flow initiation, *Canadian Geotechnical Journal*, 27, 789–801.

Keefer, D. K. (1984). Landslides caused by earthquakes, *Geological Society of America Bulletin*, 95(4), 406–421.

Keefer, D. K., and Johnson, A. M. (1983). *Earthflows: Morphology, Mobilization, and Movement*, U.S. Geological Survey Professional Paper 1264.

Keefer, D. K., Moseley, M. E., and deFrance, S. D. (2003). A 38000-year record of flood and debris flows in the Ilo region of southern Peru and its relation to El Niño events and great earthquakes, *Paleogeography, Paleoclimatology, Paleoecology*, 194, 41–77.

Keefer, D. K., Wilson, R. C., Mark, R. K., *et al.* (1987). Real-time landslide warning during heavy rainfall, *Science*, 238, 921–925.

Khalili, N., Geiser, F., and Blight, G. E. (2004). Effective stress in unsaturated soils: review with new evidence, *International Journal of Geomechanics*, 4(2), 115–126.

Kim, T.-H. (2001). Moisture-induced tensile strength and cohesion in sand. Ph.D. thesis, Deptartment of Civil, Environmental and Architectural Engineering, University of Colorado, Boulder, Colo.

Kim, T.-H., and Hwang, C. (2003). Modeling of tensile strength on moist granular earth material at low water content, *Engineering Geology* 69, 233–244.

King, G. C. P., Stein, R. S., and Lin, J. (1994). Static stress changes and the triggering of earthquakes, *Bulletin of the Seismological Society of America*, 84, 935–953.

Kirkby, M. J. (ed.) (1978). *Hillslope Hydrology*, Chichester: John Wiley & Sons.

Kirkby, M. J., and Chorley, R. J. (1967). Throughflow, overland flow and erosion, *Bulletin of the International Association of Hydrological Sciences*, 12, 5–21.

Kirkpatrick, W. M. (1957). The conditions of failure for sands, in *Proceedings of 4th International Conference on Soil Mechanics*, London, 1, pp. 172–178.

Knutson, T. R, McBride, J. L., Chan, J., *et al.* (2010). Tropical cyclones and climate change, *Nature Geoscience*, 3, 157–163.

Kool, J. B., Parker, J. C., and van Genuchten, M. Th. (1985). ONESTEP: a nonlinear parameter estimation program for evaluating soil hydraulic properties from one-step outflow experiments, *Bulletin Virginia Agricultural Experimental Station*, 85(3).

Kumar, S., and Malik, R. S. (1990). Verification of quick capillary rise approach for determining pore geometrical characteristics in soils of varying texture, *Soil Science*, 150(6), 883–888.

Kunkel, K. E., Easterling, D. R., Redmond, K., and Hubbard, K. (2003). Temporal variations of extreme precipitation events in the United States: 1985–2000, *Geophysical Research Letters*, 30, GL018052, doi:10.1029/2003GL018052.

Kunze, R. J., and Kirkham, D. (1961). Simplified accounting for membrane impedance in capillary conductivity determinations, *Soil Science Society of America Proceedings*, 26, 421–426.

Kuras, O., Pritchard, J. D. Meldrum, P. I., *et al.* (2009). Monitoring hydraulic processes with automated time-lapse electrical resistivity tomography (ALERT), *Comptes Rendus Geoscience*, 341, 868–885.

Lade, P. V. (1992). Static instability and liquefaction of loose fine sandy slopes, *Journal of Geotechnical Engineering*, 118(1), 51–71.

Lai, W. M., Rubin, D., and Krempl, E. (1978). *Introduction to Continuum Mechanics*, New York: Pergamon Press.

Lam, L., and Fredlund, D. G. (1993). A general limit equilibrium model for three-dimensional slope stability analysis, *Canadian Geotechnical Journal*, 30, 905–919.

Lane, K. S., and Washburn, S. E. (1946). Capillary tests by capillarimeters and by soil filled tubes, *Proceedings of Highway Research Board*, 26, 460–473.

Larsen, M. C., and Simon, A. (1993). A rainfall intensity-duration threshold for landslides in a humid-tropical environment, Puerto Rico, *Geografiska Annaler*, 75(A), 13–23.

Lenderink, G., and van Meijgaard, E. (2008). Increase in hourly precipitation extremes beyond expectations from temperature changes, *Nature Geoscience*, 1, 511–514.

Leshchinsky, D., and San, K. C. (1994). Pseudostatic seismic stability of slopes: design charts, *ASCE Journal of Geotechnical Engineering*, 120(9), 1514–1532.

Leshchinsky, D., and Volk, J. C. (1985). Stability charts for geotextile-reinforced walls, *Transportation Research Record 1031*, Transportation Research Board, National Research Council, Washington, D.C.: National Academy Press, pp. 5–16.

Levenberg, K. (1944). A method for the solution of certain non-linear problems in least squares, *The Quarterly of Applied Mathematics*, 2, 164–168.

Likos, W. J., and Lu, N. (2002). Filter paper technique for measurement of total soil suction, *Journal of the Transportation Research Board*, No. 1786, 120–128.

Likos, W. J., and Lu, N. (2003). An automated humidity system for measuring total suction characteristics of clays, *Geotechnical Testing Journal*, 28(2), 178–189.

Lin, M. L., and Jeng, F. S. (2000). Characteristics of hazards induced by extremely heavy rainfall in Central Taiwan – Typhoon Herb, *Engineering Geology*, 58, 191–207.

Lohnes, R. A., and Handy, R. L. (1968). Slope angle in friable loess, *Journal of Geology*, 76(3), 247–258.

Lu, N. (2008). Is matric suction stress variable? *Journal of Geotechnical and Geoenvironmental Engineering*, 134(7), 899–905.

Lu, N. (2011). Interpreting the "collapsing" behavior of unsaturated soil by effective stress principle, in *Multiscale and Multiphysics Processes in Geomechanics*, Springer Series in Geomechanics and Geoengineering, Part 3, 81–84, doi:10.1007/978-3-642-19630-0_21.

Lu, N., and Godt, J. W. (2008). Infinite slope stability under unsaturated seepage conditions, *Water Resources Research*, 44, W11404, doi:10.1029/2008/WR006976.

Lu, N., and Griffiths, D. V. (2004). Profiles of steady-state suction stress in unsaturated soils, *Journal of Geotechnical and Geoenvironmental Engineering*, 130(10), 1063–1076.

Lu, N., and Likos, W. J. (2004a). *Unsaturated Soil Mechanics*, John Wiley & Sons.

Lu, N., and Likos, W. J. (2004b). Rate of capillary rise in soils, *Journal of Geotechnical and Geoenvironmental Engineering*, 130(6), 646–650.

Lu, N., and Likos, W. J. (2006). Suction stress characteristic curve for unsaturated soils, *Journal of Geotechnical and Geoenvironmental Engineering*, 132(2), 131–142.

Lu, N., Wu, B., and Tan, C. (2007). Tensile strength characteristics of unsaturated soils, *Journal of Geotechnical and Geoenvironmental Engineering*, 133(2), 144–154.

Lu, N., Wayllace, A., Carrera, J., and Likos, W. J. (2006). Constant flow method for concurrently measuring soil-water characteristic curve and hydraulic conductivity function, *Geotechnical Testing Journal*, 29(3), 256–266.

Lu, N., Kim, T.-H., Sture, S., and Likos, W. J. (2009). Tensile strength of unsaturated sand, *Journal of Engineering Mechanics*, 135(12), 1410–1419.

Lu, N., Wayllace, A. and Godt., J. W. (2010). A hydro-mechanical model for predicting infiltration-induced landslides, in *Abstracts of Geological Society of America 2010 USGS Modeling Conference*, Bloomfield, CO, June 8–11, 2010.

Lu, N., Godt, J. W., and Wu, D. T. (2010). A closed-form equation for effective stress in unsaturated soil. *Water Resources Research*, 46, W05515, doi:10.1029/2009WR008646.

Lu, N., Şener, B., and Godt, J. (2011). Direction of unsaturated flow in a homogeneous and isotropic hillslope, *Water Resources Research*, 47, W02519, doi:10.1029/2010WR010003.

Lu, N., Şener-Kaya, B., Wayllace, A., and Godt, J. W. (2012) Analysis of rainfall-induced slope instability using a field of local factor of safety, *Water Resources Research*, 48, W09524, doi:10.1029/2012WR011830.

Lumb, P. (1975). Slope failures in Hong Kong, *Quarterly Journal of Engineering Geology*, 8, 31–65.

Malamud, B. D., Turcotte, D. L., Guzzetti, F., and Reichenbach, P. (2004). Landslide inventories and their statistical properties, *Earth Surface Processes and Landforms*, 29, 687–711.

Malvern, L. E. (1969). *Introduction to the Mechanics of a Continuous Medium*, Englewood Cliffs, N.J.: Prentice-Hall.

Martinec, J. (1975). Subsurface flow from snowmelt traced by tritium, *Water Resources Research*, 11, 496–497.

Matsui, T., and San, K. C. (1992). Finite element slope stability analysis by shear strength reduction technique, *Soils and Foundations*, 32(1), 59–70.

Maune, D. F. (2007). *Digital Elevation Model Technologies and Applications*, 2nd Edition, Bethesda, Md.: American Society for Photogrammetry and Remote Sensing, p. 655.

McCord, J. T. and Stephens, D. B. (1987). Lateral moisture flow beneath a sandy hillslope without an apparent impeding layer, *Hydrological Processes*, 1, 225–238.

McDonnell, J. J. (1990). A rationale for old water discharge through macropores in a steep, humid catchment, *Water Resources Research*, 26(11), 2821–2832.

McDonnell, J. J., and Buttle, J. M. (1998). Comment on "A deterministic-empirical model of the effect of the capillary-fringe on near-stream area runoff. 1. Description of the model" by Jayatilaka, C. J. and Gillham, R. W., *Journal of Hydrology*, 207, 280–285.

McDonnell, J. J., Freer, J., Hooper, R., *et al.* (1996). New method developed for studying flow on hillslopes, *Eos Trans. AGU*, 77, 465 and 472.

McKean, J., and Roering, J. (2004). Objective landslide detection and surface morphology mapping using high-resolution airborne laser altimetry, *Geomorphology*, 57, 331–351.

McPhaden, M. J. (2002). El Niño and La Niña: causes and global consequences, in M. C. MacCarcken and J. S. Perry, eds., *Encyclopedia of Global Environmental Change*, Vol. 1, New York: John Wiley & Sons, pp. 353–370.

Meerdink, J., Benson, C., and Khire, M. (1996). Unsaturated hydraulic conductivity of two compacted barrier soils, *Journal of Geotechnical and Geoenvironmental Engineering*, 122(7), 565–576.

Mein, R. G., and Larson, C. L. (1973). Modeling infiltration during a steady rain, *Water Resources Research*, 9(2), 384–394.

Meyer-Christoffer, A., Becker, A., Finger, P., *et al.* (2011). GPCC Climatology Version 2011 at 0.25°: Monthly land-surface precipitation climatology for every month and the total year from rain-gauges built on GTS-based and historic data. Doi: 10.5676/DWD_GPCC/CLIM_M_V2011_025.

Michalowski, R. L. (1995). Stability of slopes: limit analysis approach, in W. C. Haneberg and S. A. Anderson, eds., *Clay and Shale Slope Stability*, Geological Society of America, Reviews in Engineering Geology, 10, 51–62.

Michalowski, R. L. (2002). Stability charts for uniform slopes, *Journal of Geotechnical and Geoenvironmental Engineering*, 128 (4), 351–355.

Miller, D. J. (1991). Damage in King County from the storm of January 9, 1990, *Washington Geology*, 19, 28–37.

Miller, E. E., and Elrick, E. (1958). Dynamic determination of capillary conductivity extended for non-negligible membrane impedance, *Soils Science of America Journal*, 22, 483–486.

Minard, J. P. (1983). *Geologic Map of the Edmonds East and Part of the Edmonds West Quadrangles, Washington*, U.S. Geological Survey, Miscellaneous Field Studies Map MF-1541, scale 1:24000.

Miyazaki, T. (1988). Water flow in unsaturated soil in layered slopes, *Journal of Hydrology*, 102, 201–214.

Montgomery, D. R. and Dietrich, W. E. (1994a). Landscape dissection and drainage area-slope thresholds, in M. J. Kirkby, ed., *Process Models and Theoretical Geomorphology*, Chichester: John Wiley & Sons, pp. 221–246.

Montgomery, D. R. and Dietrich, W. E. (1994b). A physically-based model for the topographic control on shallow landsliding, *Water Resources Research*, 30, 1153–1171.

Montgomery, D. R., Dietrich, W. E., Torres, R., *et al.* (1997). Hydrologic response of a steep, unchanneled valley to natural and applied rainfall, *Water Resources Research*, 33, 91–109.

Montgomery, D. R., Sullivan, K. and Greenberg, H. M. (1998). Regional test of a model for shallow landsliding, *Hydrological Processes*, 12, 943–955

Montgomery, D. R., Schmidt, K. M., Greenberg, H. M., and Dietrich, W. E. (2000). Forest clearing and regional landsliding, *Geology*, 28(4), 311–314.

Moore, I. D., Grayson, R. B., and Ladson, A. R. (1991). Digital terrain modeling: a review of hydrological, geomorphological, and biological applications, *Hydrological Processes*, 5, 3–30.

Moreiras, S. M. (2005). Climatic effect of ENSO associated with landslide occurrence in the Central Andes, Mendoza Province, Argentina, *Landslides*, 2, 53–59.

Morgenstern, N. R. (1979). Properties of compact soils, in *Proceedings of the 6th Panamerican Conference on Soil Mechanics and Foundation Engineering*, Lima, Peru, Vol. 3, pp. 349–354.

Morgenstern, N. R., and de Matos, M. M. (1975). Stability of slopes in residual soils, in *Proceedings of the 5th Panamerican Conference on Soil Mechanics and Foundation Engineering*, Buenos Aires, Vol. 3, pp. 367–383.

Morgenstern, N. R., and Price, V. E. (1965). The analysis of the stability of general slip surfaces, *Géotechnique*, 15(4), 289–290.

Morris, S. S., Neidecker-Gonzales, O., Carletto, C., *et al.* (2002). Hurricane Mitch and the livelihoods of the rural poor in Honduras, *World Development*, 30, 49–60.

Mosely, M. P. (1979). Streamflow generation in a forested watershed, New Zealand, *Water Resources Research*, 15(4), 795–806.

Mualem, Y. (1976). Hysteretical models for prediction of the hydraulic conductivity of unsaturated porous media, *Water Resources Research*, 12(6), 1248–1254.

Muskhelishvili, N. I. (1953). *Some Basic Problems of the Mathematical Theory of Elasticity*, Leiden, the Netherlands: Noordhoof.

National Research Council (2004). *Partnerships for Reducing Landslide Risk: Assessment of the National Landslide Hazards Mitigation Strategy*, Washington D.C.: National Academy Press, p. 131.

Neiman, P. J., Ralph, F. M., Wick, G. A., Lundquist, J. D., and Dettinger, M. D. (2008a). Meteorological characteristics of overland precipitation impacts of atmospheric rivers affecting the west coast of North America based on eight years of SSM/I satellite observations, *Journal of Hydrometeorology*, 9, 22–47.

Neiman, P. J., Ralph, F. M., Wick, G. A., *et al.* (2008b). Diagnosis of an intense atmospheric river impacting the Pacific Northwest: storm summary and offshore vertical structure observed with COSMIC satellite retrievals, *Monthly Weather Review*, 136, 4398–4420.

Ng, C. W. W., and Shi, Q. (1998). A numerical investigation of the stability of unsaturated soil slopes subjected to transient seepage, *Computers and Geotetechnics*, 22, 1–28.

Ngecu, W. M., and Mathu, E. M. (1998). The El Niño-triggered landslides and their socioeconomic impact on Kenya, *Environmental Geology*, 38(4), 277–284.

O'Loughlin, C., and Ziemer, R. R. (1982). The importance of root strength and deterioration rates upon edaphic stability in steepland forests, in *Proceedings of IUFRO Workshop P.1.07-00 Ecology of Subalpine Ecosystems as a Key to Management*, 2–3 August 1982, Oregon State University, Corvallis, Ore., pp. 70–78.

Olsen, H. W., Gill, J. D., Willden, A. T., and Nelson, K. R. (1991). Innovations in hydraulic conductivity measurements, *Transportation Research Records*, 1309, Transportation Research Board, National Research Council, pp. 9–17.

Parker, R. N., Densmore, A. L., Rosser, N. J., *et al.* (2011). Mass wasting triggered by the 2008 Wenchuan earthquake is greater than orogenic growth, *Nature Geoscience*, 4, 449–452.

Pearce, A. J., Stewart, M. K., and Sklash, M. G. (1986). Storm runoff generation in humid headwater catchments 1. Where does the water come from? *Water Resources Research*, 22(8), 1263–1272.

Peck, R. B., Hansen, W. E., and Thornburn, T. H. (1974). *Foundation Engineering*, 2nd Edition, New York: Wiley.

Peixoto, J. P., and Kettani, M. (1973). The control of the water cycle, *Scientific American*, 228, 46–61.

Perkins, S. W. (1991). Modeling of regolith structure interaction on extraterrestrial constructed facilities, Ph.D. thesis, University of Colorado, Boulder, Colo.

Petley, D. N. (2008). The global occurrence of fatal landslides in 2007, in *Proceedings of the International Conference on Management of Landslide Hazards in the Asia-Pacific Region*.

Petley, D. N., Hearn, G. J., Hart, A., *et al.* (2007). Trends in landslide occurrence in Nepal, *Natural Hazards*, 43, 23–44.

Pham, H. Q., Fredlund, D. G., and Barbour, S.-L. (2005). A study of hysteresis models for soil-water characteristic curves, *Canadian Geotechnical Journal*, 42, 1548–1568.

Philip, J. R. (1957). The theory of infiltration. 1. The infiltration equation and solution, *Soil Science*, 83, 345–357.

Philip, J. R. (1969). Theory of infiltration, *Advances in Hydroscience*, 5, 215–290.

Philip, J. R. (1993). Comment on "Hillslope infiltration and lateral downslope unsaturated flow," by C. R. Jackson, *Water Resources Research*, 29(12), 4167.

Philip, J. R. (1991). Hillslope infiltration: planar slopes, *Water Resources Research*, 27(1), 109–117.

Pierson, T. C. (1983). Soil pipes and slope stability, *Quarterly Journal of Engineering Geology*, 16, 1–11.

Pollen, N., and Simon, A. (2005). Estimating the mechanical effects of riparian vegetation on stream bank stability using a fiber bundle model, *Water Resources Research*, 41, W07025, doi:10.1029/2004WR003801.

Pollen-Bankhead, N., and Simon, A. (2009). Enhanced application of root-reinforcement algorithms for bank-stability modeling, *Earth Surface Processes and Landforms*, 34, 471–480.

Pollen-Bankhead, N., and Simon, A. (2010). Hydrologic and hydraulic effects of riparian root networks on streambank stability: Is mechanical root-reinforcement the whole story? *Geomorphology*, 116(3–4), 353–352.

Pruess, K., Oldenburg, C., and Moridis, G. (2011). *TOUGH2 User's Guide, Version 2.1*, Report LBNL-43134, Lawrence Berkeley Laboratory, Berkeley, Calif.

Ragan, R. M. (1968). An experimental investigation of partial area contributions, in *Proceedings of the General Assembly, International Association of Hydrological Sciences*, Berne, Publication 76, 241–249.

Rahardjo, H., Li, X. W., Toll, D. G., and Leong, E. C. (2001). The effect of antecedent rainfall on slope stability, *Geotechnical and Geological Engineering*, 19, 371–399.

Rahardjo, H., Ong, T. H., Rezaur, R. B., and Leong, E. C. (2007). Factors controlling instability of homogeneous soil slopes under rainfall, *Journal of Geotechnical and Geoenvironmental Engineering*, 133(12), 1532–1543.

Ralph, F. M., Neiman, P. J., Wick, G. A., *et al.* (2006). Flooding on California's Russian River: role of atmospheric rivers, *Geophysical Research Letters*, 33, L12801.

Redding, T. E., and Devito, K. J. (2007). Lateral flow thresholds for aspen forested hillslopes on the Western Boreal Plain, Alberta, Canada, *Hydrological Processes*, 22(21), 4287–4300.

Reddy, J. N. (1985). *Introduction to the Finite Element Method*, New York: McGraw-Hill.

Reid, M. E. (1994). A pore-pressure diffusion model for estimating landslide-inducing rainfall, *Journal of Geology*, 102, 709–717.

Reid, M. E. (1997). Slope instability caused by small variations in hydraulic conductivity, *Journal of Geotechnical and Geoenvironmental Engineering*, 123(8), 717–725.

Reid, M. E. and Iverson, R. M. (1992). Gravity-driven groundwater flow and slope failure potential 2. Effects of slope morphology, material properties, and hydraulic heterogeneity, *Water Resources Research*, 28(3), 939–950, doi:10.1029/91WR02695.

Reid, M. E., Nielsen, H. P., and Dreiss, S. J. (1988). Hydrologic factors triggering a shallow hillslope failure, *Bulletin of the Association of Engineering Geologists*, 25(3), 349–361.

Reid, M. E., LaHusen, R. G., and Iverson, R. M. (1997). Debris-flow initiation experiments using diverse hydrological triggers, in C. L. Chen, ed., *Debris-Flow Hazards Mitigation: Mechanics, Prediction, and Assessment*, New York: American Society of Civil Engineering, pp. 1–10.

Reneau, S. L. and Dietrich, W. E. (1987). The importance of hollows in debris flows studies: examples from Marin County, California, *Reviews in Engineering Geology*, 7, 165–180.

Reneau, S. L., Dietrich, W. E., Donahue, D. J., Jull, A. J. T., and Rubin, M. (1990). Late Quaternary history of colluvial deposition and erosion in hollows, central California Coast Ranges, *Geological Society of America Bulletin*, 102, 969–982.

Richards, L. A. (1931). Capillary conduction of liquids through porous medium, *Journal of Physics*, 318–333.

Richards, S., and Weeks, L. (1953). Capillary conductivity values from moisture yield and tension measurements on soil columns, *Soil Science Society of America Journal*, 35, 695–700.

Ridley, A. M. and Wray, W. K. (1996). Suction measurement theory and practice: a state-of-the-art-review, *Proceedings 1st International Conference on Unsaturated Soils*, Paris, Vol. 3, pp. 1293–1322.

Ritsema, C. J., Dekker, L. W., Hendrickx, J. M. H., and Hamminga, W. (1993). Preferential flow mechanism in a water repellent sandy soil, *Water Resources Research*, 29(7), 2183–2193.

Roe, G. H. (2005). Orographic precipitation, *Annual Review of Earth and Planetary Sciences*, 33, 645–671.

Rogers, D. J. (1998). Mission Peak landslide, *EOS, Transactions of the American Geophysical Union*, 79(45), F266.

Rosen, M. J. (1989). *Surfactants and Interfacial Phenomena*, 2nd Edition, New York: John Wiley & Sons.

Ross, B. (1990). Diversion capacity of capillary barriers, *Water Resources Research*, 26, 2625–2629.

Rubin, J., and Steinhardt, R. (1963). Soil water relations during rain infiltration: 1. Theory, *Soil Science Society Proceedings*, 27, 246–251.

Savage, W. Z. (1994). Gravity-induced stress in finite slopes, *International Journal of Rock Mechanics and Mining Sciences & Geomechanics Abstracts*, 31(5), 471–483.

Savage, W. Z., Godt, J. W., and Baum, R. L. (2004). Modeling time-dependent areal slope stability, in landslides – evaluation and stabilization, in W. A. Lacerda, *et al.*, eds., *Proceedings of the 9th International Symposium on Landslides*, Vol. 1, London: A. A. Balkema, London, pp. 23–26.

Schmidt, K. M., Roering, J. J., Stock, J. D., *et al.* (2001). The variability of root cohesion as an influence on shallow landslide susceptibility in the Oregon Coast Range, *Canadian Geotechnical Journal*, 38(5), 995–1024.

Schofield, A., and Wroth, P. (1968). *Critical State Soil Mechanics*, London: McGraw-Hill.

Schubert, H. (1984). Capillary forces-modeling and application in particulate technology, *Powder Technology*, 37, 105–116.

Schumm, S. A. (1956). Evolution of drainage systems and slopes in badlands at Perth Amboy, New Jersey, *Geological Society of America Bulletin*, 67, 597–646.

Schuster, R. L. (1981). Effects of the eruptions on civil works and operations in the Pacific Northwest, in P. W. Lipmann and D. R. Mullineaux, eds., *The 1980 Eruptions of Mount St. Helens, Washington*, U.S. Geological Survey Professional Paper 1250, pp. 701–718.

Schuster, R. L., and Fleming, R. W. (1986). Economic losses and fatalities due to landslides, *Bulletin of the Association of Engineering Geologists*, 23(1), 11–28.

Schuster, R. L., and Highland, L. M. (2001). *Socioeconomic and Environmental Impacts of Landslides in the Western Hemisphere*, U.S. Geological Survey Open-File Report 01-0276.

Scott, P. S., Farquhar, G. J., and Kouwen, N. (1983). Hysteretic effects on net infiltration, *Advances in Infiltration*, 11(83), 163–170.

Selby, M. J. (1993). *Hillslope Materials and Processes*, 2nd Edition, Oxford: Oxford University Press, p. 430.

Shipman, H. (2004). Coastal bluffs and sea cliffs on Puget Sound, in *Formation, Evolution, and Stability of Coastal Cliffs: Status and Trends*, U.S. Geological Survey Professional Paper 1693, pp. 81–94.

Shrestha, A. B., Wake, C. P., Dibb, J. E., and Mayewski, P. A. (2000). Precipitation fluctuations in the Nepal Himalaya and its vicinity and relationship with some large scale climatological parameters, *International Journal of Climatology*, 30, 317–327.

Shroder, J. F. (1971). *Landslides of Utah*, Utah Geological and Mineralogical Survey Bulletin 90.

Sidle, R. (1992). A theoretical model of the effects of timber harvesting on slope stability, *Water Resources Research*, 28(7), 1897–1910.

Sidle, R. C. and Ochiai, H. (2006). *Landslides: Processes, Prediction, and Land Use*, American Geophysical Union Water Resources Monograph 18, Washington, D.C.

Sidle, R. C. and Swanston, D. N. (1982). Analysis of a small debris slide in coastal Alaska, *Canadian Geotechnical Journal*, 19, 167–174.

Sidle, R. C., Tsuboyama, Y., Noguchi, S., *et al.* (2000). Stormflow generation in steep forested headwaters: a linked hydrogeomorphic paradigm, *Hydrological Processes*, 14, 369–385.

Silliman, S. E., Berkowitz, B., Šimůnek, J., and van Genuchten, M. T. (2002). Fluid flow and chemical migration within the capillary fringe, *Ground Water*, 40(1), 76–84.

Silvestri, V., and Tabib, C. (1983a). Exact determination of gravity stresses in finite elastic slopes, Part I. Theoretical considerations, *Canadian Geotechnical Journal*, 20, 47–54.

Silvestri, V., and Tabib, C. (1983b). Exact determination of gravity stresses in finite elastic slopes, Part II. Applications, *Canadian Geotechnical Journal*, 20, 55–60.

Šimůnek, J., Vogel, T., and van Genuchten, M. Th. (1994). *The SWMS-2D Code for Simulating Water Flow and Solute Transport in Two-Dimensional Variably Saturated Media, Version 1.21*, Research Report No. 132, USDA-ARS U.S. Salinity Laboratory, Riverside, Calif.

Šimůnek, J., Sejna, M. and van Genuchten, M. Th. (1999). *The HYDRUS-2D Software Package for Simulating the Two-Dimensional Movement of Water, Heat, and Multiple Solutes in Variably-Saturated Media, Version 2.0*, U.S. Salinity Laboratory Agricultural Research Service, U.S. Department of Agriculture, Riverside, Calif.

Šimůnek, J., van Genuchten, M.Th., and Sejna, M. (2005). *The HYDRUS-1D Software Package for Simulating the One-dimensional Movement of Water, Heat, and Multiple Solutes in Variably-Saturated Media. Version 3.0 HYDRUS Software Series 1*, Department of Environmental Sciences, University of California, Riverside, Calif.

Šimůnek, J., van Genuchten, M. Th., and Šejna, M. (2008). Development and applications of the HYDRUS and STANMOD software packages and related codes, *Vadose Zone Journal*, 7, 587–600, doi:10.2136/vzj2007.0077.

Sinai, G., and Dirksen, C. (2006). Experimental evidence of lateral flow in unsaturated homogeneous isotropic sloping soil due to rainfall, *Water Resources Research*, 42, W12402.

Sinai, G., Zaslavsky, D., and Golany, P. (1981). The effect of soil surface curvature on moisture and yield, Beer-Sheva observation, *Soil Science*, 132(2), 367–375.

Singh, D. N., and Kuriyan, S. (2003). Estimation of unsaturated hydraulic conductivity using soil suction measurements obtained by an insertion tensiometer, *Canadian Geotechnical Journal*, 40, 476–483.

Sitar, N., and Clough, G. W. (1983). Seismic response of steep slopes in cemented soils, *Journal of Geotechnical Engineering*, 109(2), 210–227.

Skempton, A. W. (1954). The pore-pressure coefficients A and B, *Géotechnique*, 4, 143–147.

Skempton, A. W. (1960a). Significance of Terzaghi's concept of effective stress, in L. Bjerrum, A. Casagrade, R. B. Peck, and A.W. Skempton, eds., *From Theory to Practice in Soil Mechanics*, New York: Wiley, pp. 43–53.

Skempton, A. W. (1960b). Effective stress in soil, concrete, and rocks, in *Proceedings of the Conference on Pore Pressure and Suction in Soils*, London: Butterworths, pp. 4–16.

Skempton, A. W., and Hutchinson, J. N. (1969). Stability of natural slopes and embankment foundations, in *Proceedings of the 7th International Conference on Soil Mechanics and Foundation Engineering*, Mexico City, pp. 291–340.

Sklash, M. G., and Farvolden, R. N. (1979). The role of groundwater in storm runoff, *Journal of Hydrology*, 43, 45–65.

Smith, R. B. (2006). Progress on the theory of orographic precipitation, in S. D. Willett, N. Hovius, M. T. Brandon, and D. M. Fisher, eds., *Tectonics, Climate, and Landscape Evolution*, Geological Society of America Special Paper 398, Penrose Conference Series, pp. 1–16.

Smith, I. M., and Griffiths, D. V. (2004). *Programming the Finite Element*, 4th Edition, Chichester: Wiley.

Smith, R. E., with Smettem, K. R. J., Broadbridge, P., and Woolhiser, D. A. (2002). *Infiltration Theory for Hydrologic Application*, AGU Water Resources Monograph 15.

Soil Survey Division Staff (1993). *Soil Survey Manual*, Soil Conservation Service, U.S. Department of Agriculture Handbook 18.

Spanner, D. C. (1951). The Peltier effect and its use in the measurement of suction pressure, *Journal of Experimental Botany*, 11, 145–168.

Springman, S. M., Jommi, C., and Teysseire, P. (2003). Instabilities on moraine slopes induced by loss of suction: a case history, *Géotechnique*, 53, 3–10.

Srivastava, R., and Yeh, T.-C. J. (1991). Analytical solutions for one-dimensional, transient infiltration toward the water table in homogeneous and layered soils, *Water Resources Research*, 27, 753–762.

Stannard, D. I. (1992). Tensiometer: theory, construction, and use, *Geotechnical Testing Journal*, 15(1), 48–58.

Steenhuis, T. S., Parlange, J.-Y., Kung, K.-J. S. (1994). Comments on "Diversity capacity of capillary barriers" by Benjamin Ross, *Water Resources Research*, 27, 2155–2156.

Stohl, A., Forster, C., and Sodemann, H. (2008). Remote sources of water vapor forming precipitation on the Norwegian west coast at 60°N – a tale of hurricanes and an atmospheric river, *Journal of Geophysical Research*, 113, D05102.

Swan, C. C., and Seo, Y. K. (1999). Limit state analysis of earthen slopes using dual continuum/FEM approaches, *International Journal for Numerical and Analytical Methods in Geomechanics*, 23(12), 1359–1371.

Swartzendruber, D. (1963). Non-Darcy behavior and flow of water in unsaturated soils, *Soil Science Society of America Journal*, 27, 491–495.

Tarboton, D. G. (1997). A new method for the determination of flow directions and upslope areas in grid digital elevation models, *Water Resources Research*, 33, 309–319.

Taylor, D. W. (1937). Stability of earth slopes, *Journal of the Boston Society of Civil Engineers*, 24, 197–246.

Taylor, D. W. (1941). *Cylindrical Compression Research Program on Stress-Deformation and Strength Characteristics of Soils*, 7th Progress Report to U.S. Army Corps of Engineers.

ten Brink, U. S., Geist, E. L., and Andrews, B. D. (2006). Size distribution of submarine landslides and its implication to tsunami hazard in Puerto Rico, *Geophysical Research Letters*, 33, L11307.

Terzaghi, K. (1943). *Theoretical Soil Mechanics*, New York: John Wiley & Sons.

Terzaghi, K. (1936). The shearing resistance of saturated soils, in *Proceedings of the First International Conference on Soil Mechanics*, Vol. 1, 54–56.

Terzaghi, K., and Peck, R. B. (1967). *Soil Mechanics in Engineering Practice*, 2nd Edition, Hoboken, N.J.: John Wiley & Sons.

Tetens, O. (1930). Uber einige meteorologische Begriffe, *Zeitschrift Geophysic*, 6, 297–309.

Thomas, R. E., and Pollen-Bankhead, N. (2010). Modeling root-reinforcement with a fiber-bundle model and Monte Carlo simulation, *Ecological Engineering*, 36, 47–61.

Thorenz, C., Kosakowski, G., Kolditz, O., and Berkowitz B. (2002). An experimental and numerical investigation of saltwater movement in coupled saturated-partially saturated systems, *Water Resources Research*, 36(6), 1069, doi: 10.1029/2001WR000364.

Timoshenko, S., and Goodier, J. (1987). *Theory of Elasticity*, New York: McGraw-Hill.

Toda, S., Stein, R. S., Reasenberg, P. A., Dieterich, J. H., and Yoshida, A. (1998). Stress transferred by the 1995 $M\omega = 6.9$ Kobe, Japan, shock: effect on aftershocks and future earthquake probabilities, *Journal of Geophysical Research*, 103(B10), 24,543–24,565, doi:10.1029/98JB00765.

Toorman, A. F., Wierega, P. J., and Hills, R. G. (1992). Parameter estimation of hydraulic properties from one-step outflow data, *Water Resources Research*, 28, 3021–3028.

Torres, R., and Alexander, L. J. (2002). Intensity-duration effects on drainage: column experiments at near-zero pressure head, *Water Resources Research*, 38, 1240.

Torres, R., Dietrick, W. E., Montgomery, D. R., Anderson, S.P., and Loague, K. (1998). Unsaturated zone processes and the hydrological response of a steep unchanneled catchment, *Water Resources Research*, 34(8), 1865–1879.

Tóth, J. (1963). A theoretical analysis of groundwater flow in small drainage basin, *Journal of Geophysical Research*, 68, 4759–4812.

Trenberth, K. E. (1999). Conceptual framework for changes of extremes of the hydrological cycle with climate change, *Climatic Change*, 42, 327–339.

Trustrum, N. A., Gomez, B., Page, M. J., Reid, L. M. and Hicks, D. M. (1999). Sediment production, storage and output: the relative role of large magnitude events in steepland catchments, *Zeitshrift fur Geomorphologie*, Supplementband, 115, 71–86.

Tsuboyama, Y., Sidle, R. C., Hoguchi, S., and Hosoda, I. (1994). Flow and solute transport through the soil matrix and macropores of a hillslope segment, *Water Resources Research*, 30, 879–890.

Turnbull, W. J., and Hvorslev, M. J. (1967). Special problems in slope stability, *Journal of the Soil Mechanics and Foundations Division, ASCE*, 93(4), 499–528.

Uchida, T., Kosugi, K., and Mizuyama, T. (2001). Effects of pipeflow on hydrological process and its relation to landslide: a review of pipeflow studies in forested headwater catchments, *Hydrological Processes*, 15, 2151–2174.

UNESCO (1984). *Map of the World Distribution of Arid Regions*, Intergovernmental Oceanographic Commission, Paris, France.

Vanapalli, S. K., Fredlund, D. E., Pufahl, D. E., and Clifton, A. W. (1996). Model for the prediction of shear strength with respect to soil suction, *Canadian Geotechnical Journal*, 33, 379–392.

van Dam, J. C., Stricker, J. N M., and Droogers, P. (1994). Inverse method to determine soil hydraulic functions from multistep outflow experiments, *Soil Science Society of America Proceedings*, 58, 647–652.

van Genuchten, M. T. (1980). A closed-form equation for predicting the hydraulic conductivity of unsaturated soils, *Soil Science Society of America Journal*, 44, 892–898.

van Genuchten, M. Th. (1981). *Non-Equilibrium Transport Parameters from Miscible Displacement Experiments*:, Research Report No. 119, U.S. Salinity Laboratory, Riverside, Calif.

van Genuchten, M. Th., Leij, F. J., and Yates, S. R. (1991). *The RETC Code for Quantifying the Hydraulic Functions of Unsaturated Soils*, U.S. Department of Agriculture, Agricultural Research Service, Report IAG-DW12933934, Riverside, Calif.

Varnes, D. J. (1958). Landslide types and processes, in E. B. Eckel, ed., *Landslides and Engineering Practice*, Highway Research Board Special Report No. 29. pp. 20–47.

Varnes, D. J. (1978). Slope movement types and processes, in R. L. Schuster and R. J. Krizek, eds., *Landslide Analysis and Control*, Transportation Research Board Special Report 176, National Academy of Sciences, National Research Council, Washington, D.C., pp. 11–33.

Verwey, E. J. W., and Overbeek, J. Th. G. (1948). *Theory of the Stability of Lyophobic Colloids*, New York: Elsevier.

Voight, B., Janda, R. J., Glicken, H., and Douglass, P. M. (1983). Nature and mechanics of the Mount St. Helens rockslide-avalanche of 18 May 1980, *Géotechnique*, 33, 243–273.

Waldron, L. J. (1977). The shear resistance of root-permeated homogeneous and stratified soil, *Journal of Soil Science Society of America*, 41, 843–849.

Waldron, L. J., and Dakessian, S. (1981). Soil reinforcement by roots: calculation of increased soil shear resistance from root properties, *Soil Science*, 132(6), 427–435.

Walker, J., Willgoose, G., and Kalma, J. (2001). One-dimensional soil moisture profile retrieval by assimilation of near-surface measurements: a simplified soil moisture model and field application, *Journal of Hydrometereology*, 2, 356–373.

Warrick, A. W., Wierenga, P. J. and Pan, L. (1997). Downward water flow through sloping layers in the vadose zone: analytical solutions for diversions, *Journal of Hydrology*, 192, 321–337.

Watson, K. K. (1966). An instantaneous profile method for determining the hydraulic conductivity of unsaturated porous materials, *Water Resources Research*, 2, 709–715.

Wayllace, A., and Lu, N. (2012). A transient water release and imbibitions method for rapidly measuring wetting and drying soil water retention and hydraulic conductivity functions, *Geotechnical Testing Journal*, 137(1), 16–28.

Wayllace, A., Lu, N., Oh, S., and Thomas, D. (2012). Perennial infiltration-induced instability of Interstate-70 embankment west of the Eisenhower Tunnel/Johnson Memorial Tunnels, in *Proceedings of GeoCongress 2012*, ASCE.

Webster, P. J., Holland, G. J., Curry, J. A., and Chang, H. R. (2005). Changes in tropical cyclone number, duration, and intensity in a warming environment, *Science*, 309, 1844–1846.

Weiler, M., McDonnell, J. J., Tromp-Van Meerveld, I., and Uchida, T. (2005). Subsurface stormflow, in M. G. Anderson, ed., *Encyclopedia of Hydrological Sciences*, New York: John Wiley & Sons, p. 14.

Weyman, D. R. (1973). Measurements of the downslope flow of water in a soil, *Journal of Hydrology*, 20, 267–288.

Wheeler, S. J., and Sivakumar, V. (1995). An elasto-plastic critical state framework for unsaturated soil, *Géotechnique*, 45, 35–53.

Whipkey, R. Z. (1965). Subsurface stormflow from forested slopes, *International Association of Hydrological Sciences, 10th Annual Bulletin*, 2, 74–85.

Whipkey, R. Z., and Kirkby, M. J. (1978). Flow within the soil, in M. J. Kirkby, ed., *Hillslope Hydrology*, Chichester: John Wiley & Sons, pp. 121–142.

Wieczorek, G. F. (1996). Landslide triggering mechanisms, in R. L. Schuster, and R. J. Krizek, eds., *Landslide Analysis and Control*, Transportation Research Board Special Report 176, National Academy of Sciences, National Research Council, Washington, D.C., pp. 76–90.

Wieczorek, G. F., Madrone, G. and DeCola, L. (1997). The influence of hillslope shape on debris-flow initiation, in C. L. Chen, ed., *Debris-Flow Hazards Mitigation: Mechanics, Prediction, and Assessment*, New York: ASCE.

Wilson, R. C. (1997). Normalizing rainfall/debris-flow thresholds along the U.S. Pacific coast for long-term variations in precipitation climate, in C. L. Chen, ed., *Debris-Flow Hazards Mitigation: Mechanics, Prediction, and Assessment*, New York: ASCE, pp. 32–43.

Wolle, C. M., and Hachich, W. (1989). Rain-induced landslides in southeastern Brazil, in *Proceedings of the 12th International Conference on Soil Mechanics and Foundation Engineering*, Rio de Janeiro, pp. 1639–1644.

Wooten, R. M., Gillon, K. A., Witt, A. C., *et al.* (2008). Geologic, geomorphic, and meteorological aspects of debris flows triggered by Hurricanes Frances and Ivan during September 2004 in the southern Appalachian Mountains of Macon County, North Carolina (southeastern USA), *Landslides*, 5, 31–44.

Wu, W. and Sidle, R. C. (1995). A distributed slope stability model for steep forested basins, *Water Resources Research*, 31, 2097–2110.

Wu, T. H., McKinnell, W. P., III, and Swanton, D. N. (1979). Strength of tree roots and landslides on Prince of Wales Island, Alaska, *Canadian Geotechnical Journal*, 16, 19–33.

Yin, Y., Wang, F., and Sun, P. (2009). Landslide hazards triggered by the 2008 Wenchuan earthquake, Sichaun, China, *Landslides*, 6(2), 139–152.

Yu, H. S., Salgado, R., Sloan, S., and Kim, J. (1998). Limit analysis versus limit equilibrium for slope stability assessment, *Journal of Geotechnical and Geoenvironmental Engineering*, 124(1), 1–11.

Zaltsberg, E. (1986). Comment on "Laboratory studies of the effects of the capillary fringe on streamflow generation" by A. S. Abdul and R. W. Gillham, *Water Resources Research*, 22, 837–838.

Zaslavsky, D., and Sinai, G. (1981a). Surface hydrology: 1 – explanation of phenomena, *Journal of the Hydraulics Division, ASCE*, 107, 1–16.

Zaslavsky, D., and Sinai, G. (1981b). Surface hydrology: III – causes of lateral flow, *Journal of the Hydraulics Division, ASCE*, 107, 37–52.

Zhu, Y., and Newell, R. E. (1998). A proposed algorithm for moisture fluxes from atmospheric rivers, *Monthly Weather Review*, 126, 725–735.

Ziemer, R. R. (1981). Roots and the stability of forested slopes, *IAHS Publication*, 132, 343–361.

Index

Printed in the United States
By Bookmasters